Optics and Photonics: An Introduction

SECOND EDITION

Optics and Photonics:
An Introduction

Second Edition

F. Graham Smith
University of Manchester, UK

Terry A. King
University of Manchester, UK

Dan Wilkins
University of Nebraska at Omaha, USA

John Wiley & Sons, Ltd

Other Wiley Editorial Offices

John Wiley & Sons Inc., 111 River Street, Hoboken, NJ 07030, USA

Jossey-Bass, 989 Market Street, San Francisco, CA 94103-1741, USA

Wiley-VCH Verlag GmbH, Boschstr. 12, D-69469 Weinheim, Germany

John Wiley & Sons Australia Ltd, 42 McDougall Street, Milton, Queensland 4064, Australia

John Wiley & Sons (Asia) Pte Ltd, 2 Clementi Loop #02-01, Jin Xing Distripark, Singapore 129809

John Wiley & Sons Canada Ltd, 6045 Freemont Blvd., Mississauga, Ontario, Canada L5R 4J3

Wiley also publishes its books in a variety of electronic formats. Some content that appears in print may not be available in electronic
books.

Anniversary Logo Design: Richard J. Pacifico

Library of Congress Cataloging-in-Publication Data

Graham-Smith, Francis, Sir, 1923-
   Optics and photonics : an introduction. – 2nd ed. / F. Graham Smith, Terry A. King,
Dan Wilkins.
      p. cm.
   ISBN 978-0-470-01783-8 – ISBN 978-0-470-01784-5
1. Optics–Textbooks. 2. Photonics–Textbooks. I. King, Terry A. II. Wilkins, Dan, 1947-
III. Title.
   QC446.2.G73 2007
   535–dc22   2006103070

British Library Cataloguing in Publication Data

A catalogue record for this book is available from the British Library

ISBN: 9780470017838 (HB)
ISBN: 9780470017845 (PB)

Typeset in 10/12pt Times by Thomson Digital
Printed and bound in Great Britain by Antony Rowe Ltd, Chippenham, Wiltshire.
This book is printed on acid-free paper responsibly manufactured from sustainable forestry
in which at least two trees are planted for each one used for paper production.

Contents

Concave gratings. Blazed, echellette, echelle and echelon gratings. Radio antenna arrays: end-fire array shooting equally in both directions: end-fire array shooting in only one direction: the broadside array: two-dimensional broadside arrays. X-ray diffraction with a ruled grating. Diffraction by a crystal lattice. The Talbot effect.

The formation of spectral lines: the Bohr model; nuclear mass; quantum mechanics; angular momentum and electron spin. Light from the Sun and Stars. Thermal sources. Fluorescent lights. Luminescence sources. Electroluminescence.

Preface

My Design in this Book is not to explain the Properties of Light by Hypothesis, but to propose and prove them by Reason and Experiments; In order to which I shall premise the following Definitions and Axioms.

The opening sentence of Newton's Opticks, 1717

Nature and Nature's laws lay hid in night: God said, Let Newton be! and all was light.

Alexander Pope, 1688–1744.

Teaching and research in modern optics must encompass the ray approach of geometric optics, the wave approach of diffraction and interferometry, and the quantum physics of the interaction of light and matter. *Optics and Photonics*, by Smith and King (2000), was designed to span this wide range, providing material for a two-year undergraduate course and some extension into postgraduate research. The text has been adopted for course teaching at the University of Omaha, Nebraska, by our third author, Dan Wilkins, and he has contributed many improvements that have proved to be essential for a rigorous undergraduate course. The material has been rearranged to give a more logical presentation and new subject matter has been added. The text has been completely revised, many of the figures have been redrawn, and new examples have been added.

The dominant factor in the recent development of optics has been the discovery and development of many forms of lasers. The remarkable properties of laser radiation have led to a wealth of new techniques such as non-linear optics, atom trapping and cooling, femtosecond dynamics and electro-optics. The laser has led to a deeper understanding of light involving coherence and quantum optics, and it has provided new optical coherence techniques which have made a major impact in atomic physics. Not only physics but also chemistry, biology, engineering and medicine have been enhanced by the use of laser-based methods, There is now a wonderful range of new applications such as holography, optical communications, picosecond and femtosecond probes, optoelectronics, medical imaging and optical coherence tomography. Myriad applications have become prominent in industry and everyday life.

A modern optics course must now place equal emphasis on the traditional optics, dealing with geometric and wave aspects of light, and on the physics of the recent developments, usually classified as photonics. The approach in this book is to emphasize the basic concepts with the objective of developing student understanding. Mathematical content is sufficient to aid the physics description but without undue complication. Extensive sets of problems are included, devised to develop understanding and provide experience in the use of the equations as well as being thought provoking. Some worked examples are in the text, and short solutions to selected problems are given at the end of the book. Notes and full solutions for all problems are posted on a website.

We now present the book as an introduction to the essential elements of optics and photonics, suitable for a one- or two-semester lecture course and including an exposition of key modern developments. We suggest that a first course, constituting minimal core material for the subject, might comprise:

- Chapter 1 Light as waves, rays, and photons.

- Chapter 2 Geometric optics, Sections 2.1–2.7.

- Chapter 4 Periodic and non-periodic waves.

- Chapter 5 Electromagnetic waves.

- Chapter 6 Fibre optics, Sections 6.1–6.8.

- Chapter 7 Polarization.

- Chapter 8 Interference by division of amplitude, Sections 8.1–8.2.

- Chapter 12 Spectra and spectrometers.

- Chapter 15 Lasers.

Selection of further material would then depend on the intended scope of the course and its duration; for example, if time permits, we recommend these additional chapters:

- Chapter 9 Interferometry.

- Chapter 10 Diffraction, Sections 10.1–10.3.

- Chapter 11 The diffraction grating.

- Chapter 14 Holography.

Communications engineers would want to include:

- Chapter 13 Coherence and correlation.

- Chapter 16 Laser light.

- Chapter 17 Semiconductors and semiconductor lasers.

- Chapter 20 The detection of light.

Those in the biosciences could well choose the following:

- Chapter 19 Interaction of light with matter.

- Chapter 20 The detection of light.

- Chapter 21 Optics and photonics in nature.

We welcome suggestions from lecturers on such course structures; we may be contacted c/o Celia Carden, Development Editor at John Wiley & Sons Ltd, email: ccarden@wiley.co.uk.

1 Light as Waves, Rays and Photons

Are not the rays of light very small bodies emitted from shining substances?

Isaac Newton, *Opticks*

All these 50 years of conscious brooding have brought me no nearer to the answer to the question 'What are light quanta?'. Nowadays every Tom, Dick and Harry thinks he knows it, but he is mistaken.

Albert Einstein, *A Centenary Volume*, 1951.

How wonderful that we have met with a paradox. Now we have some chance of making progress.

Niels Bohr (quoted by L.I. Ponomarev in *The Quantum Dice*).

Light is an *electromagnetic wave*: light is emitted and absorbed as a stream of discrete *photons*, carrying packets of energy and momentum. How can these two statements be reconciled? Similarly, while light is a wave, it nevertheless travels along straight lines or rays, allowing us to analyse lenses and mirrors in terms of geometric optics. Can we use these descriptions of waves, rays and photons interchangeably, and how should we choose between them? These problems, and their solutions, recur throughout this book, and it is useful to start by recalling how they have been approached as the theory of light has evolved over the last three centuries.

1.1 The Nature of Light

In his famous book *Opticks*, published in 1704, Isaac Newton described light as a stream of particles or corpuscles. This satisfactorily explained rectilinear propagation, and allowed him to develop theories of reflection and refraction, including his experimental demonstration of the splitting of sunlight into a spectrum of colours by using a prism. The particles in rays of different colours were supposed to have different qualities, possibly of mass, or size or velocity. White light was made up of a compound of coloured rays, and the colours of transparent materials were due to selective absorption. It was, however, more difficult for him to explain the coloured interference patterns in thin films, which we now call Newton's rings (see Chapter 9). For this, and for the partial reflection of light at a glass surface, he suggested a kind of periodic motion induced by his corpuscles, which reacted on the particles to give 'fits of easy reflection and transmission'. Newton also realized that double refraction in a calcite crystal (Iceland spar) was best explained by attributing a rectangular

Optics and Photonics: An Introduction, Second Edition   F. Graham Smith, Terry A. King and Dan Wilkins
© 2007 John Wiley & Sons, Ltd

cross-section (or 'sides') to light rays, which we would now describe as *polarization* (Chapter 7). He nevertheless argued vehemently against an actual wave theory, on the grounds that waves would spread in angle rather than travel as rays, and that there was no medium to carry light waves from distant celestial bodies.

The idea that light was propagated as some sort of wave was published by René Descartes in *La Dioptrique* (1637); he thought of it as a pressure wave in an elastic medium. Christiaan Huygens, a Dutch contemporary of Newton, developed the wave theory; his explanation of rectilinear propagation is now known as 'Huygens' construction'. He correctly explained refraction in terms of a lower velocity in a denser medium. Huygens' construction is still a useful concept, and we use it later in this chapter.

It was not, however, until 100 years after Newton's *Opticks* that the wave theory was firmly established and the wavelength of light was found to be small enough to explain rectilinear propagation. In Thomas Young's double slit experiment (see Chapter 8), monochromatic light from a small source passed through two separate slits in an opaque screen, creating interference fringes where the two beams overlapped; this effect could only be explained in terms of waves. Augustin Fresnel, in 1821, then showed that the wave must be a *transverse* oscillation, as contrasted with the longitudinal oscillation of a sound wave; following Newton's ideas of rays with 'sides', this was required by the observed polarization of light as in double refraction. Fresnel also developed the theories of partial reflection and transmission (Chapter 5), and of diffraction at shadow edges (Chapter 10). The final vindication of the wave theory came with James Clerk Maxwell, who synthesized the basic physics of electricity and magnetism into the four Maxwell equations, and deduced that an electromagnetic wave would propagate at a speed which equalled that of light.

The end of the nineteenth century therefore saw the wave theory on an apparently unassailable foundation. Difficulties only remained with understanding the interaction of light with matter, and in particular the 'blackbody spectrum' of thermal radiation. This was, however, the point at which the corpuscular theory came back to life. In 1900 Max Planck showed that the form of the blackbody spectrum could be explained by postulating that the walls of the body containing the radiation consisted of harmonic oscillators with a range of frequencies, and that the energies of those with frequency v were restricted to integral multiples of the quantity hv. Each oscillator therefore had a fundamental energy quantum

$$\boxed{E = hv} \tag{1.1}$$

where h became known as *Planck's constant*. In 1905 Albert Einstein explained the photoelectric effect by postulating that electromagnetic radiation was itself quantized, so that electrons are emitted from a metal surface when radiation is absorbed in discrete quanta. It seemed that Newton was right after all! Light was again to be understood as a stream of particles, later to become known as photons. What had actually been shown, however, was that light energy and the momentum carried by a light wave existed in discrete units, or quanta; photons should be thought of as events at which these quanta are emitted or absorbed.

If light is a wave that has properties usually associated with particles, could material particles correspondingly have wave-like properties? This was proposed by Louis de Broglie in 1924, and confirmed experimentally three years later in two classical experiments by George Thomson and by Clinton Davisson and Lester Germer. Both showed that a beam of particles, like a light ray encountering an obstacle, could be diffracted, behaving as a wave rather than a geometric ray. The diffraction pattern formed by the spreading of an electron beam passing through a hole in a metal sheet, for example, was the same as the diffraction pattern in light which we explore in Chapter 10. Furthermore, the wavelength λ involved was simply related to the momentum p of the electrons by

$$\lambda = \frac{h}{p}. \tag{1.2}$$

The constant h was again Planck's constant, as in the theory of quanta in electromagnetic radiation; for material waves λ is the *de Broglie* wavelength. A general wave theory of the behaviour of matter, *wave mechanics*, was developed in 1926 by Erwin Schrödinger following de Broglie's ideas. Wave mechanics revolutionized our understanding of how microscopic particles were described and placed limitations on the extent of information one could have about such systems – the famous Heisenberg uncertainty relationship.

The behaviour of both matter and light evidently has dual aspects: they are in some sense both particles and waves. Which aspect best describes their behaviour depends on the circumstances; light propagates, diffracts and interferes as a wave, but is emitted and absorbed discontinuously as photons, which are discrete packets of energy and momentum. Photons do not have a continuous existence, as does for example an electron in the beam of an accelerator machine; in contrast with a material particle it is not possible to say where an individual photon is located within a light beam. In some contexts we nevertheless think of the light within some experimental apparatus, such as a cavity or a laser, as consisting of photons, and we must then beware of following Newton and being misled by thinking of photons as particles with properties like those of material particles.

Although photons and electrons have very similar wave-like characteristics, there are several fundamental differences in their behaviour. Photons have zero mass; the momentum p of a photon in equation (1.1) is related to its kinetic energy E by $E = pc$, as compared with $E = p^2/2m$ for particles moving well below light speed. Unlike electrons, photons are not conserved and can be created or destroyed in encounters with material particles. Again, their statistical behaviour is different in situations where many photons or electrons can interact, as for example the photons in a laser or electrons in a metal. No two electrons in such a system can be in exactly the same state, while there is no such restriction for photons: this is the difference between Fermi–Dirac and Bose–Einstein statistics respectively for electrons and for photons.

In the first two-thirds of this book we shall be able to treat light mainly as a wave phenomenon, returning to the concept of photons when we consider the absorption and emission of electromagnetic waves.

1.2 Waves and Rays

We now return to the question: how can light be represented by a ray? Huygens' solution was to postulate that light is propagated as a wavefront, and that at any instant every point on the wavefront is the source of a wavelet, a secondary wave which propagates outward as a spherical wave (Figure 1.1)

Each wavelet has infinitesimal amplitude, but on the common envelope where countless wavelets intersect, they reinforce each other to form a new wavefront of finite amplitude. In this way, successive positions of the wavefront can be found by a step-by-step process. The envelope[1] of the

---

[1]To define the envelope evolved after a short time from a wavefront segment, take a finite number N of wavelets with evenly spaced centres, and note the intersection points between adjacent wavelets. In the limit that N goes to infinity, the intersection points crowd together and constitute the envelope, which is the new wavefront.

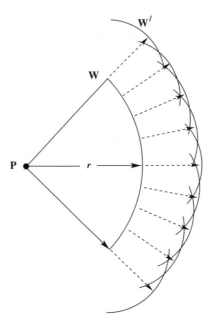

Figure 1.1   Huygens' secondary wavelets. A spherical wavefront W has originated at P and after a time t has a radius $R = ct$, where c is the speed of light. Huygens' secondary wavelets originating on W at time t combine to form a new wavefront W' at time t', when the radii of the wavelets are $c(t' - t)$

wavelets is perpendicular to the radius of each wavelet, so that the ray is the normal to a wavefront. This simple Huygens wavefront concept allows us to understand both the rectilinear propagation of light along ray paths and the basic geometric laws of reflection and refraction. There are obvious limitations: for example, what happens at the edge of a portion of the wavefront, as in Figure 1.1, and why is there no wave reradiated backwards? We return to these questions when we consider diffraction theory in Chapter 10.

Reflection of a plane wavefront W_1 reaching a totally reflecting surface is understood according to Huygens in terms of secondary wavelets set up successively along the surface as the wavefront reaches it (Figure 1.2(a)). These secondary wavelets propagate outwards and combine to form the reflected wavefront W_2. The rays are normal to the incident and reflected wavefronts. Light has travelled along each ray from W_1 to W_2 in the same time, so all path lengths from W_1 to W_2 via the mirror must be equal. The basic law of reflection follows: the incident and reflected rays lie in the same plane and the angles of incidence (i) and reflection (r) are equal.

Figure 1.2(b) shows the same reflection in terms of rays. Here we may find the same law of reflection as an example of *Fermat's principle of least time*, which states that the *time* of propagation is a minimum (or more strictly either a maximum or a minimum) along a ray path.[2] It is easy to see that the path of a light ray between the two points A and B (Figure 1.2 (b)) is a minimum if the angles i, r are equal. The proof is simple: construct the mirror image A' of A in the reflecting surface, when the line $A'B$ must be straight for a minimum distance. Any other path $AP'B$ is longer.

---

[2]This explanation of the basic law of reflection was first given by Hero of Alexandria (First century AD).

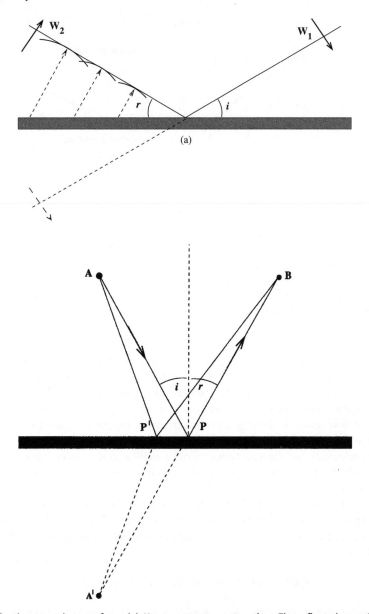

Figure 1.2   Reflection at a plane surface. (a) Huygens wave construction. The reflected wave W_2 is made up of wavelets generated as successive points on the incident plane wave W_1 reach the surface. (b) Fermat's principle. The law of reflection is found by making the path of a reflected light ray between the points A and B a minimum

Why are these two approaches essentially the same? Fermat tells us that the time of travel is the same along all paths close to an actual ray. In terms of waves this means that waves along these paths all arrive together, and reinforce one another as in Huygens' construction. When we consider periodic waves, we will express this by saying that they are *in phase*.

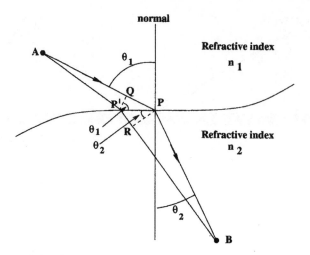

Figure 1.3   Refraction at a surface between transparent media with refractive indices n_1 and n_2. We assume the light rays and the surface normal all lie in the plane of the paper. Snell's law corresponds to a stationary value of the optical path $n_1 AP + n_2 PB$ between the fixed endpoints A, B; for small virtual variations such as shifting the point P to P', the optical path changes negligibly

The basic law of refraction (Snell's law) may be found by applying either Huygens' or Fermat's principles to a boundary between two media in which the velocities of propagation v_1, v_2 are different; as Huygens realized, his secondary waves must travel more slowly in an optically denser medium. The *refractive indices* are defined as $n_1 = c/v_1, n_2 = c/v_2$ where c is the velocity of light in free space. As we now show, the Fermat approach shown in Figure 1.3 leads to Snell's law via some simple trigonometry.

The Fermat condition is that the travel time $(n_1 AP + n_2 PB)c$ is stationary (minimum, maximum, or point of inflection); this means that for any small change in the light path of order ϵ, the change in travel time vanishes as ϵ^2 (or even faster). The distance $n_1 AP + n_2 PB$ is called the optical path. We consider a small virtual displacement of the light rays from APB to AP'B. Denote the length PP' as ϵ. By dropping perpendiculars from P and P', we create two thin triangles AP'Q and BPR that become perfect isosceles triangles in the limit of zero displacement. Fermat requires then that the change of the optical path satisfies[3]

$$n_1 QP - n_2 P'R = n_1 \epsilon \sin \theta_1 - n_2 \epsilon \sin \theta_2 = O(\epsilon^2). \tag{1.3}$$

Dividing by ϵ, and going to the limit $\epsilon = 0$, this leads directly to Snell's law of refraction:

$$n_1 \sin \theta_1 = n_2 \sin \theta_2. \tag{1.4}$$

Notice that this derivation works for a smoothly curving surface of any shape.

In Chapter 5 we show how the laws of reflection and refraction may be derived from electromagnetic wave theory.

---

[3]The notation $O(\epsilon^2)$ designates a quantity that varies as ϵ^2 in the limit of vanishing epsilon.

Figure 1.4   The light pipe. Rays entering at one end are totally internally reflected, and can be conducted along long paths which may include gentle curves

1.3   Total Internal Reflection

Referring again to Figure 1.3, and noting that the geometry is the same if the ray direction is reversed, we consider what happens if a ray inside the refracting medium meets the surface at a large angle of incidence θ_2, so that $\sin\theta_2$ is greater than n_1/n_2 and equation (1.4) would give $\sin\theta_1 > 1$. There can then be no ray above the surface, and there is *total internal reflection*. The internally reflected ray is at the same angle of incidence to the normal as the incident ray.

The phenomenon of total internal reflection is put to good use in the light pipe (Figure 1.4), in which light entering the end of a glass cylinder is reflected repeatedly and eventually emerges at the far end. The same principle is applicable to the transmission of light down thin optical fibres, but here the relation of the wavelength of light to the fibre diameter must be taken into account (Chapter 6).

1.4   The Light Wave

We now consider in more detail the description of the light wave, starting with a simple expression for a plane wave of any quantity ψ, travelling in the positive direction z with velocity v:

$$\psi = f(z - vt). \tag{1.5}$$

The function $f(z)$ describes the shape of ψ at the moment $t = 0$, and the equation states that the shape of ψ is unchanged at any later time t, with only a movement of the origin by a distance vt along the z axis (Figure 1.5). The minus sign in $(z - vt)$ indicates motion in the $+z$ direction; a plus sign would correspond to motion in the $-z$ direction. The variable quantity ψ may be a scalar, e.g. the pressure in a sound wave, or it may be a vector. If it is a vector, it may be *transverse*, i.e.

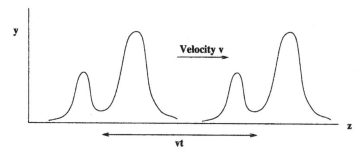

Figure 1.5   A wave travelling in the z direction with unchanging shape and with velocity v. At time $t = 0$ the waveform is $\psi = f(z)$, and at time t it is $\psi = f(z - vt)$

perpendicular to the direction of propagation, as are the waves in a stretched string, or the electric and magnetic fields in the electromagnetic waves which are our main concern. (These are the 'sides' which Newton attributed to his rays.) For most of optics it is sufficient to consider only the transverse electric field; indeed, as we shall see later, the results of scalar wave theory are sufficiently general that for many purposes we may just think of the magnitude of the electric field and forget about its vector nature.

At any one time the variation of ψ with z, i.e. the slope of the graph in Figure, 1.5, is $\partial\psi/\partial z$, and at any one place the rate of change of ψ is $\partial\psi/\partial t$. Changing to the variable $z' = (z - vt)$ and using the chain rule for partial differentiation:

$$\frac{\partial\psi}{\partial z} = \frac{\partial\psi}{\partial z'}\frac{\partial z'}{\partial z} = \frac{\partial\psi}{\partial z'} \tag{1.6}$$

$$\frac{\partial\psi}{\partial t} = \frac{\partial\psi}{\partial z'}\frac{\partial z'}{\partial t} = -v\frac{\partial\psi}{\partial z'}. \tag{1.7}$$

Similarly, the second differential of ψ with respect to z, i.e. $\partial^2\psi/\partial z^2$, which is the curvature of the graph in Figure 1.5, is related to the second differential with respect to time, i.e. the acceleration of ψ, by

$$\frac{\partial^2\psi}{\partial t^2} = v^2\frac{\partial^2\psi}{\partial z^2}. \tag{1.8}$$

This so-called one-dimensional wave equation applies to any wave propagating in the z direction with uniform velocity and without change of form.

The wave equation (1.8) may be extended to three dimensions, giving

$$\frac{\partial^2\psi}{\partial x^2} + \frac{\partial^2\psi}{\partial y^2} + \frac{\partial^2\psi}{\partial z^2} = \frac{1}{v^2}\frac{\partial^2\psi}{\partial t^2} \tag{1.9}$$

or in a more general and concise notation[4]

$$\boxed{\nabla^2\psi = \frac{1}{v^2}\frac{\partial^2\psi}{\partial t^2}.} \tag{1.10}$$

The form of the wave $f(z - vt)$ may be any continuous function, but it is convenient to analyse such behaviour in terms of *harmonic* waves, taking the simple form of a sine or cosine. (In Chapter 4 we show that any continuous function can be synthesized from the superposition of harmonic waves.) At any point such a wave varies sinusoidally with time t, and at any time the wave varies sinusoidally with distance z. The waveform is seen in Figure 1.6, which introduces the *wavelength* λ and *period* τ. At any point there is an oscillation with *amplitude* A. Equation (1.5) then becomes

$$\psi = A\sin\left[2\pi\left(\frac{z}{\lambda} - \frac{t}{\tau}\right)\right], \tag{1.11}$$

---

[4]Recall that ∇^2 is the *Laplacian operator:* $\nabla^2 = \partial^2/\partial x^2 + \partial^2/\partial y^2 + \partial^2/\partial z^2$.

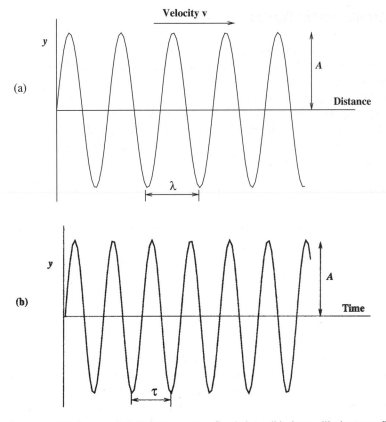

Figure 1.6   A progressive sine wave: (a) the wave at a fixed time; (b) the oscillation at a fixed point P

which is easily demonstrated to be a solution of the general wave equation (1.8) provided $\lambda/\tau = v$. The *frequency* of oscillation is $v = 1/\tau$. It is often convenient to use an *angular frequency* $\omega = 2\pi v$, and a *propagation constant* or *wave number*[5] $k = 2\pi/\lambda$. Equation (1.11) may then be written in terms of k as

$$\psi = A \sin(kz - \omega t). \tag{1.12}$$

The vector quantity $\mathbf{k} = (2\pi/\lambda)\hat{\mathbf{k}}$, where $\hat{\mathbf{k}}$ is the unit vector in the direction of $\mathbf{k}$, is also termed the *wave vector*.

Another powerful way of writing harmonic plane wave solutions of Equation (1.10) is in terms of complex exponentials

$$\psi = A \exp[\mathrm{i}(kz - \omega t)]. \tag{1.13}$$

Due to several elegant mathematical properties, including ease of differentiation and of visualization, complex functions like this can vastly simplify the process of combining waves of different amplitudes and phases, as we shall see in Chapter 4.

---

[5]Beware: the term *wave number* is also used in spectroscopy for $1/\lambda$, without the factor 2π.

1.5   Electromagnetic Waves

Although the idea that light was propagated as a combination of electric and magnetic fields was developed qualitatively by Michael Faraday, it required a mathematical formulation by Maxwell before the process could be clearly understood. In Chapter 5 we derive the electromagnetic wave equation from Maxwell's equations, and show that all electromagnetic waves travel with the same velocity in free space. There are two variables in an electromagnetic wave, the electric and magnetic fields $\mathbf{E}$ and $\mathbf{B}$; both are vector quantities, but each can be represented by the variable ψ in the wave equation (1.10). As shown in Chapter 5, they are both transverse to the direction of propagation, and mutually perpendicular. Their magnitudes[6] are related by

$$\boxed{E = vB} \qquad (1.14)$$

where v is the velocity of light in the medium. Since the electric and magnetic fields are mutually perpendicular and their magnitudes are in a fixed ratio, only one need be specified, and the magnitude and direction of the other follow. Equation (1.14) is true in general, but note that the velocity v in a dielectric such as glass is less than the free space velocity c; the *refractive index n* of the medium is

$$\boxed{n = \frac{c}{v}.} \qquad (1.15)$$

As Huygens realized, light travels more slowly in dense media than in a vacuum.

In a transverse wave moving along a direction z the variable quantity is a vector which may be in any direction in the orthogonal plane x, y. The relevant variable for electromagnetic waves is conventionally chosen as the electric field $\mathbf{E}$. The *polarization* of the wave is the description of the behaviour of the vector $\mathbf{E}$ in the plane x, y. The plane of polarization is defined as the plane containing the electric field vector and the ray, i.e. the z axis. If the vector $\mathbf{E}$ remains in a fixed direction, the wave is *linearly* or *plane* polarized; if the direction changes randomly with time, the wave is *randomly* polarized, or *unpolarized*. The vector $\mathbf{E}$ can also rotate uniformly at the wave frequency, as observed at a fixed point on the ray; the polarization is then *circular*, either right- or left-handed, depending on the direction of rotation.

Polarization plays an important part in the interaction of electromagnetic waves with matter, and Chapter 7 is devoted to a more detailed analysis.

1.6   The Electromagnetic Spectrum

The wavelength range of visible light covers about one octave of the electromagnetic spectrum, approximately from 400 to 800 nm (1 nanometre = 10^{-9} m). The electromagnetic spectrum covers a vast range, stretching many decades through infrared light to radio waves and many more decades through ultraviolet light and X-rays to gamma rays (Figure 1.7). The differences in behaviour across the electromagnetic spectrum are very large. Frequencies (ν) and wavelengths (λ) are related to the velocity of light (c) by $\lambda\nu = c$. The frequencies vary from 10^4 Hz for long radio waves (1 hertz equals

---

[6]We use the SI system of electromagnetic units throughout.

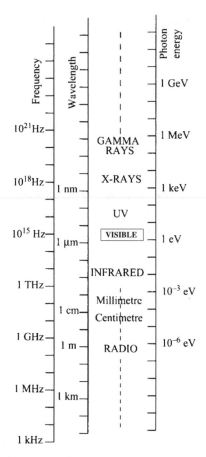

Figure 1.7   The electromagnetic spectrum

one cycle per second), to more than 10^{21} Hz for commonly encountered gamma rays; the highest energy cosmic gamma rays so far detected reach to 10^{35} Hz (4×10^{20} eV). It is unusual to encounter a quantum process in the radio frequency spectrum, and even more unusual to hear a physicist refer to the frequency of a gamma ray, instead of the energy and the momentum carried by a gamma ray photon.

Although wave aspects dominate the behaviour of the longest wavelengths, and photon aspects dominate the behaviour of short-wavelength X-rays and gamma rays, the whole range is governed by the same basic laws. It is in the optical range (waves in or near the visible range) that we most usually encounter the 'wave particle duality' which requires a familiarity with both concepts.

The propagation of light is determined by its wave nature, and its interaction with matter is determined by quantum physics. The relation of the energy of the photon to common levels of energy in matter determines the relative importance of the quantum at different parts of the spectrum: cosmic gamma rays, with a high photon energy and a high photon momentum, can act on matter explosively or like a high-velocity billiard ball, while long infrared or radio waves, with low photon energies, usually only interact with matter through classical electric and magnetic induction. We can explore these extremes in the following examples.

1. *What would be the velocity of a tennis ball, mass* 60 g, *with the same energy as a* 10^{20} eV *cosmic gamma ray photon?*

Electron volt $= 1.602 \times 10^{-19}$ J. Kinetic energy $\frac{1}{2}mv^2 = 10^{20}$ eV $= 10^{20} \times 1.6 \times 10^{-19}$ J. Velocity of 0.06 kg tennis ball is

$$v = \sqrt{\frac{2 \times 10^{20} \times 1.6 \times 10^{-19}}{60 \times 10^{-3}}} = 23 \text{ m s}^{-1} (= 83 \text{ km h}^{-1}).$$

2. *At what temperature would a molecule of hydrogen gas have, on average, the same energy as a photon of the* 21 cm *hydrogen spectral line?*

In statistical physics each degree of freedom has an average energy of $\frac{1}{2}kT$. A hydrogen molecule has 5 degrees of freedom (3 translational and 2 rotational); hence thermal energy $= \frac{5}{2}kT$. Photon energy $h\nu = hc/\lambda$, so that $T = \frac{2}{5}hc/k\lambda = 0.068$ K.

3. *What wavelength of electromagnetic radiation has the same photon energy as an electron accelerated to* 100 eV?

Photon energy $= h\nu = hc/\lambda = 100 \times 1.6 \times 10^{-19}$J. So

$$\lambda = \frac{6.63 \times 10^{-34} \times 3.00 \times 10^8}{1.6 \times 10^{-17}} = 1.24 \times 10^{-8} \text{ m} = 12.4 \text{ nm}$$

(ultraviolet light; see Figure (1.7).

4. *An X-ray photon with wavelength* 1.5×10^{-11} m *arrives at a solid. How much energy (in eV) can it give to the solid?*

$$h\nu = \frac{hc}{\lambda} = \frac{6.63 \times 10^{-34} \times 3.00 \times 10^8}{1.5 \times 10^{-11}} = 1.32 \times 10^{-14} \text{ J} = 8.3 \times 10^4 \text{ eV}.$$

The photon energy of visible light waves, ranging from 1.5 to 3 electron volts (eV), is such that quantum effects dominate only some of the processes of emission and absorption or detection. The visible spectrum contains the marks of quantum processes in the profusion of colour from line emission and in line absorption; it can also display a continuum of emission over a wide range of wavelengths, giving 'white' light, whose actual colour is determined by the large-scale structure of the continuum spectrum rather than its fine detail.

1.7   Stimulated Emission: The Laser

At the start of this chapter we remarked on the apparently complete understanding of optics at the beginning of the twentieth century. The wave nature of light was fully understood, stemming from the classical experiments of Young, Fresnel and Michelson, and substantiated by Maxwell's electromagnetic theory. Much of the content of our later chapters on interference and diffraction is derived directly from that era (with some refinements). Even Planck's bombshell announcement in 1900 that blackbody radiation is emitted by quantized oscillators, and Einstein's demonstration in 1905 of the reality of photons through his explanation of the photoelectric effect, completed rather than disturbed the picture; they had cleared up a mystery about the interchange of energy between matter and

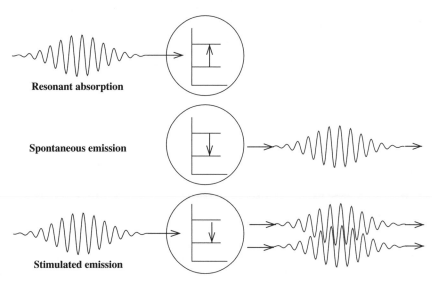

Figure 1.8   Three basic photon processes: absorption, spontaneous emission and stimulated emission. For simplicity only two energy levels are shown

electromagnetic waves. Einstein's theory of that interaction, however, contained the seed of another revolution in optics, which germinated half a century later with the invention of the laser.

Einstein in 1917 showed that there are three basic processes involved in the interchange of energy between a light wave and the discrete energy levels in an atom. All three involve a quantum jump of energy within the atom; typically in the visible region this is around 2 eV. Figure 1.8 illustrates the three basic photon processes; the processes are illustrated adopting a model with only two energy levels, although there are many more energy levels even in the simplest atom. As depicted in Figure 1.8, the first is the *absorption* of a photon which can occur when the quantum energy $h\nu$ of the photon equals the energy difference between the two levels (a *resonant* condition) and the photon falls on an atom in the lower level; the atom then gains a quantum of energy. The second is *spontaneous emission*, when an atom in the upper level emits a photon, losing a quantum of energy in the process. The third is *stimulated emission*, in which the emission of a photon is triggered by the arrival at an excited atom of another, resonant photon. This third process was shown by Einstein to be essential in the overall balance between emission and absorption. What emerged later was that the emitted photon is an exact copy of the incident photon, with the same direction, frequency and phase; further, each could then stimulate more photon emissions, leading to the build-up of a *coherent* wave which can attain a very great irradiance (or 'intensity', in old terminology).[7] The build-up requires the number of atoms in the higher energy level to exceed the number in the lower level, a condition known as population inversion, so that the rate of stimulated emission exceeds the rate of absorption. The energy supply used to create the population inversion is often referred to as a *pump*, which in Figure 1.9 is light absorbed between a ground level E_0 and level E_1. If the excitation of this level is short-lived, and it decays to a lower but longer-lived level E_2, the process leads to an accumulation

---

[7]See Appendix 1 for the definition of *irradiance* and other radiometric terms.

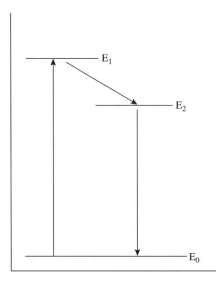

Figure 1.9 Energy levels in the three-level laser. Energy is supplied to the atom by absorption from the ground level to the excited level E_1; spontaneous emission to the long-lived level E_2 then results in overpopulation of that level. Transitions from E_2 to ground are then the stimulated emission in the laser

and overpopulation of atoms in the level E_2 compared with E_0. Stimulated emission, fed by energy from a pump, is the essential process in a *laser*. Prior to the laser, stimulated emission had been demonstrated in 1953 in the microwave region of the spectrum by Basov, Prokhorov and Townes,[8] an achievement for which they were awarded the Nobel Prize. We describe in Chapter 15 the earliest laser, due to T.H. Maiman in 1960.

The process of stimulated emission in a laser builds up a stream of identical photons, which add coherently as the most nearly ideal monochromatic light, with very narrow frequency spread and correspondingly great *coherence length* (Chapter 13). Paradoxically, lasers, which depend fundamentally on quantum processes, produce the most nearly ideal waves. Lasers have allowed the classical experimental techniques of interferometry and spectroscopy to be extended into new domains, which we explore in Chapter 9 on the measurement of length and Chapter 12 on high-resolution spectrometry.

Largely as a result of the discovery and development of lasers, a new subject of *photonics* has developed from pre-laser studies of transmission and absorption in dielectrics. Coherent laser beams easily achieve an irradiance many orders of magnitude greater than that of any thermal source, leading to very large electric fields and *non-linear* effects in dielectrics, such as harmonic generation and frequency conversion. There are many practical applications, some of which are more familiar in electronic communications, such as switching, modulation and frequency mixing. The title of this book indicates the current importance of lasers and photonics; the materials involved, including those used in non-linear optics, are included in Chapters 16 on laser light, 17 on semiconductors, 18 on light sources and 19 on detectors.

---

[8]They demonstrated a *maser* process, Microwave Amplification by the Stimulated Emission of Radiation. Note that strictly speaking this and the related laser process refer to amplification; devices which use the process in oscillators which generate microwaves and light are, however, known simply as masers and lasers.

1.8 Photons and Material Particles

As we noted in Section 1.1, the wave-like character of electrons was demonstrated in the 1920s, following the prediction by de Broglie that any particle with mass $m = E/c^2$ (where E is the total relativistic energy) and moving with velocity v has an associated wave with wavelength $\lambda = h/mv$. This association was eventually demonstrated in atoms, and even in molecules; in 1999 the wave–particle duality of the large molecule fullerene, or C_{60}, was demonstrated in a diffraction experiment by Arndt *et al.*[9]

There can be little doubt of the actual individual existence of a large particle such as a molecule of fullerene. Can we make a similar statement about the individual existence of photons? Ever since Planck and Einstein introduced quantum theory there has been a debate about the actual existence of photons as discrete objects. Light can be depicted as a ray, or as a wave; can it be thought of as a volley of photons, like a flock of birds moving from one roosting place to another? Should the wave nature of material particles, which constrains them to their behaviour in diffraction and interferometer observations, lead us to conclude that light has a similar dual nature?

Consider the classical interferometer typified by Young's double slit (Figure 1.10), which we describe in Chapter 8. Monochromatic light from the slit source passes through the pair of slits, forming an interference pattern on the screen. A detector on the screen records the arrival of individual photons, which in aggregate trace out the interference pattern, even when the intensity is so low that each recorded photon must have been the only photon present in the apparatus at any time. Through which slit did it pass? We naturally try to find out by placing some sort of detector at one or both slits, but as soon as we detect and locate the photon the interference pattern disappears. Detecting which slit the photon traverses has the same effect as forcing it to act like a localized quantum which passes through one slit at a time.

This behaviour is a simple example of the *complementarity principle* formulated by Bohr; if we know where the photon is, we cannot have an interference pattern, and if an interference pattern exists, it is impossible to specify the position of the photon. We can only observe that a photon has reached the detector, and the probability that it will arrive at any location is determined by its wave nature.

Diffraction and interference of material particles follow a similar pattern. In principle the double slit of Figure 1.10 could be demonstrating the de Broglie waves associated with a large molecule such as fullerene. Exactly the same dilemma arises: the interference pattern is observed even if only one molecule is in the apparatus at any time, but complementarity prevents us from knowing which slit the particle goes through, without destroying the interference pattern.

It has been suggested that the photon can exist in two places at once, and even that the large molecule is similarly 'delocalized'. This is better expressed by treating the wave as the basic description in both cases, and equating the probability of observing a particle or photon at a particular location to the intensity of the wave at that location. If any diffraction phenomenon is involved, the intensity pattern is determined by the *correlation* between separate wave components. If the separate components are 'de-correlated' by any process, the interference between wave components disappears. The analysis of correlation, which we present in Chapter 13, provides a unified framework for understanding diffraction both in light and in material particles. The difference, as noted in

---

[9]M. Arndt *et al.*, *Nature* **401**, 680, 1999.

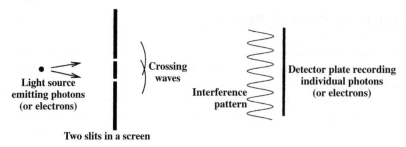

Figure 1.10   Double slit interferometer. Through which slit did each individual photon or electron go?

Section 1.1, is that a photon only exists as a quantized interchange between a field and an emitter or detector, while the individual existence of a material particle can hardly be questioned.

Problem 1.1

Gallium arsenide (GaAs) is an important semiconductor used in photoelectronic devices. It has a refractive index of 3.6. For a slab of GaAs of thickness 0.3 mm show that a point source of light within the GaAs on the bottom face will give rise to radiation outside the top face from within a circle of radius R centred immediately above the point source. Find R.

Problem 1.2

In the Pulfrich refractometer (Figure 1.11), the refractive index n of a liquid is found by measuring the emergent angle e from the prism whose refractive index is N. Show that if i is nearly $90°$

$$n \approx (N^2 - \sin^2 e)^{1/2}.$$

Problem 1.3

The angular radius of a rainbow, measured from a point opposite to the Sun, may be found from the geometry of the ray in Figure 1.12, which lies in the meridian plane of a spherical drop of water with refractive index n. The

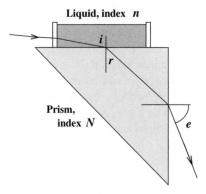

Figure 1.11   Pulfrich refractometer

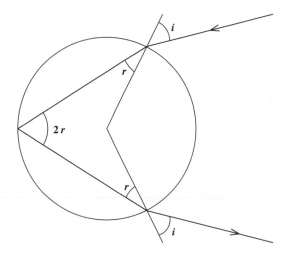

Figure 1.12 A ray refracted in the meridian plane of a spherical raindrop.

angular radius is a stationary value of the angle through which a ray from the Sun is deviated; show that it is given by

$$\cos i = \left(\frac{n^2 - 1}{3}\right)^{1/2}.$$

Note that the internal reflection is near the Brewster angle (see Section 5.4), so that the rainbow light is polarized along the circumference of the bow.

Problem 1.4
Show that the apparent diameter of the bore of a thick-walled glass capillary tube of refractive index n, as seen normally from the outside, is independent of the outer diameter, and is n times the actual diameter.

Problem 1.5
Show that the lateral displacement d of a ray passing through a plane-parallel plate of glass refractive index n, thickness t, is related to the angle of incidence θ by

$$d \approx t\theta\left(1 - \frac{1}{n}\right)$$

provided that θ is small.

Problem 1.6
If the refractive index n of a slab of material varies in a direction y, perpendicular to the x axis, show by using Huygens' construction that a ray travelling nearly parallel to the x axis will follow an arc with radius

$$n\left(\frac{dn}{dy}\right)^{-1}.$$

(Consider a sector of wavefront δy across, and compare the distances travelled in time τ by secondary waves from each end of the sector.)

Problem 1.7

Show that the geometric distance of the horizon as seen by an observer at height h metres is approximately $3.5\,h^{1/2}$ kilometres. The radius of the Earth ≈ 6000 km.

Use the result of Problem 1.6 to calculate how this is affected by atmospheric refraction, if this is due to pressure changes only with an exponential scale height of 10 kilometres. The refractive index of air at ground level is approximately $1.000\,28$.

Problem 1.8

The refractive index of solids at X-ray wavelengths is generally less than unity, so that a beam of X-rays incident at a glancing angle may be reflected, as in total internal reflection. If the refractive index is $n = 1 - \delta$ show that the largest glancing angle for reflection is $\simeq \sqrt{\delta}$. Evaluate this critical angle for silver at $\lambda = 0.07$ nm where $\delta = 5.8 \times 10^{-6}$.

2 Geometric Optics

Optics is either very simple or else it is very complicated.

Richard P. Feynman, *Lectures on Physics*, Addison-Wesley, 1963.

That y^e rays wch make blew are refracted more y^n y^e rays wch make red appears from this experim^{nt}.

Isaac Newton, *Quaestiones*.

Light, which is propagated as an electromagnetic wave, may often conveniently be represented by *rays*, which are geometrical lines along which light energy flows; the term geometric optics is derived from this concept. Rays are lines perpendicular to the *wavefronts* of the electromagnetic wave. An alternative concept is to regard the action of the various components of optical systems, such as convex and concave mirrors and lenses, as modifying a wavefront by changing its direction of travel or its curvature. This wavefront concept is useful, but the precise geometry of *ray tracing* is nevertheless essential for the detailed design of optical instruments.

We start our exposition of geometric optics by analysing the action of a thin prism and a simple lens in terms both of waves and of rays, and then develop the basic ray theory of imaging. Images are inevitably imperfect, apart from trivial cases such as images in plane mirrors; in the second part of this chapter we analyse the imperfections as various types of *aberration*.

The use of a lens as a simple magnifier, and the combination of optical components in systems such as the microscope and telescope, will be considered in the following chapter.

2.1 The Thin Prism

The wavefront concept is usefully applied to the bending of a light ray in a prism, with apex angle α and refractive index n, assuming free space[1] outside the prism. We first calculate the angle of deviation θ by applying Snell's law (equation (1.4)) to each surface in turn, and find a useful approximation for a thin prism at near-normal incidence. We then show that the wavefront approach leads directly to this approximation.

---

[1] The optical properties of free space and air are nearly the same, and are taken as identical in this chapter.

Optics and Photonics: An Introduction, Second Edition   F. Graham Smith, Terry A. King and Dan Wilkins
© 2007 John Wiley & Sons, Ltd

2.1.1   The Ray Approach

In Figure 2.1(a) the ray is incident on the first surface at angle β_1. Following the ray through the prism we have for the two refracting surfaces

$$n \sin \beta_2 = \sin \beta_1$$
$$n \sin \beta_3 = \sin \beta_4. \qquad (2.1)$$

The total deviation is $\theta = \beta_1 - \beta_2 - \beta_3 + \beta_4$. In the triangle OAB we have $\alpha = \beta_2 + \beta_3$, so that

$$\theta = \beta_1 + \beta_4 - \alpha. \qquad (2.2)$$

Figure 2.1(b) shows the results of a numerical solution of equations (2.1) and (2.2), giving θ for a prism with $\alpha = 10°$ and $n = 1.5$, with β_1 between $0°$ and $25°$. There is a minimum deviation when the ray passes symmetrically through the prism, at $\beta_1 = 7.5°$. The angle of deviation varies only between $5.02°$ and $5.23°$ over the whole range in Figure 2.1(b).

If the analysis is restricted to small values of α and β, so that to a good approximation $\sin \beta \approx \beta$, equations (2.1) become

$$\beta_1 = n\beta_2 \quad \text{and} \quad n\beta_3 = \beta_4 \qquad (2.3)$$

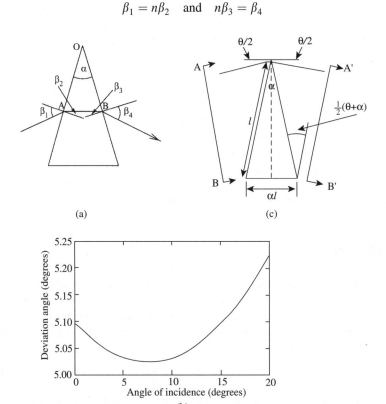

(a)

(b)

(c)

Figure 2.1   A prism with small apex angle α refracts a ray or its corresponding planar wavefront through an angle which is nearly independent of the angle of incidence, provided it is near normal. (a) The ray approach. (b) Deviation angle for a $10°$ prism over a range of angles of incidence, for $n = 1.5$. (c) The wavefront approach

and equation (2.2) becomes

$$\theta = (n - 1)\alpha. \tag{2.4}$$

In the example above the simplified equation gives $\theta = (1.5 - 1) \times 10° = 5°$, close to the correct result $\theta = 5.02°$ at minimum deviation.

2.1.2   The Wavefront Approach

We now derive equation (2.4) from the wavefront approach. In Figure 2.1(c) the incident wavefront is AB and the emergent wavefront A′ B′. The prism is arranged symmetrically, for minimum deviation, but the same argument can be applied for wavefronts over a range of angles about this position.

To calculate the angle of deviation we note that the optical paths AA′ and BB′ are equal. (Remember from Section 1.2 that this implies that the time of travel from A to A′ is the same as from B to B′.) The refracting face length of the prism is l. While the wavefront at B passes through a length $2l \sin \frac{1}{2}\alpha$ of the prism, the wavefront at A passes through a length $2l \sin \frac{1}{2}(\theta + \alpha)$ of air. The wave velocity is a factor n slower inside the prism, so that the two equal optical paths are $2nl \sin \frac{1}{2}\alpha$ and $2l \sin \frac{1}{2}(\theta + \alpha)$. At minimum deviation θ is therefore given by

$$\theta = 2 \sin^{-1}[n \sin(\alpha/2)] - \alpha. \tag{2.5}$$

As in the ray treatment, we approximate for the small-angle prism by writing the sine of an angle as the angle itself (in radian measure), and the angle of deviation θ is then given very simply by

$$\theta = (n - 1)\alpha \tag{2.6}$$

as in equation (2.4) above.

2.2   The Lens as an Assembly of Prisms

A convex lens, shown in section in Figure 2.2, is familiar as a simple hand-held magnifying glass. The lens is also shown as a series of thin prisms with apex angle increasing with distance y from the axis. As before, we assume all angles are small. If the radius of curvature of both surfaces is r, the prism angle at height y is $2y/r$ (Figure 2.2(b)) giving a wavefront deviation

$$\theta = (n - 1)2y/r. \tag{2.7}$$

As shown in Figure 2.2, a plane wavefront passing through the lens will become curved, and will converge to a focal point at a distance $f = y/\theta = r/2(n - 1)$ from the lens. This is the *focal length* of the lens. The action of the convex lens is to add a curvature[2] $2(n - 1)/r$ to the plane wavefront. Within the approximation of small angular deviation, the wavefront over the whole of the lens converges on a single focal point.

---

[2]A spherical surface with radius R is said to have a *curvature* $1/R$. A planar surface is the limiting case of a sphere with infinite radius and zero curvature.

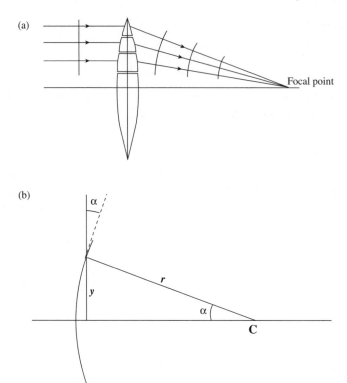

Figure 2.2 (a) A simple converging lens as an assembly of prisms. (b) The prism angle of one face of a lens at distance y off axis. $\alpha \approx \sin\alpha = y/r$

Moreover, a wavefront arriving at a different angle will converge on a different point at the same distance from the lens, i.e. in the same *focal plane*, so that the lens gives a flat image of a distant scene.

Figure 2.3(a) shows the effect of a convex lens on a diverging wavefront originating from a point source P_1 at distance u from the lens. The wavefront emerging from the lens converges on an image point P_2 at distance v from the lens. The change in curvature is related to the power of the lens.

Before proceeding further, we need to specify our sign convention for distances and angles. In geometric optics, there are two primary conventions: *real-positive* and *Cartesian*. In the first of these, which is short for 'real-positive, virtual-negative', distances along the optic axis are taken as positive for an object or image point that is real, and negative for one that is virtual. This convention is well suited to applications of Fermat's principle, or making the optical path a minimum.[3] The Cartesian convention, on the other hand, is ideal for systematic ray tracing in complex systems, i.e. those with multiple interfaces, and for this reason it is used in the matrix approach to paraxial optics (Section 2.8). The signs of coordinates and angles in the Cartesian system are explained in Figure 2.4; in addition this system specifies that if the centre of curvature of a spherical surface is on the same side as the incident light, the radius of curvature $r < 0$, and on the opposite side, $r > 0$.

---

[3]Or, more rarely, a maximum.

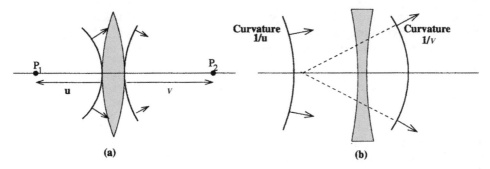

Figure 2.3   Convex (a) and concave (b) lenses changing the curvature of a wavefront. In a lens equation such as equation (2.8), the curvatures are evaluated on wavefronts immediately adjacent to the lens

We now introduce the term *vergence* for the curvature of a wavefront, using a definition which applies generally to refraction and reflection at curved surfaces. The vergence V of a wavefront emanating from (or converging to) an object (or image point) at signed distance L in a medium with refractive index n is defined as $V = n/L$; the sign of L is chosen so that vergence is positive for a converging wavefront and negative for a diverging wavefront. In Figure 2.3(a) the incident diverging wavefront has a vergence $V = 1/u$ which is negative since the object distance $u < 0$; the convex lens adds a positive vergence $2(n - 1)/r$, and the emergent wavefront with positive vergence $V' = 1/v$ converges on the image point P_2 at distance $v > 0$. The result is

$$1/v - 1/u = 2(n - 1)/r. \qquad (2.8)$$

Equation (2.8) is derived rigorously in Section 2.4. Problem 2.1 suggests a derivation based on the bending-angle approach of this section, including the case when the two surfaces have different radii.

The change in vergence imposed on the wavefront by the lens is the power P of the lens; in general for any imaging system

$$V' - V = P. \qquad (2.9)$$

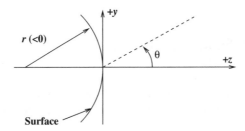

Figure 2.4   The Cartesian coordinate system in geometric optics. Light is incident from the left. The signs of distances z and y follow normal geometric convention, and anticlockwise angles are positive. A spherical surface or wavefront with centre of curvature to the left has a negative radius of curvature

A concave lens, as in Figure 2.3(b), has a negative power, so that the incident divergent wavefront in the figure becomes more divergent, i.e. it has an increased negative vergence. The action of any thin lens on the vergence of a wavefront (in air, where $n = 1$) is expressed in the *lens equation*:

$$\boxed{\frac{1}{v} - \frac{1}{u} = \frac{1}{f}.}$$ (2.10)

Example. A camera has a simple lens with a focal length $5\,\text{cm}$ ($=0.05\,\text{m}$). How far from the film must the lens be for a flower $0.7\,\text{m}$ from the lens to be in focus?

Solution. With rays diverging from the object the vergence $1/u$ and u are both negative: $u = -0.7\,\text{m}$. Solving equation (2.10) for v: $v = fu/(f + u) = 0.05(-0.7)/(0.05 - 0.7) = 0.054\,\text{m} = 54\,\text{mm}$.

We have already related the focal length of the lens to the refractive index n and the radius of curvature r of the two equally curved surfaces. A simple extension of this analysis to a thin lens with different radii of curvature r_1, r_2 (Section 2.4) gives the focal length

$$\boxed{\frac{1}{f} = (n - 1)\left(\frac{1}{r_1} - \frac{1}{r_2}\right)}$$ (2.11)

where the subscripts refer to the first and second interfaces crossed by the incident light.

For a biconvex lens such as that shown in Figure 2.3(a), $r_1 > 0$ and $r_2 < 0$; hence both surfaces add to the positive value of the power. For the diverging lens in Figure 2.3(b), the signs of r_1, r_2 are the opposite, and both contribute to a negative power.

The *power P* of a lens is defined as the inverse of its focal length, so that $P = 1/f$; measuring f in metres, the power of a lens is specified in *dioptres* ($D = m^{-1}$). If two thin lenses are placed close together or in contact their powers simply add, just as a contact lens adds to (or subtracts from) the power of the unaided eye.

Example. Consider a lens made of glass with $n = 1.5$ and $r_1 = 20\,\text{cm}$, $r_2 = -33.3\,\text{cm}$. Find its power and its focal length.

Solution. In metres: $1/f = (1.5 - 1)(1/0.20 + 1/0.333) = 4$. The power is 4 dioptres and the focal length is $0.25\,\text{m}$.

2.3   Refraction at a Spherical Surface

We now apply the concept of vergence to refraction at a single spherical surface between media with refractive indices n_1 and n_2, as in Figure 2.5. Note the sign of the radii of curvature: if the centre of curvature C is on the same side as the incident light, then $r < 0$, and on the opposite side $r > 0$.

To find the power of the refracting surface, we trace a ray from the object point P_1 to the image point P_2. Note that the labelled angles should all be considered small, so that sines and tangents are approximated by the angle itself, and the point A is taken to be not far from the axis P_1CP_2. This is

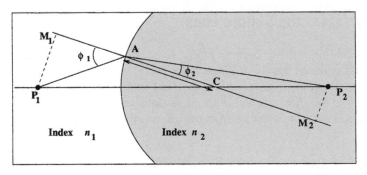

Figure 2.5   Geometry of a ray refracted at a spherical surface between media of refractive indices n_1 and n_2. P_1 and P_2 are *conjugate points*. The surface as shown has positive power since $n_2 > n_1$

the *paraxial* approximation, which applies to rays which are not far from parallel to the optic axis. Then we can take the object distance[4] $AP_1 = u < 0$ and the image distance $AP_2 = v$ as in our previous analysis of the lens. The relation between object distance and image distance is obtained by constructing perpendiculars P_1M_1 and P_2M_2 to the radial line through A, when the similar triangles P_1M_1C, P_2M_2C give the exact equation:

$$\frac{CM_1}{P_1M_1} = \frac{CM_2}{P_2M_2}. \tag{2.12}$$

Inserting the approximation that $P_1A = -u$, and $AP_2 = v$, we find

$$\frac{-u\cos\phi_1 + r}{-u\sin\phi_1} = \frac{v\cos\phi_2 - r}{v\sin\phi_2}. \tag{2.13}$$

Using the relation $n_1 \sin\phi_1 = n_2 \sin\phi_2$, and setting the cosines to unity in the paraxial approximation, this becomes:

$$\boxed{\frac{n_2}{v} - \frac{n_1}{u} = \frac{n_2 - n_1}{r} = P} \tag{2.14}$$

where P is defined as the power of the surface.

Example.   A long plastic rod of refractive index $n = 1.4$ has a radius of 1 cm and a convex spherical endface of the same radius. Where is the image of a small light bulb 10 cm from its endface?

Solution.   Using $n_2/v - n_1/u = (n_2 - n_1)/r$, we find $v = n_2[n_1/u + (n_2 - n_1)/r]^{-1}$. So $v = 1.4(1/u + 0.4/r)^{-1} = 1.4(-1/10 + 0.4)^{-1}$ cm $= 1.4$ cm$/0.3 = 4.7$ cm. With v positive, we know the rays converge to a real image point within the glass. (If v were < 0, the rays in the glass would be divergent and could be traced back to a virtual image point in the air.)

---

[4]Note the sign: this accords with the definition of vergence, and also with a Cartesian coordinate system with light travelling from left to right.

2.4 Two Surfaces; the Simple Lens

The simple thin lens in air, with two convex surfaces, is analysed by adding two equations of the form of equation (2.14) and assuming that the thickness of the lens is negligible. We give a negative sign to the second radius since the centre of curvature is to the left. For the first surface we set $n_1 = 1$ and $n_2 = n$, the refractive index of the glass, and find an image distance v_1, which becomes the object distance for the second surface. For object distance u from the lens we obtain for the first surface

$$\frac{n}{v_1} - \frac{1}{u} = \frac{n-1}{r_1} \tag{2.15}$$

and for the second surface, refracting from glass to air,

$$\frac{1}{v} - \frac{n}{v_1} = \frac{1-n}{r_2}. \tag{2.16}$$

The sum of these gives the lens equation

$$\frac{1}{v} - \frac{1}{u} = (n-1)\left(\frac{1}{r_1} - \frac{1}{r_2}\right) \tag{2.17}$$

which substantiates equation (2.11).

The power of a thin lens is the sum of the powers of the two surfaces. If the object is at infinity, v in equation (2.17) becomes the focal length f. The power is then $1/f$.

2.5 Imaging in Spherical Mirrors

Figure 2.6(a) shows the action of a spherical concave mirror M on a wavefront, illustrating the similarity with the action of a lens as in Figure 2.3. Figure 2.6(b) shows the geometry of an axial ray P_1CV and a ray at a small angle to the axis. A ray from the object at P_1 is reflected at A on the mirror surface, and reaches the image point P_2 on the axis, which is defined by the line from P_1 through the centre of curvature C. The angles θ of incidence and reflection are equal, so that the angle P_1AP_2 is bisected by the line AC. Because the angles P_1CA and P_2CA are supplementary, they have equal sines; the law of sines[5] then gives us the exact relation

$$\frac{P_1C}{P_1A} = \frac{CP_2}{P_2A}. \tag{2.18}$$

Following the vergence through this system according to equation (2.9), note that distances u and radius of curvature r are both negative, so that equation (2.18) becomes

$$\frac{r-u}{u} = \frac{r+v}{v} \tag{2.19}$$

---

[5]The law of sines asserts that in a triangle the side lengths are proportional to the sines of the opposite vertex angles.

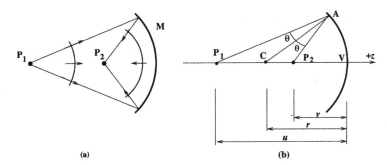

Figure 2.6 A concave spherical mirror: (a) action of the mirror on a wavefront; (b) the geometry of a paraxial ray

where we have used a paraxial approximation by writing $P_1A \approx P_1 V = -u$ and $P_2A \approx P_2V = v$. We obtain the *mirror formula*

$$\frac{1}{v} - \frac{1}{u} = -\frac{2}{r}. \tag{2.20}$$

The same equation applies for a convex mirror, having due regard for the sign convention. Equation (2.20) has the same form as equation (2.10) for a thin lens, provided we define the mirror's focal length by $f = -r/2$.

It is instructive to observe one's own image in convex and concave mirrors, especially noting the position and magnification of the image in a concave mirror as the object (the face!) is placed in front of or behind the centre of curvature. At the centre of curvature one's image is immediately in front of one's face, and so appears huge. Close to the mirror one sees a normal image, not much different from that in a plane mirror; outside the centre of curvature one sees an image not far behind the mirror, reduced in size and inverted.

Example. A shaving mirror has a concave surface on one side with a radius of curvature of 40 cm, and a plane mirror on the other side. When looking at oneself imaged in the plane side, how far from the mirror should one's face be for the image to be 30 cm away from the real face? The mirror equation is $1/v - 1/u = -2/r = 2/\infty = 0$, hence $v = u$ and one's face must be at $u = -15$ cm. Now repeat for the concave side of the mirror. You may ignore any real images.

Solution. We want our real face ($u < 0$) to form a virtual image ($v < 0$). This means $u + v = -0.3$ m, and the mirror equation is $1/v - 1/u = -2/(-0.4) = 5 \, \text{m}^{-1}$. This gives $v = u/(5u + 1)$. Substituting this into $v + u = -0.3$ gives $u^2 + 0.7u + 0.06 = 0$. This has two roots $u = -0.6, -0.1$. The first of these yields $v = 0.3$, i.e. a positive value indicating a real image. The other root $u = -0.1$ yields $v = -0.1/(1 - 5 \times 0.1) = -0.2$. We must therefore put our face 10 cm from the mirror to see its image 20 cm behind the mirror. Note that we get a real image, $v = u/(5u + 1) > 0$, whenever our real object is more than the focal length from our mirror (or $u < -(1/5)$ in this case). This is a general property of mirrors and thin lenses that converge.

2.6   General Properties of Imaging Systems

It is remarkable how well simple optical systems can work, despite the approximations that we have made in the lens theory. Even if the object point is at some distance from the axis of a simple lens, rays still converge on an off-axis image point found from the lens equation. A simple lens can therefore make an image of an extended object, in which the scale of the image is almost the same over a considerable area. The object and image planes containing an object and its image are called *conjugate planes*, and we now find the *magnification* of the image, which is the ratio between the sizes of the image and the object.

The geometric specification of a perfect optical system is that points and lines in the object space should correspond precisely to points and lines in the image space. Mathematically, the two spaces are linked by a projective transformation, and there must be a simple relation between distances in the object and image spaces. Equation (2.17) is an example of such a relation involving axial distances only. There is also a linear relationship between perpendicular distances, giving the *transverse magnification* of the system. The magnification depends of course on the positions of the conjugate planes containing the object and image.

We have so far considered only the theory of a thin spherical lens, but the same concepts can be applied to a lens whose thickness cannot be neglected, and to a multiple lens system such as those used in camera lenses (see Chapter 3). The important concept is to define planes in the system from which the axial distances should be measured. Figure 2.7 shows the location of the *principal planes* in a thick lens. These are the planes on which rays from a focal point intersect corresponding rays from a point at infinity; each focal length is measured from its corresponding principal plane. (It may also be convenient, as for example in the design of a camera, to define a *back focal length* as the distance from the back, or outgoing side, of a lens system to the focal plane.)

Example. *Unit magnification property of the principal planes.* Use ray tracing to prove that pairs of conjugate points on the principal planes are at the same height y. Given any off-axis object point in one principal plane, you will need to trace two different rays through it to locate its conjugate image point.

Solution. We apply the defining properties of the principal planes, PP_1, PP_2, shown in Figure 2.8. Ray a, passing from the first focal point through object point P on PP_1, emerges parallel to the optic axis; ray b, incident on P parallel to the axis, emerges through the second focal point. (The rays are drawn slightly separated for clarity.) On the image side, the outgoing rays intersect at point Q on PP_2. Since the rays travel parallel to the optic axis between P and Q, we see that these points are at the same height above the axis, F_1ABF_2.

The general linear relationship between distances in object and image planes becomes very simple when axial distances of object and image are measured from the focal planes, as in Figure 2.9. All distances are signed as shown: a leftward arrow signifies a negative quantity, and a rightward arrow positive. Denoting axial and transverse distances by z and y, and using subscripts 1 and 2 for the object and image spaces, the relationship deviced from similar triangles is

$$\boxed{\frac{f_1}{z_1} = \frac{z_2}{f_2} = -\frac{y_2}{y_1}.}$$

$$(2.21)$$

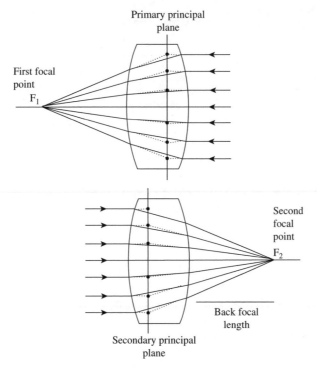

Figure 2.7   A thick lens. Rays from infinity converge on the focal points F_1, F_2. The two principal planes are located by extending each incident and outgoing ray along straight lines until they intersect. The back focal length is measured from the surface of the lens that faces away from the incident light

The *transverse* or lateral magnification is defined as $m_T = y_2/y_1$. As noted in Section 2.8, the focal lengths on the two sides are related by

$$\boxed{\frac{f_1}{f_2} = -\frac{n_1}{n_2}}$$ (2.22)

where n_1, n_2 are the refractive indices in the image and object spaces. For a system immersed in air, $n_1 = n_2$, and $f_1 = -f_2$. Equation (2.21) predicts that when $z_1 = -f_1$, $y_1 = y_2$ and $z_2 = -f_2$. This

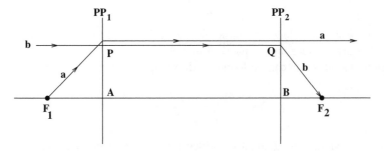

Figure 2.8   Unit magnification between principal planes

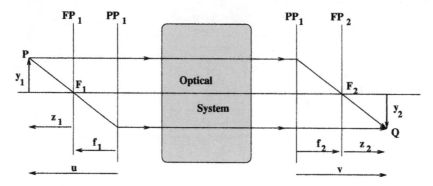

Figure 2.9 Coordinate systems for any axisymmetric, paraxial optical systems. P and Q are conjugate points. Axial distances z_1, z_2 are measured from the focal planes F_1, F_2

shows that the principal planes are conjugate planes of unit magnification: if the object point is in one, the image point is in the other and at the same height. By substituting $z_1 = 0$, we obtain $z_2 = \infty$, infinite magnification, and each focal plane has a conjugate plane at infinity. The constants f_1 and f_2 are the *principal focal lengths* of the system.

Equation (2.21) contains Newton's equation

$$z_1 z_2 = f_1 f_2. \tag{2.23}$$

Note that z_1 and z_2, like f_1 and f_2, will always have opposite signs (see Figure 2.9).

We now obtain a general Gaussian equation in place of equation (2.23), eliminating z_1, z_2 in favour of the object and image distances u, v as measured from their respective principal planes (see Figure 2.9). Substituting $z_1 = u - f_1$, $z_2 = v - f_2$ into equation (2.23), we find

$$(u - f_1)(v - f_2) = uv - vf_1 - uf_2 + f_1 f_2 = f_1 f_2. \tag{2.24}$$

This simplifies to $f_2/v + f_1/u = 1$, and insertion of $f_1/f_2 = -n_1/n_2$ from equation (2.22) converts this to the desired *Gaussian equation*:

$$\frac{n_2}{v} - \frac{n_1}{u} = \frac{n_2}{f_2} = \frac{-n_1}{f_1}. \tag{2.25}$$

It should be noted that this includes the basic equations (2.10), (2.14), (2.20) respectively for a thin lens, a single refractive surface and a spherical mirror.

A *longitudinal magnification* can be found by differentiating Newton's equation:

$$m_{\mathrm{L}} = \frac{\mathrm{d}z_2}{\mathrm{d}z_1} = -\frac{f_1 f_2}{z_1^2} = -\frac{z_2}{z_1}. \tag{2.26}$$

This indicates, for example, the amount of refocusing required when an object moves closer to a camera lens.

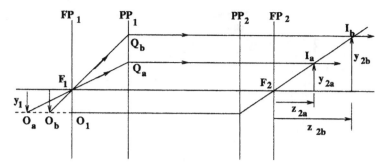

Figure 2.10 Finding the conjugate of a focal plane by ray tracing.

An *angular magnification* m_A is defined by the ratio of angles at which a ray cuts the axis in the image and object planes. It is given by the ratio of the transverse and longitudinal magnifications:

$$m_A = \frac{z_1\, y_2}{z_2\, y_1} = -\frac{f_1}{z_2} = -\frac{z_1}{f_2} \tag{2.27}$$

where we used equation (2.21) in the last two members.

Example. Consider an object point O approaching the left-hand focal plane FP_1 at a constant distance off-axis ($y_1 \neq 0$). By tracing rays through the system, make a sketch indicating that the coordinates y_2, z_2 of the image point both tend to infinity.

Solution. In Figure 2.10, we see that as the object point $(O_a, O_b, \ldots)$ moves closer to O_1 on the focal plane, the ray from O through the focal point F_1 becomes steeper; thus the image point $(I_a, I_b, \ldots)$ recedes to infinity both vertically and horizontally.

2.7   Separated Thin Lenses in Air

Many optical systems use components which are themselves made up of two or more lenses, as for example in a telescope eyepiece. The analysis of such systems by the repeated use of the simple lens formula (equation (2.17)) soon leads to tedious algebra, and it is more usual to follow a ray-tracing procedure. We now analyse the separated pair of Figure 2.11 in this way, following an incident ray parallel to the axis as it is deviated by each lens.

At some distance from the axis the lens acts like a thin prism, as in Section 2.2. From equation (2.7) the angular deviation D of the ray at a distance y from the axis of a thin lens of power P is

$$D = (n-1)y\left(\frac{1}{r_1} - \frac{1}{r_2}\right) = yP. \tag{2.28}$$

We have seen in Figure 2.1(b) that for paraxial rays the angular deviation is almost independent of the incident angle. The ray in Figure 2.11 which meets the first lens at a distance y_a from the axis, and then the second at distance y_b from the axis, has a total angular deviation given by the sum

$$D_{\text{tot}} = D_a + D_b = y_a P_a + y_b P_b. \tag{2.29}$$

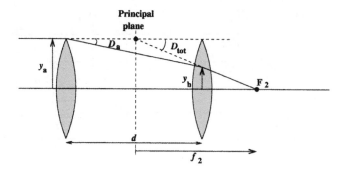

Figure 2.11 Ray tracing in a separated lens system. A ray parallel to the axis is deviated in both lenses, and crosses the axis at the focus F$_2$

As can be seen in Figure 2.11, $y_b = y_a - dD_a$. Equation (2.29) therefore becomes

$$D_{\text{tot}} = y_a(P_a + P_b - dP_aP_b). \tag{2.30}$$

The power P_{tot} of the combination, defined as f_2^{-1} where f_2 is the principal image-side focal length, is $P_{\text{tot}} = D_{\text{tot}}/y_a$. The power of the pair of lenses separated by distance d is therefore

$$P_{\text{tot}} = P_a + P_b - dP_aP_b. \tag{2.31}$$

The focal point can also be found geometrically from the figure, without recourse to tedious algebra. Note that the power of the combination is less than the sum of their individual powers, unless they are in contact, when the powers add directly.

If the space between the lenses has refractive index n, equation (2.31) should read

$$P_{\text{tot}} = P_a + P_b - \frac{d}{n}P_aP_b. \tag{2.32}$$

The powers P_a, P_b are the powers of a refractive spherical interface as in equation (2.14) (see Problem 2.6). This applies for example to thick lenses, as in the following example.

Example. Find the power of a spherical glass lens, radius R, refractive index n.

Solution. Using equation (2.14), both faces of the globe have the same power $P_a = (n-1)/R = (1-n)/(-R) = P_b$. Then equation (2.32) gives $P = 2(n-1)/R - (n-1)^2/R^2(2R/n)$, or $P = 2(n-1)/(nR)$.

2.8   Ray Tracing by Matrices

Extending the ray-tracing example of Section 2.7 to more complex multiple lens systems, such as those used in camera and microscope lenses, is conveniently achieved by a matrix method that follows a ray through a series of surfaces and the space between them.

A ray at distance z along the axis is specified by its height y above the axis and its angle θ to the axis. For definiteness, the ray is traced from an input plane to an output plane (Figure 2.12). These

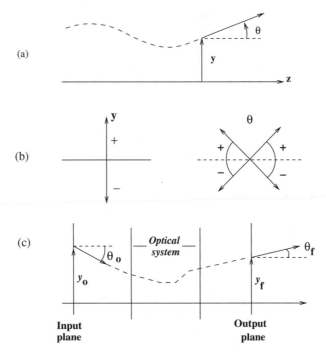

Figure 2.12   Ray tracing: (a) the height and angle of a ray are measured relative to the optic axis; (b) sign conventions for ray height and angle; (c) rays passing through the optical system are traced from input plane to output plane

planes are chosen with considerable freedom, but often the simplest choice is to place them at the outermost vertices of the system, where the optical elements intersect the optic axis. Sometimes, e.g. in a thin lens or reflecting system, a single plane may serve for both input and output.

In the paraxial approximation, we can find a 2×2 matrix M that converts the input values (y_0, θ_0) to the output values (y_f, θ_f) by matrix multiplication. But the ability to trace arbitrary rays in this fashion is not the point. The real payoff is that the existence and properties of the cardinal points follow from M; these in turn lead, for paraxial systems, to general results such as those we discussed in Section 2.6.

Figure 2.13(a) shows the progress of the ray along a distance $d = |z_2 - z_1|$ in a homogeneous medium, when the value of y increases by $d \tan \theta_1$. With the paraxial approximation $\tan \theta = \theta$, this gives the simple transformation

$$
\begin{aligned}
y_2 &= y_1 + \theta_1 d \\
\theta_2 &= \theta_1.
\end{aligned}
\tag{2.33}
$$

At a plane surface separating media with refractive indices n_1 and n_2 (Figure 2.13(b)), the transformation is

$$
\begin{aligned}
y_2 &= y_1 \\
\theta_2 &= \theta_1 \frac{n_1}{n_2}.
\end{aligned}
\tag{2.34}
$$

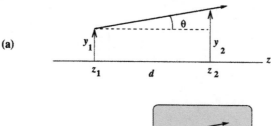

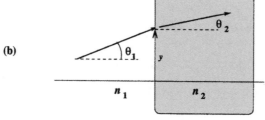

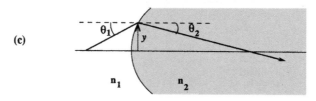

Figure 2.13   Ray tracing: (a) in a homogeneous medium; (b) at a plane surface separating media with refractive indices n_1 and n_2; (c) at a curved boundary

At a curved boundary (Figure 2.13(c)), the angular transformation follows from equation (2.14), if we put $\theta_1 = -y_1/u, \theta_2 = -y_2/v$, so that

$$y_2 = y_1$$
$$\theta_2 = \theta_1 \frac{n_1}{n_2} + \left(\frac{n_1 - n_2}{n_2}\right) \frac{y_1}{r}. \tag{2.35}$$

For a thin lens or a curved mirror, with focal length f, the transformation is

$$y_2 = y_1$$
$$\theta_2 = \theta_1 - \frac{y_1}{f}. \tag{2.36}$$

These transformations can be expressed in matrix form as

$$\begin{vmatrix} y_2 \\ \theta_2 \end{vmatrix} = \begin{vmatrix} M_{11} & M_{12} \\ M_{21} & M_{22} \end{vmatrix} \begin{vmatrix} y_1 \\ \theta_1 \end{vmatrix}. \tag{2.37}$$

In this equation the transformation matrix

$$M = \begin{vmatrix} M_{11} & M_{12} \\ M_{21} & M_{22} \end{vmatrix} = \begin{vmatrix} \alpha & \beta \\ \gamma & \delta \end{vmatrix}, \tag{2.38}$$

known as the *ray transfer matrix*, represents the action of an optical element on the ray. The advantage of this matrix representation is that a series of surfaces and spaces with ray matrices $M_1, M_2, M_3, \ldots, M_N$ is represented by a single matrix that is their product:

$$M = M_N \ldots M_3 M_2 M_1. \tag{2.39}$$

Notice that for light undergoing processes represented by the sequence 1,2,3... the matrices are multiplied in reverse order.

Table 2.1 gives examples of ray matrices corresponding to equations (2.33), (2.35), (2.36) above.

Table 2.1   Ray matrices

Optical element	Ray matrix	Notation
Uniform medium	$\begin{vmatrix} 1 & d \\ 0 & 1 \end{vmatrix}$	Distance d
Spherical interface	$\begin{vmatrix} 1 & 0 \\ (n_1 - n_2)/(n_2 r) & n_1/n_2 \end{vmatrix}$	Radius r, refractive indices n_1, n_2
Thin lens or mirror	$\begin{vmatrix} 1 & 0 \\ -1/f & 1 \end{vmatrix}$	Focal length $f = -r/2$ for mirror
		$f = [(n-1)(1/r_1 - 1/r_2)]^{-1}$

For two lenses in contact the combined ray matrix is the product[6] of their individual matrices:

$$M = \begin{vmatrix} 1 & 0 \\ -1/f_b & 1 \end{vmatrix} \begin{vmatrix} 1 & 0 \\ -1/f_a & 1 \end{vmatrix} = \begin{vmatrix} 1 & 0 \\ -(1/f_a + 1/f_b) & 1 \end{vmatrix} \tag{2.40}$$

showing that the combination acts like a single thin lens with a power $P = 1/f$ equal to the sum of the powers of the two lenses.

Notice that with all the basic optical elements shown in Table 2.1, we find the determinant $\det M = \alpha\delta - \beta\gamma = n_1/n_2$. (For the first and last cases, where n does not change, this reduces to 1.) Suppose that a light ray passing through our system encounters refractive indices $n_0, n_a, n_b, \ldots, n_y, n_z, n_f$, in that order. Let us prove

$$\det M = n_0/n_f, \tag{2.41}$$

---

[6]The product of an $m \times p$ matrix $A = |a_{ij}|$ with a $p \times n$ matrix $B = |b_{ij}|$ is an $m \times n$ matrix C with elements

$$c_{ij} = \sum_{k=1}^{p} a_{ik} b_{kj}.$$

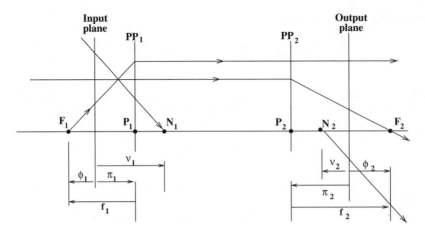

Figure 2.14 Location of the input and output planes, principal planes and the six cardinal points in an arbitrary paraxial system. Incident and transmitted rays illustrate the defining properties of the principal planes and the cardinal points. Arrows to the right or left denote, respectively, positive or negative displacements

which provides a useful check on the ray matrix. If M is the product of matrices as shown in equation (2.39), the determinants follow the rule det $M = $ det $M_N \dots$ det $M_3.$det $M_2.$det M_1. All these factors are unity except for interfaces between media; including only the latter, we can write

$$\det M = \frac{n_z}{n_f} \frac{n_y}{n_z} \frac{n_x}{n_y} \dots \frac{n_a}{n_b} \frac{n_0}{n_a} = \frac{n_0}{n_f}, \tag{2.42}$$

and so, by a series of cancellations, we verify equation (2.41).

Given a matrix M, one can show with trigonometry (Section 2.9) that the cardinal points exist and are unique. The six cardinal points are shown in Figure 2.14; they are the focal points $F_{1,2}$, the principal points $P_{1,2}$ and the *nodal points* $N_{1,2}$. The two nodal points are unique points on the axis such that any off-axis ray aimed at N_1 emerges as a conjugate ray parallel to the first and from the direction of N_2.[7] The cardinal points are located relative to the input or output planes by the signed distances given in Table 2.2.

In order to cast our generic equation (2.25) into the desired form of $V' - V = P$, we define the system's power by $P = n_f/f_2 = -n_0/f_1 = -n_f \gamma$. As an application, consider the pair of separated lenses in Section 2.7. For simplicity, take the input and output planes at the two thin lenses. Multiply right-to-left the matrices for the first lens, the space between and the second lens:

$$M = \begin{vmatrix} 1 & 0 \\ -1/f_b & 1 \end{vmatrix} \begin{vmatrix} 1 & d \\ 0 & 1 \end{vmatrix} \begin{vmatrix} 1 & 0 \\ -1/f_a & 1 \end{vmatrix} \tag{2.43}$$

---

[7]For reflective systems, nodal points still exist if we consider conjugate rays to be 'parallel' when they have equal angles: $\theta_f = \theta_0$.

to obtain

$$M = \begin{vmatrix} 1 - d/f_a & d \\ d/f_a f_b - (1/f_a + 1/f_b) & 1 - d/f_b \end{vmatrix}. \tag{2.44}$$

By inspection, we read off $\gamma = M_{21}$, and with $n_f = 1$, the total power of the system is $P_{\text{tot}} = -8 = 1/f_a + 1/f_b - d/(f_a f_b)$, which agrees with equation (2.31).

Table 2.2 Positions of cardinal points

Cardinal point	Position relative to	Displacement
F_1	Input plane	$\phi_1 = \delta/\gamma$
F_2	Output plane	$\phi_2 = -\alpha/\gamma$
P_1	Input plane	$\pi_1 = (\delta - n_0/n_f)/\gamma$
P_2	Output plane	$\pi_2 = (1 - \alpha)/\gamma$
N_1	Input plane	$v_1 = (\delta - 1)/\gamma$
N_2	Output plane	$v_2 = (n_0/n_f - \alpha)/\gamma$
F_1	Principal plane 1	$f_1 = \phi_1 - \pi_1 = (n_0/n_f)/\gamma$
F_2	Principal plane 2	$f_2 = \phi_2 - \pi_2 = -1/\gamma$

2.9 Locating the Cardinal Points

From Table 2.2, we see that the four independent variables α, γ, δ and n_0/n_f determine the locations of the cardinal points. γ expresses the power of the system, while δ and α reflect, respectively, the arbitrary positions of the input and output planes. (But where is the matrix element β? Substituting equation (2.41), namely $\alpha\delta - \beta\gamma = n_0/n_f$, we could easily rewrite the entries in the table so as to include β.)

The entries in the table can be derived from Figure 2.14 as follows.

2.9.1 Position of a Nodal Point

In Figure 2.15(a), we illustrate a ray directed at one nodal point, and its parallel conjugate ray outgoing from the direction of the other nodal point. Based on the second line of equation (2.36), the parallelism of initial and final rays requires

$$\theta_f = \gamma y_0 + \delta\theta_0 = \theta_0. \tag{2.45}$$

Substituting into this from Figure 2.14(a) the small-angle approximation $y_0 = -v_1 \tan\theta_0 = -v_1\theta_0$ gives the displacement from the input plane of the first nodal point: $v_1 = (\delta - 1)/\gamma$.

2.9.2 Position of a Focal Point

Figure 2.15(b) illustrates a defining characteristic of a focal point, i.e. that a ray extended through the object side focal point will emerge parallel to the optic axis on the image side. For such a ray, the initial angle θ_0 is any (small) angle, but the final angle θ_f, vanishes:

$$\theta_f = \gamma y_0 + \delta\theta_0 = 0. \tag{2.46}$$

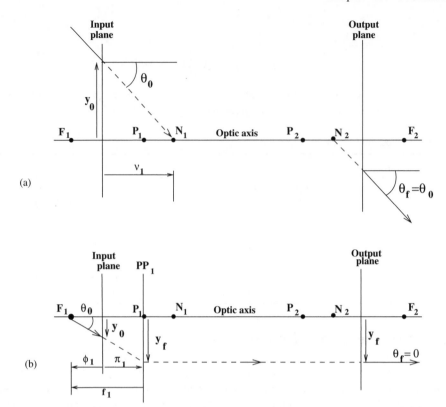

Figure 2.15 Location of (a) the nodal points, (b) a principal point

Substituting into this $y_0 = -\phi_1 \tan\theta_0 = -\phi_1\theta_0$ leads at once to the displacement from the input plane of the first focal point: $\phi_1 = \delta/\gamma$.

2.9.3 Position of a Principal Point

Figure 2.15(b) shows a ray passing through focal point F_1. Equation (2.37) gives

$$y_f = \alpha y_0 + \beta\theta_0 = y_0 + \pi_1\theta_0$$
$$\theta_f = \gamma y_0 + \delta\theta_0 = 0. \tag{2.47}$$

Notice that we have augmented the first line with the defining property of the first principal plane, i.e. that the incident focal ray has already achieved its final distance off-axis, y_f, on that plane. We know that the resulting homogeneous pair of equations

$$(\alpha - 1)y_0 + (\beta - \pi_1)\theta_0 = 0$$
$$\gamma y_0 + \delta\theta_0 = 0 \tag{2.48}$$

has a non-zero solution for (y_0, θ_0) only if the determinant of their coefficients vanishes: $(\alpha - 1)\delta - (\beta - \pi_1)\gamma = 0$. If we combine this with equation (2.41), we find the displacement from the input plane of the first principal point, $\pi_1 = (\delta - n_0/n_f)/\gamma$.

2.9.4 A Focal Length

It follows from the two preceding results that

$$f_1 = \phi_1 - \pi_1 = (n_0/n_f)/\gamma. \tag{2.49}$$

2.9.5 The Other Cardinal Points

So far we have derived four of the eight formulae in Table 2.2. The remaining four are easily derived by exploiting symmetry: the system is invariant under light ray reversal (see Problem 2.9).

Apart from the usefulness of locating the cardinal points, we note that for any paraxial system consisting of the basic optical elements mentioned above–or, equivalently, given its transfer matrix M–all six cardinal points exist[8] and are unique. This powerful result is amazing considering the infinite variety of paraxial systems one might put together from the basic elements.

2.10 Perfect Imaging

An ideal, or perfect, optical system would be one in which every point in an object space corresponds precisely to a point in an image space, being connected to it by rays passing through all points of the optical system. The optical path from any object point to its image is then the same along all rays. There is a fundamental reason, first formulated by Maxwell, why this cannot be achieved in any but the most elementary optical system. He showed that a perfect optical system can only give a magnification equal to the ratio of the refractive indices in the object and image spaces (Figure 2.16). For example, if object and image are both in air, the magnification can only be unity, which may not be very useful. A plane mirror may be perfect, but a magnifying lens cannot be.

The following demonstration of Maxwell's theorem is due to Lenz. The theorem states effectively that if two object points A_1, B_1 in a medium of refractive index n_1 give rise to image points A_2, B_2 where the refractive index is n_2, the optical paths over A_1B_1 and A_2B_2 must be equal. Suppose in Figure 2.17 the rays A_1B_1 and B_1A_1 can both pass through the optical system. They must then pass through B_2A_2 and A_2B_2 respectively. Since both optical paths from A_1 to A_2 must have the same length, and also both optical paths from B_1 to B_2, it follows that the optical paths $n_1A_1B_1$ and $n_2A_2B_2$ must be the same. Let $\Omega(AB)$ be the optical path length evaluated over a line segment AB. Since

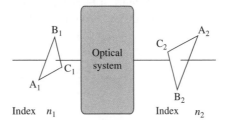

Figure 2.16 Maxwell's theorem for a 'perfect' system. Optical path lengths must be equal for corresponding parts of the object and image, so that for example $n_1A_1B_1 = n_2A_2B_2$

---

[8]We are assuming $\gamma \neq 0$. The so-called *afocal* case, where $\gamma = 0$, needs separate consideration.

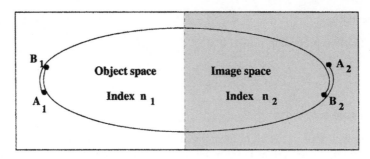

Figure 2.17   Lenz's proof of Maxwell's theorem. In a perfect system the optical paths A_1B_1 and A_2B_2 are equal

$\Omega(A_1B_1) = n_1A_1B_1$ and $\Omega(A_2B_2) = n_2A_2B_2$, we have

$$\Omega(B_1A_2B_2) = \Omega(A_1B_1A_2) - n_1A_1B_1 + n_2A_2B_2 \tag{2.50}$$

$$\Omega(B_1A_1B_2) = \Omega(A_1B_2A_2) + n_1A_1B_1 - n_2A_2B_2. \tag{2.51}$$

Subtracting the bottom equation from the top, we find $0 = 0 - 2n_1A_1B_1 + 2n_2A_2B_2$. This gives Maxwell's theorem

$$\frac{A_2B_2}{A_1B_1} = \frac{n_1}{n_2}. \tag{2.52}$$

This proof appears at first sight to be very limited, since the rays AB and BA can hardly be expected both to pass through the optical system. It may, however, be generalized by constructing a curve similar to that in Figure 2.17 but which is made up of many segments of actual rays, and integrating the whole path.

The simplest example of a perfect optical system is a plane mirror. A plane refracting surface, in contrast, only approaches perfection for rays which are nearly normal to its surface; away from the normal a bundle of rays from a single point does not form a point, or *stigmatic* image. A theoretical example of a perfect refracting system, known as the 'fish-eye' lens,[9] was invented by Maxwell; this uses an infinite spherical lens with refractive index varying with radius in such a way that all rays diverging from any point would converge on another point.

If in a more restricted system a single object point and its image point are specified, they can be connected by stigmatic rays in the optical systems of Figure 2.18. The ellipsoidal mirror has the two points as its two foci; if one point is infinitely distant, the reflector becomes the familiar paraboloid of

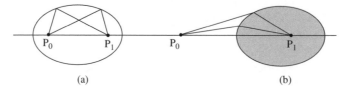

               (a)   (b)

Figure 2.18   Stigmatic imaging (a) in an ellipsoidal mirror, (b) in a refracting Cartesian oval

---

[9]See M. Born and E. Wolf, *Principles of Optics*, 2nd edn, Pergamon Press, 1980.

revolution used in reflecting telescopes and car headlights. The refracting surface is the more complicated Cartesian oval, named after Descartes.

2.11 Perfect Imaging of Surfaces

The severe restriction of the 'perfect' optical system, in which magnification can only be equal to the ratio of refractive indices in the object and image spaces, does not apply if the object points are restricted to lie on a single definite surface. This surface need not be plane, but the corresponding image points must lie on another conjugate surface if all points are to have sharp, or *stigmatic*, images.

An example of a curved but truly stigmatic imaging surface is provided by a spherical lens. Microscope objectives commonly use such a spherical lens, but with a flat face (see e.g. Figure 3.11). Figure 2.19 shows a homogeneous spherical lens, centre O, with radius a and refractive index n. A point source P_0 inside the lens is imaged at P_1 outside the lens. All rays leaving P_0 towards the left appear to diverge from a single point P_1. This is only possible when $OP_0 = a/n$ and $OP_1 = na$; the conjugate surfaces are therefore spherical.

It is, of course, not always convenient to restrict object and image surfaces to a special curve such as a sphere, but if it is required that either or both should be plane it will be necessary to abandon the requirement that the images should be strictly stigmatic. We therefore turn in the next section to the description and control of imperfections in optical images.

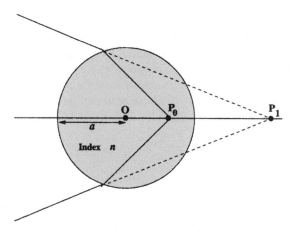

Figure 2.19 A spherical lens. All rays diverging from the point P_0 appear to diverge from the point P_1 when $OP_0 = a/n$ and $OP_1 = na$. Spheres centred on O with these radii are thus conjugate surfaces

2.12 Ray and Wave Aberrations

We have noted that a useful optical instrument can ideally only give stigmatic (sharp) images of points on a single surface, while even under this restriction the lenses or mirrors in the instrument cannot generally have simple spherical surfaces unless only a small bundle of paraxial rays is used to form the image (but note the special case of the spherical lens in Figure 2.19). In spite of this it is evident that many very useful optical instruments exist which do not conform strictly to these conditions. The quality of their images may not be ideal, but the departures from perfection, known as *aberrations*, may be tolerable for their purpose. The design of an optical system is inevitably concerned with the calculation of the various aberrations, and with their suppression below a tolerable level.

Aberration is minimized in cameras and other optical systems by the use of multiple lens systems, which we briefly describe. In astronomical telescopes a near-perfect image may be spoilt by distortions in the wavefront arriving at the telescope, due to refraction in the atmosphere; the image may then be improved by using adaptive optics in which compensating distortions are introduced into the optical system.

Aberration may be specified for any ray which contributes to the formation of a point image. The distance between an ideal image point and the intersection of the ray with the image plane is called the *ray aberration*. The total effect on the image is found by tracing sufficient rays from an object point so that the spread of intensity across the image can be found. Ray aberration therefore implies the enlargement of an ideal image point; its importance may be judged in relation to the size of the diffraction patch which is the lower limit to the size of the image of a point object below which even an ideal instrument cannot go (see Chapter 10). Alternatively the point image may be considered as the centre of a convergent wave, ideally spherical but in practice departing from sphericity; the departures are known as *wave aberrations*. The relation between ray and wave aberration is seen in Figure 2.20.

Wave aberration for an object on-axis may amount to some tens of wavelengths in a good camera lens, but it usually is less than one wavelength in an astronomical telescope. The corresponding ray aberrations may be found by geometric ray tracing rather than by analysis of wavefronts; there is, however, no need to draw a sharp distinction since both approaches lead to similar analytic results. The wave aberrations offer a clearer physical picture, as set out in the next section. Ray aberrations may be found from any pattern of wavefront aberrations by drawing ray normals from the wavefront, as in Figure 2.20. The intensity at the nominal image point is best found from the wave aberrations, since these give directly the pattern of waves which must be added to give the amplitude at the image point. The efficiency with which light is concentrated into the image point increases with decreasing wave aberration until the optical path introduced by wave aberration becomes small compared with λ.

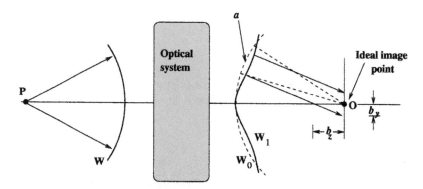

Figure 2.20   Ray and wave aberrations. A spherical wave W leaving the point P is focused by the optical system into a converging wave W_1, which departs from the ideal spherical shape W_0, centred on O. Wave aberrations are shown as a, and ray aberrations as b_y (transverse), b_z (longitudinal)

2.13   Wave Aberration On-axis – Spherical Aberration

As soon as it is admitted that a particular optical instrument, such as a camera, cannot meet the ideal of producing stigmatic images over the whole of an image, the possible range of optical designs at once becomes infinite, as does the variety of aberration patterns over the image. A simple pattern does, however, emerge from refraction or reflection at a spherical surface. Here the pattern of aberrations separates into parts which depend on the angular spread of rays from a single on-axis object, and to the width of the field containing the object (see Chapter 3 on the design of cameras). In other words, the transverse sizes of the aperture and of the object (for a given object distance) are basic.

Following Figure 2.21, the difference a in optical path between the axial ray and the ray intersecting the surface at a distance y from the axis is given by

$$a = n_1 \mathrm{AP}_1 + n_2 \mathrm{AP}_2 - n_1 |u| - n_2 |v|. \tag{2.53}$$

The distance y is taken for convenience as the chord length CA, so that the following geometrical relations hold (assuming $u < 0$ and $v > 0$):

$$\cos \psi = y/2r \quad \text{(from the isosceles triangle AOC)}$$

$$\mathrm{AP}_1 = (u^2 + y^2 - 2uy \cos \psi)^{1/2} = -u \left(1 + \frac{y^2}{u^2} - \frac{y^2}{ur} \right)^{1/2} \tag{2.54}$$

$$\mathrm{AP}_2 = (v^2 + y^2 - 2vy \cos \psi)^{1/2} = v \left(1 + \frac{y^2}{v^2} - \frac{y^2}{vr} \right)^{1/2}.$$

The wave aberration, or difference in optical path, is then found by expanding equation (2.54) as a power series in y^2 which depends on the off-axis distance y as

$$a = \frac{y^2}{2} \left[n_1 \left(-\frac{1}{u} + \frac{1}{r} \right) + n_2 \left(\frac{1}{v} - \frac{1}{r} \right) \right] - \frac{y^4}{8} \left[-\frac{n_1}{u} \left(-\frac{1}{u} + \frac{1}{r} \right)^2 \right.$$

$$\left. + \frac{n_2}{v} \left(\frac{1}{v} - \frac{1}{r} \right)^2 \right] + \text{terms of higher order in } y. \tag{2.55}$$

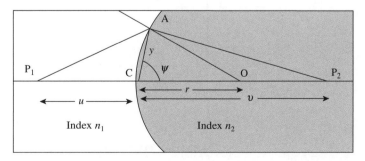

Figure 2.21   Spherical aberration at a single spherical surface. The wave aberration for a ray at a distance y from the axis is found from the small difference between the optical paths $\mathrm{P}_1\mathrm{AP}_2$ and $\mathrm{P}_1\mathrm{CP}_2$

As the radius of the aperture increases, so must the approximation in equation (2.55) be taken to higher orders in y. For paraxial rays where y is small, only the term in y^2 need be considered, and a is zero when the first half of equation (2.55) is zero. This gives the simple formula for refraction at a single surface (Equation (2.14), which is the relation between conjugate points). The second half of equation (2.55) is then the *spherical aberration* expressed as a wave aberration. The magnitude of spherical aberration increases as the fourth power of the aperture of a spherical refractor.

The ray passing through A is normal to the wavefront, so that its direction departs from the correct direction AP_2 by the angle between the ideal wavefront and the actual wavefront. This angle is found from the rate of variation of wavefront aberration a with increasing y; the angular deviation of a ray at P_2 therefore varies as $\partial a/\partial y$, i.e. as y^3 rather than y^4. The transverse ray aberration b_y increases as the cube of the aperture.

Example. Find the transverse and longitudinal spherical aberrations b_y and b_z for a ray parallel to the axis incident on a concave spherical mirror, radius of curvature R, if the ray is off-axis by a distance y (Figure 2.22). Assuming the paraxial condition $|y| \ll |R|$, expand b_y, b_z to lowest order in y. (Take all of R, y, θ, b_y, b_z as signed quantities.)

Solution. With $R < 0$,

$$CA = CF + FA = -R/2 + b_z \tag{2.56}$$

$$b_y = -b_z \tan 2\theta. \tag{2.57}$$

By dropping a perpendicular from A to CB, we see $-R = 2CA \cos\theta$ and by equation (2.56)

$$b_z = CA + R/2 = (R/2)(1 - 1/\cos\theta). \tag{2.58}$$

For small angles, $\cos\theta \simeq 1 - \theta^2/2, \tan 2\theta \simeq 2\theta$ and $\theta \simeq -y/R$ we find

$$b_z = -R\theta^2/4 = -y^2/4R \tag{2.59}$$

$$b_y = -b_z(-2y/R) = -y^3/2R^2. \tag{2.60}$$

Note that the powers of y in $b_y \propto y^3, b_z \propto y^2$ are the same as in the refractive case.

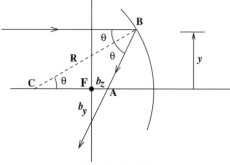

FP, the ideal image plane

Figure 2.22 Transverse and longitudinal aberrations b_y, b_z for a ray incident on a spherical mirror parallel to the axis. FP is the ideal focal plane. The mirror's centre of curvature is at C

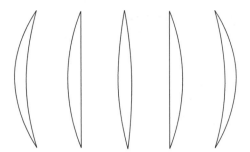

Figure 2.23 Lenses with spherical surfaces, and with the same focal lengths, but 'bent' by different amounts. Spherical aberration is minimized by using a lens shaped so that the refraction is shared roughly equally between the two surfaces; the plane or concave surface should therefore be closest to the nearer of the object and image points

Correction of spherical aberration is achieved very simply by changing the shape of the refracting surface. This can be made exactly correct for any chosen pair of conjugate points. Even if the surfaces are for simplicity constrained to be spherical, a lens may be corrected very well for spherical aberration by the 'bending' illustrated in Figure 2.23, where the surfaces are still spherical but have different radii of curvature. An exact correction requires the use of aspheric surfaces, which are frequently used to correct image distortion in optical systems. It is important to note that any correction can only apply exactly to one particular object distance, and that objects at a different distance will still suffer from spherical aberration.

Reflecting telescopes, and particularly the large reflector radio telescopes, commonly use apertures with diameters of the same order as their focal lengths. It is usual to remove the spherical aberration by making the surface a paraboloid of revolution (Figure 2.24), when the spherical aberration for an object on the axis at infinity is exactly zero. A paraboloid of revolution does not, however, form a perfect image for objects off the axis, and if it is intended to use an extended field of view in an optical or radio telescope it will be necessary to consider the off-axis aberrations, which grow more rapidly with angle for a paraboloid than for a spherical reflector. A system using a spherical mirror which avoids spherical aberration and still produces good off-axis images is used in the Schmidt

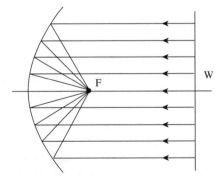

Figure 2.24 A section through a paraboloidal reflector telescope, showing rays from a distant object converging on the focus F. All optical paths from the wavefront W to the focus are exactly equal, so that there is no spherical aberration for waves from a distant object

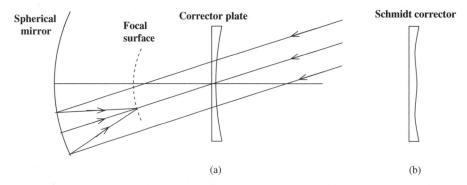

Figure 2.25 The Schmidt corrector plate (a) retards the wave in the outer parts of the aperture, removing spherical aberration. It is placed at the centre of curvature of a spherical mirror so that its effect is nearly independent of the ray inclination. A practical Schmidt plate is shown in (b); this combines the corrector with a very weak converging lens, so reducing the required overall thickness

telescope (Figure 2.25). Here a thin corrector plate located at the centre of curvature introduces a correction to the wavefront which compensates for spherical aberration. The location at the centre of curvature provides good compensation for a wide angle off-axis, although the focal surface is necessarily curved.

2.14   Off-axis Aberrations

The analysis in Section 2.13 of spherical aberration for an on-axis object point may be extended to several kinds of aberrations for an object point off-axis. The paraxial approximation involves setting $\sin \phi = \phi$ and $\cos \phi = 1$. This amounts to using the leading terms in the expansions

$$
\sin \phi = \phi - \frac{\phi^3}{3!} + \frac{\phi^5}{5!} - \cdots
$$
$$
\cos \phi = 1 - \frac{\phi^2}{2!} + \frac{\phi^4}{4!} - \cdots.
$$

$$(2.61)$$

Using the next higher order of approximation gives us the third-order, or Seidel, aberrations.

In the example that follows we will quote the wave aberration a, which is the difference of the optical path length along different paths. Ray aberrations, the deviations of rays from the ideal, paraxial image point, are then found by taking derivatives of the wave aberration.

Figure 2.26 shows several rays from an off-axis point P_1 being refracted by a single spherical interface. P_2 is the location of the ideal, paraxial image. The z axis coincides with the optic axis, the y axis is vertical, and the x axis points into the plane of the paper. Thanks to axial symmetry, rotation of the object point P_1 about the axis does not change the physical results. To make the set-up unique, we require P_1 to lie in the y, z plane. The ray P_1COP_2 through the centre of curvature O is a straight line because it strikes the interface normally. This implies that points C and P_2 also lie in the y, z plane. P_1COP_2 functions as a non-standard optic axis; relative to it, P_1 is on-axis and light emitted by it will therefore display only spherical aberration. But relative to the original axis DBOE, the description becomes more complex, and the various aberrations emerge.

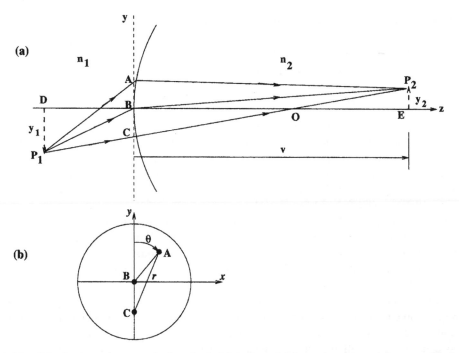

Figure 2.26   Off-axis aberration at a single spherical interface. (a) Several rays traced from an off-axis object point to its paraxial image point. Point O is the centre of curvature of the interface. (b) Appearance of the interface from the image side. An arbitrary point A is located with polar coordinates centred on the optic axis at point B. Note that the polar angle here increases clockwise from the *y* axis

If $\Omega(PQ)$ denotes the optical path over the segment PQ, it is convenient to define the wave aberration of point A relative to on-axis point B by $a(A) = \Omega(P_1AP_2) - \Omega(P_1BP_2)$. This has the advantage that $a(A)$ vanishes on-axis, where $A = B$. We can expand this in the polar coordinates of A and the image height as follows:[10]

$$a(A) = C_s\rho^4 + C_c y_2 \rho^3 \cos\theta + C_a y_2^2 \rho^2 \cos^2\theta + C_{cf} y_2^2 \rho^2 + C_d y_2^3 \rho \cos\theta. \tag{2.62}$$

The subscripts on the coefficients indicate the nature of the aberrations: s (spherical aberration), c (coma), a (astigmatism), cf (curvature of the field) and d (distortion).

Let v be the axial distance of paraxial image point P_2, and let the transverse rectangular coordinates of point A be $x = \rho \sin\theta$, $y = \rho \cos\theta$. As in Section 2.13, the angular deviation between the normals of the ideal and actual wavefronts can be found by taking a derivative of $a(A)$. Extending this to two dimensions, the transverse ray aberrations are given by

$$b_x = (v/n_2)\frac{\partial a(A)}{\partial x}$$
$$b_y = (v/n_2)\frac{\partial a(A)}{\partial y}. \tag{2.63}$$

[10]See for example F. Pedrotti and L. Pedrotti, *Introduction to Optics*, 2nd edn., Prentice Hall, 1993, sect. 5-2, and M. Born and E. Wolf, *Principles of Optics*, p. 211 *et seq.*

By evaluating these (as in Problem 2.20) one can find

$$b_x = (v/n_2)[4C_s\rho^3 \sin\theta + C_c y_2 \rho^2 \sin 2\theta + 2C_{cf} y_2^2 \rho \sin\theta]$$
$$b_y = (v/n_2)[4C_s\rho^3 \cos\theta + C_c y_2 \rho^2 (2 + \cos 2\theta) + (2C_a + 2C_{cf}) y_2^2 \rho \cos\theta \qquad (2.64)$$
$$+ C_d y_2^3].$$

For an object on-axis, $y_2 = 0$ and only the first terms, which go as ρ^3, survive. These constitute the spherical aberration already described. *Coma* is a wavefront distortion additional to spherical aberration, which only appears for object points off-axis. Rays intersect the image plane in a comet-like spread image, whose width and length increase with the square of the zonal radius ρ (Figure 2.27). The typical comatic image consists of superposed circular images, successively shifted further from the axis and focused less sharply.

Astigmatism is the result of a cylindrical wavefront aberration, which increases as the first power of ρ. The effect is unfortunately familiar in many human eyes, which show astigmatism even for objects on-axis. The focus, shown in Figure 2.27, consists of two concentrations of rays known as the focal lines, with a blurred circular region between representing the best approximation to a point focus. This is called the *circle of least confusion*.

The third term combines similar contributions from curvature of field and astigmatism. In it the wavefront has an added curvature proportional to image height squared, showing that the focal length of the lens changes for off-axis points. A flat object plane will then give a curved image surface. It is usual to find curvature still present in a lens which is corrected for astigmatism; this remaining curvature is referred to as the Petzval curvature.

Distortion represents an angular deviation of the wavefront, increasing as image height cubed. This spreads or contracts the image, destroying the linear relation between dimensions in object and image.

Since we know from the start that all aberrations cannot be eliminated from a useful optical system, it becomes a matter for choice which aberrations are the most nuisance and which can most easily be tolerated. For example, a photograph with distortion may be more displeasing to the eye than one with some blurring due to spherical aberration or coma. An astronomical photograph might on the other hand be required to show small symmetrical point images over the whole of a plate covering a large

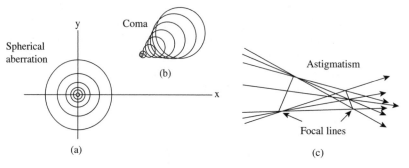

Figure 2.27 The effects of (a) spherical aberration, (b) coma and (c) astigmatism. In (a) and (b) the circles show the increasingly large images due to larger radii ρ in the optical system; in (b) these circles are displaced to form the comatic image. Patterns (a) and (b), though shown separate, actually superpose and coincide at the ideal image point, which is at the centre of the bull's eye and at the vertex of the wedge

solid angle, while it might be less important to minimize the distortion of angular scale near the edges of the plate.

We can now appreciate Feynman's remark that optics is either very simple (as in paraxial approximations) or very complicated (when a compromise must be made between conflicting aberrations). The difficult part is made easier by automatic methods of ray tracing, which can rapidly demonstrate the performance of any optical system, however complex. Many modern camera lenses use components with non-spherical surfaces, derived from computation programs which optimize performance. Such computational methods nevertheless require a performance specification and an initial outline solution, which can only be provided with a knowledge and understanding of the basic aberration theory.

2.15 The Influence of Aperture Stops

The amount of spherical aberration introduced by an uncorrected lens or reflector system varies as the cube of the lens aperture. If a large aperture is necessary, the aberration must be either tolerated or corrected, but an improvement in images can obviously be made by restricting the aperture by means of a stop. For the single purpose of restricting spherical aberration in a lens the stop would be placed against the lens itself, but the other aberrations are also affected by the stop in ways which depend on the separation of the stop from the lens. This is demonstrated in Figure 2.28, which shows a pencil of rays from an off-axis point passing through an aperture stop in front of a lens.

When the aperture stop is separated from the lens the rays from an off-axis point are constrained to pass through the outer part of the lens, as in (a). Depending on the shape of the lens, this may reduce or increase the off-axis aberrations. In (b), rays from an off-axis point reach the lens by a shorter path than in (a), and the magnification off-axis is therefore greater than in (a). Distortion can therefore be controlled by the correct positioning of the aperture stop.

2.16 The Correction of Chromatic Aberration

The power of a spherical refracting surface, radius r, is given by equation (2.14) as $(n_2 - n_1)/r$, where $n_2 - n_1$ is the difference of refractive index across the surface. So far no account has been

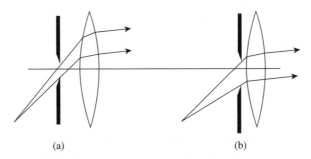

(a) (b)

Figure 2.28 The positioning of an aperture stop. In (a) the stop is spaced away from the lens, so that off-axis points are focused by the outer part of the lens. The shape of the lens may then be changed so that aberrations are reduced. In (b) the same part of the lens is used for all ray inclinations. Aberrations are then less controllable, although they will be smaller for spherical surfaces

taken of the need to focus light of a wide range of colour by the same optical system; since refractive index inevitably varies with the wavelength of the light, any optical system which depends on refraction rather than reflection will behave differently for different colours. *Chromatic aberration* is a measure of the spread of an image point over a range of colours. It may be represented either as a longitudinal movement of an image plane, or as a change in magnification, but basically it is a result of the dependence of the power of a refracting surface on wavelength. It may be compensated for by combining lenses made of different materials.

The power of a single thin lens used in air may be written as $P = (n-1)/R$, where $1/R = 1/r_1 - 1/r_2$. A small change of refractive index δn therefore changes the power by δP where

$$\delta P = P \frac{\delta n}{n-1}. \tag{2.65}$$

Varieties of optical glass differ quite widely in the way in which n varies with wavelength, so that it is possible to combine two lenses with power P_1 and P_2 in such a way that $\delta P_1 + \delta P_2 = 0$ without at the same time making the total power $P_1 + P_2 = 0$. Two colours separated in wavelength by $\delta\lambda$ will be focused together when

$$P_1 \frac{\delta n_1}{n_1 - 1} + P_2 \frac{\delta n_2}{n_2 - 1} = 0. \tag{2.66}$$

Since $\delta n/(n-1)$ has the same sign in the visible band for all glasses, this means that P_1 and P_2 must have the opposite sign, so that correcting chromatic aberration in this way reduces the power of a lens. The powers of the two components must also be inversely proportional to the value of $(dn/d\lambda)(n-1)^{-1}$ for the two glasses. The two lenses may be in contact if two surfaces have the same radii of curvature. It is advantageous to reduce the number of interfaces between glass and air, since light is lost by partial reflection at each step in refractive index. The step between the two kinds of glass is smaller than for interfaces between glass and air, but the advantage is lost unless the two lenses are cemented together using a transparent glue with a refractive index approximately the same as that for glass. Both the power and the focal plane of such an *achromatic doublet* can be made the same over a range of wavelengths, or at any two widely separated wavelengths. Outside these wavelengths, however, it will generally still suffer from some chromatic aberration.

The *dispersive power* of glass is often quoted in terms of refractive index at specific vacuum wavelengths, which have traditionally been those of the three *Fraunhofer lines* F, D and C. (These are prominent absorption lines in the solar spectrum.) Table 2.3 shows the refractive indices for representative examples of crown and flint glass.

Dispersive power Δ is defined as

$$\Delta = \frac{n_F - n_C}{n_D - 1} \tag{2.67}$$

Table 2.3   Refractive indices for crown and flint glass

Designation	Wavelength (nm)	Crown glass	Flint glass
F blue	486	1.5286	1.7328
D yellow	589	1.5230	1.7205
C red	656	1.5205	1.7076

so that typical dispersive powers of crown and flint glass are respectively 1/65 and 1/29. This definition of Δ extends the ratio $\delta n/(n-1)$ to a finite range of wavelengths. Since the deviation of a thin prism, by equation (2.6), is proportional to $(n-1)$, we can identify Δ as $\Delta = (\theta_F - \theta_C)/\theta_D$, i.e. the relative angular dispersion between the C and F lines.

2.17 Achromatism in Separated Lens Systems

The eyepieces of microscopes and telescopes often use a very simple system for achromatism, using two identical lenses separated by the focal length of one lens. The combination has a focal length which is independent of wavelength, as shown below.

At one particular wavelength the power of two lenses, separated by a distance d, is given by equation (2.31) as

$$P = P_a + P_b - dP_aP_b. \tag{2.68}$$

At a different wavelength the net change in total power is given by

$$\delta P = \delta P_a + \delta P_b - d(P_a\delta P_b + P_b\delta P_a). \tag{2.69}$$

The change in total power is zero when

$$\frac{\delta P_a}{P_aP_b} + \frac{\delta P_b}{P_aP_b} = d\left(\frac{\delta P_b}{P_b} + \frac{\delta P_a}{P_a}\right). \tag{2.70}$$

If the lenses are made of the same glass then $\delta P_a/P_a = \delta P_b/P_b$ for all wavelengths, and the achromatic condition becomes:

$$\frac{1}{P_a} + \frac{1}{P_b} = 2d \quad \text{or} \quad d = \frac{f_a + f_b}{2}. \tag{2.71}$$

The focal length of the doublet therefore is achromatic when the lenses are separated by half the sum of their focal lengths. This configuration is used in the Huygens and Ramsden eyepieces of microscopes and telescopes (see Chapter 3).

The provision of a focal length which does not vary with wavelength is not a sufficient condition to provide completely achromatic images: the position of the principal plane can still vary with wavelength, and so therefore will the position of the image. Fully achromatic doublets require further optical elements; usually each of the pair is itself made as an achromatic doublet.

2.18 Adaptive Optics

The angular resolution of large optical telescopes is usually limited by turbulence in the atmosphere, which causes random fluctuations in refractive index. Ideally the wavefront reaching the telescope from a distant point-like source is plane over the whole aperture. Turbulence disturbs the wavefront, so that it can only behave as a plane wave over a small width d instead of the whole aperture diameter D. The width of the effective telescope aperture determines the angular resolution (see Chapter 10), so that instead of angular resolution $\sim \lambda/D$ we have the larger angle $\sim \lambda/d$. Typically $d \sim 0.3\,\text{m}$, giving a resolution limited to ~ 1 arcsecond, even for the largest telescope apertures. It may seem impossible to improve on this limit, apart from observing from a telescope in space, such as the Hubble Space Telescope. Only if the atmospheric distortion can be known instantaneously, and corrected for, can the full resolution be restored. The wavefront distortions change rapidly, typically in less than 100 milliseconds, so that the measurement and correction have to be completed and repeated within this short time. How can this be achieved?

The form of the wavefront distortion can be found by a simultaneous observation of a nearby bright star, whose image will be distorted in the same way as that of the target object. For example, the wavefront across the whole aperture may be tilted, so that both objects appear to change position. The image movement can be detected if the bright star image falls on an array detector (see Chapter 20). Such a wavefront tilt can be compensated by tilting a small mirror in the optical path, near the detector. A mirror with small mass can be controlled very rapidly by a piezoelectric actuator, holding the images of both the reference star and the target object steady.

The correction of wavefront tilt is the simplest example of *adaptive optics*. Further improvements can be made by dissecting the wavefront from the reference star into a number of separate segments, and correcting each individually for tilt, using a dissected compensating mirror. Rapid measurement and computation are essential to such a scheme. Obviously such a technique is only applicable to fields of view containing a sufficiently bright reference star. An artificial star can, however, be created by shining a laser beam up through the atmosphere, when the back-scattered light from the upper atmosphere simulates a point source. Laser light tuned to sodium atoms is used, since it is scattered from sodium atoms in the upper atmosphere. A powerful laser beam can be pulsed on, and the wavefront distortion measured, in about 1 ms, well below the 100 ms within which correction and normal observation must be achieved.

Problem 2.1

Derive the thin lens equation (2.17) from the bending-angle approach (Section 2.2), as follows. Consider a lens with spherical surfaces with radius of curvature R_1 for the right-hand face, and R_2 for the right-hand face, as in Figure 2.29. O is the position of an object at distance u from the lens, and similarly I is an image at distance v from the lens. We then have for a ray travelling from O to the lens at height y and angle θ_1 to the axis a deviation $\theta = (\theta_1 + \theta_2)$ where θ_2 is the angle between the deviated ray and the optical axis. Equate this to the bending angle of equation (2.7) and obtain the thin-lens equation.

Problem 2.2

A thin mirror which is part of a spherical surface is silvered on both sides. If an object O on the concave side is reflected in it as a virtual image at O′, as a check on your sign convention show that an object at O′ will be imaged correctly as a virtual image at O.

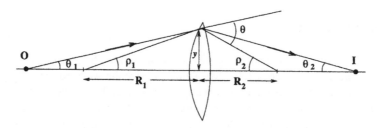

Figure 2.29   A thin converging lens (see Problem 2.1)

Problem 2.3
Two identical plano-convex lenses are each silvered on one face only, one on the plane face and the other on the convex face. Find the ratio of their focal lengths for light incident on the unsilvered side. (Hint: Can you model the system as equivalent to several thin lenses in contact?)

Problem 2.4
A lens with refractive index 1.52 is submerged in carbon disulphide, which has refractive index 1.63. What happens to its focal length?

Problem 2.5
A small fish swims along the diameter of a spherical gold fish bowl, directly towards an observer. Find how the fish's apparent position varies in terms of the bowl radius R and liquid refractive index n. Can the image be inverted?

Problem 2.6
A thick lens consists of two spherical surfaces with curvature R_1, R_2 separated by a thickness d of material of refractive index n. Show that the power P of the thick lens is given by $P = P_1 + P_2 - (d/n)P_1 P_2$ where P_1, P_2 are the powers of the two surfaces. (The argument of Section 2.7 does not work here. You can use instead the matrix algebra of Section 2.8.)

Problem 2.7
In a plane-parallel circular disc of refracting material, the refractive index $n(r)$ depends only on the distance from the axis of the disc. Following Problem 1.6, the radius of curvature of a ray nearly parallel to the axis is $R = n(\mathrm{d}n/\mathrm{d}r)^{-1}$. For $R < 0$, it curves towards the axis, and for $R > 0$, away. Suppose you are designing a converging microwave lens of radius $r = 1.0\,\text{m}$, thickness $T = 0.34\,\text{m}$, focal length $+4.9\,\text{m}$, and with refractive index 1.4 on-axis. Find the value of $n(r)$ at any radius r.

Problem 2.8
A reflecting surface giving stigmatic images of two conjugate points is a paraboloid of revolution when one of the conjugate points is on the axis and at infinity. Show that a single refracting surface between refractive indices n_1 and n_2 is similarly aplanatic for an object at infinity when it is (a) an ellipsoid of revolution, (b) a hyperboloid of revolution, depending on the sign of $n_2 - n_1$.

Problem 2.9
In geometric optics, a light ray that is reversed in direction will retrace its entire path. We know this is so because it applies to the subsidiary processes of reflection, refraction and translation. Within the context of the matrix method of Section 2.9, show that this inversion of the light path corresponds to changing the transfer matrix as follows:

$$M = \begin{vmatrix} \alpha & \beta \\ \gamma & \delta \end{vmatrix} \Rightarrow M' = \frac{n_f}{n_0} \begin{vmatrix} \delta & -\beta \\ -\gamma & \alpha \end{vmatrix} \tag{2.72}$$

Use this to locate the secondary cardinal points F_2, N_2, P_2 based on the positions already found in Section 2.9.5 for the primary cardinal points.

Problem 2.10
In the vicinity of the optic axis, left or right side, the ellipsoidal mirror (Figure 2.18) can be closely approximated by a spherical mirror. This kissing sphere is chosen of suitable radius R (focal length $f = -R/2$) to match the axial coordinate of the ellipsoid through terms of the second order in distance off-axis, and therefore reflects light almost the same as the near-axis parts of the ellipsoid. This means our standard theory of spherical mirrors will work well for paraxial rays reflecting off the ellipsoid. For definiteness, consider rays incident from the left.

If the ellipsoid has major diameter $2a$ and eccentricity e, the foci are located at distances $a(1 \pm e)$ from the vertex. Consider an object point displaced from the focus nearer the mirror by a small distance transverse to the axis.

(a) Find R and f. (b) Does the true focal point F coincide with either of the two ellipsoidal foci? (Note that for a mirror the object- and image-side focal points F_1, F_2 merge into one point F.) (c) Find the transverse magnification Y_2/Y_1.

Problem 2.11

The magnification between the object and image at the foci of the ellipsoidal mirror (Figure 2.18) might be calculated from paraxial rays reflected to either the left or the right of the foci. These evidently give different magnifications, while symmetry apparently demands unit magnification. What is wrong with this analysis?

Problem 2.12

If a thin glass filter, thickness d and refractive index n, is inserted between a camera lens and the photographic plate, show that the plate must be moved a distance $[(n-1)/n]d$ away from the lens for focus to be maintained.

Problem 2.13

The distance between an object and its image formed by a thin lens is D. The same distance is found in a second position of the lens, when it is moved a distance x. Show that the focal length of the lens is

$$f = \frac{D^2 - x^2}{4D}. \tag{2.73}$$

Show incidentally that the minimum distance between an object and its image is $4f$.

Problem 2.14

Consider a glass sphere of radius r and a narrow pencil of light parallel to the axis but off-axis by a distance αr chosen so that the refracted ray meets the opposite side of the sphere exactly on-axis. Find α in terms of the refractive index n. What are the allowed ranges of α and n?

Problem 2.15

Consider a glass sphere of radius r and refractive index n, concentric with the origin of the (y, z) coordinate system (Figure 2.30). A light ray is incident on the sphere from the negative z direction, with $y = $ constant, where

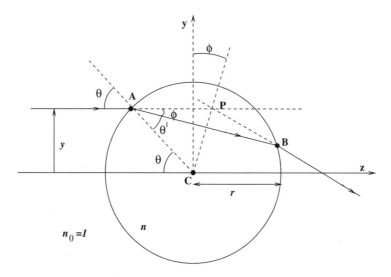

Figure 2.30   Ray tracing in a sphere

$|y|/r \ll 1$, and passes completely through the sphere. Show by ray tracing that the ingoing and outgoing rays intersect approximately on a parabola at $P(y_1, y_2) \simeq (y, (n-1)y^2/(nr))$. (For light incident from the positive z direction, only the sign of z_1 changes.) Since we can ignore the small quadratic term in the paraxial limit, conclude that both principal planes of a sphere coincide with the diameter normal to the optic axis.

Problem 2.16

Let **AB** be the vector displacement from point A to point B. Verify these identities for displacements among the cardinal points:

(i) $\mathbf{P_1N_1} = \mathbf{P_2N_2} = (n_0/n_f - 1)/\gamma$, or, equivalently, $\mathbf{N_1N_2} = \mathbf{P_1P_2}$. Interpret this result.

(ii) $\mathbf{N_1F_1} = \mathbf{F_2P_2}$.

Problem 2.17

(a) Find the ray matrix for a spherical lens of refractive index n and radius R. (b) Verify that its determinant has the correct value. (c) Find the values of the principal focal lengths. (d) Find the positions of all the cardinal points, and sketch these for $n = 1.5$. (e) Give reasoning to confirm that your result for the position of the nodal points is the unique correct answer.

Problem 2.18

One might naively assume that the cardinal points are in the same order as the input and output planes, so that an incident light wave passes F_1, N_1 and P_1 before F_2, N_2 and P_2. This exercise will illustrate that there are many simple systems where this is false and that pairs of cardinal points can appear in reverse order.

(a) As usual, let $1 = $ object side, $2 = $ image side. By definition, a parallel beam on side 1 (2) is conjugate to the focal point F_2 (F_1). Discuss the positions of the two focal points F_1 and F_2 of a diverging thin lens, and explain how it happens that F_1 and F_2 are in reverse order.

(b) Using the ray matrix, equation (2.44), for two thin lenses with focal lengths f_a, f_b separated by distance d, find the requirement on d such that the principal points will appear in reverse order. At the point of transition between reverse and normal order, what is the power of the system, and where are all the cardinal points?

Problem 2.19

A thin equiconvex lens with radii of curvature 220 mm is made of crown glass with refractive indices 1.515 and 1.508 for blue and red light respectively. Find the focal length of the lens and the axial chromatic aberration. A thin plano-concave lens of flint glass is to be used to compensate this chromatic aberration, with the concave face towards the first lens. The refractive indices of flint glass for blue and red light are 1.632 and 1.615 respectively. Find the required radius of curvature of the concave surface and the focal length of the combination.

Problem 2.20

A point source of light is at distance u from a concave spherical mirror, radius of curvature r, aperture $2h$. Following the method and approximations of Section 2.13, show that:

(a) the wave aberration a at the edge of the mirror is given by

$$a = \frac{h^4}{4r}\left(\frac{1}{u} - \frac{1}{r}\right)^2 \qquad (2.74)$$

(b) the transverse ray aberration b_y is given by

$$b_y = v\frac{da}{dh} = -\frac{h^3}{r}\left(\frac{1}{u} - \frac{1}{r}\right)^2\left(\frac{2}{r} - \frac{1}{u}\right)^{-1} \qquad (2.75)$$

(c) the longitudinal ray aberration c is given by

$$b_z = -\frac{v}{h}b_y = -\frac{h^2}{r}\left(\frac{1}{u} - \frac{1}{r}\right)^2\left(\frac{2}{r} - \frac{1}{u}\right)^{-2}. \qquad (2.76)$$

Problem 2.21

Find the difference in thickness across a Schmidt corrector plate (Figure 2.25), refractive index 1.4, used to correct the spherical aberration of the previous example without moving its focal plane.

Problem 2.22

Plane-parallel light is incident normally on the vertex of a glass hemisphere with radius 70 mm. If the refractive indices for red and blue light are 1.61 and 1.63 respectively, find the axial chromatic aberration.

3 Optical Instruments

I knew a man who, failing as a farmer,/ Burned down his farmhouse for the fire insurance/ And spent the proceeds on a telescope./ To satisfy a life-long curiosity/ About our place in the stars./ And how was that for otherworldiness?

Robert Frost (1875–1963), 'The Star Splitter'.

And besides the observations of the Moon I have observed the following in the other stars. First, that many fixed stars are seen with the spyglass that are not discerned without it; and only this evening I have seen Jupiter accompanied by three fixed stars, totally invisible because of their small mass.

Galileo Galilei, 7 January 1610.

By means of Telescopes, there is nothing so far distant but may be represented to our view; and by the help of Microscopes, there is nothing so small as to escape our enquiry; hence there is a new visible World discovered to the understanding.

Robert Hooke, *Micrographia*, 1665.

Optical imaging systems, of which the most important is the human eye, obtain information about an object or a scene in three basic ways. In the eye a complete image is formed on the retina where an array of detectors works simultaneously to send information to the brain; a conventional photographic film or digital camera has many features analogous to those of the eye. In another group the object may be dissected and scanned in sequential fashion, one piece at a time, by a single detector, as in a television camera; or there may be an array of such independent detectors, each with a simple lens, as in the multiple eyes of many insects. Finally the light from an object may be analysed to obtain its spatial Fourier components, followed by a reconstruction either mathematically or optically, as in a hologram; this will be the subject of Chapter 14.

In this chapter we deal with the typical image-forming instruments: the eye, the telescope, the microscope and the camera.

3.1 The Human Eye

The human eye is a miracle of evolution, with many parts subtly adapted to their individual purposes. The essential elements are shown in Figure 3.1. The eye is nearly spherical, about 25 mm

Optics and Photonics: An Introduction, Second Edition   F. Graham Smith, Terry A. King and Dan Wilkins
© 2007 John Wiley & Sons, Ltd

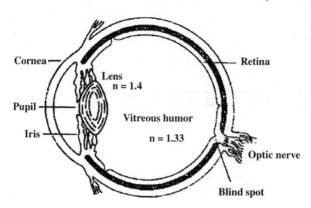

Figure 3.1 The focusing system of the human eye, as seen in horizontal section, viewed from above. Most of the refraction occurs at the front surface of the cornea, which has refractive index 1.38. The lens has a refractive index graded from 1.41 at the centre to 1.39 at the periphery, and the refractive index of the main volume is 1.34. The focal length of the lens is adjusted by tension in the surrounding ciliary muscles. The iris adjusts the aperture according to the available illumination

in diameter. The transparent front portion, the *cornea*, is more sharply curved and is covered with a tough membrane. Between the cornea and the *lens* is a liquid, the *aqueous humour*. Behind the lens is the thin jelly-like *vitreous humour*, filling the volume in front of the *retina*, on which the image is focused. The network of nerves from the sensitive cells of the retina is on the front surface of the retina; it is gathered into the *optic nerve* which passes through the retina and the *sclera*, which is the outer case of the eye. The hole in the retina may be detected as a blind spot[1] in the field of view. The *iris*, which gives individual eyes their distinctive pattern and colour, is an aperture stop; it is located in front of the lens and expands or contracts in response to the light intensity.

The eye analyses light by focusing wavefronts from different directions onto different parts of the retina. The incident wavefronts are very nearly plane; by adjusting the eye to slightly diverging wavefronts a correct focusing can be obtained for objects as close as a limiting distance D_{near}, known as the nearest distance of distinct vision. The ability of the eye to change its effective focal length to image objects over a range of distances is known as *accommodation*. In the human eye there are two focusing elements: the cornea (Figure 3.1) has a fixed power of about 40 dioptres,[2] while the lens, which is adjustable by the surrounding ciliary muscles, brings the total power to around 60 dioptres when relaxed for distant vision and around 70 dioptres when fully tensed for near vision. (In fish the adjustment is achieved by moving the lens, and in some birds it is achieved by changing the surface of the cornea.) The principal focal length within the human eye varies from about 17 mm (relaxed) to 14 mm (tensed).

The focal points (F), principal points (P) and nodal points (N) of the eye are located as shown in Figure 3.2. The two principal points almost coincide at P, as do the nodal points at N.

---

[1]With one eye closed, concentrate on one of a pair of spots about 6 cm apart on a card 20 cm away. Using the left eye, the left-hand spot will disappear if the gaze is fixed on the right-hand spot. The blind spot is located about 5 mm closer to your nose than the central axis of the eye; check that this agrees with a principal focal length of about 17 mm.

[2]Power in dioptres $= 1/f$, where the focal length f is in metres.

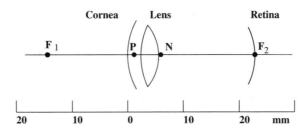

Figure 3.2 Geometric optics of the human eye. Distances are measured from the front of the cornea

The lens of the human eye operates at a focal ratio (f/D) as small as 2, but is remarkably free from spherical aberration. This is partly due to an outward gradient of refractive index from 1.41 to 1.39 within the lens while the surrounding aqueous and vitreous fluids both have an index of 1.34. Defects are, however, common, as is evident by the number of wearers of spectacles.

The common defects of short sight (*myopia*) and long sight (*hyperopia* or *hypermetropia*), are illustrated in Figure 3.3. Without correction the cornea and lens of the myopic eye bring rays from a distant object to a focus in front of the retina (myopic eyes do on the other hand have the advantage that they can focus on objects closer than D_{near}, allowing them to resolve more detail). The hyperopic eye cannot focus on close objects, and often not even on distant ones.

The power of the corneal surface may be corrected by a contact lens, which must add negative power for myopia and positive power for hyperopia. The power of the combination is the sum of the powers of the surface and the lens. The contact lens brings the focal point onto the retina, but it also changes the magnification of the image. Fortunately the brain is able to compensate for a small change in magnification, and it is only in severe cases needing correction in excess of about 8 dioptres that the effect on magnification is important.

More commonly, a lens, spaced at some distance from the eye, is used as in Figure 3.3. The spacing has an advantage: if the lens is located near the first focal point of the eye, about 16 mm in front of the

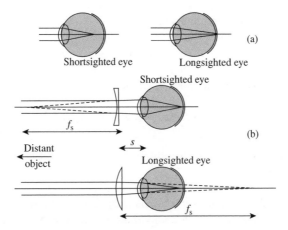

Figure 3.3 Shortsighted (myopic) and longsighted (hyperopic) eyes, showing their correction by diverging and converging spectacle lenses respectively

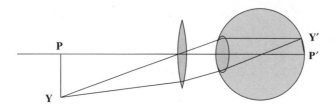

Figure 3.4   A spectacle lens located at the first focal point of the eye does not affect the magnification. The central ray from the object PY is undeviated by the lens, and forms an image P′Y′ as for a perfect eye; all other rays from Y also reach Y′

cornea, it does not affect the magnification, giving the same scale of image with and without the lens. This may be seen from Figure 3.4; the vertex ray from the off-axis object point Y passes without deviation through the spectacle lens at the focal point of the eye lens, and traverses the eye parallel to the axis reaching the retina at Y′. The position of Y' is unaffected by the spectacle lens, provided that the lens is near the front focal point.

Example.   Compare the power P_s of a spectacle lens at a distance s from the eye with the power P_c of an equivalent contact lens. Consider both kinds of spectacle lens, positive and negative.

Solution.   We use the ray diagram of Figure 3.5 for a distant object. We model the relaxed eye as a thin lens located at the cornea and with a total power $P_E = 1/f_E$. After passing through the spectacle lens, the light rays appear to the eye-equivalent lens E to diverge from, or converge towards, an object point O_E. Figure 3.5 shows that the object distance is $u = f_s - s$. For both kinds of spectacle lens, positive and negative, u has the same sign as f_s. The eye-equivalent lens E then forms an image I_E on the retina at a distance b from the cornea. For a contact lens touching the eye

$$P_C + P_E = 1/b. \tag{3.1}$$

For the spectacles, the thin-lens equation gives

$$1/v - 1/u = 1/b - 1/(f_s - s) = P_E. \tag{3.2}$$

Subtracting the latter equation from the former, we get

$$P_C = 1/(f_s - s) = P_s/(1 - sP_s). \tag{3.3}$$

Given that $f_s - s$ has the same sign as f_s, it follows that equivalent contact and spectacle lenses are both positive or both negative.

It is also common to find *astigmatism* in on-axis images, resulting from uneven curvature of the cornea. This can usually be corrected with *anamorphic* lenses, which have different powers in two perpendicular meridians.

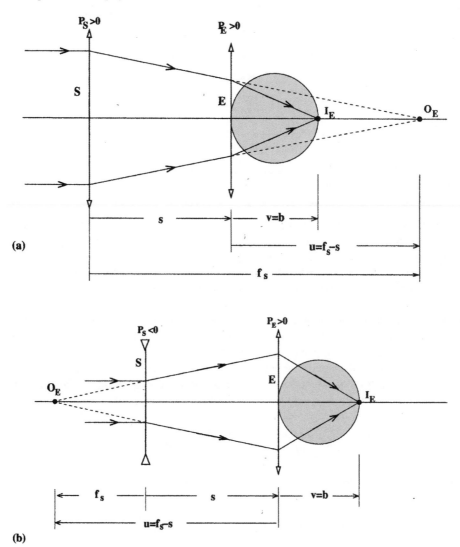

Figure 3.5 Light from a distant object is focused by a combination of a spectacle lens S and the eye, represented by a single lens E: (a) longsighted; (b) shortsighted eye. (Occasionally, as here, it is convenient to signify a thin lens with a double arrow, with arrowheads pointing outwards for a converging lens and inwards for diverging)

3.2   The Simple Lens Magnifier

The angular resolution of the eye is determined by its focal length and the separation of sensitive elements on the retina. This geometrical resolution matches well the limit of angular resolution set by diffraction at the iris, the aperture of the main part of the eye lens. In the centre part of the retina, known as the *macula*, the sensitive elements are *cones* spaced about 3 µm apart, matching the angular resolution $\sim 1'$ expected from an iris diameter of $\sim 2\,\text{mm}$ (see Chapter 10).

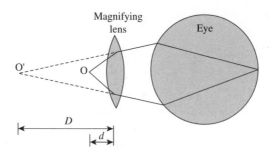

Figure 3.6 A simple lens L used as a magnifier. An object at O close to the eye can be focused by the eye as though it were at a more distant point O′

Since the angular resolution is very nearly unchangeable, it follows that the linear resolution of the unaided eye is greatest for objects as close as possible, i.e. at the near distance D_{near}; this is generally taken to be 25 cm. Closer objects are out of focus, but if the eye is aided by a convex lens an object at a very small distance can be focused, with a corresponding increase in linear resolution (Figure 3.6). In wavefront terms, the lens assists the eye by converting a wavefront which is diverging sharply into the nearly plane wave which the eye can focus unaided.

The magnification of the image depends on the position of the lens and the eye; a very large (but distorted) image can be seen if the object is near the focal plane of the lens and the eye is some distance from the lens. Normally the lens is close to the eye and the image is at the near point $D_{near} = 25$ cm. The angular magnifying power m_A of a simple lens used in this way is given by the ratio of the angular size of an object, seen as an image at the near point, to its actual angular size at the near point (Figure 3.7). If the object has height y, and using the small-angle approximation, this is the ratio

$$m_A = \frac{\theta_L}{\theta_O} = \frac{y/d}{y/D_{near}} = \frac{D_{near}}{d}. \tag{3.4}$$

If the image is viewed at the near point, we substitute image and object distances $v = -D_{near}$ and $u = -d$ into the thin-lens equation to get

$$1/v - 1/u = 1/d - 1/D_{near} = P, \tag{3.5}$$

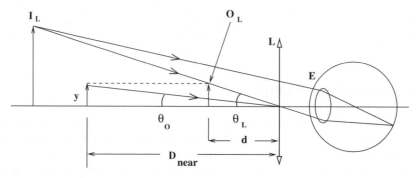

Figure 3.7 The angular magnification of an object seen, with the help of a lens, as at the near point of the eye. (For clarity, the image I_l made by the lens is shown here as more distant than the near point)

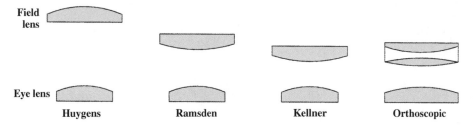

Figure 3.8   Eyepieces used in microscopes and telescopes. In the Ramsden the upper lens, known as the field lens, is at the first principal plane; the two lenses have the same focal length, which avoids chromatic aberration. The Kellner and orthoscopic eyepieces have wider fields of view; chromatic aberration is reduced by the use of different refractive indices in the doublet and triplet lenses

and therefore

$$m_A = 1 + PD_{near}. \tag{3.6}$$

A typical magnifying glass has a power of 12 dioptres, so giving

$$m_A = 1 + 12 \times 0.25 = 4. \tag{3.7}$$

When the object is placed at the focal plane of the lens, so that $d = f$, the image is seen at infinity and we find from equation (3.7) that $m_A = PD_{near}$.

Only a small magnification is available from a single lens without introducing unacceptable aberrations. When higher powers are needed, as in the eyepieces of microscopes and telescopes, some improvement is obtained from double lens systems, especially in reducing chromatic aberration (Section 2.18); examples are shown in Figure 3.8. For magnifications greater than about 10 or 20 the simple magnifier is replaced by the compound microscope.

3.3   The Compound Microscope

The simple lens magnifier of Section 3.2 provides a magnified virtual image of an object placed just within the focal plane of the lens. The eyepiece of a compound microscope (Figure 3.9) acts in this way on an object which is itself a magnified image, produced by an objective lens with very short focal length. The overall magnification, which we calculate below, is approximately the product of the two stages of magnification, amounting typically to several hundred.

The magnification of the compound microscope is calculated in two stages. First, the objective lens forms a real image; this is then magnified further by the eyepiece (Figure 3.10). If the real image is formed at a distance g beyond the focus F of the objective, whose power is P_o, then the magnification is $-P_o g$ (substitute $z_2 = g$, $f_2 = f_0$ in equation (2.21); alternatively this may be shown to be equal to v/u for a simple lens). The magnification of the eyepiece is $(1 + P_e D)$ (see equation 3.6), giving the overall transverse magnification as $-P_o g(1 + P_e D)$; here D is the (positive) distance that the virtual image of I (not shown) appears behind the eyepiece. The length g is known as the optical tube length of the microscope, since it accounts for most of the length of the instrument (see Figure 3.9).

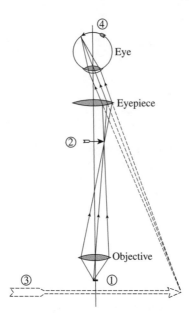

Figure 3.9   Basic optics of the compound microscope. The eyepiece is a simple magnifier focused on a real magnified image produced by the objective lens system

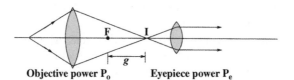

Figure 3.10   Magnification in the microscope

The large magnification of a compound microscope implies that light leaving any point in the object as a wide-angle divergent wavefront is converted into a narrow, nearly parallel, wavefront emerging from the eyepiece. The objective lens is therefore designed to collect light over a large solid angle; this is also a requirement in obtaining the maximum resolving power for detail in the object.

The design of the objective lens is crucial to the success of a microscope. Abbe showed[3] that thanks to diffraction, and ignoring aberration, the smallest distance between two points of the object that can be resolved is approximately $\lambda/(n \sin \theta)$, where θ is the maximum half angle subtended at the object by the objective, and n is the refractive index of the medium in contact with the objective lens, e.g. oil. The lens must therefore collect the spherical wavefront emerging from an object point over as wide an angle as possible, without aberrations which are easily introduced when rays traverse the lens at large angles. This is achieved in a multi-element lens in which successive meniscus lenses are used to reduce the curvature of the wavefront, as shown in Figure 3.11. The highest magnification is obtained with an oil-immersion lens, in which the space between the specimen and the first lens

---

[3]See Chapter 13.

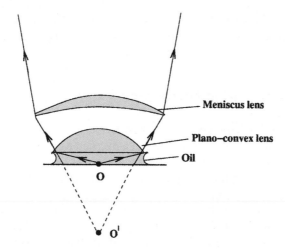

Figure 3.11 Oil-immersion microscope objective, in which a wide-angle spherical wavefront from the object O appears to diverge from the virtual object O'. The points O, O' correspond to the stigmatic points P_0, P_1 in Figure 2.19. The oil's index matches that of the first objective lens, so that the object is observed as though within a uniform sphere. The wavefront curvature is again reduced by a series of meniscus lenses, of which only the first is shown

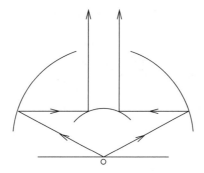

Figure 3.12 Reflecting microscope. This is a close relation of the Cassegrain telescope (Section 3.6)

surface is filled with oil with the same refractive index as the glass. The aplanatic[4] spherical surfaces of Figure 2.20 are used to form a virtual image of O'. The requirement to collect rays over a wide angle has also been met in the reflecting microscope objective of Figure 3.12; this has the advantage of avoiding the use of oil, while providing a large free space immediately above the object.

The performance of the objective in collecting light over a large angle is measured by its *numerical aperture* $NA = n \sin \theta$ where n is the refractive index of the medium in contact with the objective, e.g. oil, and θ is the half angle of the light cone entering the objective. For example, if $n = 1.515$ and $\theta = 55°$, $NA = 1.515 \sin 55° = 1.24$. In practice, numerical apertures do not exceed 1.4. Even with a large numerical aperture a microscope can only resolve detail at a scale comparable with one

---

[4]An aplanatic surface (or lens) is one on which all rays from a point source converge to a point, or stigmatic, image.

wavelength of the illuminating light. The electron microscope, which uses beams of electrons in place of rays of light, is similarly restricted in resolving power by the equivalent wavelength of the electrons and by the numerical aperture of the system.

3.4   The Confocal Scanning Microscope

A high-power conventional microscope is at a disadvantage when examining three-dimensional objects, when well-focused parts of an object are seen overlaid by confusing out-of-focus images of other parts at different depths. This becomes more confusing at larger magnifications and for larger numerical apertures, when the depth of focus becomes smaller. This disadvantage is overcome in the confocal microscope shown in Figure 3.13.

In this instrument only one point at a time is illuminated and focused to a single small electronic detector element situated behind a pinhole stop. The signals from the detector are stored and used later for a reconstruction of the image. The image at a particular depth is scanned either by moving the whole microscope with its light source detector, or more simply by moving the object, in a raster scan as in television. The scanning can be extended to different depths by refocusing. This process may appear to be elementary and slow, but the results are spectacular. As shown by the broken rays in Figure 3.13, light from planes away from the required focal plane is mainly spread outside the pinhole detector, avoiding the confusion inherent in the conventional instrument. Individual sections of the object are scanned in sequence and combined. These separate sections can be seen in Figure 3.14(a), which shows a confocal microscope scan of a portion of a compact recording disc (CD). In this microphotograph the pits which form the digital recording are 0.5 µm across, in tracks 1.6 µm apart. Many spectacular microphotographs of biological subjects, such as the cell structure in Plate 1[*], have been scanned in sections and recombined in this way.

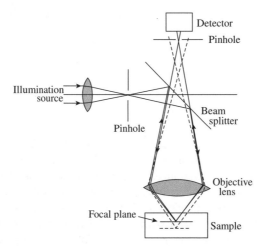

Figure 3.13   The confocal scanning microscope. The beam splitter allows the same objective lens system to be used for illumination and for focusing light onto the pinhole detector. Only light from the focal plane enters the detector; the broken lines show rays from a different depth in the sample

[*]Plate 1 is located in the colour plate section, after page 246.

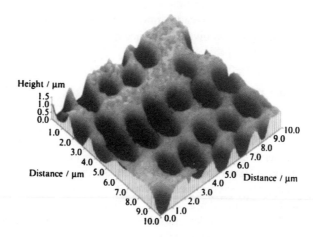

Figure 3.14 Microphotographs of a portion of a compact recording disc, using a confocal scanning microscope.

3.5 Resolving Power; Conventional and Near-Field Microscopes

The detail which can be distinguished using a microscope with high magnification is limited by the wave nature of light. This chapter is concerned with geometric optics, and not the resolving power of microscopes, but we digress to explain the need for wide-angle wavefronts at microscope objectives, and to introduce the near-field scanning optical microscope, which overcomes the limitations of conventional microscopes.

The problem of resolving power may be simply expressed as the requirement to distinguish light emitted by two similar point objects separated by a small distance, as in Figure 3.15(a). They are illuminated by the same source of light, shown as an incident plane wave from below. Each radiates a light wave in response. As in Huygens' construction (Chapter 1), straight ahead these two waves are indistinguishable. At a wide angle α, however, they may be sufficiently out of step to be distinguishable: this is the reason for designing an objective with a large numerical aperture (see

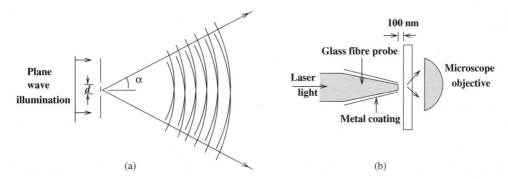

Figure 3.15 Microscope resolving power. (a) In a conventional microscope, light waves from two adjacent sources are only distinguishable if they are collected over a wide angle; even at the largest numerical apertures the resolution is no better than one wavelength. (b) The evanescent field within one wavelength of an object can be probed by a scanning near-field optical microscope (SNOM); a resolution of $\lambda/10$ is achievable

Section 3.3 above). Anticipating the consideration of interference in Chapter 8, this gives an approximate value for the minimum resolvable distance d as

$$d \approx \lambda/2n \sin \alpha, \tag{3.8}$$

which is inevitably no better than about one wavelength.

This argument does not apply to the wave fields very close to the two point objects. Within a distance of around one wavelength the electric field is dominated by components which decay rapidly with distance and have no effect on the normal microscope. In the *scanning near-field optical microscope*, or SNOM, a very fine fibre optic probe can be scanned across this near field, and is in practice able to distinguish detail less than one-tenth of the wavelength across. As shown in Figure 3.15(b), the scanning probe is usually the light source, and the detector is a conventional microscope feeding a photomultiplier detector. (For an opaque object, the same probe can be used both for illuminating the object and as a detector; this requires the use of a directional coupler in the fibre, as described in Chapter 6.) The mechanical requirements of the SNOM are severe: the active tip of the probe must be only some tens of nanometres across, and it must be located and maintained at a distance of less than 100 nanometres from the surface of the object.

3.6   The Telescope

When the eye attempts to distinguish details of a distant object, it is attempting to separate nearly plane waves which are inclined at small angles to each other. The limit of resolution can only be improved by using an instrument which increases the angular separations of a range of plane waves. This is the action of a telescope.

In its most familiar use, of viewing distant objects, the telescope converts parallel incident rays from the object to parallel outgoing rays, which can be focused by a relaxed normal eye. In this case, the object and image points are both at infinity. But parallel outgoing rays signify that the object point coincides with a principal focal point of the system; likewise for the image point. In other words, the principal focal points are both at infinity. Table 2.2 shows that an *afocal* system, i.e. one with $f_1 = f_2 = \infty$, has a matrix with element $M_{21} = \gamma = 0$.

If a plane wave at a small angle θ_1 to the axis of a telescope is to emerge as a plane wave at a larger angle θ_2, the refractive index at both ends being the same, then we will show that the width of the wavefront is reduced in the ratio θ_1/θ_2. Consider the wavefronts entering and leaving a telescope, as shown in Figure 3.16. This shows the simple *astronomical* telescope, using two convex lenses with long and short focal lengths f_o and f_e; these lenses are called the *objective* and the *eyepiece*.

The wavefront enters the telescope with width w_1. It is at an angle θ_1 to the axis of the telescope, so that the difference in optical path l across the wavefront in the diagram is $l = w_1 \theta_1$ (where a small-angle approximation may be used). This path difference l is preserved as the wavefronts traverse the telescope, so that the difference in angle as the wavefronts leave the telescope is determined by l and the new width w_2 of the wavefront. Again for small angles, Figure 3.16 shows that the ratio of widths is the ratio f_o/f_e of the focal lengths. The angular magnification is therefore:

$$\frac{\theta_2}{\theta_1} = \frac{w_1}{w_2} = \frac{f_o}{f_e}. \tag{3.9}$$

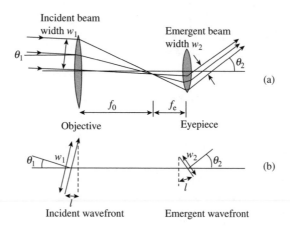

Figure 3.16   The action of a simple telescope, with convergent objective and eyepiece lenses, focal lengths f_o and f_e. (a) A pencil of parallel rays from a distant source enters at angle θ_1 and emerges at angle θ_2. The magnification of the telescope is $\theta_2/\theta_1 = f_o/f_e$. (b) The widths of the wavefront as it enters and emerges are w_1 and w_2. The optical paths l are identical; the angles are small in practice, so that $l = w_1\theta_1 = w_2\theta_2$

Figure 3.16 shows the simplest form of astronomical telescope, using two convex lenses with long and short focal lengths f_o and f_e. As in a compound microscope, the eyepiece can be regarded as a lens magnifier which is used to view an image formed by the objective. Again for small angles, the similar triangles in Figure 3.16(a) of light rays between the lenses show that the ratio of the widths is the ratio f_o/f_e of the focal lengths. Note that in any two-element telescope set for direct viewing, as illustrated here, the distance between the lenses equals $f_o + f_e$; this also holds when the eyepiece is a negative lens.

Practical arrangements for telescopes giving angular magnification are shown in Figure 3.17, which includes many of the conventional varieties of telescope. For each the figure shows the reduction of the width of a plane wavefront. (In some optical systems, notably in laser optics, a telescope may be used in reverse to expand rather than contract the area of a wavefront.) The emerging wavefront may be observed directly by eye, or it may be focused by a camera onto a photographic film or an array detector; the eyepiece may then become part of the camera. The angular magnification of all these arrangements is given by the ratio of the widths of the plane wavefronts entering and leaving the telescope; this is numerically equal to the ratio of the focal lengths of the two optical elements, either lenses or mirrors, which form the objective and eyepiece elements of the system.

A telescope also has the advantage over the eye that it can gather a larger area of plane wavefront, so that a point source of light becomes more easily visible. It is most important not to confuse this increase in sensitivity with the question of the visibility of a uniformly bright object with a finite size: the surface of the Moon is no brighter as seen through a telescope, while stars which are effectively point sources of light may easily be seen through a telescope even if they are invisible to the naked eye. If the object is already resolved in angle by the eye, then its visibility is related to its *luminance*, which is essentially the visible power emitted per unit area into unit solid angle. The luminance as seen through the telescope cannot be greater than the original, according to a basic theorem of photometry; there can in practice only be a loss of luminance in a telescope, due to partial reflection at lens surfaces or to incomplete reflection at a mirror surface.

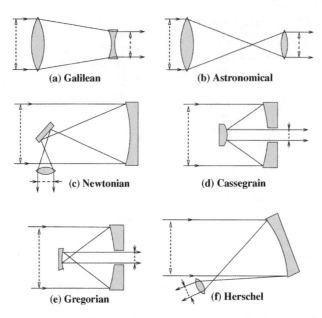

Figure 3.17   The reduction in width of a wavefront in various types of telescope. The telescopes are shown adjusted for direct viewing of the emergent beam; the emergent wave could instead be made convergent, focusing an image on a photographic plate

Example. A simple astronomical telescope has an objective with focal length $f_o = 30$ cm. What should be the focal length f_e and diameter D of the second lens to give a magnification of 15 and an angular field of view $2°$ in diameter?

Solution. Magnification $= f_o/f_e = 15 = 30/f_e$. Hence $f_e = 2$ cm. Field of view is determined by the second lens acting as a field stop. Hence by tracing rays from the boundary of the object straight through the vertex of the objective and converting the angle to radians, we find $D/(f_o + f_e) = 2\pi/180$ giving $D = 11$ mm.

3.7   Advantages of the Various Types of Telescope

The types of telescope in Figure 3.17 are distinguished mainly as reflecting or refracting by the use either of mirrors or of lenses for the objective,[5] and by the use of second elements with positive or negative power. Further elements, such as a camera lens or an eyepiece, may be added to focus the emergent wavefront. Systems such as the Galilean telescope and the Cassegrain telescope have the advantage that they are shorter than the corresponding instruments using second elements with

---

[5]A telescope or microscope system which uses only lenses, such as the Galilean, is referred to as *dioptric*, and with mirrors only as *catoptric*; a combination of lenses and mirrors, as in the Schmidt telescope (Figure 2.25), is a *catadioptric* system.

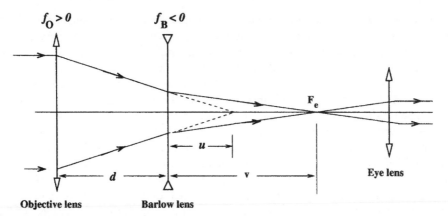

$f_O > 0$ $f_B < 0$

F_e

Eye lens

Objective lens   Barlow lens

Figure 3.18 Ray trace for a Keplerian telescope with Barlow diverging lens inserted. (Distances are not to scale)

positive power (astronomical and Gregorian). They may, however, be unsuitable for terrestrial survey and position measurement because they have no real image plane at which a reference scale or graticule can be placed.

Increasing the magnification of the simple telescope in Figure 3.16 involves either reducing the eyepiece focal length, which may introduce aberrations, or increasing the objective focal length, which may make the telescope too long. An alternative is to introduce a diverging lens before the primary focus, as shown in Figure 3.18. This lens, introduced by Peter Barlow in 1834, reduces the vergence of the wavefront, and increases the magnification with only a small extension of the telescope. A Barlow lens giving acceptable magnification ×2 to ×4 is often introduced in small telescopes used by amateurs. Larger magnifications would introduce unacceptable aberrations.

The extra magnification provided by a Barlow lens may be calculated with the help of Figure 3.18. Let the Barlow lens have a negative focal length f_B and be separated by distance d from an objective of focal length f_O. The focal point of the objective is a distance $u = f_O - d$ behind the Barlow lens. By equation (2.31), the combined focal length of the objective plus Barlow is

$$f_{com} = [1/f_O + 1/f_B - d/(f_O f_B)]^{-1} = (f_B + f_O - d)^{-1} f_O f_B = f_O f_B/(f_B + u). \qquad (3.10)$$

The angular magnification of the telescope is now to be computed according to equation (3.9), but with this combined focal length in place of f_O; it is therefore larger than the magnification without the Barlow by a factor

$$m_B = f_B/(f_B + u) = |f_B|/(|f_B| - u). \qquad (3.11)$$

It is evident that when u is near $|f_B|$, a small displacement of the Barlow lens can lead to a large change in m_B (see Problem 3.7).

Most reflector telescope systems are axially symmetric, and consequently the secondary tends to obstruct part of the aperture. The Herschel system uses the whole of the aperture without obstruction, but at the cost of using the primary off-axis; it is easier to control aberrations with the more symmetrical mirror arrangements Figure 3.17 (c), (d) and (e).

The control of aberrations has already been discussed in Chapter 2, but the detailed application to a full telescope system becomes complicated. Modern astronomical telescopes commonly use a

Cassegrain system, and the control of aberrations may entail the addition of a Schmidt corrector plate in the beam as it enters the telescope, or a corresponding asphericity of both primary and secondary.

A lens has the great advantage over a mirror that a small distortion due to gravity or uneven temperature has no first-order effect on the optical path through it, whereas if part of a mirror bends forward by an amount z it shortens the optical path by $2z$. If nearly perfect images are required, in which all optical paths are near equal, this means that the mounting of a large mirror must be considered much more carefully than that of a lens. On the other hand, a mirror has no inherent chromatic aberration, and very little light is lost at a reflection. The largest telescopes, and some of the best survey theodolites, use mirror systems. A mirror must, of course, be used for wavelengths where no good lens can be made, as at infrared, radio or X-ray wavelengths.

Most mirrors for optical telescopes use a thin silver or aluminium film evaporated onto glass, or preferably onto a ceramic with near-zero thermal coefficient of expansion. Mirrors for astronomical telescopes must be as large as possible to obtain sufficient light-gathering power; some of the largest have been cast in a single piece up to 8 metres in diameter, as for the Gemini telescopes in Hawaii and Chile. The mirrors of the Keck telescopes on Hawaii are even larger, with diameter 10 metres; these are built up of hexagonal elements mounted to produce an almost complete single mirror. Orbiting space telescopes are necessarily smaller; the diameter of the Hubble Space Telescope is 2.4 metres, but it has of course the tremendous advantage of avoiding the effects of absorption and random refraction in the atmosphere. For all these large-telescope mirrors the surface profile must be accurate to a small fraction of a wavelength; this is extremely demanding both in manufacture and in the support systems which maintain the shape in use. Errors may be hard to rectify: the Hubble Space Telescope was launched with serious spherical aberration due to a faulty test procedure, and the wavefront entering the cameras and spectrometers had to be corrected subsequently by special optical systems.

Radio telescopes use a simple metal surface, fabricated or polished so that the surface profile is correct within a small fraction of a wavelength. X-ray telescopes present a different problem: the only efficient reflector is a polished metal surface at a grazing angle of incidence. The Wolter telescope of Figure 3.19 uses a section of a paraboloid which is only slightly tapered, followed by a second reflector element which is part of a hyperboloid (the combination reduces off-axis aberrations, giving a wider field of view). X-ray telescopes in spacecraft use Wolter telescopes, often with several concentric reflector systems so as to increase the effective collecting area. The aperture is typically 1 metre in diameter, and the focal point is several metres beyond the reflector system. Electronic detector arrays, such as the charge-coupled detector arrays described in Chapter 20, are used to obtain remarkably detailed images of the X-ray emission of energetic astronomical objects such as active galactic nuclei.

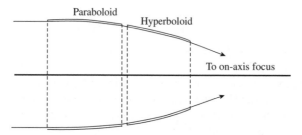

Figure 3.19 The Wolter X-ray telescope. The grazing incidence reflecting elements are sections of a paraboloid followed by a section of a hyperboloid

The attainment of the theoretical angular resolving power of telescopes (given approximately by the ratio wavelength/diameter, see Chapter 10) depends on a number of factors. For infrared and longer wavelengths the full theoretical resolution may be achieved, even for the largest astronomical telescopes. In the visible region, however, atmospheric effects typically limit the resolution to around 0.3 arcseconds, while for X-ray telescopes the limitation is the accuracy of the mirror surfaces.

3.8   Binoculars

The binocular telescope, or 'binoculars', as used by bird-watchers and amateur astronomers, must be one of the most widely used forms of telescope; for many people, using both eyes gives a considerable improvement in the perception even of diffuse objects. Binoculars comprise a pair of refracting telescopes, with objective lenses some centimetres across and with eyepieces large enough to allow normal vision of the magnified scene. Each telescope is basically an astronomical telescope, with internally mounted prisms to correct the inversion of the image. Let us imagine that we are to design a binocular telescope for general use.

We know already that the magnification of a telescope focused for object and image at infinity is given by the ratio f_o/f_e between the focal lengths of objective and eyepiece. A magnification of $\times 8$ is common for binoculars: a hand-held instrument does not usually have a magnification greater than about $\times 8$ or $\times 10$, otherwise the image cannot be held sufficiently steady without a tripod mounting. We also know that the total amount of light entering the instrument is determined by the aperture of the objective, and that this affects the visibility of point sources of light; a large-diameter objective is therefore important. We now discuss the factors which determine the field of view of the binoculars, what sorts of lenses we must use, and what determines the diameter of the eyepiece.

The eye is especially sensitive to chromatic aberration, which has the effect of colouring the edges of objects away from the axis. Binocular objectives must therefore be carefully corrected; they are therefore made as cemented achromatic doublets (Section 2.17). In the eyepiece a single cemented achromatic pair is insufficient to control other aberrations over a wide field of view; practical eyepieces usually consist of a separated pair, one of which is itself a cemented doublet. Figure 3.20(a)

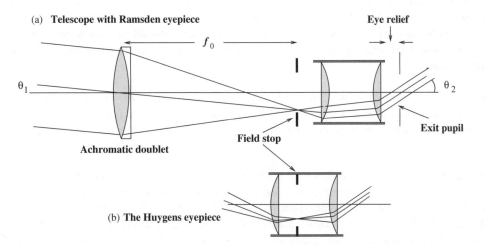

Figure 3.20   Astronomical telescope system, as used in binoculars

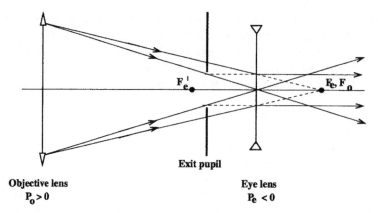

Figure 3.21 The exit pupil of a Galilean telescope is the image of the objective by the eyepiece and can be located by tracing several convenient rays

shows a simple pair in the form known as the Ramsden eyepiece. In another type of eyepiece, due to Huygens and shown in Figure 3.20(b), the primary image falls inside the eyepiece, where a graticule or cross-hair can be mounted; this is useful in a survey instrument such as a theodolite. (Note the different arrangement of the front plano-convex lens, following the advantage of splitting the refractive power evenly between the surfaces: see Figure 2.23.)

An important part of eyepiece design concerns the position of the eye. The rays from a point source in Figure 3.20 cross the axis beyond the eyepiece; at this point they fill the *exit pupil* of the system. The eye is placed at the exit pupil, which is separated from the eye lens by a distance known as the *eye relief.*

The exit pupil is defined as the image of the aperture stop as viewed through the eyepiece. In the Galilean telescope the exit pupil is inside the telescope, where the eye cannot be placed (Figure 3.21); this is a disadvantage of the Galilean telescope, as it results in a reduced field of view.

The optimum size of the exit pupil is determined by the size of the pupil of the eye. If the exit pupil is smaller than the eye pupil then the *eye* is used inefficiently since only part of the eye pupil is illuminated. If the exit pupil is larger than the eye pupil some light is wasted and the *telescope* is being used inefficiently. In practice the exit pupil should be somewhat larger than the eye pupil so that the exact position of the eye pupil is not too critical; the binoculars are then easier to use.

Example. A ray of light from a star makes an angle of 0.01 radians with the axis of a simple telescope whose objective has a focal length of 50 cm and an eyepiece of focal length 2 cm. Calculate the distance D beyond the eyepiece where the ray crosses the axis of the telescope. (This is the *eye relief.*)

Solution. With the help of an undeflected ray through its vertex, we see that the objective focuses the light at $0.01 \times 50 = 0.5$ cm from the axis, and the light reaches the eyepiece at $0.01 \times 52 = 0.52$ cm from the axis. The angular magnification is $50/2 = 25$, so the light leaves the eyepiece at angle $0.01 \times 25 = 0.25$ radians and crosses the axis at $D = 0.52/0.25 = 2.08$ cm. (Note that this distance, although calculated for a specific angle of starlight, does not depend on angle. The general expression is $D = (f_o + f_e)/m_A = f_e(1 + 1/m_A)$, where m_A is the angular magnification).

The focal plane of eyepieces of the Ramsden type is in front of and close to the first lens. A real image of any object at infinity exists at this point, so that in this case the angular width of the field of view is determined by the aperture of the first lens of the eyepiece. An aperture which limits the field in this way is known as a *field stop*; the first lens is therefore often called the field lens. The angular width of the field of view is the diameter of the field lens divided by the focal length of the objective.[6] A long-objective focal length f_o therefore gives a large magnification but a small field of view. We might therefore expect to see large magnifications obtained instead by using a small-eyepiece focal length f_e. However, as we shall see, the diameter of the eyepiece is fixed by other considerations, and reducing f_e becomes difficult without introducing aberrations (recall that aberrations of a lens tend to increase with the ratio of diameter to focal length).

Let us assume that the angular magnification is fixed at the comfortable limit of $\times 10$, and find the diameters and focal lengths which must be used for the eyepiece and objective. We shall find that all of these depend on our requirements for the *field of view*. Consider again the rays entering the eye at the exit pupil in Figure 3.20. The angular spread of rays at this point is the angular width of the field of view multiplied by the magnification; it is therefore a large angle, often about 50°. The eye lens must be larger than the exit pupil to accommodate these rays, and the field lens must be somewhat larger again. The field lens must therefore be at least 15 mm in diameter; the size of the real image at this point must be the same size, as this lens constitutes the field stop. The field of view, which in this example would be 50° divided by the magnification of 10, is 5°, and accordingly $5° = 0.087$ rad $= 15\,\mathrm{mm}/f_0$ which gives $f_0 = 17\,\mathrm{cm}$. The focal length of the eyepiece is therefore one-tenth of this, i.e. 1.7 cm.

Finally, from equation (3.9), the diameter D of the objective must, for a magnification m_A, be m_A times the diameter of the exit pupil, which is usually about 4 mm to match the pupil of the eye. The objective is therefore 40 mm in diameter. We have reached the specification in the form usually quoted: these binoculars would be specified as 10×40.

Two remaining problems are solved simultaneously by the use of a pair of prisms, as seen in Figure 3.22(a). These invert the image, so that it appears upright, and they fold the light path so that the total length is much less than $f_o + f_e$, the standard length of an afocal telescope. A more compact form is the roof prism shown in Figure 3.22(b). A Galilean arrangement would of course provide an upright image without prisms, but without folding the light path the telescope length would be too great for anything more than the small magnification used in opera glasses.

3.9 The Camera

Astronomical research is seldom conducted by looking through a telescope: instead an image is formed on a photographic plate or detector array (Chapter 20), or it may be focused on the slit of a spectrograph. The telescope then becomes a camera, which is an artificial eye; the photographic plate is the retina and the lens of the eye is the primary lens or mirror of the telescope. A camera is usually focused on an object at a distance large compared with the focal length of the lens; the linear size of the image is then given directly by the product of the focal length of the lens and the angular size of the object.

---

[6] In some systems (see for example Problem 3.4), one should divide the diameter of the field lens by the distance from the objective to the field lens.

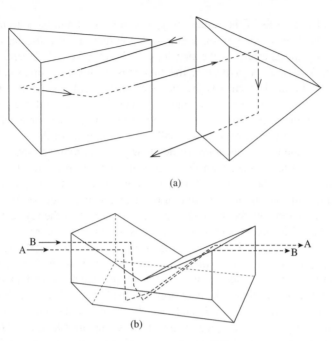

(a)

(b)

Figure 3.22   (a) A pair of erecting prisms, as used in the binocular telescope. (b) The roof prism, which performs the same function and is more compact

Similarly a camera may be arranged to focus on very near objects, when it becomes a photomicroscope. Photographic and television cameras are often provided with interchangeable sets of lenses with a range of focal lengths, so that the scale of a picture may be selected according to the required angular resolution; alternatively a 'zoom' lens may be used which is an adjustable compound lens whose focal length can be varied over a range which may be as large as five to one. Small cameras commonly have a lens with focal length about 40 millimetres; an astronomical telescope may have a focal length of 10 metres or more, so as to provide a sufficiently large linear scale on the photographic plate. Even with a focal length of 10 metres an angle of 1 second of arc corresponds to only 0.05 millimetres at the focal plane; since diffraction images smaller than 1 second of arc are obtainable in large telescopes a stellar image usually has a microscopic scale on the image plane. The effective focal length may be adjusted by any of the devices of Figure 3.17; we have already seen how these affect the angular scale of a pattern of plane waves. It is in fact only necessary to change the position of the secondary lens or mirror to obtain a real image at any desired distance. For example, the Galilean telescope may be converted into the telephoto lens (Figure 3.23) by moving the secondary away from the primary. The advantage over the use of a single objective lens is that a long focal length is available without a corresponding and inconveniently long distance between the first lens and the photographic plate.

The objective lens of a camera with a wide field of view, which is required of most modern cameras, usually consists of four or more elements. An example is shown in the single lens reflex camera of Figure 3.24; this design, known as the Tessar, is widely used. The two main elements are cemented pairs designed to correct for achromatism, while the outer lenses provide correction for geometrical aberrations. In this camera the viewfinder uses the same lens, viewing the field via a mirror which hinges out of the light path when the film is exposed. The image on a translucent screen is seen upright through a reversing prism.

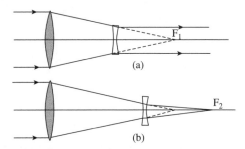

Figure 3.23 Comparison of Galilean telescope (a) and telephoto lens (b). In the telescope the diverging secondary lens is placed so that the foci of the two lenses coincide at F_1; in the telephoto lens the secondary is moved so that a real focus F_2 is located on a photographic plate

A telescope or camera photographing an extended object produces an image in which we need to know the amount of radiant power (or flux) falling on the device per unit area, which is called *irradiance* (see Appendix 2). This depends on the light flux leaving unit solid angle of the source in the direction of the observer, i.e. on the *radiant intensity* of the source. The irradiance of the image is proportional to the aperture area, but it also varies inversely an s the area of the image, which is itself proportional to the square of the focal length. The intensity on the photographic plate therefore varies as the ratio $(f/D)^{-2}$, where f/D is the familiar 'focal ratio', or F-number, of a camera lens; it is the ratio of focal length f to aperture diameter D. Many compact cameras incorporate a *zoom* lens, which will adjust the focal length over a range of two or three. The effect on the field of view is presented to the photographer by adjusting the viewfinder in synchronism.

The depth of focus, which is the range of object distance over which the image is effectively in focus, depends on the F-number. In Figure 3.25 a point object at a distance u_0 forms a point image at P, distance v from the lens, while a closer point object at u_1 forms an image at $v + \delta v$. The converging rays form a blurred image at P with diameter d; if this is small enough the object is still effectively in focus. The lens diameter is D, so by simple proportion

$$\frac{d}{\delta v} = \frac{D}{v}. \tag{3.12}$$

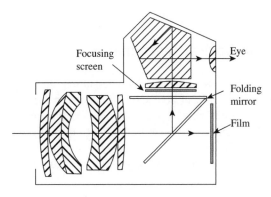

Figure 3.24 A single lens reflex (SLR) camera, showing the multiple element lens and the viewfinder arrangement

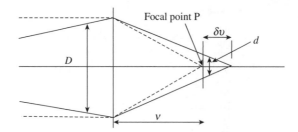

Figure 3.25   Geometrical construction for the depth of focus

The case for a camera focused on infinity is dealt with in Problem 3.5. More generally, and provided that δ_v is small, we can find the depth of field δu by differentiating the lens equation

$$\frac{1}{v} - \frac{1}{u} = \frac{1}{f}$$

$$\frac{\delta u}{u^2} = \frac{\delta v}{v^2}. \tag{3.13}$$

After some algebra, this gives

$$\delta u = \frac{Fu(u+f)d}{f^2} \tag{3.14}$$

where $F = f/D$. For example, a camera with $F = 2.5$ and focal length 5 cm focused on an object at 2 m distance (i.e. $u = -2$ m), using film with an acceptable blurring diameter of 50 μm, will be in focus for objects 19 cm in front of or behind the 2 m position.

A modern compact automated camera (Plate 2[*]) conceals from the user many sophisticated design features. Object distance is measured by an infrared rangefinder, and luminance is measured by a photometer, followed by automatic focusing, exposure and aperture adjustments. Digital cameras use electronic array detectors such as the CCD (Chapter 20), with their own complex circuitry, offering possibilities of enhanced sensitivity and spectral range and with image detail comparable with that of the photographic plate.

3.10   Illumination in Optical Instruments

The discussion of the compound microscope started by assuming that wavefronts left the object over a wide range of angles. The object may of course be illuminated naturally by diffuse light, but this is often insufficient. Extra illumination must be provided, and for efficiency and good angular resolution the light must be encouraged to leave the object in the right range of directions. This is achieved for transparent objects by the use of a *condenser*, which may be a concave mirror or a lens system, as in Figure 3.26. No great optical quality is required, since only a rough image of a diffuse source of light

---

[*]Plate 2 is located in the colour plate section, after page 246.

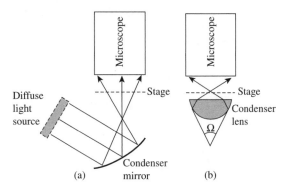

Figure 3.26 Condenser systems for a microscope: (a) concave mirror; (b) lens system. A short focal length is needed to collect light over a large solid angle Ω and to cover a wide range of angles as it enters the microscope

need be formed on or near the objective plane of the microscope. The total light entering the microscope depends on the solid angle over which the condenser collects the light; the condenser must therefore have a short focal length both for this reason and so that the object plane is well illuminated by light traversing it over a wide range of angles.

A similar problem is encountered in projection systems and enlargers; there the requirement is to obtain as much light as possible through a system consisting of a transparency, a projection lens and a screen; the transparency may be the key element of a digital cinema projector. The illumination of the transparent object must be even, but there is no requirement for illumination over a wide range of angles. Figure 3.27(a) shows the way in which light from a small source traverses a projection system: it is important not to confuse this diagram with the more conventional ray diagram of Figure 3.27(b) which is concerned with the image on the screen of a point on the transparent object. This image is formed by a narrow pencil of rays within the light paths of Figure 3.27(a).

The condenser lens of a projector need not be an accurate high-quality component. For the familiar 'overhead' projector a stepped lens is used (Figure 3.28); this is a thin sheet of glass or plastic

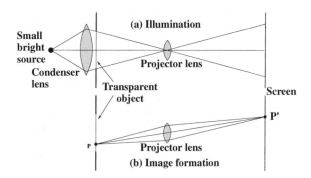

Figure 3.27 Illumination in a projector system. (a) The action of the condenser lens is to collect light emerging from a small bright source over a wide angle, providing an even illumination over the transparent object, and concentrating the light as it passes through the projector lens. (b) The projector lens forms an image P' of each part P of the object in a narrow pencil of rays determined by the illumination

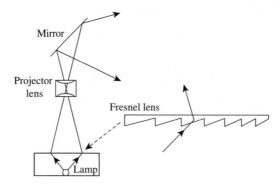

Figure 3.28   Overhead projector. The light from the lamp is concentrated into the projector lens by the stepped lens plate, known as a Fresnel lens

with an embossed array of prisms.[7] The equivalent simple lens which it replaces would be an impossibly thick and massive piece of glass, while any imperfections in the stepped lens are unimportant.

Problem 3.1

The microscopist Antoni van Leeuwenhoek (1632–1723) used a single lens to obtain magnifications up to ×200 or more. Find the diameter of a spherical glass bead which would give such a magnification. (Van Leeuwenhoek apparently fabricated biconvex lenses of such a small diameter.) (Hint: You will need the thick-lens power formula given in Problem 2.6.)

Problem 3.2

Show that the angular magnification M_A of the astronomical telescope in Figure 3.17 can be measured by placing a scale across the objective lens, and measuring the transverse magnification (< 1) of this scale in the image formed by the eyepiece.

Problem 3.3

The immersion technique, in which a liquid fills the space between a microscope objective and a slide cover glass, can give improved image brightness by allowing more light to enter the objective. Calculate the improvement which can be obtained when the objective in air accepts a cone of half angle 30° and the liquid and glass both have refractive index 1.5.

Problem 3.4

Compare the fields of view of the Galilean and astronomical (Keplerian) telescopes of Figure 3.17, in the following example. The objective diameters are both 2 cm, with focal lengths 20 cm, and the eyepiece diameters are both 1 cm, with focal lengths −10 and +10 cm respectively. Show that the magnifications are +2 and −2 respectively, and the fields of view are 1/10 and 1/30 radians respectively. (Note that in this case the eyepiece itself is assumed to play the role of the field stop.) Show that the exit pupil for the Keplerian telescope is outside the eye lens.

---

[7]The stepped lens is known as the Fresnel lens after its inventor, who was the first to use the principle in lighthouse lenses.

Problem 3.5

A camera focused on infinity has a depth of focus depending on the focal ratio F, the focal length f and the acceptable image diameter d. Show that objects beyond a distance u_1 are in focus, where $u_1 = f^2/Fd$. Find this distance for $F = 2.5, f = 5\,\text{cm}, d = 50\,\mu\text{m}$.

If the camera is focused on u_1, what is the nearest object in focus? (Use the approximate analysis of Section 3.9)

Problem 3.6

For binoculars specified as 8×40, with objective focal length 15 cm and $6°$ field of view, what are (a) the magnification of a distant object, (b) the focal length of the eyepiece, (c) the diameter of the exit pupil, d) the size of the field stop?

Problem 3.7

Given a Barlow lens of focal length $-75\,\text{mm}$, find the magnitude and the direction of the displacement needed to change the magnification m_B from $\times 2$ to $\times 4$.

Problem 3.8

(a) Show that if we add a Barlow lens of magnification m_B at distance d from an objective of focal length f_O, the focal length is increased by $\Delta z = (m_B - 1)u$, where $u = (f_O - d)$. (b) If the Barlow has focal length $f_B = -6\,\text{cm}$, how much does the telescope length change if m_B is varied from 2 to 4?

4 Periodic and Non-periodic Waves

Fourier, Jean Baptiste Joseph (*1768–1830*), *French mathematician ... born Auxerre ... son of a tailor ... soon distinguished himself as a student, and made rapid progress, delighting most of all, but not exclusively, in mathematics.*

Encyclopaedia Britannica, 9th edn, 1898.

Although light is emitted and absorbed in photons, which are discrete packets of energy and momentum, the propagation of all electromagnetic radiation is determined by its wave nature; geometric optics, described in terms of rays, is an approximation. The propagation of light, and in particular its behaviour in interference and diffraction, must be described by the wave theory that we develop in this chapter. We consider first a wave in any quantity, which we designate as ψ, which might be the pressure in a sound wave, the height of a wave in the sea, or the amplitude of the electric or magnetic field of a light wave.

Although the plane wave $\psi = f(z - vt)$, progressing in the positive z direction with velocity v, may have any wave shape and will keep that same shape as it progresses, it is both convenient and physically meaningful to concentrate on the simple harmonic waveform or sinusoidal wave introduced in Chapter 1:

$$\psi = A \cos k(z - vt). \tag{4.1}$$

Using the angular frequency ω and adding an arbitrary phase ϕ the wave becomes

$$\boxed{\psi = A \cos(kz - \omega t + \phi).} \tag{4.2}$$

Note that when $\phi = -\pi/2$ the wave becomes the sine wave of equation (1.12).

The constants in equation (4.2) are the *angular frequency* ω (which is $2\pi v$ where v is the *frequency*), the *wave number k* and the *phase* ϕ. A cycle of oscillation occurs at time intervals of one *period* $= 2\pi/\omega$, and at distance intervals of one *wavelength* $\lambda = 2\pi/k$.

The velocity of the wave is $v = \omega/k$ and its form at any time is a simple sine or cosine wave along the z axis. The phase term ϕ determines the position of the cosine wave at $t = 0$ (Figure 4.1). Adding $-\pi/2$ to ϕ makes the cosine wave into a sine wave, moving it along the z axis by a quarter wavelength.

Optics and Photonics: An Introduction, Second Edition   F. Graham Smith, Terry A. King and Dan Wilkins
© 2007 John Wiley & Sons, Ltd

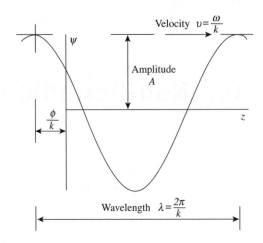

Figure 4.1   The progressive cosine wave $\psi = A\cos(kz - \omega t + \phi)$ at t $= 0$

We will also write the same wave in complexified form as an exponential function

$$\tilde{\psi} = A\exp[\mathrm{i}(kz - \omega t + \phi)] \tag{4.3}$$

which lends itself well to analytical work, although it represents less obviously the same wave and will require explanation in this chapter.

However it is expressed, the simple harmonic wave is a type of wave that is easily recognizable, as for example in a sound wave composed of a single pure note, or the monochromatic light from a laser. Familiarity with the various ways of representing and visualizing simple harmonic waves is essential to understanding the behaviour of light.

In this chapter we introduce the representation of a simple harmonic wave mathematically by complex exponential functions and graphically by a rotating vector known as a *phasor*. We show how simple harmonic waves are added, taking account of phase; this is at the heart of interference and diffraction phenomena in optics. We then show how any waveform can be built up by the addition of simple harmonic motions, using Fourier synthesis, or separated into component parts by Fourier analysis.

4.1   Simple Harmonic Waves

Any non-trivial solution $\psi = \psi(z, t)$ of the one-dimensional wave equation

$$\frac{\partial^2 \psi}{\partial z^2} - \frac{1}{v^2}\frac{\partial^2 \psi}{\partial t^2} = 0 \tag{4.4}$$

(where v is the propagation velocity) is a wave, by definition. For example, substitution of $\psi = A\cos(kz - \omega t + \phi)$ shows it will be a solution provided k and ω satisfy $v = \omega/k$. Sinusoidal functions such as this represent an idealized limit of certain quasi-monochromatic waves often found in nature. Since equation (4.4) is linear, if we are given any two solutions ψ, ψ', any combination of the form $\tilde{\psi} = a\psi + b\psi'$, where a, b are arbitrary complex constants, is also a solution. Suppose we

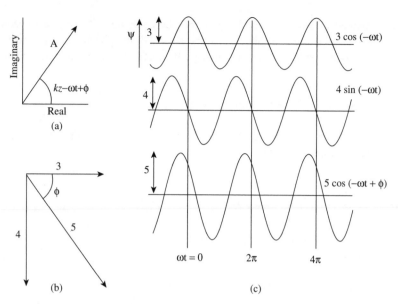

Figure 4.2 (a) A general phasor of amplitude A at an angle of $(kz - \omega t + \phi)$ to the horizontal axis; (b) the addition of phasors at $z = 0$ and $t = 0$ representing $3\cos(kz - \omega t)$ and $4\sin(kz - \omega t) = 4\cos(kz - \omega t - \pi/2)$ give a resultant phasor of length 5 at angle ϕ, representing $5\cos(kz - \omega t + \phi)$ where $\phi = -\tan^{-1}(4/3)$; (c) the two waves and their resultant as a function of ωt for $z = 0$

have a real-valued monochromatic wave of the form $\psi = A\cos\theta$, where A and $\theta = kz - \omega t + \phi$ are both real. Taking $\psi' = A\sin\theta$, $a = 1$ and $b = i$, we have

$$\tilde{\psi} = A\cos\theta + iA\sin\theta = A(\cos\theta + i\sin\theta) = A\exp(i\theta), \qquad (4.5)$$

a particularly handy, complex-valued solution. (In the last member, we have used Euler's famous theorem on the expansion of $\exp(i\theta)$.) This wave, the complexified form of ψ, has an *amplitude A* and *phase θ*. The real and imaginary parts of $\tilde{\psi}$ are given by $A\cos\theta = \mathrm{Re}(\tilde{\psi})$, $A\sin\theta = \mathrm{Im}(\tilde{\psi})$. Figure 4.2 (a) shows $\tilde{\psi}$ plotted in the complex plane with $\mathrm{Re}(\tilde{\psi})$, $\mathrm{Im}(\tilde{\psi})$ as rectangular coordinates. We see that $\tilde{\psi}$ acts as a two-dimensional vector of amplitude A, and at angle θ (measured anticlockwise from the $\mathrm{Re}(\tilde{\psi})$ axis). Such a vector, which embodies the amplitude and phase of a real oscillation, is called a *phasor*. When $\theta = kz - \omega t + \phi$, and z is held fixed, this phasor will rotate clockwise as time increases. In other words, instead of oscillating sinusoidally as did the original wave, the phasor has a much simpler motion: it rotates uniformly, its endpoint describing a circle.

The exponential form and the phasor provide a simple representation of a phase change. Expanding $\psi = A\cos(kz - \omega t + \phi)$, we find

$$\psi = A\cos\phi\cos(kz - \omega t) - A\sin\phi\sin(kz - \omega t), \qquad (4.6)$$

a rather complicated mixing of cosines and sines. Contrast this with the exponential form

$$\tilde{\psi} = A\exp[i(kz - \omega t + \phi)] = A\exp(i\phi)\exp[i(kz - \omega t)]. \qquad (4.7)$$

Here the complex wave keeps its original form and is merely multiplied by a complex constant; correspondingly the phasor is simply rotated by the phase shift.

Superposition also becomes much simpler with the phasor picture. Phasors obey vector addition in the complex plane. An especially simple case is one where all the phasors being superposed have the same frequency ω; as a result, they all rotate rigidly together, and their resultant has a constant magnitude.

Example. Use phasors to evaluate the sum of $\cos\theta$ and $\cos(\theta + \alpha)$.

Solution. The real quantity to be evaluated is $\psi = \cos\theta + \cos(\theta + \alpha)$. Write the complex version of this and merge the separate terms into one:

$$
\begin{aligned}
\tilde{\psi} &= \exp(i\theta) + \exp[i(\theta + \alpha)] = \exp(i\theta)[1 + \exp(i\alpha)] \\
&= \exp(i\theta)\exp(i\alpha/2)[\exp(-i\alpha/2) + \exp(i\alpha/2)] \\
&= \exp[i(\theta + \alpha/2)]2\cos(\alpha/2).
\end{aligned}
\tag{4.8}
$$

We require the real part of this, which is $2\cos(\alpha/2)\cos(\theta + \alpha/2)$.

Example. Sketch a diagram showing phasors that represent the following oscillations: $2\sin\omega t$, $3\cos\omega t$, $4\cos(\omega t + \pi/4)$. By evaluating their real (R) and imaginary (I) components at $t = 0$, find the amplitude and phase of the phasor representing the sum of all three.

Solution. Note that $2\sin\omega t$ can be written as $2\cos(\omega t - \pi/2)$. The three phasors are $2\exp[i(\omega t - \pi/2)]$, $3\exp(i\omega t)$ and $4\exp[i(\omega t + \pi/4)]$. At $t = 0$, their sum is $A_R = 3 + 2\sqrt{2} = 5.828$, $A_I = -2 + 2\sqrt{2} = 0.828$, giving $A = (A_R^2 + A_I^2)^{1/2} = 5.887$ with phase $\tan^{-1}(A_I/A_R) = 8.09°$.

Note that a phasor can represent any periodically varying quantity, which might itself be a scalar, such as pressure in a sound wave, or a vector, such as the vector components E_z or B_x of an electromagnetic wave. The phasor is a representation of the amplitude and phase of a harmonic wave. When two or more waves at the same frequency are superposed, each with its own amplitude and phase, the way they add depends on their relative phases: the *interference* between them may give an increased or decreased amplitude. This is done algebraically by adding the complex magnitudes, but it may be pictured by adding the corresponding phasors as vectors to produce a single phasor representing the combination in amplitude and phase.

To summarize this important concept, we can represent a harmonic wave in amplitude and phase by a complex number, the *complex magnitude*; summing the complex magnitudes for a combination of waves gives the complex magnitude for the combined wave.

The intensity of the wave (i.e. the energy flux, also known as the 'irradiance' in optics) can be found by taking the time average of the square of the wave. A constant multiplicative factor will be needed which will depend on the details of the physical system, e.g. mechanical, electromagnetic, etc., and the system of units: for present purposes we ignore this factor.[1] For the wave $\psi = A\cos\theta$, where θ is some linear function of the time, we would calculate $I = \langle A^2\cos^2\theta\rangle_{\text{avg}} = A^2/2$; again we will ignore the factor of one-half, and set $I = A^2$. Using the complex conjugate of the complexified wave,

---

[1]In Section 5.5, where we discuss energy flow in electromagnetism, we specify this factor precisely.

or $\tilde{\psi}^* = A\exp(-i\theta)$, we have

$$I = A^2 = \tilde{\psi}\tilde{\psi}^* = |\tilde{\psi}|^2. \tag{4.9}$$

We will see in Section 4.10 that any continuous waveform can be represented as the sum of simple cosine and sine waves. If the waveform is *periodic*, repeating at equal intervals of time with basic frequency v, these will be a series of *harmonics* at frequencies which are integral multiples of v. If the waveform is not truly periodic, but changes with time (and therefore with distance), then it can be constructed from a continuous spectrum of sinusoidal waves. The mathematical link between a waveform and its components is by way of Fourier analysis.

4.2   Positive and Negative Frequencies

As we have just seen, a typical form of a phasor is

$$\boxed{\tilde{\psi} = A\exp[i(kz - \omega t + \phi)].} \tag{4.10}$$

The *exponential frequency term* $\exp(-i\omega t)$ represents a point at unit distance from the origin in the complex plane, rotating clockwise around the origin $\omega/2\pi = v$ times per second. Similarly an exponential frequency term with a plus sign rotates the phasor in the opposite direction, continually adding phase rather than subtracting from it. Both signs are equally good mathematical representations of the same oscillation $\cos\omega t$, which is the real part both of $\exp(i\omega t)$ and of $\exp(-i\omega t)$. Why then do we bother with negative frequencies, when positive frequencies are equally useful? If we want to represent an arbitrary function $f(t)$ by exponential components, we need frequency terms with both positive and negative signs. A simple example is a cosine wave, where

$$\cos\omega t = \frac{1}{2}[\exp(i\omega t) + \exp(-i\omega t)] \tag{4.11}$$

while a sine wave is represented by

$$\sin\omega t = \frac{1}{2i}[\exp(i\omega t) - \exp(-i\omega t)]. \tag{4.12}$$

A real-valued wave with intermediate phase, i.e. with both cosine and sine components, can be represented by a sum such as

$$a\cos\omega t - b\sin\omega t = \frac{1}{2}(a + ib)\exp(i\omega t) + \frac{1}{2}(a - ib)\exp(-i\omega t). \tag{4.13}$$

The two components now have *complex amplitudes* which are complex conjugates.

Extending this to a spectrum with a range of frequency components, a function $f(t)$ may be written as

$$\boxed{f(t) = \int_{-\infty}^{+\infty} A(\omega)\exp(i\omega t)d\omega} \tag{4.14}$$

where $A(\omega)$ is the complex amplitude at frequency ω.

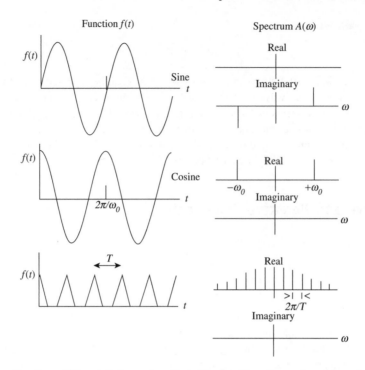

Figure 4.3   Three functions $f(t)$ and their spectra $A(\omega)$. The functions themselves ($\sin \omega_0 t, \cos \omega_0 t$, chopped triangle) are real, but their spectra have in general both real and imaginary parts. Note that the lines of various heights in the spectra represent delta functions with various multiplying factors. The cosine's spectrum has only real components, and the sine's has only imaginary components. We assume that the chopped triangular function keeps the same form eternally. It is then an even function, so its spectrum is real and consists of delta functions uniformly spaced but varying in magnitude

Note that for $f(t)$ to be real, $A(-\omega) = A^*(\omega)$, and vice versa. The real and imaginary parts of the complex $A(\omega)$ together form the *complex spectrum* of the function $f(t)$, and may be plotted on two graphs, one for the real and one for the imaginary part. Figure 4.3 shows the spectra of three periodic waves in this way. Notice that there is no need always to plot the negative frequency half of these spectra, because of the conjugate property.

We need to relate the discrete spectra of equations (4.11) and (4.12), and the continuous spectrum of equation (4.14). For this we introduce the *delta function*, named after the physicist P.A.M. Dirac, who introduced it in the context of quantum mechanics. The Dirac delta function $\delta(x)$ can be regarded loosely as an infinitely narrow, peaked function, with peak at $x = 0$ and with unit area. It is usually defined as follows:

$$
\begin{aligned}
\delta(x) &= 0 \quad x \neq 0 \\
\int_{-a}^{b} \delta(x)\mathrm{d}x &= 1 \quad a, b > 0.
\end{aligned}
\tag{4.15}
$$

It follows that for any continuous function $f(x)$

$$
\int_{x_0-a}^{x_0+b} f(x)\delta(x - x_0)\mathrm{d}x = f(x_0).
\tag{4.16}
$$

Some useful properties of the delta function can be found in Section 4.14. (See also Problem 4.2.)

We now see how to write down the amplitude of a sine or cosine function. Setting $A(\omega) = (1/2)[\delta(\omega - \omega_0) + \delta(\omega + \omega_0)]$ in equation (4.14) yields $f(t) = \cos(\omega_0 t)$, and setting $A(\omega) = (2\mathrm{i})^{-1}[\delta(\omega - \omega_0) - \delta(\omega + \omega_0)]$ gives $\sin(\omega_0 t)$. A function $f(t)$ that is an arbitrary discrete sum of harmonic functions can be handled in a similar fashion (Problem 4.3).

4.3   Standing Waves

The simplest example of two waves with the same frequency adding with varying phase is given by two cosine waves travelling in opposite directions adding to give an interference pattern of *standing waves*. This may be seen in water waves reflected from a pond wall, or heard in sound waves; it is often conspicuous in VHF radio (the FM band) where standing wave patterns inside a room may be explored by moving a portable receiver. Two waves with equal amplitudes travelling in the directions $+z$ and $-z$ add as

$$\psi = A\cos(kz - \omega t) + A\cos(-kz - \omega t). \tag{4.17}$$

It is a useful exercise to write this as the sum of exponential terms $A\exp[-\mathrm{i}(\omega t \pm kz)]$, obtaining

$$\begin{aligned}\tilde{\psi} &= A\exp(-\mathrm{i}\omega t)[\exp(-\mathrm{i}kz) + \exp(\mathrm{i}kz)] \\ &= 2A\cos kz\exp(-\mathrm{i}\omega t).\end{aligned} \tag{4.18}$$

Here the complex quantity $\tilde{\psi}$ has amplitude $2A\cos kx$ and phase ωt. This is a wave with the same phase everywhere at a given time, but with an amplitude varying with position z. Figure 4.4 shows the envelope pattern of the standing wave, with the actual displacement ψ at intervals of one-sixteenth of the period, i.e. at phase intervals of $\pi/8$. The amplitude of the superposition varies along the z axis as the relative phase of the two component waves changes. If the oscillations are in phase at $z = 0$, the phase of one wave increases and the phase of the other decreases as kz, as indicated in Figure 4.5. The phase reference is the phase of the oscillation at $z = 0$. Equation (4.18) shows that the sum of the two

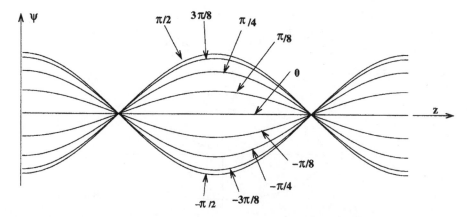

Figure 4.4   The oscillation in a standing wave pattern, with successive plots at intervals of one-sixteenth of the period. The curves at phase $\pi/2$ and $-\pi/2$ constitute the envelope of the oscillation

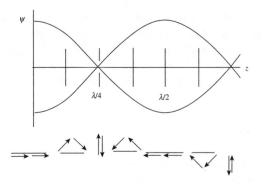

Figure 4.5   The envelope of the standing wave pattern formed by two sinusoidal waves of equal amplitudes, in phase at $z = 0$ and travelling in opposite directions. The phasor diagrams show the waves in phase at $z = 0$, with the resultant amplitude, which oscillates in a straight line, depending on the relative phase as z increases

waves gives a maximum at $z = 0$, falling as $\cos kz$ to zero at $z = \lambda/4$ and increasing to a maximum again at $z = \lambda/2$. The successive minima and maxima are called *nodes* and *antinodes*.

The pattern of standing waves provides a simple example of the phasor representation of amplitude and phase. Figure 4.5 shows the phasors for the two waves at intervals of $\lambda/8$ along the z axis. Starting at an antinode with two waves in phase at $z = 0$ and moving to larger values of z, the phase difference increases in steps of $\pi/4$. The sum of the two vectors decreases to zero to give the first node, and then increases to give the next antinode at $z = \lambda/2$; here the phasor is seen to be rotated through angle π compared with the first antinode, i.e. there is a phase change of π.

The standing wave pattern for waves of unequal amplitude does not have zero amplitude at the nodes. Figure 4.6 shows the envelope of the standing wave pattern, with the phasor diagrams at

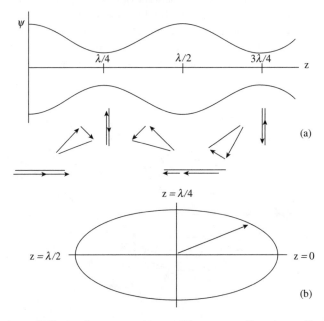

Figure 4.6   The envelope of the standing wave pattern, with corresponding phasor diagrams (a), for waves of unequal amplitude. The resultant phasor traces out an ellipse (b)

intervals of $\lambda/8$ along the z axis. The phasor representing the standing wave then traces an ellipse as it varies along the x axis. The major and minor axes represent amplitudes at antinodes and nodes; for equal amplitudes, as in Figure 4.5, the ellipse degenerates into a straight line.

4.4 Beats Between Oscillations

The addition of two oscillations with slightly different frequencies gives the effect of beating, which is familiar in sound waves. This is closely analogous to the standing wave patterns of the previous section, with the relative phases of two oscillations varying with time rather than with distance. The addition of two sinusoids, one of which is smaller in amplitude and which has a slowly increasing phase relative to the first, is shown in Figure 4.7(a). Here the phase of the larger oscillation is taken as the reference phase, so that the phasor representing the smaller oscillation rotates; the tip of the phasor representing the sum oscillation traces a small circle as the relative phase changes. Although the intensity[2] of the sum varies sinusoidally, the phase does not, as shown in Figure 4.7(b) and (c).

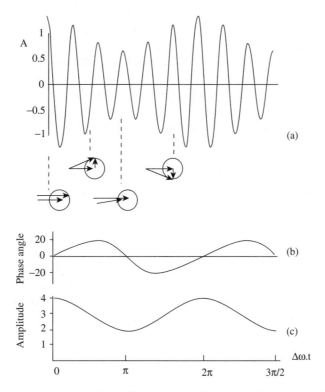

Figure 4.7 (a) The sum of two sinusoidal oscillations with different amplitudes and with a slowly changing relative phase corresponding to slightly different frequencies, showing phasor diagrams; (b) the phase variations relative to the phase of the larger oscillation, shown together with (c), the amplitude variations

---

[2]Denoting the sum by $y = \exp(i\omega t)[1 + a\exp(i\Delta\omega t)]$, the intensity $yy^* = [1 + a\exp(i\Delta\omega t)][1 - a\exp(i\Delta\omega t)] = 1 + a^2 + 2a\cos\Delta\omega t$.

The phase reference has been taken as one of the two oscillations, with the phase of the other increasing uniformly with time. One oscillation then has angular frequency ω, and the other has $\omega + \Delta\omega$, and the beat frequency is $\Delta\omega/2\pi$. When the amplitudes are equal, a more convenient phase reference may be taken as that of an oscillator with frequency half-way between these two, so that we are adding angular frequencies $\omega - \Delta\omega/2$ and $\omega + \Delta\omega/2$. The time variation of the phasors now looks like the spatial variation in the standing wave pattern of Figure 4.5. For equal amplitudes the phase of the resultant is now constant for half a period, reversing at the instants of zero amplitude.

4.5   Similarities Between Beats and Standing Wave Patterns

The common use of the phasor diagram to illustrate the phenomena of beats and of standing waves demonstrates their underlying similarity. Beats are variations of amplitude with time at one point, whilst standing waves are variations of amplitude with position at one time. Both phenomena can in fact be produced simultaneously in very simple circumstances. In Figure 4.8 two sources of sinusoidal waves S_1, S_2 are separated by a distance of several wavelengths, so that the distances S_1X, S_2X to a point X may differ by anything from zero to several wavelengths. Such a situation may occur, for example, with two sources of sound, or with two radio transmitters. At X the two waves add, with a phase relation that depends on the relative phase of the sources S_1 and S_2, and on the difference $S_1X - S_2X$ expressed in terms of the wavelength λ. Beats can now be produced at X by keeping the geometric arrangement fixed, and transmitting two different frequencies from S_1 and S_2. Alternatively, the two transmitters can be set to the same frequency, and arranged to transmit exactly in phase. Then if $(S_1X - S_2X) = n\lambda$ the waves will arrive in phase at X, while if at another point X' the path difference $(S_1X' - S_2X') = (n + \frac{1}{2})\lambda$ the waves will arrive there out of phase. There is therefore a pattern of waves resulting from the interference of the waves, with maxima and minima of amplitude following a simple geometric pattern.

Now let the phase of S_1 change slowly with respect to S_2. The result can be described in two ways: either the interference pattern is moving, or at each point there is a beat between the two transmitters. The physical situation can be reversed, so that X represents a single source or transmitter and S_1 and S_2 represent two receivers connected together in such a way that the relative phase of the two waves they receive determines the sum of their signals. This situation occurs in optical and radio interferometers, such as the Michelson stellar interferometer. The radio interferometer, as used in radio astronomy, collects waves from a single radio source in two separate antennas, adding them in a single receiver. As the Earth rotates, a celestial source moves across the interferometer, the path difference changes, and the receiver output varies. Alternatively, in the addition of the two waves an extra phase difference can be inserted deliberately, and if this increases steadily with time, the rate of variation can be adjusted to compensate for the rotation of the Earth.

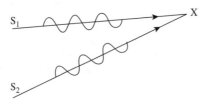

Figure 4.8   Waves from the two sources S_1, S_2 reach X by different path lengths. As X moves, it explores an interference pattern: alternatively, if S_1 and S_2 transmit different frequencies beats will be heard at X

The basic equivalence in these examples is that an addition or subtraction of phase linearly with time in a sinusoidal oscillation is equivalent to a change of frequency.

4.6   Standing Waves at a Reflector

Standing wave patterns can most easily be demonstrated by arranging for the total reflection of a plane wave. Following a classic experiment by O. Wiener in 1890, G. Lippmann carried out in 1891 a striking demonstration of the standing waves of light reflected by a mirror. He used a photographic plate with very fine grain, backed with a layer of mercury to act as a smooth reflector. A plane wave of monochromatic light, falling on the plate, formed a standing wave which could be seen in the developed film by cutting a section at a very shallow angle. Lippmann also showed that a plate exposed and developed in this way could be used for colour photography, since, as explained below, light was reflected selectively according to its wavelength from the planes of silver left in the emulsion. Coloured holographic images, which we describe in Chapter 14, are based on the same concept of selective reflection from a three-dimensional negative, made by interference between light beams within a photographic emulsion. In an emulsion of $20\,\mu m$ thickness, since the interference planes are separated by $\lambda/2$, some 80 interference planes are formed, depending on the wavelength. This array acts as a selective reflective filter for white light. A similar phenomenon of selective reflection is found in X-ray diffraction at planes of atoms within a crystal (Chapter 11).

The Lippmann demonstration can now be repeated much more easily by using radio waves with wavelength of a few centimetres; but it had a particular historical importance in that it showed not merely the wave pattern but the way in which the pattern was related to the reflecting surface. The two standing wave patterns in Figure 4.9 show the patterns obtained at two different kinds of boundaries. This difference is well known in wind instruments: the resonant oscillations in an organ pipe have a node at the end of the pipe for closed pipes, and an antinode for open ones. For electromagnetic waves the relevant boundary conditions are determined by the dielectric and magnetic properties of the material at the boundary. In 1891 it was still interesting to prove that a metal surface, being an excellent conductor, would determine that there would be essentially zero electric field at the surface, giving the pattern of Figure 4.9(a). The Lippmann films showed this clearly, giving layers of silver starting one-quarter wavelength above the surface. The pattern of

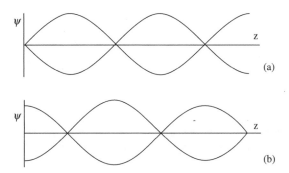

Figure 4.9   Envelope of the standing wave patterns due to reflections at different types of boundary: (a) zero amplitude at the boundary; (b) maximum amplitude at the boundary

Figure 4.9(a) is produced by two waves out of phase by π radians at the surface, while in Figure 4.9(b) the two waves are in phase at the surface. A property of the surface, the reflection coefficient (see Section 5.3), determines the relation between the incident and the reflected waves.

The concepts of interference in space and in time are well illustrated in the Doppler radar system used for measuring the speed of moving aeroplanes or automobiles, which we discuss after setting out the theory of the Doppler effect itself.

4.7 The Doppler Effect

The Doppler effect is familiar as a change in pitch of a sound as the source or observer moves. It applies over the whole of the electromagnetic spectrum, but for light in particular it is important on physical scales from atomic to cosmic. On the atomic scale we shall be concerned with the spread in frequency of spectral lines due to the thermal velocities of atoms or molecules in a gas (Chapter 12), and on the cosmic scale we observe the *redshift* of spectral lines from galaxies receding, in effect,[3] with velocities comparable with the velocity of light. We shall show how the simple theory of the Doppler effect can be refined to take account of such large velocities by incorporating the theory of special relativity.

From the point of view of a stationary observer, a source emitting v_0 waves in 1 second and moving away from the observer with velocity v will expand the v_0 waves to a distance $(c + v)$ where c is the velocity of the waves, as in Figure 4.10. The frequency will therefore be seen by the observer as

$$v_{\text{obs}} = \frac{v_0}{1 + v/c}. \tag{4.19}$$

Similarly an observer moving away from a stationary source with velocity v will receive a frequency decreased by the rate at which the observer covers wavelengths of distance. The observed frequency is therefore

$$v_{\text{obs}} = v_0 - \frac{v}{\lambda_0} \quad \text{or}$$

$$v_{\text{obs}} = v_0 \left(1 - \frac{v}{c}\right). \tag{4.20}$$

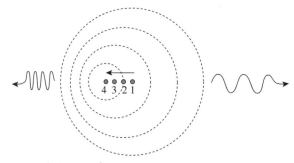

Figure 4.10 The Doppler effect. A moving source emits a periodic wave, represented by the broken circles. The waves are bunched together in the direction of motion and spread out behind

---

[3]Standard cosmological theory holds the more sophisticated view that galaxies are locally at rest but they ride on an expanding 'fabric' of spacetime.

These two equations (4.19) and (4.20) are nearly identical for small velocities, but they differ increasingly at large velocities. For light, however, unlike sound, there can be no difference between motion of the source and motion of the observer, which must give the same Doppler shift. The two equations are reconciled by a relativistic correction, which results from the different ways in which the source and the observer measure frequency. Ticks of a moving clock, as compared with an identical clock at rest, are prolonged, hence the clock goes slower. According to special relativity, any oscillator at frequency v_0, moving with velocity v relative to a stationary observer, will appear to the observer to be oscillating at a lower frequency v_1,

$$v_1 = v_0 \left(1 - \frac{v^2}{c^2}\right)^{1/2}. \tag{4.21}$$

Substituting this for v_0 in equation (4.19), the observed frequency becomes

$$v_{\text{obs}} = v_0 \left(\frac{1 - v/c}{1 + v/c}\right)^{1/2}. \tag{4.22}$$

The same result is obtained for the moving observer, since as seen by the stationary source the observer's clock goes slow by the same relativistic factor. In terms of wavelength, since $\lambda = c/v$

$$\lambda_{\text{obs}} = \lambda_0 \left(\frac{1 + v/c}{1 - v/c}\right)^{1/2}. \tag{4.23}$$

The full relativistic formula is essential in the context of astronomical measurements of distant galaxies, which may be receding with velocities approaching the velocity of light. The wavelengths λ_{obs} of spectral lines have been observed to be *redshifted* from their original wavelength λ_0 by a factor of more than 7, corresponding to a velocity $v = 0.96c$.

4.8 Doppler Radar

In a Doppler radar (Figure 4.11), one form of which is the radar used by police for measuring the speed of traffic, a transmitter T sends out a constant sinusoidal wave, say with wavelength

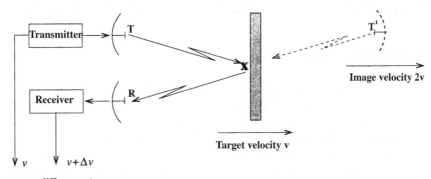

Figure 4.11 A Doppler radar system. The reflected wave from a target receding with velocity v is at a frequency lower by $2v/\lambda$. This is measured as a beat against the transmitted frequency

$\lambda = 3\,\text{cm}$ (frequency $v = 10\,\text{GHz}$). The reflected wave is received at R and added to part of the transmitted wave; the relative phase of these two waves is determined by the distance TXR. If the target X moves with velocity v, this distance changes at a rate $2v$, and, for $v \ll c$, the beat frequency from equation (4.22) will be $v_0(2v/c) = 2v/\lambda_0$. For wavelength $3\,\text{cm}$ and a velocity $v = 50\,\text{km}$ per hour $\simeq 14\,\text{m\,s}^{-1}$, the beat frequency is $926\,\text{Hz}$; this may be used to give a direct reading of the velocity of the target.

Another way of looking at the same problem is also shown in Figure 4.11, where the signal reaching the receiver may be considered to have originated in an image T′ of the transmitter, which moves with velocity $2v$ away from the radar. The beat is now between the frequency v_0 and the Doppler shifted frequency $v_0(1 - 2v)/c$, giving a beat frequency $2v/\lambda_0$ as before.

Example. We can give a treatment of Doppler radar that is exact for all velocities. In special relativity theory, two successive transformations by velocity v yield a net velocity $2v/(1 + v^2/c^2)$. This is the correct velocity of the transmitter's image T′ relative to the receiver. Use this to obtain the exact observed frequency and the beat frequency.

Solution. Substituting this net velocity for v in equation (4.22), we find

$$v_{\text{obs}} = v_0 \left(\frac{1 + v^2/c^2 - 2v/c}{1 + v^2/c^2 + 2v/c} \right)^{1/2} = v_0 \left(\frac{1 - v/c}{1 + v/c} \right) \tag{4.24}$$

and

$$v_{\text{beat}} = v_0 - v_{\text{obs}} = v_0 \frac{2v}{(c + v)} = \frac{2v}{\lambda_0(1 + v/c)}. \tag{4.25}$$

Notice that in the limit as v approaches c, the beat frequency is half that which we obtained for the low-velocity case.

Doppler radar measurements give velocity, not distance. Distance, or range,[4] is measured by the time of flight of a reflected radar pulse, which travels at the group velocity (Section 4.16).

4.9   Astronomical Aberration

Astronomers are familiar with an effect in the propagation of light which is related to the Doppler shift but which is purely geometrical and which does not depend on wavelength or frequency. If a source of light appears to be in a certain direction, and if the observer then starts to move *transverse* to this direction, what change does the observer see? The change is easily assessed for slower wave motions, such as water waves seen from a moving boat, since we have only to compound the observer's motion with the wave motion, as in Figure 4.12. A vector difference of the two velocities

---

[4]*Radar* is an acronym for RAdio Detection And Ranging.

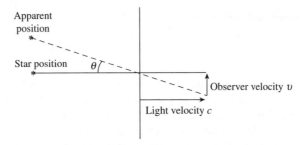

Figure 4.12 Astronomical aberration. An observer O moves transverse to a light wave with velocity c. The vector difference of their velocities gives an angular shift $\theta \approx v/c$

$\vec{v}_{\text{wave}} - \vec{v}_{\text{boat}}$ gives the effective motion of the waves past the boat. As for the Doppler frequency shift, in the previous section, the calculation should include a relativistic correction if the velocities are large, but the simple vector sum for velocity v transverse to light with velocity c gives an angular shift θ given by

$$\tan \theta = \frac{v}{c}. \tag{4.26}$$

Figure 4.13 shows that vector velocity addition is unacceptable because it produces a variable vacuum velocity for light, but that according to special relativity where light speed is always equal to c, the correct solution is

$$\tan \theta = \frac{v}{c} \left(1 - \frac{v^2}{c^2} \right)^{-1/2} \tag{4.27}$$

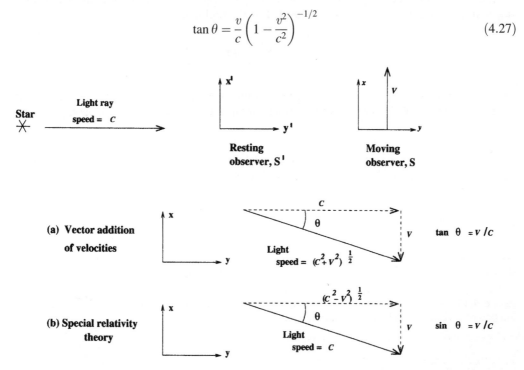

Figure 4.13 Astronomical aberration. (a) Non-relativistic theory predicts the wrong speed for light. (b) Relativistic theory gives the correct light speed

which reduces to the simple relation

$$\sin \theta = \frac{v}{c}. \tag{4.28}$$

The effect is observed as a periodic shift in the position of stars as the Earth follows its orbit round the Sun, known as *astronomical aberration* and illustrated in Figure 4.12. The Earth's orbit around the Sun with a velocity $v = 30\,\mathrm{km\,s^{-1}}$ causes a maximum aberration angle of $\theta = 20$ seconds of arc for any star in the sky. Astronomical aberration was discovered by James Bradley in 1725, when he was attempting to measure stellar parallax, a perspective effect in which the position of a star varies according to the position of the Earth in its orbit rather than to its velocity. The discovery of aberration was important in establishing both the finite velocity of light and the orbital motion of the Earth.

4.10   Fourier Series

A simple harmonic oscillation with an infinite extent in time and space is an idealized concept; in practice we deal with waves covering a range of frequencies and also travelling in various directions. To handle these cases we need to add a *spectrum* of waves distributed in frequency and in angle. Both frequency spectra and angular spectra are conveniently expressed in Fourier terminology, which we explore first in the domain of waveform and frequency.

Figure 4.14 shows how a periodic square waveform may be built up from a *harmonic series*[5] of sine waves; only the odd harmonics are needed, with amplitudes $4/\pi n$:

$$f(t) = \frac{4}{\pi} \sum_{n=1,3,5} \frac{1}{n} \sin\left(\frac{2\pi n t}{T}\right). \tag{4.29}$$

Only a small number of harmonics are needed to produce a recognizable square wave, although in theory a sharply defined square wave requires an infinite series. Note that all the components are in

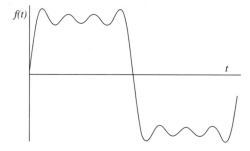

Figure 4.14 A periodic square wave built up from a harmonic series of sine waves, including only the fundamental with the third and fifth harmonics

---

[5]Until now we have used the term *harmonic* to mean a single angular frequency ω. A harmonic series is a sum of terms with frequencies $n\omega$, where n is any integer.

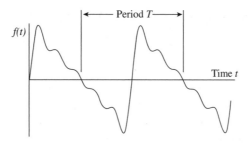

Figure 4.15   Sawtooth wave built up from Fourier components, including only the fundamental with the second to fifth harmonics

phase at the rising edge of the square wave. A different periodic wave, the sawtooth wave of Figure 4.15, can be constructed from sine components with amplitudes $a_n = 2/\pi n$ where n is an integer:

$$f(t) = \frac{2}{\pi} \sum_{n=1}^{5} \frac{1}{n} \sin\left(\frac{2\pi nt}{T}\right). \tag{4.30}$$

The more components that are added, the sharper is the sawtooth waveform.

The square wave in Figure 4.14 and the sawtooth in Figure 4.15 were constructed from sine wave harmonics. Choosing a different time origin would require a change of phase for each harmonic, which is equivalent to the introduction of cosine wave components.

A general theorem due to Fourier states that any *periodic* function $f(t)$ which repeats with period T can be expressed in terms of a constant plus a harmonic series of sine and cosine waves as

$$f(t) = \frac{1}{2}a_0 + \sum_{n=1}^{+\infty} a_n \cos\left(\frac{2\pi nt}{T}\right) + \sum_{n=1}^{+\infty} b_n \sin\left(\frac{2\pi nt}{T}\right). \tag{4.31}$$

Building up a periodic waveform as in Figures 4.14 and 4.15 is a process of *Fourier synthesis*, which consists of adding together a fundamental frequency component and harmonics of various amplitudes. The derivation of the frequencies and amplitudes of the components of a periodic waveform is *Fourier analysis*. This is achieved for the harmonic series of equation (4.31) as follows. The coefficients a_n and b_n are obtained by multiplying equation (4.31) by $\cos(2\pi nt/T)$ and $\sin(2\pi nt/T)$ respectively and integrating over one period, any interval, $t_1 \le t \le t_1 + T$:

$$a_n = \frac{2}{T}\int_0^T f(t) \cos\left(\frac{2\pi nt}{T}\right) dt, \quad b_n = \frac{2}{T}\int_0^T f(t) \sin\left(\frac{2\pi nt}{T}\right) dt. \tag{4.32}$$

Note that any constant term in $f(t)$ appears as $\frac{1}{2}a_0$, while b_0 is always zero.

In the exponential notation the Fourier series in equation (4.31) becomes

$$f(t) = \frac{1}{2}a_0 + \frac{1}{2}\sum_{n=1}^{\infty}(a_n - ib_n)\exp\left(i\frac{2\pi nt}{T}\right) + \frac{1}{2}\sum_{k=1}^{\infty}(a_k + ib_k)\exp\left(-i\frac{2\pi nt}{T}\right). \tag{4.33}$$

Note that we have replaced the dummy suffix n by k in the second summation. If we now write $k = -n$ we get

$$f(t) = \sum_{n=-\infty}^{\infty} A_n \exp\left(i\frac{2\pi nt}{T}\right) \tag{4.34}$$

where $A_n = \frac{1}{2}(a_n - ib_n)$, provided that for $n < 0$ we define $a_n = a_{-n}$ and $b_n = -b_{-n}$. Note that if $f(t)$ is a real-valued function, then A_n is the complex conjugate of A_{-n}.

So far we have considered functions in time, and Fourier harmonic series involving time and frequency. But we can also have periodic functions in space, such as the diffraction gratings described in Chapter 11. Any periodic function in the space domain with period L may be represented by the Fourier series

$$f(x) = \frac{1}{2}a_0 + \sum_{n=1}^{n=+\infty} a_n\cos\left(\frac{2\pi nx}{L}\right) + \sum_{n=1}^{+\infty} b_n\sin\left(\frac{2\pi nx}{L}\right). \tag{4.35}$$

The coefficients are then

$$a_n = \frac{2}{L}\int_0^L f(x)\cos\left(\frac{2\pi nx}{L}\right)dx,$$

$$b_n = \frac{2}{L}\int_0^L f(x)\sin\left(\frac{2\pi nx}{L}\right)dx. \tag{4.36}$$

This spatial form of the Fourier series and Fourier transform occurs in many areas of optics and photonics.

4.11   Modulated Waves: Fourier Transforms

Beats and standing waves are simple examples of *modulated waves*, whose amplitude varies periodically with time or space. Fourier theory also allows us to analyse non-periodic, or *aperiodic*, modulation in the same way. A short burst of waves, such as a pulse of laser light, can be considered as the sum of waves with a continuous range of frequencies rather than a single frequency. In general, both periodic and aperiodic modulation of a cosine wave can be expressed in terms of a *modulating function* $g(t)$, so that the wave is $g(t)\cos(2\pi v_1 t)$. If the modulating function $g(t)$ is a sinusoid, the wave can be decomposed into two components with frequencies above and below v_1; these would for example be the two frequencies producing a beat at the modulating frequency. If the wave is not fully modulated, so that the amplitude of $g(t)$ does not go to zero, there are also components at $\pm v_1$; the spectrum of the modulated wave then consists of a *carrier* and two *sidebands*.

Fourier analysis is a general technique for relating the form of a variable function to its spectrum; for example, it relates the spectrum of a sound's intensity to its actual waveform, and it gives a precise description of the wavelength (or frequency) components in a pulse of laser light. The relation, which was set out in equation (4.31) for the discrete harmonic components of a periodic wave, must now be extended to include a continuous spectrum and aperiodic modulation.

The relation between a variable $f(t)$ and its frequency spectrum $F(v)$ is given by the integrals

$$f(t) = \int_{-\infty}^{+\infty} F(v)\exp(2\pi ivt)dv, \ f(v) = \int_{-\infty}^{+\infty} f(t)\exp(-2\pi ivt)\,dt \tag{4.37}$$

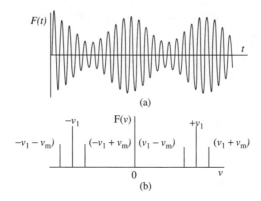

Figure 4.16 Cosinusoidal modulated wave (a) and its spectrum (b). The line at $v = 0$ is the vertical axis. As in Figure 4.3, the vertical lines at $v \neq 0$ represent delta functions

These integrals in the time domain express the one-to-one relation between a waveform and the infinite set of frequency components that constitute its spectrum. In the space domain, both $1/\lambda$ and the wave number $k = 2\pi/\lambda$ play roles similar to frequency. Using k, the spatial Fourier integrals corresponding to Equation (4.37) take the compact form

$$f(x) = \frac{1}{2\pi} \int_{-\infty}^{\infty} F(k)\exp(idx)dk, \quad F(k) = \int_{-\infty}^{\infty} f(x)\exp(-ikx)dx \qquad (4.38)$$

These *Fourier transforms* apply equally well to non-periodic as to periodic variables. We apply them first to the modulated wave $g(t)\cos(2\pi v_1 t)$, whose spectrum is found from equation (4.38) as follows:

$$\begin{aligned} F(v) &= \int_{-\infty}^{+\infty} g(t)\cos(2\pi v_1 t)\exp(-2\pi ivt)dt \\ &= \int_{-\infty}^{+\infty} g(t)\frac{1}{2}\{\exp[-2\pi i(v - v_1)t] + \exp[-2\pi i(v + v_1)t]\}dt \qquad (4.39) \\ &= \frac{1}{2}G(v - v_1) + \frac{1}{2}G(v + v_1) \end{aligned}$$

where $G(v)$ is the Fourier transform of the modulating function $g(t)$.

The simplest example is the spectrum of a cosinusoidally modulated wave, as in Figure 4.16(a). Suppose that we start with a 'carrier wave' of form $\cos(2\pi v_1 t)$ and this is modulated by $g(t) = a + b\cos(2\pi v_m t)$. The full spectrum of the unmodulated wave (the carrier wave) has components at v_1 and $-v_1$; the spectrum of the modulating cosine wave at any frequency v_m similarly has components at $\pm v_m$. The resulting spectrum of the modulated wave, evaluated with help of equation (4.51), is shown in Figure 4.16(b); there are now sidebands separated from the original components by v_m. This result is as expected from the consideration of beating between two cosine waves, which are now seen as the sidebands on either side of the carrier.

4.12 Modulation by a Non-periodic Function

In Figure 4.17 the carrier oscillation $\cos(2\pi v_1 t)$ is confined by a time-limited modulation function $g(t)$, which is shown as either a *Gaussian* or the abrupt *top-hat* function. Following equation (4.39)

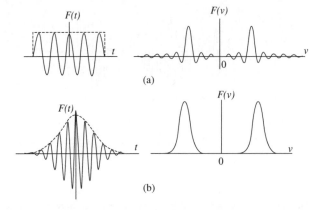

Figure 4.17   A wave whose duration is limited by (a) a 'top-hat' function and (b) a Gaussian

the spectrum of either of these time-limited waves is found from the spectrum of the modulating function. A top-hat function with height h and width b, centred on the origin at $t = 0$, is written

$$g(t) = h \text{ for } \left(-\frac{1}{2}b < t < +\frac{1}{2}b\right)$$

$$g(t) = 0 \text{ elsewhere.}$$

(4.40)

The Fourier integral equation (4.38) then gives the spectrum

$$G(v) = \int_{-b/2}^{+b/2} h \exp(-2\pi i v t) \mathrm{d}t$$

$$= \frac{h}{2\pi i v} \left[\exp\left(+2\pi i v \frac{b}{2}\right) - \exp\left(-2\pi i v \frac{b}{2}\right)\right]$$

$$= hb \frac{\sin\psi}{\psi} = hb \text{ sinc } \psi, \quad \text{where } \psi = \pi v b.$$

(4.41)

This Fourier transform of a top-hat function is a *sinc function*. (We shall encounter the sinc function again in Section 10.1 on diffraction at a single slit.) Figure 4.17 shows the full spectrum of the time-limited wave, with positive and negative components each with the shape of a sinc function. The width of the sinc function is inversely proportional to the width of the top-hat.

   The sinc function is frequently encountered in physics and in communication engineering. We shall see later that the spectrum of a waveform abruptly started and stopped has an intrinsic width inversely proportional to the length of the wavetrain; it also has sidebands which extend on either side of the main spectral component. A smooth modulation, avoiding the abrupt start and stop, has a wider main spectral component but lower sidelobes. The most important example of such a smooth modulating function is a *Gaussian*. The Gaussian function in Figure 4.17 is written

$$g(t) = h \exp\left(-\frac{t^2}{\sigma^2}\right).$$

(4.42)

Evaluation by the same process, using the identity

$$\int_{-\infty}^{+\infty} \exp(-a^2 x^2) \mathrm{d}x = \sqrt{\pi}/a \ldots (a > 0)$$

(4.43)

gives the spectrum

$$G(v) = h\sigma\sqrt{\pi}\exp(-\pi^2 v^2 \sigma^2). \tag{4.44}$$

The transform of a Gaussian function is therefore another Gaussian, whose width $1/\pi\sigma$ is inversely proportional to the original width[6] σ. Applying the general relation, equation (4.39) gives the spectra shown in Figure 4.17. We see that the Gaussian modulation of an oscillation gives Gaussian spectral lines, whose width is inversely proportional to the duration of the wavetrain. Similar analyses may be applied to a wave limited in space, showing that the spectrum of a wave group, such as the group of waves associated with a single particle or photon, depends on the length of the wave group. (For spatial analysis of waves, the variable analogous to frequency used to express the spectrum is the wave number, $k = 2\pi/\lambda$.) Extreme examples of short wave groups are found in pulsed light from lasers, where the pulse may be only a few wavelengths long, and consequently has a very wide spectral range (Chapter 16).

Gaussian modulating functions are also encountered in the lateral spread of concentrated light beams, and especially those from lasers. The lateral spread of the beam is related by a Fourier transform to the angular divergence of the beam, which is similarly described by a Gaussian function (Section 16.2).

4.13   Convolution

We now state the convolution theorem which enables us to find the Fourier transform of a further class of functions, those which are obtainable by convolving together two functions, say $f(t)$ and $g(t)$. The convolution $C(t)$ of two functions $f(t)$ and $g(t)$ is defined by the equation

$$C(t) = \int_{-\infty}^{+\infty} f(\tau)g(t - \tau)\mathrm{d}\tau = \int_{-\infty}^{\infty} f(t - \tau)g(\tau)\mathrm{d}\tau. \tag{4.45}$$

This equation is often written symbolically as

$$C(t) = f(t) \star g(t). \tag{4.46}$$

The convolution equation is useful in Fourier analysis of any function, whether it is of time, distance or angle. It occurs naturally in the response of any optical instrument such as a telescope or spectrometer which is intended to 'resolve' light according to either direction or wavelength; any such instrument has a limited resolving power which inevitably modifies and degrades the image or spectrum. For example, a photograph of a point source taken by an astronomical telescope appears to show the light originating from an extended source, which is the result of diffraction. Let a cross-section of this apparent source have a brightness distribution $g(\theta)$, called the *blurring function* or *point spread function*. Then a photograph of an object which has an actual brightness distribution $f(\theta)$ has a blurred image made by a convolution of the two functions $f(\theta)$ and $g(\theta)$. This blurred image $h(\theta)$ at θ is made up of contributions from a range of angles covered by the blurring function.

---

[6]The 'standard deviations' are smaller than these 'widths' by a factor of $\sqrt{2}$. They are, respectively, $\sigma/\sqrt{2}$ for $g(t)$ and $(\sqrt{2}\pi\sigma)^{-1}$ for $G(v)$. But their product is still a constant.

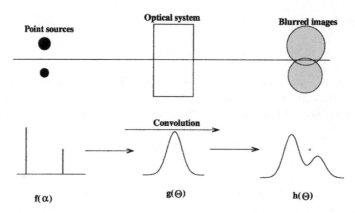

Figure 4.18 Response of an optical system. The image of a sharply defined object has been blurred by the point spread function

The contribution from a point α, where the true brightness is $f(\alpha)$, is proportional to the blurring function centred on the point θ, i.e. $g(\theta - \alpha)$. The resultant at θ is the integral over α:

$$h(\theta) = \int_{-\infty}^{+\infty} f(\alpha)g(\theta - \alpha)\mathrm{d}\alpha = \int_{-\infty}^{+\infty} f(\theta - \alpha)g(\alpha)\mathrm{d}\alpha. \tag{4.47}$$

This is a convolution of the source function $f(\alpha)$ with the point spread function $g(\theta)$. Notice that for an ideal system without any blurring, $g(\theta - \alpha) = \delta(\theta - \alpha)$, and $h(\theta) = f(\theta)$. Figure 4.18 illustrates the effect of blurring on the image of a geometric object, showing also the point spread function.

We now write the Fourier transforms of the functions $f(t)$, $g(t)$, $h(t)$ as $F(v)$, $G(v)$, $H(v)$; using the definitions of convolution and Fourier transform it can easily be shown that if $h(t) = f(t) \star g(t)$ then

$$\boxed{H(v) = F(v)G(v).} \tag{4.48}$$

This is the convolution theorem, which may be stated as follows:

The Fourier transform of the convolution of two functions is the product of their individual transforms.

We will apply the convolution theorem in diffraction theory (Chapter 10), where the functions $f(t)$ etc. are replaced by amplitude distributions across a diffraction aperture; their transforms then represent the corresponding diffraction patterns.

4.14 Delta and Grating Functions

When a single pulse becomes infinitely narrow, its transform becomes infinitely wide. It is convenient to describe an infinitely narrow function as a *delta function* (see Section 4.2). We may for example describe the envelope of a very short pulse of light travelling along an optical fibre as an approximation to a delta function, implying that it has an almost infinitely wide spectrum and can be used in a communication circuit with a very wide bandwidth. The broadening of the pulse as it travels, or in the detector circuits, can be regarded as a series of convolution processes, with a corresponding reduction in bandwidth and limitation of usefulness of the communication circuit.

The delta function itself may be thought of as a limiting case of an ordinary function, such as a Gaussian, with unit area but zero width:

$$\delta(x) = \lim_{a \to \infty} \sqrt{\frac{a}{\pi}} \exp(-ax^2).$$ (4.49)

The delta function $\delta(x - x_0)$ is an infinitely short function centred on x_0, and with unit area. The convolution of any reasonable function $f(x)$ with a delta function takes the form

$$\int_{-\infty}^{+\infty} f(x)\delta(x_0 - x)dx = f(x_0).$$ (4.50)

A convenient integral representation is

$$\delta(x) = (2\pi)^{-1} \int_{-\infty}^{\infty} \exp(ixy)dy.$$ (4.51)

Notice that if b is any non-zero real constant, a change of variable in equation (4.51) gives the useful identity $\delta(bx) = |b|^{-1}\delta(x)$.

An infinite set of uniformly spaced delta functions is known as the *grating function* (or *comb* or *shah function*). For positions $x_n = nx_0$ the grating function is

$$\sum_{n=-\infty}^{+\infty} \delta(x - x_n).$$ (4.52)

This may be regarded as a periodic function comprising an infinite series of delta functions: it can be shown that its Fourier transform is another grating function

$$F(k) = \int_{-\infty}^{+\infty} \sum_n \delta(x - x_n)\exp(-ikx)dx = \sum_n \exp(-ikx_n) = \frac{2\pi}{x_0} \sum_n \delta\left(k - \frac{2\pi n}{x_0}\right).$$ (4.53)

and the latter has spikes spaced apartly $2\pi/x_0$. The periodicities in the grating function and its transform are related inversely. Grating functions are used in the theories of the diffraction grating (Chapter 11) and of pulsed laser light (Chapter 16).

4.15   Autocorrelation and the Power Spectrum

The full description of a spectrum must contain the amplitude and the phase of all components. However, it is often only necessary to consider intensity (or power, or radiance, for example), which is proportional to the square of the amplitude and contains no phase information. This intensity distribution may also be expressed as a spectrum and is usually referred to as the *power spectrum*. The power is a real quantity; for a harmonic component $F(v)$ which may be a complex quantity, we obtain the power by multiplying by the complex conjugate[7] $F^{\star}(v)$.

---

[7] The complex conjugate of $A + iB$ is $A - iB$, and the product is $A^2 + B^2$.

If the signed magnitude $A(t)$ of a time-varying quantity is convolved with itself, the result is the inverse Fourier transform of its power spectrum. Although this follows from Section 4.13, the result is so valuable that we set out a proof as follows.

Convolving a function with itself, or self-convolution without the reversed sign seen in equation (4.45), is also known as an autocorrelation. We define the autocorrelation function Γ_{11} as

$$\Gamma_{11}(\tau) = \int_{-\infty}^{\infty} A(t+\tau)A^*(t)\mathrm{d}t. \tag{4.54}$$

Now take the Fourier transform of this using equation (4.38) and integrate first with respect to τ and then over t:

$$\int \Gamma_{11}(\tau)\exp(-2\pi i\nu\tau)\mathrm{d}\tau = \int_\tau \int_t A(t+\tau)A^*(t)\exp(-2\pi i\nu\tau)\mathrm{d}\tau\mathrm{d}t$$

$$= \int_\tau \int_t A^*(t)\exp(2\pi i\nu t)A(t+\tau)\exp[-2\pi i\nu(t+\tau)]\mathrm{d}\tau\mathrm{d}t$$

$$= \int_t \{A(t)\exp(-2\pi i\nu t)\}^*\mathrm{d}t.F(\nu)$$

$$= F^*(\nu).F(\nu) \tag{4.55}$$

where all integrals extend to $\pm\infty$. This is the power spectrum, or power spectral density. Lastly, performing the inverse Fourier transform of equation (4.37) on both sides of equation (4.55),

$$\Gamma_{11}(t) = \int_{-\infty}^{\infty} F^*(\nu)F(\nu)\exp(2\pi i\nu t)\mathrm{d}\nu. \tag{4.56}$$

This is known as the *Wiener–Khintchine theorem*, which is particularly useful in finding the width and structure of narrow spectral lines from measurement of amplitude fluctuations (see Chapters 12 and 13). Equation (4.55) is conveniently remembered as follows: *the Fourier transform of the amplitude autocorrelation is the power spectral density.*

4.16   Wave Groups

Modulated waves, and in particular wave groups such as those of Figure 4.17(b), are of great importance in many branches of physics. In view of this we now consider modulated waves in a simple physical way, so as to illuminate the mathematical results of the Fourier approach. We start with two wave components only. The addition of two waves travelling in the $+x$ direction with equal amplitude a but slightly different angular frequencies $\omega \pm \Delta\omega/2$ and wave numbers $k \pm \Delta k/2$ is expressed as

$$y = a\exp\left\{i\left[\left(\omega+\frac{\Delta\omega}{2}\right)t - \left(k+\frac{\Delta k}{2}\right)x\right]\right\} + a\exp\left\{i\left[\left(\omega-\frac{\Delta\omega}{2}\right)t - \left(k-\frac{\Delta k}{2}\right)x\right]\right\}$$

$$= a\exp[i(\omega t - kx)]\left\{\exp\left[i\left(\frac{\Delta\omega t}{2}-\frac{\Delta k x}{2}\right)\right] + \left[\exp-i\left(\frac{\Delta\omega t}{2}-\frac{\Delta k x}{2}\right)\right]\right\}$$

$$= 2ia\sin\left(\frac{\Delta\omega t - \Delta k x}{2}\right)\exp[i(\omega t - kx)]. \tag{4.57}$$

The exponential term is a wave at the centre frequency, and the sine term is a slower modulation of the wave in time at angular frequency $\Delta\omega/2$ and in space with wave number $\Delta k/2$. The real part of equation (4.57) is

$$y_{\text{real}} = -2a \sin\left(\frac{\Delta\omega\, t - \Delta k\, x}{2}\right) \sin(\omega t - kx). \tag{4.58}$$

The wave at the centre frequency moves as before with a velocity

$$v = \frac{\omega}{k}. \tag{4.59}$$

This is known as the *phase velocity* of the group. The modulation moves with a different velocity, such that $\sin(\Delta\omega t - \Delta kx)$ is constant; this is known as the *group velocity* v_{g}, given by the ratio

$$v_{\text{g}} = \frac{\Delta\omega}{\Delta k}. \tag{4.60}$$

Any pair in a group of waves can be analysed in this way, so we may deduce that the whole group will move with the same velocity as the sinusoidal modulation pattern. In the limit, a group must be considered as an infinite series of waves all with angular frequencies and wave numbers near ω and k. The group velocity v_{g} is then the derivative

$$\boxed{v_{\text{g}} = \mathrm{d}\omega/\mathrm{d}k.} \tag{4.61}$$

Note the distinction between group velocity and the phase velocity ω/k. On a graph of ω versus k, they correspond respectively to instantaneous slope (v_{g}) and average slope (v_{p}). Since $\mathrm{d}\omega(k)/\mathrm{d}k$ may vary with k, the derivative in equation (4.61) should, for greatest accuracy, be evaluated at some k_0 near the middle of the spectrum.

The limitation of the duration of a wave, or the limitation of its extent in space, requires the superposition of an infinite series. Two waves differing by Δv in frequency reinforce over a time $\tau \approx 1/\Delta v$, which is the time between successive beat minima. A group lasting for time τ must consist of sinusoidal waves spread over a range $\Delta v \approx 1/\tau$, so that they are in phase during the time τ, and outside this time their relative phases become large enough that they cancel each other out by destructive interference. Similarly a limitation of a group of waves to a spatial extent of length L implies that the group contains a range of wavelengths such that the component waves become out of step outside the group. By analogy with $\Delta v \approx 1/\tau$, the requirement is $\Delta(1/\lambda) \approx 1/L$ or $\lambda^2/\Delta\lambda \approx L$; if $L = n\lambda$ then the range of $\Delta\lambda$ is given by $\lambda/\Delta\lambda \approx n$.

The first measurement of the velocity of light was achieved in 1675 by Roemer, who timed the orbital motion of the four Galilean satellites of Jupiter, and found a delay which depended on the varying distance of Jupiter from the Earth. Later and more accurate determinations by Michelson and by Bergstrand timed the passage of light over a terrestrial path, using either a pulse of light or a sinusoidal modulation. The discussion above shows us that all these methods in principle measure the group velocity. In contrast, the phase velocity can be determined from a measurement of the wavelength of light whose frequency is known by comparison with harmonics of a standard oscillator. Then we define $v_p = v\lambda$. (Since the velocity of light is now regarded as a fundamental constant, this determination is, in modern terms, a means of relating standards of time and length.)

4.17   An Angular Spread of Plane Waves

The wave groups discussed in previous sections are limited in extent only along the direction of travel. A wave packet describing a particle, or a simple light beam, must also have a limited extent laterally. Can this also be regarded as a result of superposing plane waves? The longitudinal extent of a wave group is governed by the range of wavelengths of the plane waves constituting the group: we now show that the lateral extent is determined by a spread in wave *directions* rather than by a spread in *wavelength*.

Consider first the addition of two plane waves, with velocity c and wavelength λ, crossing at an angle 2θ, as in Figure 4.19. Along the broken lines in this figure the two waves add in phase, making a wave progressing at velocity $c \sec \theta$. This resultant wave pattern shows a cosine variation of amplitude *across* the wavefront, i.e. perpendicular to the direction of propagation, with zero amplitude half-way between the maxima on the broken lines. For small θ, the maxima are separated by a distance λ/θ. If we now add more pairs of waves, with the same wavelength but crossing at different values of θ, we add to the resultant wave pattern further cosine components with different scales λ/θ. Following the same idea as in the longitudinal limitation of the wave group, we see that these different scales of lateral variation can add to produce a wave limited in space transverse to the direction of propagation.

A wavefront limited in this way to a lateral extent D requires a range of crossing waves with angles from zero to λ/D. The required distribution of wave amplitude with angle depends on the shape of the distribution of amplitude across the wavefront: following the example of the wave group we may expect the relation to be given again by a Fourier transform. This is explored in more detail in Chapter 13.

Example. A harmonic plane wave propagating in any direction has the complex representation of the form $\exp[i(\mathbf{k} \cdot \mathbf{r} - \omega t)]$. where $\mathbf{k} = (2\pi/\lambda)\,\hat{\mathbf{k}}$ is the wave vector, and $\hat{\mathbf{k}}$ is the unit vector giving the direction of propagation. Use this form to:

(a) find the resultant from superposing two complex plane waves that cross at angle 2θ, as shown in Figure 4.19, and

(b) find the spacing in y and z of the wave maxima.

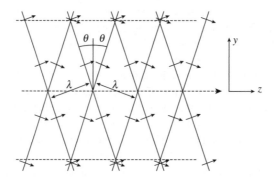

Figure 4.19   Two plane waves crossing at angle 2θ. The waves add in phase along the broken lines, which are spaced by $\lambda/(2\sin\theta)$, and are always in antiphase half-way between the broken lines

Solution. (a) The two propagation vectors that cross at angle 2θ have components $\mathbf{k}_\pm = (0, \pm k_y, k_z) = (0, \pm k \sin\theta, k \cos\theta)$. We evaluate the resultant:

$$\begin{aligned}
\tilde{\psi} &= \exp[i(\mathbf{k}_+ \cdot \mathbf{r} - \omega t)] + \exp[i(\mathbf{k}_- \cdot \mathbf{r} - \omega t)] \\
&= [\exp(ik_y y) + \exp(-ik_y y)]\exp[i(k_z z - \omega t)] \\
&= 2\cos(k\sin\theta y)\exp[i(k\cos\theta z - \omega t)].
\end{aligned} \tag{4.62}$$

(b) From the preceding, we see that the wave goes through a complete cycle when a coordinate changes by

$$\begin{aligned}
\Delta y &= 2\pi/(k\sin\theta) = \lambda/\sin\theta \text{ or} \\
\Delta z &= 2\pi/(k\cos\theta) = \lambda/\cos\theta.
\end{aligned} \tag{4.63}$$

Problems in Fourier Analysis

Problem 4.1

Suppose a function $f(x)$ has Fourier expansion as in equation (4.14). Prove the statement in Section 4.2 that if $f(t)$ is real, then $A^*(\omega) = A(-\omega)$. (Hint: You can assume that if two functions are equal, $f(t) = g(t)$, then their Fourier amplitudes are also equal, $A_f(\omega) = A_G(\omega)$.)

Problem 4.2
Some properties of the delta function

In what follows, assume a, b are positive constants, and that $f(x)$ is any continuous function:

(a) In the context of equation (4.14), what is the amplitude $A(\omega)$ that generates $\delta(t)$?

(b) Prove that $b(x)$ is an even parity function in the sense that

$$\int_{-a}^{b} f(x)[\delta(x) - \delta(-x)]dx = 0. \tag{4.64}$$

(c) Use integration by parts to prove that

$$\int_{-a}^{b} f(x)\delta'(x)dx = -f'(x) \tag{4.65}$$

where $\delta' = d/dx$.

(d) Prove that for any non-zero constant A

$$\delta(Ax) = |A|^{-1}\delta(x) \tag{4.66}$$

in the sense that $\int_{-a}^{b} f(x)\delta(Ax)dx = |A|^{-1}\int_{-a}^{b} f(x)\delta(x)dx$.

Problem 4.3

Given a complex-valued function of the form $(f)t) = \sum_{n=-\infty}^{n=\infty} A_n \exp(i\omega_n t)$, write down the amplitude $A(\omega)$ that corresponds to it according to equation (4.14).

Problem 4.4

Suppose that a real function $F(t)$ is of even or odd parity, $F(t) = \pm f(t)$, where the upper (lower) sign represents the even (odd) parity case. Prove that its frequency spectrum $F(v)$, as given in Section 4.11, is real for even parity and imaginary for odd parity.

Problem 4.5

The negative half-cycles of a sinusoidal waveform $E = E_0 \cos \omega t$ are removed by a half-wave rectifier. Show that the resulting wave is represented by the Fourier series

$$E = E_0 \left(\frac{1}{\pi} + \frac{1}{2} \cos \omega t + \frac{2}{3\pi} \cos 2\omega t - \frac{2}{15\pi} \cos 4\omega t + \ldots \right).$$

Show that a full-wave rectifier, which inverts the negative half-cycles, has an output

$$E = E_0 \left(\frac{2}{\pi} + \frac{4}{3\pi} \cos 2\omega t - \frac{4}{15\pi} \cos 4\omega t + \text{ even harmonics} \right).$$

Problem 4.6

A Gaussian function with height h and standard deviation σ is $f(t) = h \exp[-(t^2/2\sigma^2)]$. Show that its Fourier transform is

$$F(v) = (2\pi)^{1/2} \sigma h \exp(-2\pi^2 v^2 \sigma^2).$$

(You will require the integral $\int_{-\infty}^{\infty} \exp(-x^2) \mathrm{d}x = \pi^{1/2}$.)

Problem 4.7

Show that the Fourier transform $F(v)$ of an isosceles triangular function, centred on $t = 0$, with height h, base width b, is

$$F(v) = \frac{hb}{2} \frac{\sin^2 \psi}{\psi^2} \quad \text{where} \quad \psi = \frac{\pi b}{2} v.$$

Problem 4.8

A decaying wave train is represented by

$$f(t) = a \exp\left(-\frac{t}{\tau}\right) \exp(i\omega_0 t).$$

Show that the Fourier transform of $f(t)$ is

$$F\left(\frac{\omega}{2\pi}\right) = \frac{a}{1/\tau + i(\omega - \omega_0)}$$

and hence that the energy spectrum for ω close to ω_0 is given by

$$\left| F\left(\frac{\omega}{2\pi}\right) \right|^2 = \frac{a^2}{(1/\tau)^2 + (\omega - \omega_0)^2}.$$

Problem 4.9

(a) Find the spectrum $F(v)$ for each of the functions $\cos(2\pi v_0 t)$ and $\sin(2\pi v_0 t)$.

(b) The function illustrated in Figure 4.16(a) has the form $g(t) \cos(2\pi v_1 t)$, where $g(t) = a + b \cos(2\pi v_m t)$ is a real function. Find its spectrum. Convince yourself that your spectrum agrees with Figure 4.16(b). This is an amplitude-modulated wave. In the case depicted, $g(t)$ is the more slowly varying factor ($v_m \ll v_1$) and can therefore be described as "modulating"; but the spectrum obtained is valid regardless of that limitation.

Problem 4.10

Find the spectrum of the frequency-modulated wave

$$f(t) = A\cos(pt + B\cos qt)$$

when B is small.

(Hint: Expand the waveform as a power series in $B\cos qt$, and neglect B^2.) What distinguishes this spectrum from the amplitude-modulated spectrum of problem 4.9b?

Physics Problems

Problem 4.1

Evaluate the sum $y = \sin(kx - \omega t) + \sin(kx + \omega t + \alpha)$. Manipulate the complexified form $\tilde{y}$, without the help of trigonometric identities, to show that $y = 2\sin(kx + \alpha/2)\cos(\omega t + \alpha/2)$.

Problem 4.2

Show that the energy in the sum of two oscillations is equal to the sum of their individual energies, provided that they differ in frequency and a suitable time average is taken.

Problem 4.3

Demonstrate the equivalence of the following expressions for group velocity v_g, when phase velocity $v = c/n$:

$$v_g = \frac{d\omega}{dk} = \frac{c}{n + \omega dn/d\omega} = v - \lambda\frac{dv}{d\lambda}.$$

Problem 4.4

A plane wave propagates in a dispersive medium with phase velocity v given by

$$v = a + b\lambda$$

where a and b are constants. Find the group velocity.

Show that any pulse modulated waveform will reproduce its shape at times separated by intervals of $\tau = 1/b$, and at distance intervals of a/b. (Hint: Consider any pair of component waves separated in wavelength by $\Delta\lambda$ as in Section 4.16.)

Problem 4.5

Calculate the group velocity for the following types of waves, given the variation of phase velocity v with wavelength λ:

(a) Surface water waves controlled by gravity: $v = a\lambda^{1/2}$.

(b) Surface water waves controlled by surface tension: $v = a\lambda^{-1/2}$.

(c) Transverse waves on a rod: $v = a\lambda^{-1}$.

(d) Radio waves in an ionized gas: $v = (c^2 + b^2\lambda^2)^{1/2}$.

Problem 4.6

The refractive index for electromagnetic waves propagating in an ionized gas is given by

$$n^2 = 1 - \omega_p^2/\omega^2$$

where ω_p, the plasma frequency, is determined by the density of the gas. Show that the product of the group and phase velocities is c^2.

Problem 4.7 The relativistic Doppler effect

A signal received from an oscillator with frequency v moving in a space vehicle with velocity v directly away from an observer has an apparent frequency v' where

$$v' = v\left(\frac{1 - v/c}{1 + v/c}\right)^{1/2}.$$

Find the difference between this and the non-relativistic value $v'_{nr} = v(1 - v/c)$ for an oscillator at 6 GHz moving at $6\,\text{km s}^{-1}$ in the line of sight.

Find also the transverse Doppler frequency shift $v - v_t$ for a velocity of $6\,\text{km s}^{-1}$ in the line of sight where

$$v_t = v(1 - v^2/c^2)^{1/2}.$$

Problem 4.8

Find the exact change in frequency in the Doppler radar problem by compounding two Doppler shifts: that from sender to target, and that from target to receiver. Assume the target recedes from the sender–receiver at speed v.

Problem 4.9

Although a good approximation at low speeds, strict vector addition of velocities is impossible within the context of special relativity because it would lead to changes in the observed velocity of light. Instead, velocities add in a non-linear way. Suppose, as illustrated in Figure 4.13, the two inertial frames have corresponding axes parallel, and the moving observer, S, has velocity v'_x relative to the resting observer, S'. Special relativity tells us that the velocity of any particle transforms according to

$$v_x = \frac{v'_x - v}{1 - vv'_x/c^2} \tag{4.67}$$

$$v_y = \frac{v'_y(1 - v^2/c^2)^{1/2}}{1 - vv'_x/c^2}. \tag{4.68}$$

Use this to find the velocity components and speed of a light ray incident parallel to the $+y'$ axis as seen by the moving observer, S.

Problem 4.10

In the solar spectrum the same Fraunhofer line at 600 nm appears at wavelengths differing by 0.004 nm at the pole and at the edge of the disc near the equator. Find the velocity at the equator, and deduce the rotation period, given that the Sun's distance is 500 light-seconds and that it subtends an angle of $32'$ at the Earth.

Problem 4.11

The Crab Pulsar emits a precisely periodic pulse train whose frequency is close to 30 Hz. It lies in a direction close to the ecliptic plane, in which the Earth orbits round the Sun. Calculate the peak-to-peak variation in the observed pulse frequency due to the Earth's annual motion given that the Sun's distance from the Earth is 500 light-seconds.

Problem 4.12

The mean radius of the orbit of the Earth is 1.5×10^{11} m. Find the amplitudes of astronomical parallax and aberration for a star at a distance of 10 light-years situated in the plane of the orbit. What is the phase relation between these two periodic motions?

Problem 4.13

A ray of light falls at angle of incidence i on a mirror surface moving normally to its surface with a velocity v small compared with c. Use Huygens' construction to show that the angle of reflection r differs from i by approximately $(2v/c)i$ for small i.

5 Electromagnetic Waves

The ether, this child of sorrow of classical mechanics.

Max Planck, quoted by Jean-Pierre Luminet in *Black Holes.*

Light is always propagated in empty space with a definite velocity c which is independent of the state of motion of the emitting body.

A. Einstein.

The wave theory of light, which was applied so successfully in the nineteenth century to the phenomena of propagation, interference and diffraction, was naturally thought of in the same way as water waves and sound waves, which were obviously waves in a medium. Maxwell showed that light was an electromagnetic wave. But what was the medium through which light propagated? The ether, as it was called, had no observable properties. Attempts to detect it by measuring the motion of the Earth through it all failed, and it became clear that the description of electromagnetic waves did not depend in any way on the existence of the ether. Maxwell's equations, which are the basis of our understanding of electromagnetic waves, are relations between electric and magnetic fields, and not between these fields and some all-pervading medium.

In this chapter we first show how electromagnetic waves may be derived from the fundamental laws of electricity and magnetism, as formulated in Maxwell's equations.[1] We then consider the flow of energy in an electromagnetic wave, and what happens when an electromagnetic wave meets a boundary, where it may be partly reflected and partly transmitted, depending on the materials at the boundary, the angle of incidence of the wave and its polarization.

What happens to photons at a partially reflecting boundary? The question is meaningless for an individual photon: if light is regarded as a stream of photons, wave theory gives the *probability* that photons will be reflected or transmitted. The transport of energy by the photons averages to that of the classical electromagnetic wave, and the momentum associated with a photon leads to a radiation pressure at an interface between media. These are examples of the dual nature of light; only the quantum picture, however, can account for the wavelength shift of Compton scattering or for the spectrum of *blackbody radiation*, which we consider at the end of this chapter.

---

[1]We use electromagnetic SI units throughout.

Optics and Photonics: An Introduction, Second Edition   F. Graham Smith, Terry A. King and Dan Wilkins
© 2007 John Wiley & Sons, Ltd

5.1   Maxwell's Equations

We start with the set of four equations, known as *Maxwell's equations*, which encapsulate the basic laws of classical electrodynamics.[2] They relate electric and magnetic fields to two different kinds of sources: first by charges and currents, and second through induction, in which a changing magnetic field induces an electric field and a changing electric field induces a magnetic field. Both the variables in an electromagnetic wave, the electric and magnetic fields **E** and **B**, are vector quantities, and we use vector notations throughout. We confine our analysis to isotropic and homogeneous materials, and mainly to non-conducting materials with linear properties. The full Maxwell's equations in vector form[3] are

$$\operatorname{div} \mathbf{D} = \rho, \qquad \operatorname{div} \mathbf{B} = 0$$
$$\operatorname{curl} \mathbf{E} = -\frac{\partial \mathbf{B}}{\partial t}, \qquad \operatorname{curl} \mathbf{H} = \mathbf{J} + \frac{\partial \mathbf{D}}{\partial t}. \tag{5.2}$$

Here ρ is the free charge density and **J** the free current density; in most of optics ρ and **J** are zero and the medium is non-magnetic. The vector fields **D** and **H** are needed for material media that show electric and magnetic polarization in the presence of external fields. In this book, we deal mainly with linear isotropic media where $\mathbf{D} = \epsilon \mathbf{E}$ and $\mathbf{B} = \mu \mathbf{H}$; ϵ is the dielectric constant of the medium, and μ is the magnetic permeability. In vacuum, the permittivity ϵ and (magnetic) permeability μ reduce to $\epsilon_0 = 8.854 \times 10^{-12}\,\mathrm{F\,m^{-1}}$ (farad per metre, the so-called electric constant) and to $\mu_0 = 4\pi \times 10^{-7}\,\mathrm{H\,m^{-1}}$ (henry per metre, the magnetic constant); ϵ and μ are conveniently expressed by their values relative to vacuum, namely $\epsilon_r = \epsilon/\epsilon_0$, $\mu_r = \mu/\mu_0$. The relative permittivity, ϵ_r, is also known as the dielectric constant.

An electromagnetic field tends to polarize any medium it permeates, producing an instantaneous distribution of electric and magnetic dipoles. The electric dipole moment per unit volume is called the *electric polarization* and equals $\mathbf{P} = \mathbf{D} - \epsilon_0 \mathbf{E} = \epsilon_0 \chi_e \mathbf{E}$. The magnetic dipole volume density is called the *magnetization* **M** and is given by $\mu_0 \mathbf{M} = \mathbf{B} - \mu_0 \mathbf{H} = \mu_0 \chi_m \mathbf{H}$. (We meet the polarization again in Chapters 16 and 19, where it plays major roles in the theories of light propagation and scattering.)

---

[2]See for example I.S. Grant and W.R. Phillips, *Electromagnetism*, 2nd edn, John Wiley & Sons, Ltd, 1990.
[3]In Cartesian coordinates the divergence and curl of a vector **F** are

$$\operatorname{div} \mathbf{F} = \nabla \cdot \mathbf{F} = \frac{\partial F_x}{\partial x} + \frac{\partial F_y}{\partial y} + \frac{\partial F_z}{\partial z} \tag{5.1}$$

$$\operatorname{curl} \mathbf{F} = \nabla \wedge \mathbf{F}$$
$$= \hat{\mathbf{x}}\left(\frac{\partial F_z}{\partial y} - \frac{\partial F_y}{\partial z}\right) + \hat{\mathbf{y}}\left(\frac{\partial F_x}{\partial z} - \frac{\partial F_z}{\partial x}\right) + \hat{\mathbf{z}}\left(\frac{\partial F_y}{\partial x} - \frac{\partial F_x}{\partial y}\right)$$

where $\hat{\mathbf{x}}, \hat{\mathbf{y}}, \hat{\mathbf{z}}$ are unit vectors in the x, y, z directions.

In terms of the **E** and **B** fields, the four Maxwell's equations within a uniform medium become

$$\operatorname{div} \mathbf{E} = \frac{\rho}{\epsilon}, \ \operatorname{div} \mathbf{B} = 0$$

$$\operatorname{curl} \mathbf{E} = -\frac{\partial \mathbf{B}}{\partial t}, \ \operatorname{curl} \mathbf{B} = \epsilon\mu \frac{\partial \mathbf{E}}{\partial t} + \mu \mathbf{J}.$$

(5.3)

In a non-conducting material ($\mathbf{J} = 0$) with no free charge ($\rho = 0$)

$$\operatorname{div} \mathbf{E} = 0, \ \operatorname{div} \mathbf{B} = 0$$

$$\operatorname{curl} \mathbf{E} = -\frac{\partial \mathbf{B}}{\partial t}, \ \operatorname{curl} \mathbf{B} = \epsilon\mu \frac{\partial \mathbf{E}}{\partial t}.$$

(5.4)

The last two equations in (5.4) are Faraday's law of electromagnetic induction and the complementary law of magneto-electric induction introduced by Maxwell.

The properties of electromagnetic waves involve the interaction between the two fields expressed in the two laws of induction. We now eliminate one of the fields by combining the last two equations in (5.4). Taking the curl of both sides of the third Maxwell equation,

$$\operatorname{curl} \operatorname{curl} \mathbf{E} = -\operatorname{curl} \frac{\partial \mathbf{B}}{\partial t}.$$

(5.5)

Since

$$-\operatorname{curl} \frac{\partial \mathbf{B}}{\partial t} \equiv -\frac{\partial}{\partial t} (\operatorname{curl} \mathbf{B})$$

(5.6)

we can use the fourth equation to give

$$\operatorname{curl} \operatorname{curl} \mathbf{E} = -\epsilon\mu \frac{\partial^2 \mathbf{E}}{\partial t^2}.$$

(5.7)

Using the operator identity

$$\operatorname{curl} \operatorname{curl} \ \equiv \ \operatorname{grad} \operatorname{div} - \nabla^2$$

and noting that $\operatorname{div} \mathbf{E} = 0$ from the first Maxwell equation, we obtain

$$\boxed{\nabla^2 \mathbf{E} = \epsilon\mu \frac{\partial^2 \mathbf{E}}{\partial t^2}.}$$

(5.8)

A similar derivation for **B** yields

$$\boxed{\nabla^2 \mathbf{B} = \epsilon\mu \frac{\partial^2 \mathbf{B}}{\partial t^2}.}$$

(5.9)

These are the wave equations for an unattenuated electromagnetic field at any frequency and travelling in any direction. Comparison with the general wave equation (see Chapter 1)

$$\nabla^2 \psi = \frac{1}{v^2} \frac{\partial^2 \psi}{\partial t^2}$$

(5.10)

gives the wave propagation velocity

$$v = \left(\frac{1}{\epsilon\mu}\right)^{1/2}.$$ (5.11)

All electromagnetic waves in free space $(\epsilon_r = \mu_r = 1)$ travel with the same speed, which is a fundamental constant usually given the symbol c. For a medium with permittivity ϵ and permeability μ the wave velocity is

$$v = \frac{1}{\sqrt{\epsilon\mu}} = \frac{c}{\sqrt{\epsilon_r\mu_r}}.$$ (5.12)

Of the factors ϵ_r and μ_r which depend on the medium, the dielectric constant is usually the more important, since it is unusual to encounter light waves in media where μ_r differs appreciably from unity. In a dielectric, the ratio of the velocity in free space and the velocity in the medium is defined as the *refractive index n* of the medium. Hence for a dielectric

$$n = \frac{c}{v} = \sqrt{\epsilon_r}.$$ (5.13)

The velocity in free space $c = (\epsilon_0\mu_0)^{-1/2}$ was evaluated by Maxwell using laboratory electrical measurements for ϵ_0 and μ_0. He obtained the velocity $3 \times 10^8\,\mathrm{m\,s^{-1}}$, in remarkable agreement with the measured speed of light. This led him to conclude that light was an electromagnetic disturbance which propagated according to the laws of electromagnetism.

5.2   Transverse Waves

We stated in Chapter 1 that light is an electromagnetic wave with fields $\mathbf{E}$ and $\mathbf{B}$ oscillating *transversely* to the direction of propagation. We now show that the transverse nature of light follows directly from electromagnetic theory.

The wave equation (5.8) can represent waves of any frequency and any form, and it may be expressed in any system of coordinates. In Cartesian coordinates the vector field $\mathbf{E}$ has components

$$\mathbf{E} = \hat{\mathbf{x}}E_x + \hat{\mathbf{y}}E_y + \hat{\mathbf{z}}E_z$$ (5.14)

where $\hat{\mathbf{x}}, \hat{\mathbf{y}}, \hat{\mathbf{z}}$ are unit vectors in the x, y, z directions. These three components can be independent solutions of equation (5.8). For a plane wave $\hat{\mathbf{x}}E_x$ travelling in the z direction in free space

$$\frac{\partial^2 E_x}{\partial z^2} = \frac{1}{c^2}\frac{\partial^2 E_x}{\partial t^2}.$$ (5.15)

This has the general solution

$$E_x = f(z - ct) + g(z + ct)$$ (5.16)

representing the superposition of two waves of any form travelling in the $\pm z$ directions. Note that in free space all electromagnetic waves, with whatever waveform, travel with the same velocity. In the modern system of units (since 1983), the velocity of light in free space is a defined quantity set at $299\,792\,458\,\mathrm{m\,s^{-1}}$ exactly.[4]

Can there be a solution representing a longitudinally polarized wave? Consider a plane wave travelling in the z direction, with

$$\mathbf{E} = \hat{\mathbf{z}} E_0 \cos(\omega t - kz).$$

As it is a plane wave there is no variation of the field $\mathbf{E}$ in the x and y directions; in the expansion of div $\mathbf{E} = 0$

$$\frac{\partial E_x}{\partial x} + \frac{\partial E_y}{\partial y} + \frac{\partial E_z}{\partial z} = 0. \tag{5.17}$$

The first two terms are zero, so that E_z must be independent of z and no such progressive wave can exist.

The $\mathbf{B}$ field is at right angles to the $\mathbf{E}$ field. Assuming that $\mathbf{E}$ is along the x axis as in equation (5.16), this follows from the third Maxwell equation curl $\mathbf{E} = -\partial\mathbf{B}/\partial t$, where, thanks to $\hat{\mathbf{y}} \partial E_x/\partial z$, the only non-zero component of $\partial\mathbf{B}/\partial t$ is along the y axis. Both $\mathbf{B}$ and $\mathbf{E}$ are transverse to the direction of propagation, constituting a so-called TEM wave, in contrast to the TE and TM waves encountered for example in fibre optics (Chapter 6). Those are not plane waves, and there may be components along the direction of propagation. The ratio B_y/E_x when only one wave is present may in general be found from partial differentiation of equation (5.16). The relation between $\mathbf{E}$ and $\mathbf{B}$ is encapsulated in vector notation:

$$\mathbf{B} = \frac{1}{c}\hat{\mathbf{k}} \wedge \mathbf{E} \quad \text{(in free space)}. \tag{5.18}$$

Here $\hat{\mathbf{k}}$ is the *propagation vector*, which is a unit vector in the direction of propagation of the wave. Thus for the above example, with a wave moving in the $+Z$ direction, $E_x = cB_y$. In a dielectric, or any isotropic non-conducting medium, the relation becomes

$$\boxed{\mathbf{B} = \frac{1}{v}\hat{\mathbf{k}} \wedge \mathbf{E}} \tag{5.19}$$

where, as before,

$$\boxed{v = \frac{c}{\sqrt{\epsilon_r \mu_r}} = \frac{c}{n}} \tag{5.20}$$

and $n \geq 1$. Since the wave speed is less than the speed in free space, for a given frequency the wavelength in a medium, $\lambda = v/\nu$, is less than that in free space. The wave speed in the medium, and the refractive index, may vary with frequency; this is called *dispersion*. The spreading of colours in light refracted by a prism is due to dispersion by the glass in the prism.

---

[4]See Chapter 9. The speed of light $c = (\epsilon_0\mu_0)^{-1/2}$ is a fundamental constant with the defined value of $299\,792\,458\,\mathrm{m\,s^{-1}}$. In SI units the magnetic constant is given the value $\mu_0 = 4\pi \times 10^{-7}\,\mathrm{H\,m^{-1}}$, and it follows that the electric constant is $\epsilon_0 = 8.854\,188 \times 10^{-12}\,\mathrm{F\,m^{-1}}$.

5.3   Reflection and Transmission: Fresnel's Equations

Snell's law, discussed in Chapter 1, relating the angles of incidence and refraction as a ray enters or leaves a refracting medium, tells only part of the story. A wave encountering a boundary between media with different refractive indices n_1, n_2 will not only be refracted, but also be partly reflected. The ratios of the amplitudes of the reflected and transmitted waves to that of the incident wave are known as the amplitude *reflection* and *transmission* coefficients, r and t. The *Fresnel* equations for an electromagnetic wave express the way in which these coefficients depend on the angles of incidence (θ_1) and refraction (θ_2), and on the polarization of the wave.

In Figure 5.1 the reflected and refracted rays are shown for two cases of plane polarization, when **E** is (a) in the plane of incidence and (b) perpendicular to it. At the boundary, where the three rays meet, there must be a match between the components of the electric and magnetic fields on either side of the interface. Based on the second and third equations of (5.2), the boundary conditions to be met[5] by **E** and **B** are

(i) the component of the electric field parallel to the boundary, and

(ii) the component of the magnetic field perpendicular to the surface,

which must be the same on either side of the boundary. The subscripts $\parallel$ and $\perp$ refer to the orientation of the electric field vector; for $\parallel$ it is parallel to the plane of incidence (the plane containing the propagation vector $\hat{\mathbf{k}}$ and the normal to the interface), and for $\perp$ it is perpendicular.

In Figure 5.1(a) the surface component of the electric fields of the incident ray and transmitted rays are $E_i \cos \theta_1$ and $E_t \cos \theta_2$. The reflected ray is at angle $\pi - \theta_1$, so that the surface component is $-E_r \cos \theta_1$. The first boundary condition is therefore

$$E_i \cos \theta_1 - E_r \cos \theta_1 = E_t \cos \theta_2. \tag{5.21}$$

For the polarization shown in Figure 5.1(a) the magnetic fields are parallel to the interface, giving a second boundary condition[6]

$$B_i + B_r = B_t. \tag{5.22}$$

Since the magnitudes of E and B are related by

$$B = \frac{n}{c} E \tag{5.23}$$

where n is the refractive index of the medium, equation (5.22) gives

$$n_1 E_i + n_1 E_r = n_2 E_t. \tag{5.24}$$

---

[5]See for example Grant and Phillips, *Electromagnetism*, 2nd edn, 1990, p. 392 *et seq.*
[6]Since our media are assumed non-magnetic, $\mu = \mu_0$, the continuity of the component of **H** parallel to the surface, which is implied by equation (5.2), carries over to **B**.

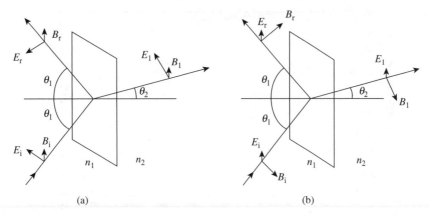

Figure 5.1 Reflected and refracted rays at a boundary. The directions of the vector fields are shown for (a) **E** in the plane of incidence (∥) and (b) **E** normal to the plane of incidence (⊥)

Combining equations (5.21) and (5.24) gives the reflection and transmission coefficients $r_\parallel$ and $t_\parallel$ which are defined as ratios of *amplitudes*:

$$
\begin{aligned}
r_\parallel &= \left(\frac{E_r}{E_i}\right)_\parallel = \frac{n_2 \cos\theta_1 - n_1 \cos\theta_2}{n_2 \cos\theta_1 + n_1 \cos\theta_2} \\
t_\parallel &= \left(\frac{E_t}{E_i}\right)_\parallel = \frac{2n_1 \cos\theta_1}{n_2 \cos\theta_1 + n_1 \cos\theta_2}.
\end{aligned}
\tag{5.25}
$$

A similar analysis for **E** perpendicular to the plane of incidence (Figure 5.1(b)) gives

$$
\begin{aligned}
r_\perp &= \left(\frac{E_r}{E_i}\right)_\perp = \frac{n_1 \cos\theta_1 - n_2 \cos\theta_2}{n_1 \cos\theta_1 + n_2 \cos\theta_2} \\
t_\perp &= \left(\frac{E_t}{E_i}\right)_\perp = \frac{2n_1 \cos\theta_1}{n_1 \cos\theta_1 + n_2 \cos\theta_2}.
\end{aligned}
\tag{5.26}
$$

Using Snell's law $n_1 \sin\theta_1 = n_2 \sin\theta_2$, the amplitude reflection and transmission coefficients can be expressed in terms of angles only:

$$
\begin{aligned}
r_\parallel &= \frac{\tan(\theta_1 - \theta_2)}{\tan(\theta_1 + \theta_2)}, \qquad t_\parallel = \frac{2 \sin\theta_2 \cos\theta_1}{\sin(\theta_1 + \theta_2)\cos(\theta_1 - \theta_2)} \\
r_\perp &= -\frac{\sin(\theta_1 - \theta_2)}{\sin(\theta_1 + \theta_2)}, \qquad t_\perp = \frac{2 \sin\theta_2 \cos\theta_1}{\sin(\theta_1 + \theta_2)}.
\end{aligned}
\tag{5.27}
$$

Figure 5.2(a) is a typical plot of these reflection (r) and transmission (t) coefficients, for an air/glass boundary with $n_2 = 1.5$ (where $n_1 \approx 1$).

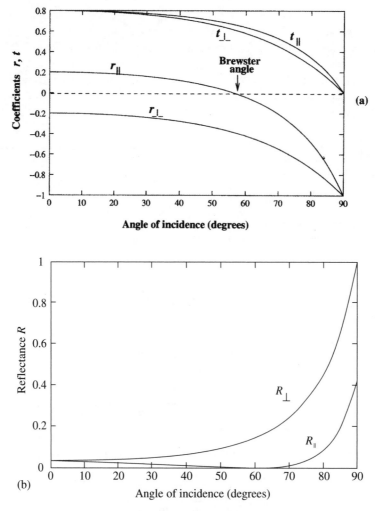

Figure 5.2   (a) Reflection $(r_\parallel, r_\perp)$ and transmission $(t_\parallel, t_\perp)$ coefficients for light incident on an air/glass boundary with refractive index $n = 1.50$. (b) Reflectance coefficients $R_\parallel, R_\perp$

Note that these coefficients are for amplitudes. The flow of energy across a surface, known as the *irradiance* (see Appendix 2), is proportional to the square of the amplitude, so that the *reflectance R* is r^2, (See Figure 5.2(b)). The *transmittance T* is found from

$$T = \frac{n_2}{n_1} \left(\frac{\cos \theta_2}{\cos \theta_1} \right) t^2, \tag{5.28}$$

where the extra factor of n_2/n_1 accounts for power flow within a medium, and the geometric factor is due to the lateral compression of the wavefront (see Section 5.5 below). It is a useful exercise to check that $R + T = 1$ for both polarizations.

It will be seen from equation (5.27) that $r_\parallel$ goes through zero when $\theta_1 + \theta_2 = \pi/2$ (since $\tan(\pi/2) = \infty$), and that it changes sign. At this point the angle of incidence is known as the *Brewster angle*, shown in Figure 5.2. The change of sign indicates a phase reversal. Light reflected at the Brewster angle becomes completely linearly polarized, with the electric vector normal to the plane of incidence. It is this behaviour that makes polaroid glasses useful in reducing the glare of light reflected off a wet road, and in allowing fishermen to see into a lake despite the reflection of the bright sky in its surface. Similarly, a glass plate at the Brewster angle is completely transparent for light with the electric vector parallel to the plane of incidence; this is used in the windows of gas lasers to avoid reflection losses.

For normal incidence the magnitudes[7] of the reflection and transmission coefficients are independent of polarization, becoming simply

$$r = r_\parallel = -r_\perp = \frac{n_2 - n_1}{n_1 + n_2}$$
$$t = t_\parallel = t_\perp = \frac{2n_1}{n_1 + n_2}.$$

(5.29)

The reflectance R and transmittance T are then

$$R = r^2 = \left(\frac{n_2 - n_1}{n_1 + n_2}\right)^2$$
$$T = \frac{n_2}{n_1}t^2 = \frac{4n_1 n_2}{(n_1 + n_2)^2}.$$

(5.30)

The reflectance loss of 4% at normal incidence for a typical air/glass surface with $n_2 = 1.5$ becomes a serious problem in the multi-component lenses of optical instruments such as cameras and telescopes. The losses can, however, be halved by coating the surface with a transparent layer with a lower refractive index $(n_1 n_2)^{1/2}$, as may be verified with the help of equation (5.30) (see Problem 5.3). Further improvement can be achieved in very thin coated layers, through the effects of thin-film interference between reflections from the front and back of the coating (see Chapter 8). The reflectance can be reduced to zero for a chosen wavelength if the layer is made a quarter wavelength thick.

5.4   Total Internal Reflection: Evanescent Waves

In Chapter 1 we saw that a ray meeting a boundary between media with higher and lower refractive indices at a large angle of incidence may be totally reflected; this is referred to as total internal reflection. In this case there are two important extensions required to the Fresnel theory. The geometric ray approach merely shows total reflection, and makes no distinction between reflection at a dielectric and at a metallic surface. The boundary conditions are, however, quite different, since for the metallic conductor the tangential electric field is zero, while there is no such restriction on the tangential field at a dielectric surface. There are two consequences: first, there is an extension of the

---

[7]The opposite signs may be understood from the geometry of Figure 5.1.

field across the boundary into the medium of lower refractive index, and, second, there is a phase shift in the reflected wave.

The wave field outside the dielectric boundary is an *evanescent wave*, whose amplitude falls exponentially with distance from the boundary. This field contains energy and transports it parallel to the boundary but not normal to it. The presence of this evanescent field is important in fibre optics, where light is confined to a thin glass fibre by total internal reflection. The energy flow is not confined to the core of the fibre, but extends to a cladding of lower refractive index glass into which according to geometric optics it cannot penetrate. No energy is lost by the evanescent wave unless there is absorption in the medium in which it is travelling. The cladding must therefore be thick enough to accommodate the evanescent wave, and it must also, like the core, be made of low-loss material.

The analysis of reflection coefficients now involves the matching at the boundary of the evanescent wave to the incident and reflected waves. The reflection coefficients[8] then contain an imaginary component. Writing $n = n_2/n_1$ and eliminating $\theta_2 = \theta_t$ with the help of Snell's law, equations (5.27) yield

$$
\begin{aligned}
r_\parallel &= \frac{n^2 \cos\theta_i - i(\sin^2\theta_i - n^2)^{1/2}}{n^2 \cos\theta_i + i(\sin^2\theta_i - n^2)^{1/2}} = \exp\left(i\phi_\parallel\right) \\
r_\perp &= \frac{\cos\theta_i - i(\sin^2\theta_i - n^2)^{1/2}}{\cos\theta_i + i(\sin^2\theta_i - n^2)^{1/2}} = \exp(i\phi_\perp).
\end{aligned}
\tag{5.31}
$$

These equations have been cast in a form suitable for the case of total internal reflection, where $\sin\theta_i > n$ and the reflection coefficients are complex numbers of unit modulus. In this case, the reflectance takes the form $R = |r|^2$, and we see that the reflection is indeed total: $R = 1$.

The phase change on reflection $\phi(\theta)$ is found from these reflection coefficients:

$$
\tan\frac{\phi_\parallel}{2} = -\frac{(\sin^2\theta_i - n^2)^{1/2}}{n^2 \cos\theta_i}
\tag{5.32}
$$

$$
\tan\frac{\phi_\perp}{2} = -\frac{(\sin^2\theta_i - n^2)^{1/2}}{\cos\theta_i}.
\tag{5.33}
$$

Figure 5.3 shows the phase change for a glass/air interface where $n = 1.5$. Note that the difference $\phi_\parallel - \phi_\perp$ reaches $-45°$, so that the polarization of a linearly polarized ray with both parallel and perpendicular components can be changed substantially on reflection (see Chapter 7). This phase change on reflection can be used to produce circularly polarized light from plane polarized light.

5.5   Energy Flow

The total energy per unit volume u contained in a system of electric and magnetic fields in an isotropic medium is[9]

$$
u = \frac{1}{2}(\mathbf{D} \cdot \mathbf{E} + \mathbf{B} \cdot \mathbf{H}).
\tag{5.34}
$$

[8]See Born and Wolf, *Principles of Optics*, 6th edn, p. 48.
[9]See for example Grant and Phillips, *Electromagnetism*, 2nd edn, 1994, p. 383.

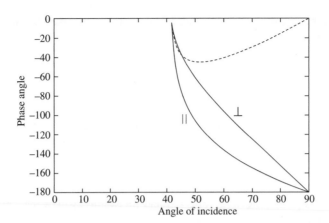

Figure 5.3   The phase change at total internal reflection in a glass/air interface when $n = 1.5$, for parallel and perpendicular polarizations. The broken line shows the difference between them

Thus the energy density in a combination of electric and magnetic fields with magnitudes E and B may be written as $\epsilon E^2/2 + B^2/2\mu$. In a rapidly varying harmonic wave, we must take the average over a whole cycle. The energy is proportional to the square of the fields, so that for any wave component such as $\mathbf{E} = \hat{\mathbf{x}} E_0 \sin(k(z - vt))$ the average square of the field is $\frac{1}{2} E_0^2$ where E_0 is the field amplitude. The mean energy density u is therefore

$$\bar{u} = \frac{1}{4} \left(\epsilon E_0^2 + B_0^2/\mu \right). \tag{5.35}$$

Since $B_0 = E_0/v$ and $v = (\epsilon\mu)^{-1/2}$, the two terms are equal and the energy density may be written as

$$\boxed{\bar{u} = \frac{1}{2} \epsilon E_0^2.} \tag{5.36}$$

The average energy crossing unit area per unit time in the z direction is the product $\bar{S} = v\bar{u}$:

$$\bar{S} = \frac{1}{2} v\epsilon E_0^2 = \frac{1}{2} E_0^2 \sqrt{\frac{\epsilon}{\mu}} = \frac{1}{2} cn\epsilon_0 E_0^2. \tag{5.37}$$

The last member assumes a non-magnetic medium where $n = \sqrt{\epsilon_r}$.

In free space $\bar{S} = \frac{1}{2} \epsilon_0 c E_0^2 = \frac{1}{2} E_0^2/Z_0$, where $Z_0 = (\mu_0/\epsilon_0)^{1/2} = \mu_0 c$ has the dimensions of resistance; it is known as the *impedance of free space*. Substitution of the values of ϵ_0, μ_0 in SI units gives $Z_0 = 376.73$ ohms (often quoted as 377 ohms). Using a root mean square of the field $E_{\text{rms}} = (\overline{E^2})^{1/2} = (E_0^2/2)^{1/2}$ in volts per metre, the energy flow is $(E_{\text{rms}}^2/377)\,\text{W}\,\text{m}^{-2}$ (see Problems 5.2(iii) and 5.6).

The energy flow is a vector known as the *Poynting vector* **S**. Electromagnetic theory shows that in terms of the magnetic intensity **H**, its generic and instantaneous value is

$$\mathbf{S} = \mathbf{E} \wedge \mathbf{H}. \tag{5.38}$$

For a plane wave in the direction of the unit wave vector $\hat{\mathbf{k}}$, and in a medium with permittivity ϵ and permeability μ, the time-averaged Poynting vector is

$$\bar{\mathbf{S}} = \frac{1}{2}E_0^2\sqrt{\frac{\epsilon}{\mu}}\hat{\mathbf{k}}. \tag{5.39}$$

Since optical frequencies are so high ($\nu_{opt} \sim 10^{15}$ Hz), most detectors of optical radiation will respond to the cumulative effect of many cycles. The time average of the magnitude of $\mathbf{S}$ is known as the *irradiance* $I = \bar{S}$. When referred to visible light and calibrated to the response of the human eye, it is called *illuminance* (see Appendix 2).

We are now in a position to return to the Fresnel transmittance issue and derive equation (5.28). Consider those portions of the incident, reflected and transmitted waves that intersect the interface between the two media in a common footprint of area A_0. If I_i and A_i are the irradiance and cross-sectional area of the incident beam, it carries a power I_iA_i; but since the incident ray is tilted at angle θ_i from the normal, the area of the incident beam is foreshortened: $A_i = \cos\theta_iA_0$. Analogous formulae hold for the other two beams. Since the dielectrics are non-conducting, there are no free surface currents to create ohmic dissipation. Conservation of energy then requires that all incident power emerges in the reflected or transmitted beams:

$$I_i\cos\theta_iA_0 = I_r\cos\theta_rA_0 + I_t\cos\theta_tA_0. \tag{5.40}$$

Inserting the irradiances from the last member of equation (5.37),

$$n_1E_{0i}^2\cos\theta_i = n_1E_{0r}^2\cos\theta_r + n_2E_{0t}^2\cos\theta_t. \tag{5.41}$$

Dividing through by the left side, with $\theta_r = \theta_i$, $r = E_{0r}/E_{0i}$ and $t = E_{0t}/E_{0i}$ gives

$$1 = r^2 + \frac{n_2\cos\theta_t}{n_1\cos\theta_i}t^2. \tag{5.42}$$

We can identify the first term on the right (reflected power/incident power) as the reflectance R, and the second term (transmitted power/incident power) as the transmittance T.

5.6   Photon Momentum and Radiation Pressure

The reality of assigning a discrete momentum to a photon was demonstrated by A. Compton in 1923. He investigated the scattering of monochromatic X-rays by the electrons in a block of paraffin. An X-ray photon in collision with an electron will change direction, as in Figure 5.4, and transfer part of its energy and momentum to the recoiling electron (see *Compton scattering*, Section 19.11).

The X-ray photon leaves the scatterer with energy reduced by an amount depending on the angle of scatter. Taking account of the conservation both of momentum and of energy, the increase in wavelength $\lambda' - \lambda$ of the photon at the collision can be found from the dynamics of the collision[10]

$$\lambda' - \lambda = \frac{h}{m_ec}(1 - \cos\phi). \tag{5.43}$$

---

[10]See for example F.H. Read, *Electromagnetic Radiation*, John Wiley & Sons, 1980, p. 230.

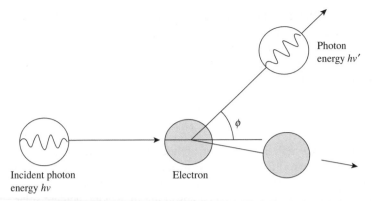

Figure 5.4 The Compton effect. An X-ray photon with energy $h\nu$ (wavelength λ) collides with an electron, loses energy and momentum, and emerges deviated through angle ϕ and with reduced energy $h\nu'$ (wavelength λ')

In equation (5.43), the constant $h/m_e c = 2.43 \times 10^{-12}$ m is known as the *Compton wavelength of the electron*; it is 2×10^5 times shorter than the wavelength of visible light. Subsequent experiments detected the individual recoil electrons in the Compton effect, but the measurement of the wavelength shift was in itself sufficient to establish the reality of this corpuscular behaviour of a photon, i.e. that photons behave like billiard balls.

When electromagnetic radiation meets a boundary between two media it exerts a pressure known as *radiation pressure*. This pressure is related to the flow of momentum in the radiation, and it is therefore most easily understood by considering the radiation in terms of photons. The momentum p carried by a photon is $p = h/\lambda$. The flux of photons, i.e. the number N crossing unit area per unit time, is obtained from the time-averaged Poynting vector, or irradiance I, divided by the photon energy:

$$N = \frac{I}{h\nu}. \tag{5.44}$$

If all the photons are incident normally from air, and are absorbed at the surface, the radiation pressure P is given by Newton's second law, as is the rate of absorption of momentum:

$$P = N\frac{h}{\lambda} = \frac{I}{c} = \epsilon_0 \overline{E^2} \tag{5.45}$$

where $\overline{E^2} = \frac{1}{2}E_0^2$ is the mean square field in the radiation. For total reflection the momentum transfer is doubled, and correspondingly the pressure is doubled because the direction of the photon is reversed: $P = 2\epsilon_0\overline{E^2} = \epsilon_0 E_0^2$.

Radiation pressure is, of course, explicable in purely classical terms. In a reflection at a conductor, the radiation field E acts on charge carriers to produce a current, and the B field acts on the induced current to give a Lorentz force which is directed into the conductor. Since B is proportional to E, the pressure is proportional to ϵE_0^2 as in equation (5.45).

Circularly polarized radiation carries an inherent angular momentum, so that in addition to radiation pressure there is also a torque on any refracting or reflecting surface which it encounters. All photons, of any energy, have an intrinsic angular momentum[11] $\hbar = h/2\pi$; this is aligned in the

---

[11]See for example Read, *Electromagnetic radiation*, p. 36.

direction of travel for RH circular polarization, and in the reverse direction for LH circular polarization. No torque is experienced in random polarization, in which there are equal numbers of LH and RH photons, or in linear polarization, in which the LH and RH photons are equal in number and also correlated. With the number flux of equation (5.44), the maximum rate of transfer of angular momentum per unit area to an absorber is

$$J = \frac{h}{2\pi} \frac{I}{h\nu} = \frac{I}{\omega}. \tag{5.46}$$

The practical use of the radiation pressure of laser light on individual atoms and other small particles is described in Section 16.7. Even at the distance of the Earth, the pressure of solar radiation may be important for artificial satellites, and may be used for accelerating low-mass satellites by the use of solar sails. The pressure on a solar panel absorbing the whole incident solar energy at Earth's distance from the Sun $(1.4\,\mathrm{kW\,m^{-2}})$ is $4.7 \times 10^{-6}\,\mathrm{N\,m^{-2}}$; on a completely reflecting solar sail this value is double. Note that the force on 1 square metre of sail equals the gravitational force on a mass of half a milligram on Earth.

5.7   Blackbody Radiation

The quantized nature of radiation has a profound effect on the spectrum of thermal radiation, and we end this chapter by considering the spectrum of electromagnetic radiation from a blackbody.

A blackbody is one that completely absorbs any radiation of any wavelength incident upon it. The intensity and spectrum of radiation from a blackbody are then characteristic only of its temperature. The concept of blackbody radiation is usually illustrated in terms of radiation inside an isothermal enclosure, inside which radiation from the walls is balanced by absorption. A small hole in the surface of the blackbody enclosure gives access to the radiation, like a peephole into an oven. The hole will absorb all radiation from outside, and therefore acts as a blackbody. The radiation within the enclosure reaches an equilibrium in which emission balances absorption, and the small sample of the radiation which emerges from the hole is the blackbody radiation. We need to relate the spectrum and the intensity (the irradiance) of this radiation to the temperature of the enclosure.

Consider first the classical pre-1900 view of radiation and absorption, in which each small range of frequencies is continuously emitted and absorbed by an oscillator consisting of an electron in a resonant system. Each oscillator has an average energy kT, and according to classical electromagnetic theory it radiates energy at a rate proportional to kT and to ν^2. It must absorb at the same rate if the radiation is in equilibrium with its surroundings. The calculation of the equilibrium intensity involves the relation of the absorption cross-section of the oscillator to its rate of radiation, but the essential point is that the equilibrium intensity of the radiation is also proportional to $kT\nu^2$. The exact relation is the Rayleigh–Jeans formula[12]

$$u(\nu)\mathrm{d}\nu = \frac{8\pi kT}{c^3} \nu^2 \, \mathrm{d}\nu \tag{5.47}$$

where $u(\nu)\mathrm{d}\nu$ is defined as the energy per unit volume in a frequency range $\mathrm{d}\nu$.

---

[12]See for example F. Mandl, *Statistical Physics*, 2nd edn, John Wiley & Sons, 1988, Ch. 10.

The problem with this classical calculation is the factor v^2, which gives an intensity increasing indefinitely with frequency, which is obviously physically impossible. The radiation from an electric heater, for example, is concentrated in the red and infrared, and not in the ultraviolet. The solution to this dilemma was found by Planck (see Chapter 1), who abandoned the assumption that all oscillators would have an average energy of kT, and introduced an apparently arbitrary assumption that the energy of any oscillator at frequency v could only exist in discrete units of hv, where Planck's constant $h = 6.626 \times 10^{-34}$ J s. This *quantization* gives the oscillator an average energy not of kT but kT multiplied by the factor

$$\frac{hv}{kT} \left[\exp \left(\frac{hv}{kT} \right) - 1 \right]^{-1}.$$

The energy density of the blackbody radiation spectrum in the frequency range dv then becomes

$$u(v)dv = \frac{8\pi hv^3}{c^3} \frac{dv}{[\exp(hv/kT)-1]}. \tag{5.48}$$

This is the Planck radiation formula for the energy density within a blackbody.

The irradiance I of a blackbody is related to the energy density u by considering the energy flowing out of a unit area hole in a blackbody cavity. Within the cavity the flow is uniform in direction over solid angle 4π. Outside, the flow at angle θ to the normal through solid angle $d\Omega$ is $uc \cos \theta d\Omega/4\pi$. In direction θ, ϕ, where ϕ is the azimuth angle, $d\Omega = \sin \theta d\theta d\phi$, so that the irradiance is $I = \int_0^{2\pi} \int_0^{\pi/2} uc \cos \theta \sin \theta d\theta d\phi/4\pi = uc/4$.

The Planck formula for irradiance is therefore

$$I(v)dv = \frac{2\pi hv^3}{c^2} \frac{dv}{[\exp(hv/kT) - 1]}. \tag{5.49}$$

The concept of quantized oscillators in the walls of a cavity was later replaced by quantization of resonant modes of electromagnetic waves within the cavity, but the theory is otherwise unchanged. The effect of the Planck term is seen in the solid line of Figure 5.5, where the unmodified Rayleigh–Jeans curve, shown as a broken line, indicates a spectrum increasing indefinitely at high frequencies. Note that the Rayleigh–Jeans formula may be sufficiently nearly correct to be used at low frequencies when hv/kT is small; see Problem 5.9.

The blackbody spectrum defined as a function of wavelength is

$$u(\lambda)d\lambda = \frac{8\pi hc}{\lambda^5} \frac{1}{[\exp(hc/\lambda kT) - 1]} d\lambda, \tag{5.50}$$

and becomes $I(\lambda)d\lambda$ when multiplied by $c/4$. This is plotted in Figure 5.6 for a range of temperatures. The peak in each curve at λ_{max} is near the wavelength at which $hc/\lambda = kT$, so that the product $\lambda_{max}T$

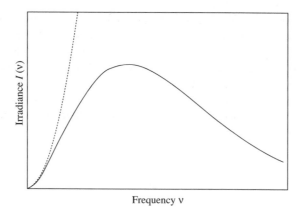

Figure 5.5 The blackbody radiation curve. The broken curve shows the dependence expected without quantization

is a constant. This gives Wien's law:[13]

$$\lambda_{max}T = 2.897 \times 10^{-3} \text{m K}.$$ (5.51)

It is interesting to note that Wien's law was formulated before quantization was introduced by Planck. It turns out that thermodynamic arguments alone can establish both Wien's law and another fundamental radiation law due to Stefan. This concerns the total energy integrated over a blackbody spectrum, and is unaffected by quantization. *Stefan's law*, found experimentally in 1879 and derived from thermodynamics by Boltzmann in 1884, states that the total power radiated by a blackbody over all wavelengths is proportional to the fourth power of the temperature, giving

$$I(T) = \int_0^\infty I(v)\mathrm{d}v = \sigma T^4$$ (5.52)

where I is the total power radiated per unit area, and $\sigma = 5.67 \times 10^{-8} \text{ W m}^{-2} \text{ K}^{-4}$ is the *Stefan–Boltzmann* constant.[14]

---

[13]Wien's law may also be stated in terms of frequency v; it is easily derived in this form from equation (5.50) by writing Planck's formula as

$$I(q) = \frac{2\pi k^3 T^3}{c^2 h^2} \frac{q^3}{\exp(q) - 1}$$

where $q = hv/kT$ and differentiating with respect to q. The result is $v_{max} = 2.82kT/h$. Note that this calculation relates to the maximum per unit *frequency*, while equation (5.51) refers to a maximum per unit *wavelength*, which occurs at occurs at wavelength $\lambda_{max} = ch/4.965kt = 0.568c/v_{max}$.

[14]Strictly speaking, the power leaving unit area of a surface is known as the radiant exitance, M_e, but physically it is very close to irradiance so we here denote it as such.

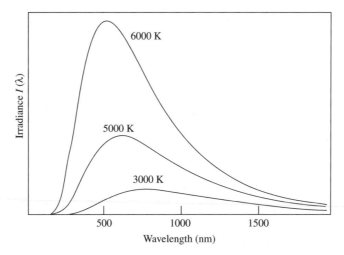

Figure 5.6 Blackbody radiation; intensity plotted against wavelength for different temperatures

Very closely connected to the irradiance $I(t)$ is the energy density $u(T)$ found by integrating equation (5.48) over all frequencies. Using $q = hv/kT$, and the identity

$$\int_0^\infty \{q^3/[\exp(q) - 1]\}\mathrm{d}q = \pi^4/15 \tag{5.53}$$

we find

$$u(T) = \frac{8\pi h}{c^3} \int_0^\infty \frac{v^3 \mathrm{d}v}{\exp(hv/kt) - 1} = \frac{8\pi(kT)^4}{h^3 c^3} \int_0^\infty \frac{q^3 \mathrm{d}q}{\exp(q) - 1} = \frac{8\pi^5 k^4}{15 h^3 c^3} T^4. \tag{5.54}$$

In other words, the energy density of blackbody radiation has the form

$$u(T) = aT^4 \tag{5.55}$$

where

$$a = (8\pi^5/15)k^4(hc)^{-3} = 7.57 \times 10^{-16} \mathrm{J\,m^{-3}\,K^{-4}}. \tag{5.56}$$

Since we have just shown that $I = cu/4$, the constants in equations (5.52) and (5.55) are related by $\sigma = ca/4$.

The most perfect blackbody radiation curve ever observed is that of the cosmic microwave background radiation, which is a relic of the concentrated thermal radiation which filled the early Universe soon after the Big Bang. As the Universe expands, reducing the energy concentration in this radiation, the radiation cools but its spectrum remains that of a blackbody. At the present state of expansion the temperature of this radiation is 2.73 K, giving a spectrum peaking near 1 millimetre wavelength. The spectrum was measured with remarkable precision from above the Earth's atmosphere, with a spectrometer on the COBE satellite (Figure 5.7).

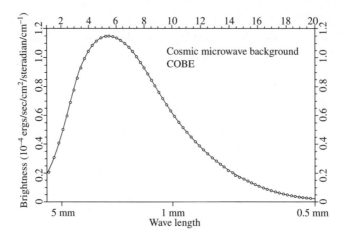

Figure 5.7 The blackbody spectrum of the microwave cosmic background radiation. The observational data from the COBE satellite fit precisely on the theoretical curve for a temperature of 2.73 K. (Mather J.C. *et al.*, 1994, *Astrophys. J.*, **420**, 439)

Wien's law gives a useful guide to the spectral range at which any hot body radiates most efficiently, even if it is not a perfect blackbody. The spectrum of solar radiation is a good approximation to that of a blackbody at 6000 K; the peak at 500 nm comes within the visible spectrum, coinciding with the range of wavelengths that can penetrate the Earth's atmosphere and to which our eyes are sensitive. X-rays originate in hotter places, with temperatures of order 10^6 K; in astronomy most such sources are ionized gas clouds, such as the outer part of the solar atmosphere.

Problem 5.1

The general one-dimensional wave equation $\partial^2 y/\partial z^2 - (1/v^2)\partial^2 y/\partial t^2 = 0$ is just like equation (5.15) but allows for a wave speed v that may differ from c, the vacuum speed of light. Find out by inspection which of the following are solutions (real- or complex-valued) of the wave equation, and when they are, give their wave speed v. Note that singular and divergent solutions are allright so long as they are well defined over at least some finite range of z and t.

(a) $y = \tan^7(z - 3t)$

(b) $\tilde{y} = \exp[i(a^2 z^2 - 2abzt + b^2 t^2)]$.

(c) $y = 5\cos(z - 2t) + 8\sin(z + 3t)$.

(d) $y = \ln(z^2 - 25t^2)$.

(e) $y = \exp[-a^2(z - t)^2 - b^2(z + t)^2]$

(f) $y = \sin[1/(z + 2t)^3]$.

Problem 5.2

(i) A slab of GaAs crystal, used in a laser, has refractive index $n = 3.6$. What fraction of the energy of radiation generated in the slab and incident normally on the top face is reflected? What is the transmittance for radiation from outside entering the slab at normal incidence?

(ii) Two glass slabs, with refractive indices 1.5 and 1.3, are glued together with a thick layer of transparent material with refractive index 1.4. Show that the light lost by reflection is approximately halved compared with a direct contact between the slabs.

(iii) A light wave in glass with refractive index 1.5 has a transverse electric field amplitude of $10\,V\,m^{-1}$. What is the associated magnetic field and the energy density?

(iv) At what wavelengths are the maximum output of radiation from blackbodies at temperatures $3\,K$, $20°C$, $5800\,K$?

(v) The average irradiance of solar radiation at the Earth is $1.4\,kW\,m^{-2}$. Most is absorbed; calculate the total force on the whole of the Earth. The mean radius of the Earth is $6.4 \times 10^6 m$.

Problem 5.3

A film with refractive index n_f is placed between two media with indices n_1 and n_2. (a) For light incident normally, passing from 1 to 2, find the value of n_f that maximizes the net transmittance, T_{1f2}, and determine this optimal value. (Hint: Instead of maximizing T_{1f2} itself, it is easier to maximize its natural logarithm.) (b) Compare the minimal reflective loss $(1 - T_{1f2})$ with $(1 - T_{12})$, the value it would have in the absence of the film for the two cases: (i) $n_1 = 1.44, n_2 = 1.69$; and (ii) an air–diamond interface, $n_1 = 1.00, n_2 = 2.40$.

Problem 5.4

What fraction of light is reflected at the surface of a lens with refractive index 1.5? Show how this may be reduced by a suitable surface coating.

Problem 5.5

Compare the solar radiation pressure on the Earth (see Problem 5.2(v) above) with the gravitational attraction of the Sun, and find the radius of a sphere with density the same as the mean density of the Earth $(5.5\,g\,cm^{-3})$ for these forces to balance. (Hint: The gravitational force can be found from the period of the Earth's orbit and its distance $1.5 \times 10^{11}\,m$ from the Sun.)

Problem 5.6

Following Section 5.5, estimate the electric field amplitude due to normal illumination from a desk lamp. Assume it converts some 2% of its wattage to light.

Problem 5.7

A $1\,kW$ laser beam has a cross-sectional diameter of $5\,mm$. Calculate the irradiance and the amplitudes of the electric and magnetic fields.

Problem 5.8

Consider two monochromatic electromagnetic waves of the same frequency. Under what circumstances of polarization can they add so that the irradiance of the sum is always equal to the sum of their two separate irradiances?

Problem 5.9

Two plane waves exactly in phase combine to form a wave with double amplitude, i.e. with quadruple power. Where does the extra energy come from?

Problem 5.10

In Section 5.7 we state that a (one-dimensional) simple harmonic oscillator with frequency v and in thermal equilibrium at temperature T radiates energy at a rate proportional to kTv^2. In Section 18.1 we show that $\bar{P}$, the average power radiated by a Hertzian dipole, goes as $\omega^4 x_0^2$, where $\omega = 2\pi v$ and x_0 is the amplitude of the oscillation. Reconcile these two statements.

Problem 5.11

Show that Planck's formula for blackbody radiation goes over to the Rayleigh–Jeans formula in the low-frequency limit. (Note that for $|x| \ll 1$, $\exp(x) \simeq (1 + x)$.) Determine the frequency below which the Rayleigh–Jeans formula applies for the 3 K cosmic background radiation.

Problem 5.12

Any mass M compressed into a sphere of radius $R_{bh} = 2GM/c^2$ is dense enough to become a black hole, a region of spacetime with gravity so intense that no particles or radiation can escape. The cosmic microwave background (CMB) with a temperature of 2.73 K fills space uniformly. Find out whether the CMB is dense enough to turn the Observable Universe into a black hole. (Note that the mass density of blackbody radiation is $u/c^2 = aT^4/c^2$, where $a = 7.57 \times 10^{-16}\,\mathrm{J\,m^{-3}\,K^{-4}}$, and that the Observable Universe has a radius of $R \approx 10^{10}$ light-years (lyr), where $1\,\mathrm{lyr} = 9.46 \times 10^{15}$ m).

Problem 5.13

What is the weakest incident photon that can lose two-thirds of its energy when Compton scattering off an electron? Give its energy in eV. By reference to Figure 1.7, tell what kind of photon it is.

Problem 5.14

A standard formula to calculate the flux of any scalar quantity Q (mass, charge, number of particles, etc.) through a chosen area is

$$\text{Flux of } Q = \Delta Q/(\Delta t \Delta A) = (\Delta Q/\Delta V) \times \bar{V}_n. \tag{5.57}$$

Here t is time, V is volume, and $\bar{V}_n$ is the mean component of velocity normal to the area. Let us consider the energy flux of blackbody radiation escaping from a peephole in the cavity. The radiation is isotropic, which means that the net flux is zero; there are equal and opposite fluxes that cancel each other out. Assume the hole is small enough that it does not disturb the radiation a small distance inside the cavity. But as we approach the hole, the inward-moving photons vanish (no cavity photons are entering from outside) and only the outward-moving ones remain. These latter photons are the ones with a positive component of V_n. Our flux formula becomes

$$I(t) = \Delta U/(\Delta t \Delta A) = (\Delta U/\Delta V)_{\text{out}} \times \bar{V}_n = \frac{1}{2}u(T)\bar{V}_n. \tag{5.58}$$

Since outward-moving photons near the hole represent half of the photons, we have set the relevant energy density equal to half the total. By evaluating $\bar{V}_n$, verify that the constants in equations (5.52) and (5.55) are related by $\sigma = ca/4$. (Hint: Properly oriented spherical coordinates make the calculation easier.)

6 Fibre and Waveguide Optics

... beauty draws us with a single hair.

Alexander Pope, *The Rape of the Lock.*

If hairs be wires.

William Shakespeare, *Sonnets.*

The transmission of light along a curved dielectric cylinder was the subject of a spectacular lecture demonstration by John Tyndall in 1854. His *light pipe* was a stream of water emerging from a hole in the side of a tank which contained a bright light. The light followed the stream by total internal reflection at the surface of the water. Light pipes made of flexible bundles of glass fibres are now routinely used to illuminate internal organs in surgical operations in the *fibrescope* (or *endoscope*) which also transmits an image back to the surgeon. The overwhelmingly important use of glass fibres is, however, to transmit modulated light over large distances for communications.

Electrical cables and radio have largely been replaced by optical fibres in long-distance terrestrial communications. Hundreds of thousands of kilometres of fibre optic cables are now in use, carrying light modulated at high frequencies, providing the large communication bandwidths needed for television and data transmission. The techniques which made this possible are the subject of this chapter. These techniques involve the manufacture of glass with very low absorption of light, the development of light emitters and detectors which can handle high modulation rates, and the fabrication of very thin fibres which preserve the waveform of very short light pulses. An essential development has been the *cladding* of fibres with a glass of lower refractive index, which prevents the leakage of light from the surface.

Optical fibres are also useful in short communication links, especially where electrical connections are undesirable. They also offer remarkable opportunities in computer technology and in laboratory instrumentation such as interferometers and a variety of optical fibre sensors.

We start by discussing the propagation of a light ray by internal reflection in a light pipe, and show how this approach may be developed into the concept of waves guided inside a dielectric slab or along a thin fibre. The cylindrical geometry of a fibre, and the light-confining feature of a fibre, which is a step or a gradient in refractive index, both need special consideration. Propagation in a light fibre

Optics and Photonics: An Introduction, Second Edition   F. Graham Smith, Terry A. King and Dan Wilkins
© 2007 John Wiley & Sons, Ltd

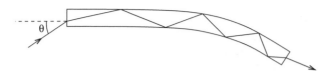

Figure 6.1   A light pipe, showing total internal reflection of a light ray

may be *dispersive*, so that different wavelengths travel at different speeds; we show how the effects of dispersion are calculated and how they may be compensated for. We then briefly describe some of the many applications of fibres outside the field of communications.

6.1   The Light Pipe

The transmission of light along a glass rod depends on the total internal reflection of a ray reaching the surface at a glancing angle, i.e. at a high angle of incidence (Section 1.3). A *light pipe* in the form of a glass rod can be used to conduct light round corners (Figure 6.1), provided that the corners are not so sharp that the complementary angle of incidence[1] θ falls above the critical angle defined by $\cos\theta_{\text{crit}} = \sin\theta_{\text{crit}}^{\text{nl}} = (n_2/n_1)$.

A bundle of thin glass fibres, usually coated with glass or plastic with lower refractive index, can transmit an optical image, as in the surgeon's endoscope. If the fibre ends are aligned in a plane, the light distribution across them will be reproduced at the other end of the bundle. Tapering the fibres along the bundle can be used to reduce or enlarge an image; rearranging them can compensate for distortions introduced in other parts of an optical system. A random rearrangement of the fibres within a bundle can be used to "scramble" an image; the opposite rearrangement, using the bundle in the reverse direction, will then restore the image. Fibre optic bundles occur naturally in both the animate and the inanimate world. The retina of the human eye, and many other eyes, has an assembly of rods and cones which transmit light from the surface of the retina to light-sensitive cells. In many insects, the transmission is sensitive to polarization, providing the information which insects use for navigation. Natural inanimate fibre optics is found in the crystalline material known as ulexite, which is a fibrous form of borax.

To deal with the propagation of light by the thin glass fibres used in communications a different approach is necessary; now the diameter of the fibre is comparable with the wavelength of light, and we must consider the light as a wave which is guided by the fibre. The ray concept is useful in understanding refraction at the ends of a fibre, and to some extent in considering propagation within it and reflection at the boundary; however, the wave theory is essential for understanding the field distribution within the fibre. The configuration of the wave inside the fibre is constrained by conditions at the boundary of the fibre; we must also consider the extension of the wave field outside the boundary, into the cladding of the core fibre.

---

[1]In fibre optics it is conventional to designate the angle between a ray or wavenormal and the fibre axis as θ, which is the complement of the angle θ^{nl} measured from the normal and used in the analyses of refraction by Snell and by Fresnel; see Chapters 1 and 5.

6.2 Guided Waves

We develop the theory of the propagation of light along a cylindrical fibre in three stages. The requirement is to find wave configurations within the fibre which are solutions of Maxwell's equations, and which conform to the *boundary conditions*, i.e. the physical conditions at the surface of the core of the fibre. The first stage is to apply Maxwell's equations (Chapter 5) to a wave confined to a parallel-sided slab of dielectric. There are two major differences from free space propagation: there can be components of both **E** and **B** fields in the direction of propagation, and only a limited number of wave patterns between the faces of the slab, known as modes, can propagate between the faces of the slab. The allowable mode patterns depend on the thickness of the slab and on the boundary conditions.

The effect of the boundary conditions is easiest to understand if the faces of the slab are perfectly conducting metal slabs; this is close to the practical case of waveguides for centimetric and millimetric radio waves. The second stage is to consider a slab guide bounded by a step in dielectric constant; the boundary conditions are then more complicated and there is a component of the wave outside the surface of the slab. These two stages allow us to understand the fundamental characteristics of guided waves, and in particular their field patterns and their velocities. The geometry then needs to be adapted to the more complex mathematics of cylindrical rather than rectangular symmetry.

Maxwell's equations *in free space* (equations (5.4)) are

$$\text{div } \mathbf{E} = 0 \quad \text{div } \mathbf{B} = 0 \tag{6.1}$$

$$\text{curl } \mathbf{B} = \epsilon\mu \frac{\partial \mathbf{E}}{\partial t} \quad \text{curl } \mathbf{E} = -\frac{\partial \mathbf{B}}{\partial t}. \tag{6.2}$$

As we have seen in Chapter 5, these lead to the wave equations

$$\nabla^2 \mathbf{E} = \epsilon\mu \frac{\partial^2 \mathbf{E}}{\partial t^2} \tag{6.3}$$

$$\nabla^2 \mathbf{B} = \epsilon\mu \frac{\partial^2 \mathbf{B}}{\partial t^2}. \tag{6.4}$$

The electric field may be expressed in Cartesian components:

$$\mathbf{E} = \hat{\mathbf{x}}E_x + \hat{\mathbf{y}}E_y + \hat{\mathbf{z}}E_z \tag{6.5}$$

where $\hat{\mathbf{x}}, \hat{\mathbf{y}}, \hat{\mathbf{z}}$ are unit vectors in the x, y, z directions.

The separate field components each obey the wave equation, so that the y component obeys

$$\nabla^2 E_y = \epsilon\mu \frac{\partial^2 E_y}{\partial t^2}. \tag{6.6}$$

For a plane wave in the z direction E_y does not vary in directions x or y, and equation (6.6) reduces to

$$\frac{\partial^2 E_y}{\partial z^2} = \epsilon\mu \frac{\partial^2 E_y}{\partial t^2} \tag{6.7}$$

which represents waves of any form $E_y = E_0 f(z - vt)$ where the velocity $v = (\epsilon\mu)^{-1/2}$. In free space the corresponding magnetic field is in the x direction; both fields are transverse to the direction of propagation.

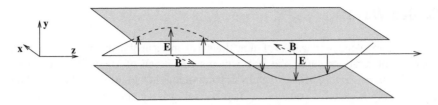

Figure 6.2 A waveguide formed by two conducting plates, showing the simplest propagating mode (the TEM mode)

The same wave will propagate along the slab guide shown in Figure 6.2, since it conforms to the boundary conditions at the conducting walls, where the tangential component of the electric field and the normal component of the magnetic field are required to be zero at the walls. Hence the electric field must be perpendicular to the walls. The wave velocity $v = \omega/k$ is the velocity of light in the medium between the plates. This mode is referred to as the transverse electric and magnetic, or TEM, mode. We now find other modes which will propagate in the slab guide. As in equation (6.7) the wave travels in the z direction, but the electric and magnetic fields are now constant only in the x direction. Setting $\partial/\partial x = 0$, equations (6.2) reduce to

$$\frac{\partial E_z}{\partial y} - \frac{\partial E_y}{\partial z} = -\frac{\partial B_x}{\partial t} \tag{6.8}$$

$$\frac{\partial E_x}{\partial z} = -\frac{\partial B_y}{\partial t} \tag{6.9}$$

$$\frac{\partial E_x}{\partial y} = \frac{\partial B_z}{\partial t} \tag{6.10}$$

$$\frac{\partial B_z}{\partial y} - \frac{\partial B_y}{\partial z} = \epsilon\mu\frac{\partial E_x}{\partial t} \tag{6.11}$$

$$\frac{\partial B_x}{\partial z} = \epsilon\mu\frac{\partial E_y}{\partial t} \tag{6.12}$$

$$\frac{\partial B_x}{\partial y} = -\epsilon\mu\frac{\partial E_z}{\partial t}. \tag{6.13}$$

Two sets of solutions emerge from this array. Equations (6.9), (6.10) and (6.11) contain only E_x together with B_y and B_z; these form solutions in which the electric field has no components in the direction of propagation, but the magnetic field does; in contrast equations (6.8), (6.12), and (6.13) contain only B_x together with E_y and E_z; these form solutions in which the magnetic field has no components in the direction of propagation, but the electric field does. These two sets of solutions represent *transverse electric* (TE) and *transverse magnetic* (TM) modes respectively. We now describe the field patterns in the individual modes.

Based on Fourier analysis, we consider harmonic waves as the basic modes into which any wave within the guide can be decomposed. Of course, only those harmonic waves are allowed that satisfy the appropriate boundary conditions. Each mode has a simple field pattern which varies sinusoidally across the guide. This can conveniently be regarded as the combination of two crossing plane waves with certain allowed values of wave vectors $\mathbf{k}_{\perp}$. Consider first a pair of waves with electric vector in the x direction, and with vectors $\mathbf{k}_\pm$ making angles $\pm\theta$ with the z direction:

$$\mathbf{E_1} = \hat{\mathbf{x}}E_0 \exp[i(\omega t - kz\cos\theta + ky\sin\theta)] \tag{6.14}$$

$$\mathbf{E_2} = -\hat{\mathbf{x}}E_0 \exp[i(\omega t - kz\cos\theta - ky\sin\theta)]. \tag{6.15}$$

The sum of these is

$$\mathbf{E} = \hat{\mathbf{x}}\,2i\sin(ky\sin\theta)E_0\exp[i(\omega t - kz\cos\theta)]. \tag{6.16}$$

The boundary condition is that $E_x = 0$ at both plates, at $y = 0$ and $y = b$. This is achieved if the angle θ is chosen to give

$$kb\sin\theta = n\pi \tag{6.17}$$

where n is an integer. There may be several pairs of waves with different values of θ which satisfy this criterion, provided that

$$n \le \frac{kb}{\pi}. \tag{6.18}$$

Each pair constitutes an allowable wave pattern, or *mode*, which can propagate independently along the guide in the z direction. The propagation constant k_g along the guide is $k\cos\theta$; substituting for θ from equation (6.17) we have

$$k_g = \left(k^2 - \frac{n^2\pi^2}{b^2}\right)^{1/2}. \tag{6.19}$$

These modes are TE modes; note that there are components of the magnetic field in the direction of propagation. In the TM modes, the magnetic field is wholly transverse and there is a component of the electric field in the z direction. The modes are designated TE_n and TM_n according to their mode number n. Equation (6.16) shows that the wave velocity v_p of each mode is

$$v_p = \frac{\omega}{k_g} \tag{6.20}$$

where the subscript p indicates the *phase* velocity in contrast to the *group* velocity (see Chapter 4). From equation (6.19) and recalling that in non-magnetic media, the separate harmonic waves of equations (6.14), (6.15) both have phase velocity $c/\sqrt{\epsilon_r} = \omega/k$,

$$v_p = \frac{c}{\sqrt{\epsilon_r}}\left(1 - \frac{n^2\pi^2}{k^2b^2}\right)^{-1/2}. \tag{6.21}$$

Let us discuss the simple case of a vacuum. We see that the phase velocity is greater than the free space velocity c, and that it depends on the wave number k. The group velocity v_g is given by

$$v_g = \frac{d\omega}{dk_g} \tag{6.22}$$

which from differentiating equation (6.19) with respect to ω, and using $d\omega/dk_g = (dk_g/d\omega)^{-1}$, is

$$v_g = c\frac{k_g}{k}. \tag{6.23}$$

The group velocity, as might be expected, is always less than c; the product $v_p v_g = c^2$.

The complete analysis of metallic waveguides must also involve boundaries in the x direction, to form a rectangular waveguide.

6.3   The Slab Dielectric Guide

A wave may be guided along a dielectric slab, such as a sheet of glass, provided that it is bounded by a material of smaller refractive index. The analysis is similar to that for the guide with conducting plates, but there are different boundary conditions to consider. The wave amplitude does not fall to zero at the boundary, and there is a component of the field beyond the boundary. We follow the same procedure of analysing pairs of crossing waves, each allowable pair constituting a propagating mode. It is convenient, however, to consider the pair of waves as a ray which is reflected to and fro between the boundaries of the slab, as in Figure 6.3.

There must be total internal reflection at the boundary. From Snell's law (Chapter 1) this means that the angle of incidence must be larger than the critical angle (see Chapter 5), so that the ray angle must be closer to the axis than θ_{crit} given by

$$\cos\theta_{crit} = \frac{n_2}{n_1} \tag{6.24}$$

where n_1, n_2 are the refractive indices inside and outside the slab. The pair of crossing waves which constitute a mode is now represented as a single ray which is reflected to and fro across the guide, as in Figure 6.4. After the two reflections shown in Figure 6.4 the ray CD must have the same phase as the incident ray AB, so that it constitutes the single wavefront of equation (6.14). The twice-reflected ray has travelled an extra distance,[2] and in contrast to reflection at the

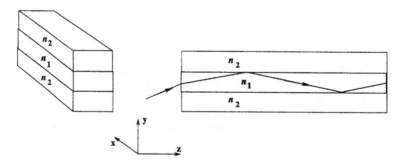

Figure 6.3   Dielectric slab waveguide, showing total internal reflection at the interface between refractive indices n_1 and n_2

---

[2]In Figure 6.4 the angle θ is shown larger than usual, to help visualize the geometry. Note the similarity to the analysis of the plane-parallel plate in Chapter 9.

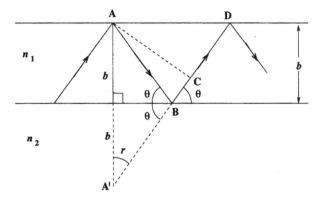

Figure 6.4 The path difference between reflected rays in a dielectric guide. A′ is the image of point A as if reflected in the lower surface of the guide. By congruent triangles, AA′ = 2b and A′B = AB

conducting plate there is also a phase change $\phi(\theta)$ at each reflection. The extra path AB + BC for the reflected ray is found from

$$AB + BC = A'B + BC = 2b \cos r = 2b \sin \theta. \tag{6.25}$$

The rays arriving at A and C must be in phase, as they lie on the same wavefront. This gives a phase condition, including $2\phi(\theta)$ for the two reflections:

$$2b \sin \theta + \lambda_1 \frac{2\phi(\theta)}{2\pi} = N\lambda_1. \tag{6.26}$$

Aside from the additional term $\phi(\theta)$, this is similar to equation (6.17); λ_1 now refers to the wavelength λ_0/n_1 in the dielectric. N is again a mode number; there is a set of ray directions which can propagate, and each has its own group velocity. Equation (6.26) is a general condition for a propagating mode. Its solution is best approached by numerical methods; note that $\phi(\theta)$ depends on the polarization of the wave as well as the angle of incidence.

6.4 Evanescent Fields in Fibre Optics

The electric field does not fall to zero at the boundary of the dielectric slab, although the components of the propagating wave are totally internally reflected and, following equation (6.24), there is no refracted ray propagating away from it. The wave amplitude must therefore fall to zero in the y direction. The wave outside the slab is an *evanescent wave*[3] (Figure 6.5); we show that the amplitude of this evanescent wave decays exponentially with distance y.

Consider a refracted wave transmitted across the boundary when the grazing– or off-surface–angle θ_1 is more than the critical angle; θ_1 is the angle that the wave vector, i.e. the incident ray, makes with

---

[3]*Evanescent* is fleeting or vanishing, from evanesce: to fade away.

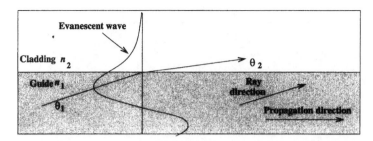

Figure 6.5 Cross-section of the electric field pattern E_y in a multi-mode dielectric guide, showing the penetration of an evanescent wave into the cladding

the surface of the slab (Figure 6.6). As in Section 5.4, the amplitude of the refracted wave is E_t, and it propagates at angle θ_2 to the surface as a wave E_2 with the form (compare equation (6.14))

$$E_2 = E_t \exp[ik_2(z\cos\theta_2 - y\sin\theta_2)], \tag{6.27}$$

where we omit the factor $\exp(-i\omega t)$. Since from Snell's law $n_1\cos\theta_1 = n_2\cos\theta_2$,

$$\sin\theta_2 = \sqrt{1 - \frac{n_1^2}{n_2^2}\cos^2\theta_1} \tag{6.28}$$

we can write equation (6.27) in terms of the angle of incidence as

$$E_2 = E_t \exp(ik_2)\left(\frac{n_1}{n_2}z\cos\theta_1 - y\sqrt{1 - \frac{n_1^2}{n_2^2}\cos^2\theta_1}\right). \tag{6.29}$$

For a ray beyond the critical angle, the square root term becomes imaginary, and we can write

$$\sqrt{1 - \frac{n_1^2}{n_2^2}\cos^2\theta_1} = \pm i\alpha. \tag{6.30}$$

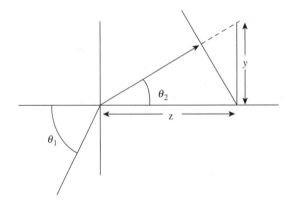

Figure 6.6 Geometry of the refracted wave

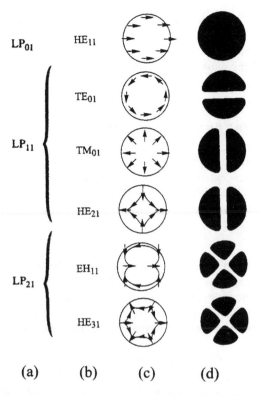

Figure 6.7   Electric field patterns in the three lowest order linearly polarized modes propagating in a circular cross-section waveguide. The modes are designated (a) in terms of linear polarization and (b) in terms of TE_{lm} and TM_{lm} transverse modes for meridional rays and hybrid modes HE_{lm} and EH_{lm} for skew rays; (c) shows the electric field distributions, and (d) the electric field intensity distributions. (From J.M. Senior, *Optical Fiber Communications*, Prentice Hall, 1992)

The wave outside the boundary (Figure 6.5) is now seen to be the evanescent wave[4]

$$E_T = E_t \exp(-\alpha k_2 y) \left[\exp\left(ik_2 \frac{n_1}{n_2} z \cos \theta_1 \right) \right]. \tag{6.31}$$

The wave propagates along the boundary, matching the guided wave inside the boundary, while the amplitude decays exponentially in the y direction. The exponential decay constant αk_2 is the inverse of the *penetration depth* ξ, given by

$$\xi^{-1} = k_2 \left(\frac{n_1^2}{n_2^2} \cos^2 \theta_1 - 1 \right)^{1/2}. \tag{6.32}$$

The evanescent wave can penetrate a significant distance into the cladding of an optical fibre, which must be thick enough for the amplitude to fade away almost to zero. In a silica fibre with

---

[4]The solution with an exponentially *growing* wave is clearly unphysical.

$n_1/n_2 = 1.01$, for an angle close to the critical angle, and for a wavelength of $1.3\,\mu m$, the value of ξ is about $2\,\mu m$.

A considerable fraction of the wave energy travelling along an optical fibre is transmitted in the evanescent wave. The cladding must therefore be a glass with as high an optical quality as that of the fibre core, so as to avoid transmission losses.

6.5   Cylindrical Fibres and Waveguides

The main principles of a slab or rectangular waveguide may be applied to a cylindrical dielectric guide without modification except for the detailed pattern of the propagating field. As before, we require monochromatic solutions of Maxwell's equations in the form of the propagating fields

$$\mathbf{E} = \mathbf{E}_0(\rho, \phi)\exp(-i\beta z)$$
$$\mathbf{B} = \mathbf{B}_0(\rho, \phi)\exp(-i\beta z) \tag{6.33}$$

where a common factor of $\exp(i\omega t)$ is understood. The analysis requires the wave equations for $\mathbf{E}$ and $\mathbf{B}$ to be written in a cylindrical coordinate system, with ρ, ϕ replacing x, y. Solutions which satisfy the boundary conditions resemble those for the slab guide, showing discrete modes which will propagate for any given free space wavelength, provided it is less than a critical wavelength. The field patterns for each mode are in the form of Bessel functions rather than the sine and cosine functions in the rectangular guide.

Figure 6.7 shows some of the field patterns in the simplest case of a cylindrical dielectric waveguide. In each mode there must be a small component of electric or magnetic field along the axis, and the modes are often distinguished in terms of the field component which is wholly transverse as TE_{lm} or TM_{lm} modes, which have meridional[5] travelling rays with radially symmetric field distributions. The two-dimensional cross-section of the cylindrical waveguide requires two integers l, m to designate the modes. The integer numbers indicate the number of azimuthal (circumferential) nodes (l) and the number of radial nodes (m) that are in the field pattern. The TE modes have zero axial electric field ($E_z = 0$) and the TM modes have zero axial magnetic field ($H_z = 0$). Skew ray propagation leads to hybrid modes HE_{lm} and EH_{lm} in which E_z and H_z are non-zero, with the designation depending on whether the axial H or E field makes the dominant contribution to the transverse field. The set of modes can be approximated by a set of linearly polarized (LP_{lm}) modes. In the cylindrical guide the modes are designated according to the order of the Bessel function describing the field pattern, as shown in Figure 6.7. In a rectangular guide each mode is designated according to the number of cycles across the guide; the simplest mode in the rectangular waveguide is designated TE_{01}.

In a practical fibre, with dielectric cladding, the field extends into the cladding as an evanescent wave. The detailed field configuration depends on the form of the interface between the fibre and the cladding; solutions can be obtained for a step change in refractive index, but in practice there are advantages in a more gradual transition of refractive index, the graded index involving a more complicated analysis.

We have already seen how the boundary conditions of dielectric slab guides affect the field patterns as compared with those of waveguides with conducting walls. The differences are more important in

---

[5]A meridional ray is one that lies in a plane containing the axis of the fibre. Any non-meridional ray is called *skew*, and does not pass through the fibre axis.

optical fibres, since they are usually clad with a dielectric with refractive index n_2 which is only slightly below n_1, the refractive index of the guide core itself. This has the advantage that rays at a large angle to the direction of propagation are not then internally reflected, and only low-order modes are propagated (see equation (6.18)); in the limit, there is only one propagating mode. Such a *single mode fibre* is particularly important in long-distance communications, where the difference in group velocity between modes is a severe disadvantage. Single mode fibres must have a core diameter less than a few wavelengths, and a small change in refractive index from core to cladding. While single mode fibres provide large bandwidth, e.g. for optical communications, the multi-mode fibre has a larger core radius than the single mode fibre, so that it is easier to launch light into the fibre and connections between similar fibres can more readily be made.

Exact analysis of the maximum diameter for a single mode fibre is tedious, but an approximate analysis for a slab dielectric is easily done in terms of a ray in the slab at the limiting angle θ_{crit} for internal reflection. Then $n_1 \cos \theta_{crit} = n_2$. If there are N half wavelengths in the wave pattern across a slab with thickness b, then, based on equation (6.26) with $\phi(\theta_{crit}) = 0$, which follows from equations (5.32) and (5.33), $b \sin \theta_{crit} = N\lambda_1/2$, where λ_1 is the guide wavelength λ_0/n_1, giving N as

$$N = \frac{2b}{\lambda_0}(n_1^2 - n_2^2)^{1/2}. \tag{6.34}$$

In practice the two refractive indices differ by only a small amount and may be written as n and $n + \Delta n$, so that $(n_1^2 - n_2^2)^{1/2} \simeq (2n\Delta n)^{1/2}$. The maximum thickness b_{max} of a slab which carries only a single mode is

$$b_{max} = \frac{\lambda_0}{2(2n\Delta n)^{1/2}}. \tag{6.35}$$

A similar but more complicated analysis for cylindrical fibre guides yields the useful parameter known as the *V number*, which for a step-index guide with radius a is

$$V = \frac{2\pi a}{\lambda_0}(n_1^2 - n_2^2)^{1/2}. \tag{6.36}$$

The analysis shows that if this V number is less than 2.405, only a single mode can propagate. This occurs for a wavelength $\lambda_0 \geq 2\pi a(n_1^2 - n_2^2)^{1/2}/2.405$. The fibre is known as single mode or monomode. This requires very thin fibres: for example, if $n_1 - n_2$ is 20 % of n_1, the maximum diameter for a single mode fibre is only 4 vacuum wavelengths. For a cylindrical waveguide the number 2.405 corresponds to the first zero of the Bessel function which is the solution of the wave equation for the fundamental mode. For large values of V the total number of modes M (including both polarizations) for a step-index fibre is

$$M \simeq \frac{\pi^2}{2}\left(\frac{2a}{\lambda_0}\right)^2 (n_1^2 - n_2^2) = \frac{V^2}{2}. \tag{6.37}$$

The total number of modes is proportional to (fibre diameter/free space wavelength)^2. A step-index fibre with a radius $a = 25\,\mu m$, $n_1 = 1.520$, $n_2 = 1.505$ and operating at a vacuum wavelength of $2\,\mu m$ is able to propagate about 140 modes. Modes become unguided or cut off when the mode field in the cladding changes from being an evanescent field to a real field carrying power. For a field varying as

$\exp[-\mathrm{i}(\omega t - \beta_{lm}z)]$ with propagation constant β_{lm} for the (lm) mode, the mode becomes cut off when $\beta_{lm} = \beta_c$, where β_c is the propagation constant in the cladding.

6.6   Numerical Aperture

Although the rays in an optical fibre are at a small angle to the axis, they spread to a wider angle as they emerge at the end of the fibre. This is important in matching light detectors to fibres, where for efficient light collection the angle accepted by a detector should be approximately the same as the emergent light cone of the guide; similarly, in injecting light into a fibre efficiently the light cone from the source should match the acceptance angle of the fibre. The numerical aperture determines the maximum acceptance angle of the fibre. For a fibre with refractive index n_1 and cladding with refractive index n_2, the largest angle θ_{crit} within the fibre is shown in Figure 6.8, where the refracted ray is along the surface. Applying Snell's law to a meridional ray (a ray crossing the axis of the cylinder),

$$\sin \theta_i^{\mathrm{nl}} = \cos \theta_{\mathrm{crit}} = \frac{n_2}{n_1}. \tag{6.38}$$

A ray at this limiting angle enters a plane face at the end of the slab at angle θ_a to the normal, as shown in Figure 6.8. Then if the refractive index outside the fibre is n_0

$$n_0 \sin \theta_a^{\mathrm{nl}} = n_1 \sin \theta_{\mathrm{crit}} = n_1 \left[1 - \left(\frac{n_2}{n_1} \right)^2 \right]^{1/2}. \tag{6.39}$$

All rays inside the acceptance angle θ_a^{nl} will propagate within the slab.

The limited acceptance cone is usually expressed in terms of the *numerical aperture* (NA), which characterizes a cone of rays in any optical instrument, defined as $\mathrm{NA} = n_0 \sin \theta_a^{\mathrm{nl}}$. In this case

$$\mathrm{NA} = (n_1^2 - n_2^2)^{1/2}. \tag{6.40}$$

The numerical aperture determines for light the maximum acceptance angle of the fibre. A useful simplification can be made when the relative refractive index difference $\Delta = (n_1 - n_2)/n_1$ is small:

$$\mathrm{NA} \simeq n_1 (2\Delta)^{1/2}. \tag{6.41}$$

Typically $n \approx 1.4$, and the fractional step $\Delta \sim 1\%$, giving $\mathrm{NA} \sim 0.2$ and an acceptance cone with half angle around $10°$.

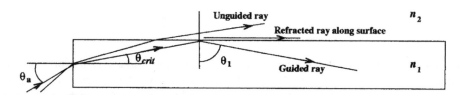

Figure 6.8   The acceptance angle for light entering a dielectric guide

6.7 Materials for Optical Fibres

The most important requirement for the glass in an optical fibre is a low transmission loss.

Transmission loss is usually measured in decibels per kilometre (dB km^{-1}), as in communications engineering.[6] A slab of ordinary silica glass usually has a loss much greater than 100 dB km^{-1}; this is due to absorption by impurities, particularly metallic ions such as iron, chromium and copper. Pure silica glass has remarkably low losses, below 1 dB km^{-1} at infrared wavelengths between 1.0 and 1.8 microns. Beyond those wavelengths the losses increase sharply (Figure 6.9). In the visible region the principal losses are due to elastic Rayleigh scattering from inhomogeneities frozen into the glass; this gives a loss increasing as λ^{-4} (see Chapter 19). There are also losses in the ultraviolet region due to electronic transitions. The increase in absorption at longer infrared wavelengths is the residual effect of vibrational states of the lattice and absorption bands such as that at 9.2 microns, due to a resonance in Si–O bonds. At higher transmitted powers additional losses may result from stimulated Brillouin and Raman scattering; these are inelastic scattering processes in which the scattered light undergoes a change in wavelength (see Chapter 19).

Within the window between 1.0 and 1.8 microns there is an appreciable rise in attenuation centred on 1.38 microns. This is related to water dissolved in the glass; the resonance actually occurs in the hydroxyl ion (OH^{-}) at 2.7 microns with a second harmonic at 1.38 microns. However, the practical situation is that there are two low absorption bands in silica glass, at 1.3 and 1.55 microns, the longer wavelength band having attenuation loss of down to 0.2 dB km^{-1}.

Losses in optical fibres may also be due to geometric imperfections introduced in the manufacturing process, and from sharp bends which the guided waves may not be able to follow. The critical condition is that the guided wave in the outer part of the cladding should not be required to travel at a speed greater than the velocity of light in that medium. The allowable radius of a bend depends on the mode and the difference in refractive index at the core interface; typically the losses are small for a radius greater than around 30 mm.

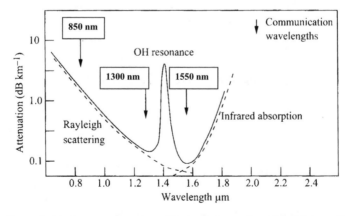

Figure 6.9 Transmission loss as a function of free space wavelength in high-quality silica glass

---

[6]Loss in decibels is $10\log_{10}$(ratio of input power to output power). It is useful to remember that 10 dB is a factor of 10, 3 dB is close to a factor of 2, and 1 dB loss is approximately 20%.

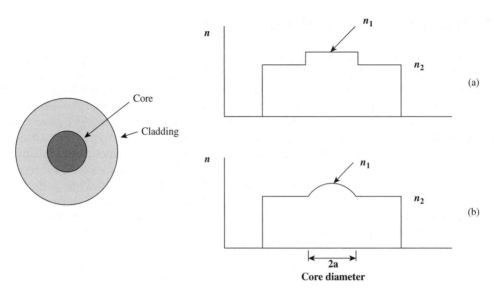

Figure 6.10   Refractive index profiles of (a) step and (b) graded refractive index fibres

The simple cladding of a fibre with a different material results in the stepped refractive index profile of Figure 6.10(a). There is, however, an important advantage in fibres manufactured with a gradient of refractive index, decreasing from the axis to join the lower refractive index of the cladding, as shown in Figure 6.10(b). Such a graded-index (GRIN) fibre can be made by allowing the cladding to diffuse into the fibre, but the manufacturing techniques which we describe below allow a more precise control of the refractive index profile. The advantage of a graded-index fibre, as described below, is that the difference in velocity of the allowable modes is minimized.

An example of a graded-index fibre is one in which the refractive index has a parabolic radial dependence. In this case for a fibre with a core radius a

$$n(r) = n_1[1 - (r/a)^2]^{1/2}. \tag{6.42}$$

Here n_1 is the refractive index on the axis. A more general graded-index function is of the form

- Core: $n(r) = n_1[1 - 2\Delta f(r/a)]^{1/2}$ for $r < a$.

- Cladding: $n(r) = n_1[1 - 2\Delta]^{1/2} = n_c$ for $r > a$.

The quantity $\Delta = (n_1^2 - n_2^2)/2n_1^2 \approx (n_1 - n_2)/n_1$ and, over $0 \leq r \leq a$, $f(r/a)$ increases monotonically from $f(0) = 0$ to $f(1) = 1$.

Expressing $f(r/a) = (r/a)^\alpha$ describes an α profile, with the parabolic profile for $\alpha = 2$ and the step-index profile for $\alpha = \infty$. Among all the α profiles the parabolic case $\alpha = 2$ is distinguished by its ability to nearly eliminate the modal dispersion. A light ray in a fibre with parabolic $n(r)$ oscillates sinusoidally across the axis (see Problem 6.3).

Analytical solutions for the propagation of EM waves in the cylindrically symmetrical dielectric waveguide in the form of specified functions can only be obtained for the step-index fibre and the parabolic graded-index profile.

For the α profile the number of modes that can propagate is

$$M \approx \left(\frac{\pi^2}{2}\right)\left(\frac{\alpha}{\alpha+2}\right)\left(\frac{2a}{\lambda_0}\right)^2(n_1^2 - n_2^2). \tag{6.43}$$

Then for $\alpha = 2, M \approx V^2/4$, half that for the step-index fibre.

The absorption of silica increases for wavelengths greater than 2.2 μm. Other fibres have been developed for infrared transmission based on specialist glasses; these include fluoride, germanium dioxide, chalcogenide and crystalline halide glasses. Because of their greater attenuation compared with silica these are suitable only for short-distance applications such as optical fibre sensors. Optical fibres may be made from polymer materials in which the step-index or graded-index core is an acrylic resin and the cladding is a fluorinated polymer. Although the transmission losses of typically 50–150 dB km^{-1} are greater than for silica fibre, they can be made with large core diameters up to 1 mm. These are able to provide bandwidths in the range of 10 MHz km (step-index) to 500 MHz km (graded-index). A multi-mode step-index fibre made up of a hybrid of a silica core and a polymer cladding (PCS) fibre provides lower attenuation than the polymer core fibre. An important application of fibre optics is in the transmission of high-power laser radiation from the laser to its point of application over short or long distances. The silica fibre is suitable to transmit wavelengths over 200 nm to 2.0 μm, particularly for the 1.06 μm Nd:YAG laser. Specialist fibres have been developed for longer wavelengths, including the 10.6 μm CO$_2$ laser, but with more severe limitations on their power handling capability. Typical optical power delivery applications of fibres are in robotic laser welding in the automotive industry and in laser surgery.

6.8 Dispersion in Optical Fibres

Fibre optic communication systems usually use pulses of light. A typical train of pulses might be transmitted as in Figure 6.11(a), and after travelling for a large distance might appear as in Figure 6.11(b). In this figure the amount of pulse spreading is close to the limit which would still allow the signal to be decoded. Pulses will start to merge if they are separated by less than the temporal width $\Delta\tau$ acquired through dispersion. This limits the communication bandwidth to a maximum of $v_{max} = 1/\Delta\tau$.

Pulse spreading is inevitable in multi-mode fibres, although its effect can be reduced in graded-index fibres. There are, however, important effects in the single mode fibres used for long-distance communications, due to the spread in travel times over the wavelength band of the pulsed light. This may not matter if a narrow wavelength band is used, as in a laser light source, but if several adjacent

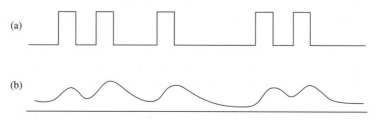

Figure 6.11 The effect of dispersion in travel time on a train of pulses

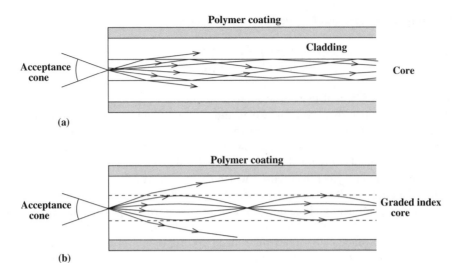

Figure 6.12   Light paths in two types of fibre: (a) step index, (b) graded index

spectral channels are used, or a wideband LED source, it is important to minimize the differences in travel time. We first examine the spread in travel time in a multi-mode fibre.

A simple way of appreciating this effect is illustrated in Figure 6.12, where the two rays represent two different modes; as we saw in the analysis of a slab guide, the higher the order of the mode, the larger the inclination of the equivalent rays to the axis. Here light pulses travel along the axis at velocity c/n_1, while an oblique ray at angle θ to the axis only progresses at the projected velocity $c\cos\theta/n_1$. For a step-index fibre, equations (6.38) and (6.40) allow the difference $\delta\tau$ in travel time between rays on-axis and rays at the maximum allowed angle, over a length L, to be expressed in terms of the numerical aperture NA as

$$\delta\tau = \frac{Ln_1}{c}\left(\frac{1}{\cos\theta_{\text{crit}}} - 1\right) \simeq \frac{Ln_1\Delta}{c} \simeq \frac{L(\text{NA})^2}{2n_1c}. \tag{6.44}$$

The bandwidth of the step-index fibre is limited by intermodal dispersion to about $1/\delta\tau = c/(Ln_1\Delta)$; for $\Delta \approx 10^{-3}$, this is about 100 MHz km/L. This demonstrates how the information-carrying capacity, which is proportional to the bandwidth, deteriorates with increasing length. In a graded-index (GRIN) fibre this difference in travel time is reduced or eliminated; the path of the more oblique ray, although longer, is mainly in glass with a lower refractive index, and the increased speed compensates for the extra path length (Figure 6.12). The sinuous path followed by a meridional ray in a GRIN fibre is the subject of Problems 6.3 and 6.4 at the end of this chapter.

Although multi-mode dispersion is significantly reduced in GRIN fibres, in practice it confines the use of multi-mode fibres to comparatively short or narrowband communication links and local area networks. Long-distance communications must use single mode fibres, where dispersion effects are smaller.

There are two distinct causes for wavelength dispersion in travel time in a single mode fibre; these are respectively the *material dispersion* (the intrinsic dispersion of the glass) and *waveguide dispersion*, which is inherent in the waveguide geometry. The travel time for a pulse in the fibre is determined by the group velocity

$$v_{\mathrm{g}} = \frac{\mathrm{d}\omega}{\mathrm{d}k}. \tag{6.45}$$

Consider first the effect of material dispersion, due to the variation of refractive index n with free space wavelength λ_0. The group velocity v_{g} is c/n_{g}, where n_{g} is the *group refractive index* given as a function of wavelength by differentiation as follows:

$$\omega = \frac{2\pi c}{\lambda_0}, \quad \frac{\mathrm{d}\omega}{\mathrm{d}\lambda_0} = -\frac{2\pi c}{\lambda_0^2} \tag{6.46}$$

$$k = \frac{2\pi n}{\lambda_0}, \quad \frac{\mathrm{d}k}{\mathrm{d}\lambda_0} = 2\pi\left(-\frac{n}{\lambda_0^2} + \frac{1}{\lambda_0}\frac{\mathrm{d}n}{\mathrm{d}\lambda_0}\right) \tag{6.47}$$

$$\frac{\mathrm{d}\omega}{\mathrm{d}k} = \frac{\mathrm{d}\omega}{\mathrm{d}\lambda_0}\frac{\mathrm{d}\lambda_0}{\mathrm{d}k} = \frac{c}{n - \lambda_0 \mathrm{d}n/\mathrm{d}\lambda_0} \tag{6.48}$$

giving

$$n_{\mathrm{g}} = n - \lambda_0\frac{\mathrm{d}n}{\mathrm{d}\lambda_0}. \tag{6.49}$$

The difference in travel time $\Delta\tau_{\mathrm{mat}}$ for light pulses centred at two wavelengths separated by $\Delta\lambda_0$, for a length L of fibre, is

$$\Delta\tau_{\mathrm{mat}} = \frac{L}{c}\frac{\mathrm{d}n_g}{\mathrm{d}\lambda_0}\Delta\lambda_0 \tag{6.50}$$

giving

$$\Delta\tau_{\mathrm{mat}} = \Delta(L/v_g) = -\frac{L}{c}\lambda_0\frac{\mathrm{d}^2 n}{\mathrm{d}\lambda_0^2}\Delta\lambda_0. \tag{6.51}$$

Derived from equation (6.51), the *material dispersion parameter*

$$\frac{1}{L}\frac{\Delta\tau_{\mathrm{mat}}}{\Delta\lambda_0} = -\frac{\lambda_0}{c}\frac{\mathrm{d}^2 n}{\mathrm{d}\lambda_0^2}$$

is quoted in units $\mathrm{ps\,nm}^{-1}\,\mathrm{km}^{-1}$.

The group velocity dispersion in transmission delay $(1/L)(\mathrm{d}\tau/\mathrm{d}\lambda_0)$ is shown in Figure 6.13 for wavelengths near 1 micron in silica glass. Fortunately the dispersion is very small at wavelengths close to the transmission band at 1.3 microns; this band has therefore been preferred for long-distance communications with a broad bandwidth. Techniques are, however, available for removing the effect of dispersion, and the low-loss band at 1.5 microns is also now in general use for links with bandwidths of several gigahertz. The importance of dispersion delay may be illustrated by considering a fibre optic cable at 0.85 microns, where Figure 6.13 gives a comparatively large delay of $98\,\mathrm{ps\,nm}^{-1}\,\mathrm{km}^{-1}$. An LED source (see Chapters 17 and 18) at this wavelength might have a

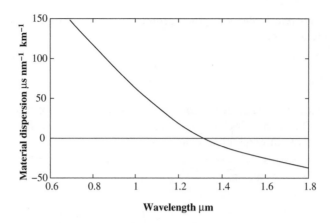

Figure 6.13   The dispersion in group velocity as a function of wavelength

spectral width of 50 nm, so that the dispersion in delay would be 5 ns per kilometre. In a communications link of 1000 km this would spread a narrow pulse to a width of 5 μs, limiting the bandwidth to about 0.2 MHz.

The second cause of dispersion in a single mode fibre is waveguide dispersion. This arises as a result of dependence of the group velocity on the ratio between the core radius and the wavelength. An exact analysis is complex, since the effect depends on the refractive index profile. For a step-index

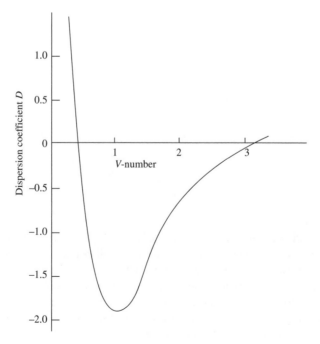

Figure 6.14   The waveguide dispersion coefficient D as a function of V-number

fibre the result is usually expressed as a delay $\Delta\tau_w$ related to the parameter V, introduced in Section 6.5 above, by[7]

$$\Delta\tau_w = \frac{L}{c}\left(\frac{\Delta\lambda_0}{\lambda_0}\right)(n_2 - n_1)DV \qquad (6.52)$$

where the dimensionless coefficient D is a function of the V-number, as shown in Figure 6.14. For a single mode fibre, assuming a source wavelength $\lambda_0 = 1\,\mu m$ and linewidth $\Delta\lambda_0 = 1\,nm$, the contribution to pulse broadening from waveguide dispersion is $\Delta\tau_w/L \simeq 2\,ps\,km^{-1}$. For silica and a wavelength $\lambda > 1.3\,\mu m$ the sign of waveguide dispersion is opposite to that for material dispersion. Then a dispersion-shifted fibre can be fabricated in which the zero-dispersion wavelength is able to be moved to a wavelength near $1.55\,\mu m$ where, as seen in Figure 6.9, the fibre loss is a minimum.

The importance of these various effects on the bandwidth of long-distance communications has prompted much analysis and experimentation. The results may be expressed as a product of bandwidth and fibre length. A typical step-index fibre bandwidth is less than 100 MHz for a length L of 1 km, due to multi-mode propagation. In a GRIN fibre, where the effects of multi-mode propagation are reduced, the bandwidth may be increased typically to 1 GHz km/L. The performance of single mode fibres can achieve in excess of 3 GHz km/L.

A single fibre can be used to carry simultaneously several signal channels on different optical wavelengths, giving an increased overall signal bandwidth. This technique of *wavelength division multiplexing* (WDM) requires optical filters at the transmitter and receiver.

6.9   Dispersion Compensation

Comparison of Figures 6.9 and 6.13 shows that dispersion is comparatively large at one of the wavelength bands with the lowest losses, i.e. at $1.55\,\mu m$. This band can nevertheless be exploited for long-distance broad-bandwidth communication by the use of compensating devices, which introduce a delay with equal and opposite dispersion. The delay is introduced by diverting the light signal into a short reflecting fibre whose effective length varies rapidly with wavelength.

Selective reflection of light in a narrow wavelength band can occur in an optical fibre if a periodic structure can be created along the length of the fibre. The effect is similar to the selective reflection of X-rays by a crystal lattice, which we analyse in Chapter 11. The periodic structure is an artificially constructed cyclic variation of refractive index, with a half-wavelength period and extending for many wavelengths. Figure 6.15 shows a typical plot of reflection coefficient against wavelength for such a structure. The maximum reflection occurs when the small reflected waves from each peak in refractive index add exactly in phase. The selectivity of the reflection, and the resemblance to Bragg reflection of X-rays in crystals, lead to the name *fibre Bragg gratings* for such devices, which are used as wavelength-selective reflection filters.

The next stage is to vary the spacing linearly along the fibre, so that different wavelengths are reflected at different distances. This then becomes a dispersive element, in which the travel time in a return journey depends on wavelength (Figure 6.16). The variation in reflection wavelength, and therefore in frequency, has become known as a *chirp*, and the device as a *chirped Bragg grating* (by analogy with the high-pitched sound emitted by some birds and bats which

---

[7]See A.H. Cherin, *An Introduction to Optical Fibers*, McGraw-Hill, 1985, p. 103.

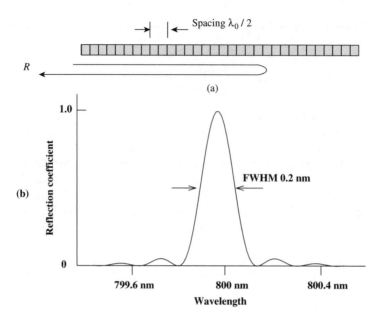

Figure 6.15   Bragg grating. (a) Periodically varying refractive index in a fibre. (b) The wavelength-dependent reflection coefficient

is accompanied by an increase in pitch). The magnitude of dispersion-induced delay can be made to match that of tens of kilometres of normal fibre in a device less than a metre in length overall.

Dispersion compensation may be achieved by the insertion of a length of fibre which has an equal and opposite dispersion–length product to that of the transmission fibre. The increase in the

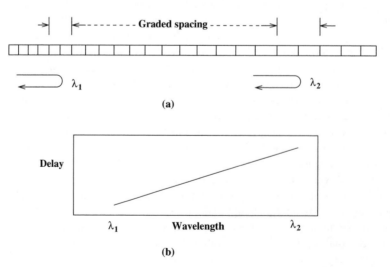

Figure 6.16   Dispersion compensation. (a) Grating with periodicity varying along the fibre. (b) The variation with wavelength in travel time for a return journey

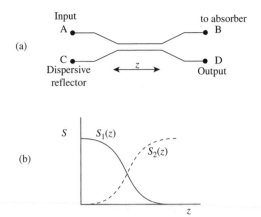

Figure 6.17 (a) Directional coupler used to connect a dispersive reflector into a fibre optic communication system. (b) The exchange of light signal between the coupled light guides. The signal S_1 entering at port A is divided between the two fibres as $S_1(z)$, $S_2(z)$, as a sinusoidal function of the length z

length of the fibre introduces additional loss which may be compensated by extra fibre amplification.

Such a dispersive element can be inserted in a fibre optic communication link by using a *directional coupler* shown diagrammatically in Figure 6.17. The coupler consists of two light guides running close together so that their evanescent fields overlap. A signal in one of the two guides is then progressively transferred to the other as shown in Figure 6.17(b). The length of the coupler is chosen so that half the signal from the input port A is transferred into the dispersive reflector at port C; the other half is lost in an absorber on port B. The returning signal from C, now compensated for dispersion, is then split again, half returning to the input fibre at A and half proceeding to the detector by the fourth port D. This is the required 'de-dispersed' signal.

Methods of imposing a periodic variation of refractive index along the dispersive fibre element are shown in Figure 6.18. A small but sufficient change in index (of order 1 in 10^4) can be induced in germanium-doped silica by subjecting the glass to a very intense flash of ultraviolet light. The periodic structure is created by forming an interference pattern within the fibre core from two coherent laser light beams incident at an angle to the fibre axis, as shown in Figure 6.18(a) (compare Figure 4.19). This is a form of holographic writing (Chapter 14). Ultraviolet light with $\lambda_{UV} \sim 240\,\text{nm}$ is required for the writing beams since the glass of the fibre is photosensitive in this region; this may be obtained by harmonic generation from a longer wavelength laser, or it may be directly produced by a pulsed ultraviolet excimer laser (Chapter 15). The design Bragg wavelength is dependent on λ_{UV} and on the angle between the interfering beams, and can be varied from λ_{UV} to longer wavelengths. The regularly spaced element in Figure 6.15 is used as a selective filter; by using a curved wavefront as in Figure 6.18b, the technique is extended to make the gradient in fringe spacing required for the dispersive element.

The holographic method of writing the grating, using a wavefront amplitude beam splitter to create the two beams, is technically difficult since the paths of the two writing beams must be kept constant to a fraction of a wavelength. An alternative method illustrated in Figure 6.18(c) is to use a form of transmission diffraction grating, termed a phase mask, which has surface relief acting as diffracting elements. This is placed a short distance from the fibre, so that an interference pattern develops between the two first-order diffracted beams at the fibre core.

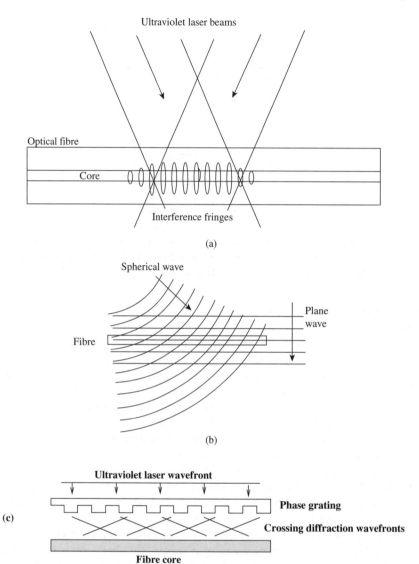

Figure 6.18 Creating the periodic variation in refractive index from an interference pattern: (a) with uniform spacing; (b) a curved wavefront providing a graded spacing; (c) a phase mask in contact with the fibre

6.10.   Modulation and Communications

Modulation of amplitude, phase, polarization or frequency of light in a fibre allows information to be encoded onto and, after transmission, extracted from the beam. Modulation of amplitude (or irradiance) is used in optical communications for analogue or digital encoding of information. It is also used in fibre optic sensors, to generate pulsed illumination, for mode locking of fibre lasers (Chapter 16) and in pulsed range-finding or LIDAR systems. Amplitude (or irradiance) modulation is

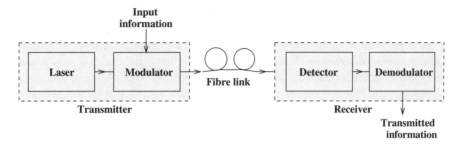

Figure 6.19 Fibre optic communication link

usually achieved by modulating the light source, e.g. semiconductor laser or LED, externally to the fibre. Direct modulation of lasers is described in Chapter 17.

The very high frequency of visible and near-infrared light $\sim 10^{14}$ Hz gives the potential for communication channels of very high information-carrying capacity, greater by a factor of about 10^4 than for microwave and radio wave frequencies. The optical fibre provides a transmission medium immune to environmental degradation and with gigahertz bandwidths. An outline of a fibre optic communication link is shown in Figure 6.19. In the transmitter information is impressed on a laser or LED beam by modulation. The modulated output is transmitted by the fibre to a receiver, most usually a photodiode, that generates an electric current in response to incident light (see Chapter 20). At the receiver the signal is amplified and demodulated to provide the output signal. Typically the signal is pulsed at a defined rate, known as the *bit rate*; an 'on' pulse represents a digital '1' and an 'off' pulse represents a '0'. The information-carrying capacity is determined by the bit rate, which is limited by the rate at which the signal can be switched between '1' and '0'. Following Section 4.12, Fourier analysis tells us that this in turn is determined by the bandwidth of the light signal.

At the very high bit rates used in fibre communications, it is essential to transmit sufficient photons per bit; random fluctuations in number within a single bit must be small enough that a '1' signal does not fall below a threshold and appear as a '0'. The fluctuations in photon number are random, and there is always a finite probability of such an error; thus for a data rate of 1 Gbit s^{-1} about 1000 photons per bit are required to achieve a bit error rate of 10^{-9}. In addition to the statistical random nature of the photodetection mechanism, the required photon number per bit is dependent on the electronic receiver sensitivity; since this is degraded by noise in the detector and amplifier, the photon number must be increased accordingly. In long-distance optical fibre communications the signal is steadily attenuated but can be restored using the erbium-doped fibre amplifier (EDFA) described in Chapter 15. A short length of EDFA can be spliced into the transmission fibre, with in-line optical isolators to prevent back reflection into the laser source; each such amplifier provides about 20 dB gain in the wavelength band 1530–1570 nm. Long-distance communication systems have been standardized internationally to have channel data rates of 155 and 622 Mbit s^{-1} and 2.5, 10 and 40 Gbit s^{-1}, to be followed in the future by a 160 Gbit s^{-1} system.

6.11 Fibre Optical Components

Devices which generate, amplify, control and detect light in fibre optic systems are described in many texts such as those listed in Appendix 5. Fibre optical components may be grouped into active and passive components. Active components require an external power source or signal to function: these include lasers, amplifiers, detectors, modulators, frequency shifters and polarization controllers. Passive components include connectors, couplers, directional couplers, filters, reflectors, isolators, polarizers and polarization retarders. Joins between fibres can be made with low loss down to 0.1 dB, and these may be permanent or demountable. A permanent joint is termed a splice and may be formed by fusing the fibre glass or by glueing. Demountable joints are formed by connectors, either by bringing two fibres in close proximity (butt joint) or by a lens arrangement to image one fibre end onto the other. Fibre beam splitters and combiners (such as the directional coupler described in Section 6.9 above) may be made by bringing the cladding of two fibres in close contact over a length of a few millimetres. The fibres are then fused by heating, while drawing the softened fibre to make a taper. Fibre optic switches selectively direct optical signals between different fibres. The switching

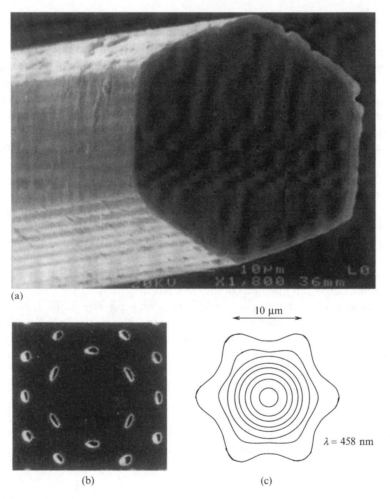

Figure 6.20   (a) Scanning electron micrograph of a cleaved end face of a large mode area photonic crystal fibre. The fibre shown here as a core diameter of 22.5 μm and a relative air hole diamter $d/\Delta = 0.11$, and is monomode at all wavelengths $\lambda > 458$ nm at least. (b) The central hole pattern. (c) Contour map of the near field irradiance distribution for the guided mode in the fibre shown in (a) at a wavelength of $\lambda = 458$ nm. The contours are plotted at 10% intervals in the modal field intensity distribution (J. C. Knight, University of Bath)

can be classified as optomechanical, electronic or photonic. Optomechanical switches include a mechanical movement of a component such as the fibre, prism or lens to deflect the beam. Electronic switches use an electro-optic effect and photonic switches use electro-optic or acousto-optic switches in an integrated optics crystal of lithium niobate ($LiNbO_3$). The fibre Bragg grating described in Section 6.9 for dispersion compensation has numerous other applications which utilize the high reflection coefficient and low insertion loss.

6.12 Hole-Array Light Guide; Photonic Crystal Fibres

The regularly spaced variation in refractive index along the length of a fibre, which is used in the Bragg filter (Section 6.9), may be extended to two or three dimensions to make a *photonic crystal lattice*. If the spacing between the refractive index discontinuities is comparable with the wavelength, the propagation of light waves within the lattice is subject to conditions similar to those of X-rays propagating in a crystal lattice, as described in Chapter 11. An example with practical use in fibre optics is a regular hexagonal array of airholes along the length of an otherwise uniform silica glass fibre. Such an array can be fabricated with a single missing hole, as seen in the micrograph of the cross-section (Figure 6.20). The intact hexagonal region then acts as a light guide, in which light is trapped as it is in the core of a conventional fibre.

For light waves travelling in the direction of the fibre axis, the array of holes has the effect of lowering the refractive index. The central hexagon therefore acts as the core, and the surrounding array as the cladding, as in the conventional guide where the difference in refractive index is achieved by chemical doping. The behaviour of the hole-array fibre depends on the ratio of the diameter d to the spacing Δ of the holes (Figure 6.21), but if this ratio is less than about 0.2 light will be propagated

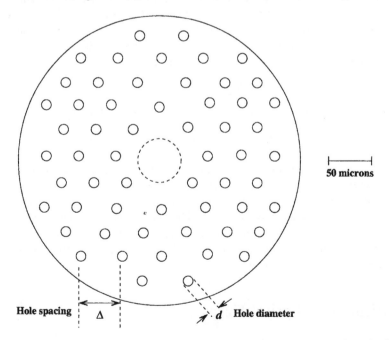

Figure 6.21 Photonic crystal fibre. The small circles represent airholes running the length of the fibre. The central broken circle is the fibre core, which may be solid or hollow

in only a single fundamental mode; this applies over a wide range of wavelengths. An illustration of the distribution of irradiance across the core is shown in Figure 6.20.

The two main types of photonic crystal fibre are illustrated in Figure 6.21. The index-guiding type described above has a solid core surrounded by cladding containing the array of airholes, while the air-guiding type has a hollow core. In this case the periodic pattern of airholes creates conditions in which confinement and guiding is provided by Bragg reflection.

An important advantage of such a fibre is that a single mode propagation can be achieved over a large cross-section of the core, so that the energy density can be very much lower than in the conventional fibre, avoiding the non-linear effects associated with the transmission of higher powers. Single mode operation has been demonstrated in a fibre with core diameter up to 50 free space wavelengths.

6.13   Optical Fibre Sensors

Optical fibre sensors have the general characteristics of high sensitivity, nonelectrical method of operation, immunity to electromagnetic interference, low power consumption, small size and weight and may readily be multiplexed. This has led to a remarkably widespread range of applications in the measurement of temperature, pressure, current and voltage, magnetic field, strain, chemical composition, position, movement and vibration, rotation, acoustic waves, microparticle sizing and fluid flow.

The main parameter which is exploited in fibre optic sensors is propagation time, as measured by the phase of an emergent wave. Light travelling in a fibre of length L undergoes a phase delay of $\phi = kn_{e}L$, where $k = 2\pi/\lambda_0$ and n_e is the effective refractive index of the core. This may be written $\phi = \beta L$ with $\beta = 2\pi n_e/\lambda_0$ being the propagation constant. The effective refractive index is the ratio of the propagation constant of light in a vacuum to that propagating in the LP_{01} modes (see Section 6.5). A change in ϕ can be related to a change in the fibre length or to a change in the fibre propagation constant, which might for example be induced by stress:

$$\Delta\phi = \beta\Delta L + L\Delta\beta. \tag{6.53}$$

The main contribution to $\Delta\beta$ is from a change in the refractive index, $\Delta\beta = \partial\beta/\partial n.\Delta n$. Then

$$\Delta\phi = \beta\Delta L + L\partial\beta/\partial n.\Delta n = 2\pi/\lambda_0(n\Delta L + L\Delta n). \tag{6.54}$$

Many sensors exploit a change in length L induced by stress; for example, the fibre can be wrapped tightly round a piezoelectric cylinder, which expands when a voltage is applied.

Changes $\Delta\phi$ in emergent phase are measured by comparison with a light wave from the same source propagated down an undisturbed fibre path. The two waves are combined in an interferometer, such as the Michelson, Mach–Zehnder and Sagnac interferometers, described in Chapter 9, in which fibres replace the open paths. It is possible using such an interferometer to measure phase changes equivalent to about 10^{-6} of a wavelength.

6.14   Fabrication of Optical Fibres

A crude glass fibre is easy to make by heating and softening the centre of a glass rod and pulling the ends apart. A useful fibre, with a constant diameter and consisting of a core and cladding, needs a more sophisticated technique. Two methods are available: drawing from a preformed thick rod, which already contains the core and cladding, and drawing from a concentric double crucible in which the two components are separately melted. The temperature (1700°C to 2000°C) at which pure silica has a workable viscosity is considerably greater than that for common glass (around 1000°C). In the double crucible method the properties of the core and cladding glasses must be reasonably well matched so that they can flow together and not be under stress when they cool. A controlled gradient of refractive index, which is essential in most applications, is obtained by drawing from a 'preform' which already contains the graded index.

A preform with graded index can be made by diffusing various dopants such as GeO_2 and P_2O_5 into silica glass. Both these dopants increase the refractive index; typically an addition of 10% raises the index from 1.46 to 1.47. The addition of fluorine lowers the refractive index. The dopants can be added by gas deposition, termed chemical vapour deposition, onto the inside of a tube of pure silica; the tube is collapsed later by melting to create the preform. The 'hole-array' fibre described in the previous section is drawn from a preform which is assembled from thin hexagonal rods, which themselves have been drawn from larger diameter hollow tubes. A single solid rod is packed into the centre to form the core. The preform is held at the top of a long fibre-pulling tower; when the preform end is heated it may be drawn down to a fibre and collected on a drum.

Drawing the fibre must be controlled so that the diameter is maintained within about ±2%. After the main drawing process an extra coating of some plastic material is added to protect the fibre. All these processes are adaptable to continuous operation, which typically runs for some days at a rate of up to $1\,m\,s^{-1}$, i.e. more than 80 km per day.

Problem 6.1

(i) Light from the end of an optical fibre in air forms a patch of light radius 3 cm on a screen 10 cm away. Find the numerical aperture. If the core refractive index is 1.5, find the fractional step in index $(n_1 - n_2)/n_1$ between the core and the cladding.

(ii) A single mode optical fibre has core diameter 4 μm and step in index $\Delta n/n = 2\%$. Using equation (6.36) find the minimum wavelength inside the core, λ_0/n_1, which will propagate in a single mode.

(iii) If the loss in a fibre is $0.5\,dB\,km^{-1}$ and there is an added loss of 1 dB at joints which are 10 km apart, find the necessary interval between amplifiers when the transmitter power is 1.5 mW and the detector level is 2 μW.

(iv) From equation (6.44) find the dispersion τ in propagation time for a fibre 5 km in length with ungraded index $n = 1.5$ and $\Delta = 1\%$. What bit rate B_T could be used in this length if $B_T = 1/2\tau$?

(v) The effect of material dispersion on pulse travel time in a fibre depends on the bandwidth of the light source. Find the dispersion $\Delta\tau_{mat}/L$ (i) for an LED with bandwidth 20 nm and (ii) for a laser with bandwidth 1 nm when using a glass fibre at $\lambda = 0.85$ μm where $\lambda^2 d^2 n/d\lambda^2 = 0.025$.

Problem 6.2

Suppose every meridional ray in a GRIN fibre follows successive arcs of a circle, with the centre of the circle displaced by distance r_1 from the axis. A complete oscillation about the axis occurs in length Λ. The refractive index on-axis is n_1. (a) Find the refractive index profile $n(r)$ for the fibre, (b) prove that r_1 is the same for all rays,

(c) find the arc's radius a, (d) find the turning point distance from the axis r_t, and (e) find the ray's axis-crossing angle ϕ_1.

Problem 6.3

In Problem 1.6 of Chapter 1, we asked the reader to derive an approximate equation $1/R = n^{-1}dn/dy$ for the radius of curvature of a light ray in a stratified optical medium (i.e. one in which the refractive index n depends only on one coordinate y). An axial plane through a fibre is nothing but a two-dimensional stratified medium in which the refractive index depends on the radial coordinate. In the following problems, we develop the exact expression for the curvature of meridional rays.

We let the y, z plane lie in the axial plane of interest, with the z axis the fibre's axis and the y axis along the radial direction (but, unlike a radius, y can go negative). The refractive index has the form $n = n(y)$.

(a) In classical mechanics, a one-dimensional system moving along the trajectory $q = q(t)$ can be described by a Lagrangian function $L = L(q, q', t)$ where $q' = dq/dt$. Requiring that the least action $S = \int L(q, q', t)dt$ be unchanged to first order in small virtual variations of $q(t)$, while holding the endpoints (q_1, t_1) and (q_2, t_2) fixed, leads to the Euler–Lagrange equation $d/dt(\partial L/\partial q') - \partial L/\partial q = 0$.

We can describe a ray in the y, z plane by $z = z(y)$, provided we limit ourselves to a segment of the path where y changes monotonically. Following Fermat's principle, express the optical path between two fixed endpoints (ct, where t is the travel time) in the form of an action-like integral, $ct = \int_{y_1}^{y_2} L(z, z', y)dy$, where $z' = dz/dy$. Integrate the Euler–Lagrange equation to prove that along the ray

$$n \cos \theta = \text{constant} \tag{6.55}$$

where θ is the angle the ray makes with the z axis.

(b) Suppose you modelled a planar slab as a series of thin layers parallel to the x, z plane, in each of which $n(y)$ is a constant. Explain how you would arrive at the relation (6.55) above by a simple argument.

(c) Suppose that a ray passes $y = 0$ at a positive angle θ_0, and moves toward positive y. If $n(y)$ decreases continuously as y increases, show that the ray will tip over to smaller angles. What condition must the refractive index satisfy in order that the ray will reach a turning point at some $y = y_t$, where it will move parallel to the z axis, and then curve back to $y = 0$ again?

(d) Prove that, regardless what form $n(y)$ has, no ray can describe an arc of a circle equal to or exceeding a semicircle.

Problem 6.4

(a) As in the previous problem, consider a fibre where y, z is an axial plane, the z axis is the fibre's axis, and the refractive index varies in the radial direction as $n = n(y)$. By using the result (6.55), find a differential formula for curvature of a meridional ray in that plane. The curvature of the ray is defined by $1/R$, where R is the instantaneous radius of the arc. Show that the curvature is equal to $d\theta/ds$, where s is the element of the arc length and θ the ray's angle from the z axis. Derive an exact result for $1/R$ in terms of θ, n and $dn(y)/dy$.

(b) Use the preceding to determine the curvatures for the cases of $\theta = 0$, and $\theta = \pm \pi/2$ (assuming dn/dy is finite everywhere). Explain why the result for $\theta = 0$ is surprising if we consider only the motion of a single ray.

Problem 6.5

Suppose every meridional ray in a GRIN fibre follows a sinusoidal path of wavelength Λ. Given n_1, r_t, and Λ, defined as in Problem 6.2, find (a) the refractive index profile $n(r)$ and (b) the ray's axis-crossing angle θ_1.

7 Polarization of Light

I will found my enjoyments on the affections of the heart, the visions of the imagination, and the spectacle of nature.

Etienne Louis Malus, born Paris, 23 July 1775, the discoverer of the polarization of light.

Michael Faraday...*'magnetised a ray of light'*.

Linearly polarized light is a surprisingly common phenomenon in everyday circumstances. It can be detected by the use of the 'polaroid' material in glasses used, especially by motorists, to reduce glare in bright sunlight. Polaroid transmits light which is plane polarized in one direction only, and absorbs light polarized perpendicular to this direction. Light reflected from any smooth surface, such as a wet road or a polished table top, is partially linearly polarized; this is easily demonstrated by rotating the polaroid glass, which gives a change in brightness according to the change in angle between the plane of polarization and the transmission axis of the polaroid. Complete polarization is found for reflection at a particular angle of incidence, the 'Brewster angle' (Section 5.3). The light of the blue sky, which is sunlight scattered through an angle, is also noticeably polarized. Insects such as honey bees can detect the polarization of the sky, and use its direction in relation to the Sun for navigation.

Circular polarization is less easily observed, but it is important in several phenomena concerned with the propagation of electromagnetic waves in anisotropic media, e.g. in the propagation of light in crystals such as quartz, and in some liquids such as sugar solution.

In this chapter we show how any state of polarization in a wave can be expressed in terms of elementary components, either plane or circularly polarized, and how the state of polarization may be changed by transmission through optically active materials.

7.1 Polarization of Transverse Waves

In the previous chapter we showed that light is a transverse electromagnetic wave. The *polarization* of the wave is the description of the behaviour of the vector $\mathbf{E}$ in the plane x,y, perpendicular to the direction of propagation z.

Optics and Photonics: An Introduction, Second Edition   F. Graham Smith, Terry A. King and Dan Wilkins
© 2007 John Wiley & Sons, Ltd

The plane of polarization is defined as the plane containing the propagation vector, i.e. the z axis, and the electric field vector.[1] The plane of polarization need not be constant at any point on the ray, but if the vector **E** does remain in a fixed direction, the wave is said to be *linearly* or *plane* polarized. If the direction of **E** changes randomly with time, the wave is said to be *randomly* polarized, or unpolarized. The vector **E** can also rotate uniformly in the plane x, y at the wave frequency, as observed at a fixed point on the ray; the polarization is then *circular*, either right- or left-handed depending on the direction of rotation. A combination of plane and circularly polarizations produces elliptical polarization. A *partially polarized* light wave can have a combination of polarized and unpolarized components.

It is convenient to consider a polarized wave as the sum of components E_x and E_y on the orthogonal axes x and y; these are two independent plane polarized waves with individual amplitudes and phases. The vector addition of these two components can produce any state of polarization of the actual electric field, depending on the relative phase of the two oscillations. If the two oscillations are in phase, the successive vector sums are as in Figure 7.1(a). The resultant is a vector at a constant angle to the x axis. Two plane polarized waves have combined to produce another wave which is also plane polarized.

If the two oscillations are in quadrature, so that their values of ϕ in equations of the form $E = a\cos(\omega t - kz + \phi)$ differ by $\pi/2$, the successive vector additions follow Figure 7.1(b). Here a is the amplitude on the x axis, and b the amplitude on the y axis. The resultant now rotates in the plane x, y, following a circle if the two amplitudes a, b are the same, or an ellipse if $a \neq b$. If the x oscillation is phase advanced on the y oscillation the rotation is anticlockwise; if it is retarded the rotation is clockwise. The two plane polarized waves have combined to produce a wave which is elliptically polarized.

We now show how any state of polarization in a ray of light can be described in terms of elementary components of plane polarized light. It is necessary to distinguish first between the polarized and unpolarized components of the ray. Note that when we consider states of polarization we are adding field components such as E_x and E_y, while for unpolarized light these components have a randomly changing phase relation and only the mean square of the amplitude is significant; it is the mean square of the amplitude which is proportional to the irradiance of the ray.

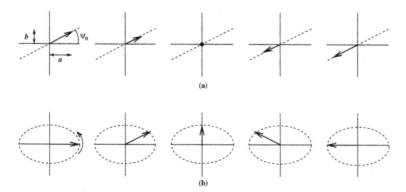

Figure 7.1   Vector addition of two oscillating electric fields, at successive moments through one half-cycle. The two fields have unequal amplitudes and are mutually perpendicular: in (a) they are in phase and in (b) they are in quadrature. The resultant oscillation is linearly polarized in (a) and elliptically polarized in (b)

---

[1]It is to be emphasized that it is the plane of vibration of the vector **E** which is taken to define the plane of polarization; there is ambiguity and confusion over this in some of the older literature.

We now analyze the polarized component of a wave propagating along the z axis in terms of linear components[2] with any phase difference ϕ:

$$E_x = a\cos(\omega t - kz)$$
$$E_y = b\cos(\omega t - kz + \phi). \tag{7.1}$$

If $\phi = 0$, E_x and E_y combine vectorially to give a resultant field **E** with magnitude $(a^2 + b^2)^{1/2}$, and at an angle ψ_0 to the x axis given by

$$\tan\psi_0 = \frac{b}{a}. \tag{7.2}$$

If $\phi = \pi/2$ and $a = b$, the wave is circularly polarized, and the vector **E** describes a uniform clockwise circular motion in space; the handedness reverses when $\phi = -\pi/2$ (see Problem 7.2). In optics[3] the hand is defined looking back towards the source of the ray, when the electric field vector in any one plane rotates clockwise for a right-handed circular polarization. In this case, when the thumb of the right hand points back towards the source, the fingers will curl in the clockwise direction. Figure 7.2 shows that in a right circularly polarized wave, at a fixed moment in time the tip of the vector **E** describes a right-handed screw in space.

More generally, adding orthogonal vector oscillations when ϕ is not zero or $\pi/2$ produces an elliptical polarization whose major axis does not lie along x or y. We add the real parts of the oscillations, with components E_x and E_y:

$$\frac{E_x}{a} = \cos\omega t, \quad \frac{E_y}{b} = \cos(\omega t + \phi)$$
$$= \cos\omega t\cos\phi - \sin\omega t\sin\phi. \tag{7.3}$$

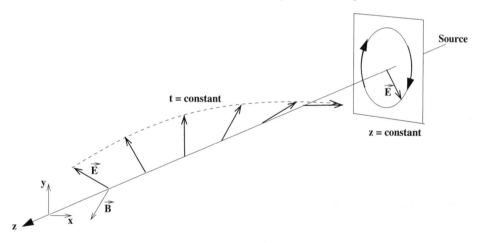

Figure 7.2 A right circularly polarized wave moving in the z direction

---

[2]The reader should be careful not to confuse the two dimensional plot of (E_x, E_y) with a phasor plot. In particular, a complex representation of (E_x, E_y) would require two separate phasors, one each for E_x and E_y.

[3]Unfortunately, this is the opposite of the convention for radio waves, where the handedness is defined looking *along* the direction of propagation.

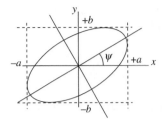

Figure 7.3 Elliptically polarized oscillation, combining linearly polarized oscillations on the x and y axes, with amplitudes a and b, and with an arbitrary phase difference

Eliminating the time factor,

$$\frac{E_y}{b} = \frac{E_x}{a}\cos\phi - \left[1 - \left(\frac{E_x}{a}\right)^2\right]^{1/2}\sin\phi, \tag{7.4}$$

$$\frac{E_y^2}{b^2} + \frac{E_x^2}{a^2} - \frac{2E_xE_y}{ab}\cos\phi - \sin^2\phi = 0. \tag{7.5}$$

This equation describes an ellipse, as in Figure 7.3. At any time the resultant field vector **E** reaches to a point on the ellipse, moving round as time progresses. The ellipse is contained in a rectangle with sides $2a$ and $2b$. The position of the major axis of the ellipse, at an angle ψ to the x axis, is found as follows.

The amplitude $E_x^2 + E_y^2$ is maximum on the major axis. Therefore at this point

$$E_x\,dE_x + E_y\,dE_y = 0. \tag{7.6}$$

Also from equation (7.5), for a fixed value of ϕ:

$$\left(\frac{E_x}{a^2} - \frac{\cos\phi}{ab}E_y\right)dE_x + \left(\frac{E_y}{b^2} - \frac{\cos\phi}{ab}E_x\right)dE_y = 0. \tag{7.7}$$

On the major axis we have $\tan\psi = E_y/E_x$, so that combining equations (7.6) and (7.7)

$$\frac{1}{a^2} - \frac{\cos\phi}{ab}\tan\psi = \frac{1}{b^2} - \frac{\cos\phi}{ab}\cot\psi. \tag{7.8}$$

Since $\tan\psi - \cot\psi = -2\cot 2\psi$, we find:

$$\tan 2\psi = \frac{2ab\cos\phi}{a^2 - b^2}. \tag{7.9}$$

The ratio of the maximum and minimum axes of the ellipse may be found by rotating the coordinate axes through the angle ψ. The axial ratio may in this way be shown to be given by R in

$$\frac{R}{1 + R^2} = \frac{ab}{a^2 + b^2}\,|\sin\phi\,|. \tag{7.10}$$

7.2 Analysis of Elliptically Polarized Waves

By a suitable choice of amplitudes, a, b, and relative phase ϕ, the vibration ellipse of Figure 7.3 may be set with any axial ratio and with the major axis at any position angle. It follows that any elliptical oscillation or elliptically polarized wave may be analyzed mathematically into two linearly polarized components at right angles, using axes at any angle to the major axis of the ellipse. The relative phase depends on this position angle; it is only important to note that the two components are in quadrature if they are aligned with the major and minor axes of the ellipse. The analysis may be done experimentally by using a device which will change the relative phases of two orthogonal components of a ray by a known amount. A particularly useful device is the *quarter-wave plate*, a transparent slice of anisotropic crystalline material in which the wave velocity differs between two perpendicular directions by such an amount that one component takes a quarter period longer to propagate than the other (see Section 7.8). This, in combination with a polaroid analyzer, can be used for analysis of elliptically polarized light by turning the axes of the quarter-wave plate until a position is found where the emergent light is fully plane polarized.

A combination of an analyzer and a quarter-wave plate can also be used to determine the state of polarization of an arbitrarily polarized wave. For this purpose it is convenient to think of the most general state of polarization as a combination of elliptical and random polarization. The procedure is as follows:

1. Using the analyzer discussed above, the amount of plane polarized light can be determined by rotating this analyzer. The remaining light when the analyzer is set to admit a minimum irradiance may be circularly or randomly polarized.

2. The quarter-wave plate is inserted before the analyzer, and the orientations of both are changed independently to produce a minimum irradiance. Elliptically polarized light will give zero irradiance in these circumstances, since the quarter-wave plate when properly oriented will turn it into plane polarized light which will be rejected by the analyzer. Any remaining light at minimum irradiance must have a random polarization.

This procedure is used in the *ellipsometer*, an instrument designed to measure certain characteristics of a surface by observing its polarizing effect on reflected light. This has a particular importance in measuring the thickness and composition of thin deposited films, in which the polarizing effect is wavelength dependent (see the discussion on thin films in Chapter 8).

7.3 Polarizers

Light from most sources is unpolarized. It can be converted into fully polarized light by the removal of one component, usually either plane or circularly polarized.

A simple example is the wire grid polarizer used originally for radio waves, but which can be demonstrated to work for infrared light at about 1 µm wavelength. This is simply a parallel grid of thin conducting wires whose diameter and spacing are small compared with the wavelength. In such a grid only the electric field component perpendicular to the wires can exist; for the component polarized parallel to the wires the grid acts as a reflecting plane. Only the plane of polarization perpendicular to the grid is transmitted (Figure 7.4). Such devices are called *polarizers* when they are used to create polarized light, or *analyzers* when they are used to explore the state of polarization, as

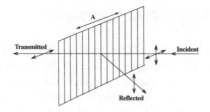

Figure 7.4   A wire grid polarizer. The wires are spaced at less than a wavelength apart; light polarized parallel to the wires is reflected. The grid's transmission axis A is normal to its wires

in the previous section. The action of a linear analyzer on a linearly polarized wave is shown in Figure 7.5. Light from a linear polarizer with transmission axis A_1 is incident normally on a linear analyzer with axis A_2 at angle θ to the plane of polarization defined by A_1. The amplitude of the transmitted wave is reduced by the factor $\cos\theta$, giving *Malus's law* for the irradiance I

$$I(\theta) = I_0 \cos^2 \theta. \tag{7.11}$$

The metallic wire grid has an analogue in the aligned molecular structure of a polaroid sheet. This is a stretched film of polyvinyl alcohol containing iodine; the iodine is in aligned polymeric strings which absorb light polarized parallel to the direction of alignment. A general class of crystals, including the well-known material tourmaline, have the same property of selectively absorbing one plane of polarization. These materials are often referred to as *dichroics*.[4]

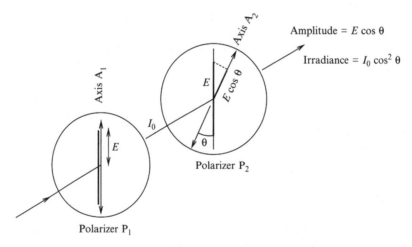

Figure 7.5   Malus's law. Light from a polarizer with transmission axis vertical falls on a linear analyzer with axis at angle θ. The irradiance is reduced by $\cos^2 \theta$

[4]The term dichroic originated in mineralogy, where it referred to the different *colours* of two polarized rays emerging from birefringent crystals. The colours arise from selective absorption; if the absorption over a large wavelength range is much larger in one polarization than the other, the material is dichroic in the sense used in optics.

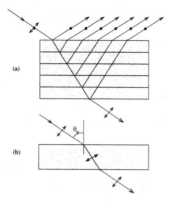

Figure 7.6 Reflection and transmission at the Brewster angle θ_p: (a) reflection at a pile of plates; (b) transmission at a Brewster window

Light can also be made linearly polarized by reflection at the Brewster angle from a dielectric surface (Chapter 5). The reflection coefficient is, however, often inconveniently low; this can be overcome by the *pile-of-plates* polarizer shown in Figure 7.6(a), where reflections from multiple layers of glass or other dielectric add to give almost complete reflection. The high transmission coefficient for the other polarization at the Brewster angle is used to make perfect non-reflecting windows, as shown in Figure 7.6(b). Such *Brewster windows* are used in gas lasers (Chapter 15), where light from the laser cavity makes repeated passes through windows placed in front of mirrors at either end of the cavity. The emerging laser light is usually fully linearly polarized in a plane determined by the Brewster windows.

7.4   Liquid Crystal Displays

The familiar digital displays of pocket calculators and wrist watches employ a form of electro-optic modulator shown in Figure 7.7. The active element is a liquid crystal, in which long organic molecules align naturally parallel to a liquid–glass interface, but can be realigned by an electric field. The natural alignment is determined by conditions at the surface, so that a twisted structure can be produced in a thin cell, as in Figure 7.7(a). The effect is now to rotate the plane of polarization, as in the optical activity of a quartz crystal (Section 7.9), and in the arrangement of an LCD the reflected light from a mirror behind the cell is rejected by a polarizer. Application of an electric field rearranges the molecules as in Figure 7.7(b), removing the spiral structure and allowing light to pass both ways through the cell.

An illustration of the use of the liquid crystal cell as a reflective display device is shown in Figure 7.8. The cell LC is placed between two polarizers, which are aligned to correspond to the directions of the molecular ordering on the two surfaces, and in front of a mirror. Incident light polarized by the first polarizer has its polarization direction rotated by the cell, passes through the second polarizer, is reflected by the mirror and again passes through both polarizers. With no electric field on the cell the image therefore appears bright. When a field is applied the direction of polarization of light is not rotated, light cannot travel in either direction through the cell and the image appears dark.

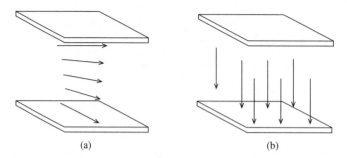

Figure 7.7   Molecular alignment in an LCD. The cell is typically about 5 µm thick: (a) with no field between the electrodes, the molecules align with the surface structures, which are arranged to give a twisted molecular structure; (b) an electric field aligns the molecules and removes the twist

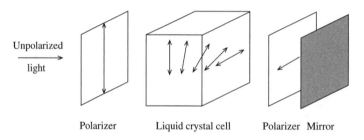

Figure 7.8   An element of an LCD using reflected light

7.5   Birefringence in Anisotropic Media

In an *anisotropic* medium, and in particular many transparent crystals, the phase velocity of light varies with crystal orientation. The refractive index is then not a single number, but a quantity which varies with direction; it may be represented by a surface such as an ellipsoid. A further complication is that the refractive index in any one direction may be a function of the state of polarization of the light wave, so that a ray entering a crystal with random polarization will be split into two components which will be refracted differently. The medium is then said to be *doubly refracting* or *birefringent*. The effect (Figure 7.9) in calcite (Iceland Spar) was described by Newton: 'If a piece of this crystalline Stone be laid upon a Book, every Letter of the Book seen through it will appear double, by means of a double refraction' (*Opticks*, Book 3). Newton also recognized that the two rays differed in some intrinsic geometric property. He said they had 'sides'; we now say they are plane polarized.

The refractive index of a crystal depends generally on the direction of polarization in relation to the crystal structure. Not all crystals behave in this way; those with a highly symmetric form, such as sodium chloride, are not birefringent. Birefringence is also observed as a difference between the propagation of two hands of circular polarization, as in quartz crystals and in solutions of some *optically active* substances such as sugar (Section 7.9 below).

The crystal structure of calcite ($CaCO_3$) has a single axis of symmetry, which coincides with an *optic axis*; this is also an axis of symmetry for the refractive index surface. (Calcite is a *uniaxial* crystal; other crystals have more complex symmetries and their birefringence is correspondingly

Figure 7.9   Double refraction in Iceland Spar. (©Andrew Alden, geology.about.com., reproduced with permission of the author)

more complex.) A point source of unpolarized light within a calcite crystal will generate two wavefronts, as shown in Figure 7.10. One is spherical, and is known as the *ordinary wave*, or *o-wave*; the other forms an oval,[5] and is known as the *extraordinary wave*, or *e-wave*. (Note that these are wavefronts; distances from the central point source are proportional to the phase velocity, $v = c/n$, and thus vary inversely with the refractive index, n. The refractive index surface for the extraordinary ray is a prolate, not oblate, ellipsoid. The difference in refractive indices $n_e - n_o$ is negative for calcite, which is classified as *negative uniaxial*.) It is their different (orthogonal) polarizations relative to the crystal structure that distinguish the two waves, causing them to interact differently with the molecules and thus to propagate at different velocities. In the o-wave the electric vector is everywhere normal to the optic axis, and in the e-wave it has a component parallel to the optic axis.

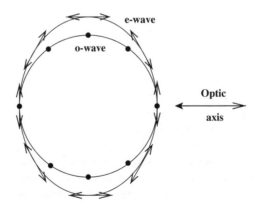

Figure 7.10   Birefringence in a uniaxial crystal: ordinary and extraordinary wavefronts radiating from a point source in the crystal. The electric field of the e-wave is shown by the double-headed arrows; the polarization of the o-wave is out of the plane of the diagram. For propagation perpendicular to the optic axis the refractive index depends on the orientation of the vector **E** in relation to the axis; both waves travel at the same velocity along the axis

Table 7.1   Indices of refraction at $\lambda_0 = 589.3$ nm

Substance	n_o	n_e
Isotropic		
rock salt (NaCl)	1.544	
sylvite (KCl)	1.4900	
fluorite/fluorspar (CaF_2)	1.434	
Uniaxial		
calcite/calcspar ($CaCO_3$)	1.658	1.486
quartz (SiO_2)	1.544	1.553
rutile (TiO_2)	2.613	2.909

Refractive indices of several crystalline substances are shown in Table 7.1.

7.6   Birefringent Polarizers

An unpolarized ray incident on a face of a calcite crystal will in general be refracted into two rays, propagating in different directions within the crystal, and with orthogonal plane polarizations. This separation is used in various forms of *birefringent polarizer*. In the *Nicol prism*, made of calcite (Figure 7.11), the two rays are separated at a layer of transparent cement within the calcite, arranged so that one of these rays is removed by total internal reflection. The single emergent ray is accurately linearly polarized. Figure 7.11 also shows the more commonly used *Glan–Foucault prism* in which there is no deviation at the first face, and the transmitted ray is undeviated overall. The space between the two prisms is usually filled with air (the *Glan–air polarizer*); polarization selection then requires simply that the prism angle θ is related to the two refractive indices n_o and n_e by

$$\frac{1}{n_o} < \sin\theta < \frac{1}{n_e} \tag{7.12}$$

so that the ordinary ray alone will be removed by total internal reflection. An increased field of view is obtained by cementing the prisms together (the *Glan–Thompson polarizer*), but the air-spaced version can handle larger irradiances, as is often required in high-powered laser systems.

In the Wollaston prism (Figure 7.12) the optic axes of the two components are orthogonal, as shown. The two polarized rays are both transmitted, but they are separated by a sufficient angle for them to be treated individually; for example, they may go to separate photoelectric detectors. Such devices are used in optical telescopes for measuring the plane polarized component of starlight. The advantage over the Nicol and Glan–Foucault prisms is symmetry: both components are transmitted through similar paths in the crystal, and any absorption is the same for both.

---

[5]This wavefront surface resembles an ellipsoid, but is actually a fourth-degree oval. However, the corresponding wave vector (**k**) surface is an ellipsoid.

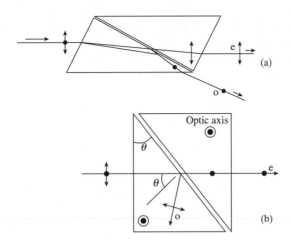

Figure 7.11 (a) Nicol and (b) Glan–Foucault prisms. Selective reflection in the Nicol prism is obtained by using a transparent cement between the parts of the calcite crystal with refractive index 1.52, intermediate to the index for the e-ray (1.49) and the index of the o-ray (1.66). In the Glan–Foucault polarizer the two prisms are spaced by an air gap. The calcite prisms require an angle $\theta = 38° - 42°$

7.7 Generalizing Snell's Law for Anisotropic Materials

Cnsider a monochromatic, plane wave incident on the flat face of a transparent crystal. Boundary conditions for the incident, reflected and transmitted waves require that $\exp[i(\mathbf{k} \cdot \mathbf{r} - \omega t)]$ is the same for all three waves. (We assume that the origin of coordinates is located in the interface.) But since the frequency is the same for all three waves, this implies that

$$\mathbf{k}_i \cdot \mathbf{r} = \mathbf{k}_r \cdot \mathbf{r} = \mathbf{k}_t \cdot \mathbf{r}, \tag{7.13}$$

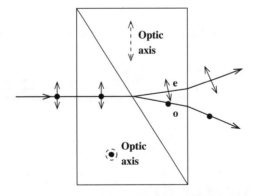

Figure 7.12 Wollaston prism. The two parts are made of a birefringent material such as quartz (a positive uniaxial substance, with $n_e - n_o > 0$), with the optic axis in the two directions orthogonal to the incident ray. The e- and o-rays are separated at the interface. In quartz the refractive indices for rays normal to the optic axis are $n_o = 1.544$ and $n_e = 1.553$ (refractive indices vary with wavelength: these values are for the Fraunhofer sodium D-line at 589 nm; see Table 2.1). Calcite is also used; it has a larger difference in refractive indices, and separates the rays by a larger angle

where **r** is any displacement within the surface. Since the three wave vectors have the same vector component along the surface, we deduce that the reflected and transmitted rays lie in the plane of incidence, which is defined by $\mathbf{k}_i$ and the surface normal. Suppose the refractive index changes from n_1 to n_2 (possibly anisotropic). If each **k** makes an angle θ to the normal, then since $k = \omega n/c$, equation (7.13) reduces to

$$n_1 \sin \theta_i = n_1 \sin \theta_r = n_2 \sin \theta_t. \tag{7.14}$$

The first two terms give $\theta_r = \theta_i$, the law of reflection. The final term provides a generalization of Snell's law valid for anisotropic materials.

Consider light entering a uniaxial crystal. The last member of equation (7.14) can refer equally well to the o- or the e-wave. For the latter case, n_2 becomes a variable function of direction, and solving equation (7.14) for the transmitted angle is often non-trivial.

Example. Consider a uniaxial crystal cut parallel to its optic axis. Light is incident on the crystal at an angle $\theta (= \theta_i)$ in the plane containing the optic axis. We shall find the angle $\phi (= \theta_t)$ at which the e-wave is transmitted. For definiteness, let the z axis be normal to the surface, and the y axis lie along the optic axis. Note that in the plane of incidence, the e-wave has a variable value n_ϕ of its refractive index and this lies along an ellipse. This has the equation

$$(n_y/n_o)^2 + (n_z/n_e)^2 = (n_\phi \sin \phi/n_o)^2 + (n_\phi \cos \phi/n_e)^2 = 1, \tag{7.15}$$

where n_y, n_z are the Cartesian coordinates of the ellipse, and n_o refers to the ordinary ray (not to vacuum). Show that

$$\tan \phi = \frac{n_o}{n_e} \frac{\sin \theta}{\sqrt{n_o^2 - \sin^2 \theta}}. \tag{7.16}$$

Solution. Equation (7.14) tells us that $\sin \theta = n_\phi \sin \phi$. Solving equation (7.15) for $\sin \phi$, we get

$$\sin^2 \theta = n_\phi^2 \sin^2 \phi = \frac{\sin^2 \phi}{(\sin^2 \phi/n_o^2 + \cos^2 \phi/n_e^2)} = \frac{n_o^2 \tan^2 \phi}{(\tan^2 \phi + n_o^2/n_e^2)}. \tag{7.17}$$

Solving this for $\tan \phi$ yields equation (7.16).

7.8   Quarter- and Half-Wave Plates

The polarization state of light can also be analyzed using components sensitive to circular and other states of polarization; many of these components depend on phase changes during propagation in anisotropic media rather than on selective refraction or absorption.

Consider the propagation of plane polarized light incident normally on a parallel-sided thin slab of crystal such as calcite, cut so that the optic axis is in the plane of the slab (Figure 7.13). The component of the wave with electric vector parallel to the optic axis travels faster than the perpendicular component (assuming $n_e < n_o$), thereby defining fast and slow axes in the slab; these are the e- and o-waves (for extraordinary and ordinary) introduced in the previous section. These two components are in phase as they enter the slab, but the e-wave travels faster and a phase difference δ grows as they travel. If the two refractive indices are n_e and n_o, the phase difference after a distance d is

$$\delta = \frac{2\pi}{\lambda_0}(n_e - n_o)d \quad \text{radians,} \tag{7.18}$$

where λ_0 denotes the vacuum wavelength.

Crystal slabs giving $|\delta| = \pi/2$ and π are known as *retarders*, either quarter-wave or half-wave plates, respectively. We have already shown in Section 7.2 how a quarter-wave plate can be used in the analysis of polarized light. These components have important uses in manipulating polarization in optical systems.

The amplitudes of the two components are $A\cos\theta$ and $A\sin\theta$, where θ is the angle between the incident plane of polarization and the optic axis and A is the amplitude in the incident ray. Combining these two again with phase difference δ produces a different state of polarization (Figure 7.14); for $\delta = \pi/2$ this is an ellipse with a principal axis along the optic axis, while for $\delta = \pi$ the polarization is again plane but rotated by angle 2θ. In the particular case where $\theta = 45°$ the ellipse becomes a circle, and circularly polarized light is produced; the opposite hand of circular polarization is obtained when $\theta = 135°$. For $\delta = \pi$ and $\theta = 45°$, the plane of polarization is rotated by 90°. The successive changes in polarization are shown in Figure 7.15.

7.9 Optical Activity

In many anisotropic media the refractive index is different for the two hands of *circular* polarization. This form of birefringence, known as *chirality*, has an important effect on plane polarized light: a beam of linearly polarized light passing through such an optically active medium

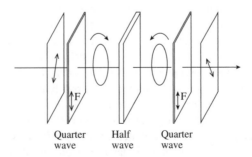

Quarter wave Half wave Quarter wave

Figure 7.13 A plane polarized wave at 45° passes through a quarter-wave plate and becomes circular; the hand is reversed by a half-wave plate; and the orthogonal plane is produced by a second quarter-wave plate. The fast axis is indicated by F

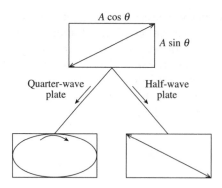

Figure 7.14 A plane polarized wave at angle θ to the fast direction of a quarter-wave and a half-wave plate, converted into an elliptical and a plane polarized wave respectively

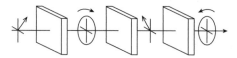

Figure 7.15 Changes of polarization through a series of quarter-wave plates

emerges with its plane of polarization rotated through an angle proportional to the path length in the medium. The same effect occurs in the propagation of linearly polarized radio waves through the ionized hydrogen of interstellar space, where the birefringence is due to the interstellar magnetic field (see Chapter 19).

The common characteristic of these particular media is that light of different hands of circular polarization travels with different velocities, so that if the two hands are propagated together, their phase relation changes progressively along the line of sight. The addition of two circularly polarized oscillations or waves, of equal amplitudes but with opposite hands, results in a plane polarized wave whose plane depends on the relative phase of the two circular oscillations. The difference in propagation velocity therefore results in a rotation of the plane of polarization, as observed for example in the propagation of plane polarized light in sugar solution.

The double refraction, or *birefringence*, of a crystal depends on anisotropy in its structure. In some crystals, notably crystalline quartz (but not fused quartz), and in the molecules of many organic substances such as sugar, the molecular structure is a helix. The refractive index for circularly polarized light then depends on the relation between the hand of polarization and the hand of the spiral structure. The phenomenon is useful both in the manipulation of polarization and in elucidating the molecular structure of so-called *optically active* materials.

A plane polarized ray traversing an optically active crystal must then be thought of as the combination of two circularly polarized rays, which travel at different speeds. Their relative phases, which determine the position angle of the linear polarization, change along the ray path, and the plane rotates.

The rate of rotation of the plane of polarization in quartz, for light propagated along the optic axis, is 21° per millimetre. In liquids the rotation is normally less, but for the so-called *liquid crystals*, which are liquids in which molecules are partially oriented as in a crystal lattice, the rotation may be very much larger. Cholesteric liquid crystals, in which the molecules have a helical structure, have rotations up to $40\,000°$ per mm. These are used in LCDs (Section 7.4).

7.10   Formal Descriptions of Polarization

The analysis of fully polarized light in terms of orthogonal components, either linear or circular, lends itself to a simple mathematical formulation in terms of *Jones vectors*. In this analysis, the orthogonal linear components of the electric field which determine the state of polarization are written in column form

$$\begin{bmatrix} E_{0x} \\ E_{0y} \end{bmatrix} = \begin{bmatrix} A\exp(i\phi_x) \\ B\exp(i\phi_y) \end{bmatrix}$$

where the non-negative numbers A, B are the magnitudes of the complex amplitudes E_{0x}, E_{0y} and ϕ_x, ϕ_y are their phases.[6]

The Jones vector is a simplified and normalized form of this. For example, for $A = B$ and $\phi_x = \phi_y$, corresponding to a linear polarization at $45°$, it is written

$$\begin{bmatrix} 1 \\ 1 \end{bmatrix}.$$

A phase difference appears as an exponential; for example, a circular polarization in which the y component leads in time by $90°$ is written

$$\begin{bmatrix} 1 \\ \exp(-i\pi/2) \end{bmatrix}$$

or simply

$$\begin{bmatrix} 1 \\ -i \end{bmatrix}.$$

The Jones vector of the sum of two *coherent* light beams is the sum of their individual Jones vectors.

The advantage of this formulation is that devices such as polarizers and wave plates can be specified by simple 2×2 matrices, the *Jones matrices*, and their operation on a polarized wave is found by matrix multiplication. For example, the product

$$\begin{bmatrix} a_{11} & a_{12} \\ a_{21} & a_{22} \end{bmatrix} \begin{bmatrix} E_{0x} \\ E_{0y} \end{bmatrix}$$

gives the modified field components

$$\begin{aligned} E'_{0x} &= a_{11}E_{0x} + a_{12}E_{0y} \\ E'_{0y} &= a_{21}E_{0x} + a_{22}E_{0y}. \end{aligned} \tag{7.19}$$

---

[6]The variable part of the phasors that we have factored out is assumed to have the form $\exp[i(kz - \omega t)]$.

In equations (7.19), the unprimed and primed components represent the components of the complex electric field, respectively, before and after passing through a given device. The reader must understand that, in contrast to orthogonal rotation of axes, *the x and y axes are invariant*; they do not change from "before" to "after".

As an illustration, consider a polarizer with a transmission axis in the x, y plane with an arbitrary direction given by the unit vector $\hat{p}$. If a plane wave with electric field amplitude $\boldsymbol{E}_0$ is incident normally on this, the electric field vector along the transmission axis is $(\boldsymbol{E}_0 \cdot \hat{p})\hat{p}$. The projections of this onto the x and y axes yield the components E'_{0x} and E'_{0y}, which can be expanded in terms of the original electric field:

$$\boldsymbol{E}'_{0x} = (\boldsymbol{E}_0 \cdot \hat{p})\hat{p} \cdot \hat{x} = (E_{0x}\hat{x} \cdot \hat{p} + E_{0y}\hat{y} \cdot \hat{p})\hat{p} \cdot \hat{x}$$
$$\boldsymbol{E}'_{0y} = (\boldsymbol{E}_0 \cdot \hat{p})\hat{p} \cdot \hat{y} = (E_{0x}\hat{x} \cdot \hat{p} + E_{0y}\hat{y} \cdot \hat{p})\hat{p} \cdot \hat{y}.$$

(7.20)

The polarizer's Jones matrix can be read off as

$$\begin{bmatrix} (\hat{p} \cdot \hat{x})^2 & \hat{p} \cdot \hat{x}\hat{p} \cdot \hat{y} \\ \hat{p} \cdot \hat{y}\hat{p} \cdot \hat{x} & (\hat{p} \cdot \hat{y})^2 \end{bmatrix}.$$

For example, if the transmission axis is rotated by 45° from the x axis, $\hat{p} = (1/\sqrt{2})(\hat{x} + \hat{y})$, the Jones matrix is

$$\frac{1}{2}\begin{bmatrix} 1 & 1 \\ 1 & 1 \end{bmatrix}.$$

The action of a series of components can be found from matrix multiplication, giving a single 2×2 matrix to represent the whole system. Matrices representing some of the polarizers and retarders dealt with in this chapter are tabulated below.

Jones matrices

Linear polarizers:

$$\text{horizontal}\begin{bmatrix} 1 & 0 \\ 0 & 0 \end{bmatrix} \quad \text{vertical}\begin{bmatrix} 0 & 0 \\ 0 & 1 \end{bmatrix} \quad 45° \; \frac{1}{2}\begin{bmatrix} 1 & 1 \\ 1 & 1 \end{bmatrix}.$$

Circular polarizer:

$$\text{right-hand} \; \frac{1}{2}\begin{bmatrix} 1 & i \\ -i & 1 \end{bmatrix} \quad \text{left-hand} \; \frac{1}{2}\begin{bmatrix} 1 & -i \\ i & 1 \end{bmatrix}.$$

Polarization plane rotator:

$$\text{rotation angle } \beta \quad \begin{bmatrix} \cos\beta & -\sin\beta \\ \sin\beta & \cos\beta \end{bmatrix}.$$

Phase retarders: F is the fast axis of a quarter-wave plate (QWP),

$$\text{QWP, F vertical} \qquad \exp(i\pi/4)\begin{bmatrix} 1 & 0 \\ 0 & -i \end{bmatrix}$$

$$\text{QWP, F horizontal} \qquad \exp(i\pi/4)\begin{bmatrix} 1 & 0 \\ 0 & i \end{bmatrix}.$$

Example. Use the Jones method to find the result when a horizontal linear polarizer acts on: (a) a wave polarized in the x,y plane at angle β to the x axis; and (b) circularly polarized waves of either hand. In each case, compare the initial and final irradiance, proportional to $|E_{0x}|^2 + |E_{0y}|^2$.

Solution. (a) Horizontal linear polarizer acts on rotated linearly polarized wave:

$$\begin{bmatrix} 1 & 0 \\ 0 & 0 \end{bmatrix} \begin{bmatrix} \cos\beta \\ \sin\beta \end{bmatrix} = \begin{bmatrix} \cos\beta \\ 0 \end{bmatrix}.$$

The result is a wave polarized along the x axis, but with irradiance reduced by a factor of $\cos^2\beta$.
(b) Horizontal linear polarizer acts on circularly polarized wave:

$$\begin{bmatrix} 1 & 0 \\ 0 & 0 \end{bmatrix} \begin{bmatrix} 1 \\ \mp i \end{bmatrix} = \begin{bmatrix} 1 \\ 0 \end{bmatrix}.$$

The final wave is linearly polarized along the x-axis, but with half the original irradiance of the incident wave.

The Jones vectors apply only to fully polarized light. For partial polarization the appropriate analysis uses the *Stokes parameters*, which are functions of irradiance rather than fields. If the irradiance is measured through four different analyzers, (i) passing all states (but transmitting only half of each), (ii) and (iii) linear analyzers with axes at angles $0°$ and $45°$, (iv) a circular analyzer, which measure respectively I_0, I_1, I_2, I_3, the Stokes parameters are

$$\begin{aligned}
S_0 &= 2I_0 \\
S_1 &= 2I_1 - 2I_0 \\
S_2 &= 2I_2 - 2I_0 \\
S_3 &= 2I_3 - 2I_0.
\end{aligned} \tag{7.21}$$

For partially polarized light with polarized and unpolarized components of irradiance I_p and I_u we define the *degree of polarization P* as

$$P = \frac{I_p}{I_u + I_p} = \frac{(S_1^2 + S_2^2 + S_3^2)^{1/2}}{S_0}. \tag{7.22}$$

The Stokes parameters for two *incoherent* light beams are the sum of their individual Stokes parameters.

Example. What are the polarization states of the two independent (incoherent) light beams with Stokes parameters $(1, -1, 0, 0)$ and $(3, 0, 0, -2)$, and of their sum?

Solution. The first is fully linearly polarized (vertically, i.e. at $90°$, $P = 1$); the second is partially left-hand circularly polarized ($P = 0.67$) and their sum $(4, -1, 0, -2)$ is partially elliptically polarized (long axis vertical) with $P = \sqrt{5}/4 = 0.56$.

Consider, for example, a wave of the form given by equation (7.1). Allowing the amplitudes a, b and phase ϕ to be slowly varying with time (relative to the wave period $2\pi/\omega$), the wave becomes quasi-monochromatic, and can be polarized or unpolarized, depending on the time dependencies. The Stokes parameters (apart from a multiplicative constant) reduce to

$$
\begin{aligned}
S_0 &= \langle a^2 \rangle + \langle b^2 \rangle \\
S_1 &= \langle a^2 \rangle - \langle b^2 \rangle \\
S_2 &= \langle 2ab \cos \phi \rangle \\
S_3 &= \langle 2ab \sin \phi \rangle.
\end{aligned}
\tag{7.23}
$$

The brackets stand for time averages over an observation period of many cycles.

Example. Evaluate P of equation (7.22), and confirm it has the values expected for:

(a) a fully polarized wave, where the amplitudes and phase are constant;

(b) a completely unpolarized wave, where the phase varies randomly and $\langle E_x^2 \rangle = \langle E_y^2 \rangle$.

Solution. (a) With no need for the time averages, one finds that

$$
S_0^2 = S_1^2 + S_2^2 + S_3^2 = (a^2 + b^2)^2;
$$

hence $P = 1$.

(b) $\langle a^2 \rangle = \langle b^2 \rangle$ and the random change of phase leads to $S_1 = S_2 = S_3 = 0$, so that $P = 0$.

7.11   Induced Birefringence

Some isotropic materials can be made birefringent by an external electric or magnetic field. The effects can be understood at the atomic or molecular level, as in the permanently birefringent materials.

In the *Kerr effect*, discovered in 1875 by J. Kerr, the birefringence is induced in many solids, liquids and gases by an electric field transverse to the light ray. As in a uniaxial crystal, quarter-wave and half-wave plates can be created, although a cell several centimetres long may be needed in practice. The difference in refractive indices is related to the field E and the vacuum wavelength λ_0 by

$$
n_e - n_o = \lambda_0 K E^2
\tag{7.24}
$$

where K is the *Kerr constant* for the substance. The *Kerr cell* is used to modulate a ray of plane polarized light. A cell about 10 cm long containing nitrobenzene (which has a large value of the Kerr constant) becomes a half-wave plate when a transverse field of around $10 \, \text{kV cm}^{-1}$ is applied. If the incident polarization is at 45° to the field, the emergent beam is rotated by 90° and it can be transmitted by an analyzer set at 90° to the original plane, as in Figure 7.16(a). The Kerr cell can therefore be used as an electrically operated light switch.

The *Pockels effect* is a birefringence induced in a crystal by a longitudinal electric field. The classes of crystal which show this effect are also piezoelectric. Among the many exotic crystals developed specially for a large Pockels effect are barium titanate and potassium dideuterium phosphate (known as KD*P). A Pockels cell, as shown in Figure 7.16(b), acts in a similar way to the Kerr cell; it is,

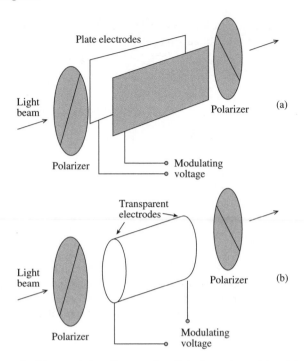

Figure 7.16   (a) A Kerr cell; (b) a Pockels cell. In each a light beam is modulated by an electric field which induces birefringence, rotating the plane of polarization

however, more compact and is widely used for electrical modulation and switching of light beams in communications systems.

The *Faraday effect* (Figure 7.17) is induced optical activity, in which a longitudinal magnetic field can induce a rotation of the plane of polarization in an isotropic material such as glass. The angle of rotation ψ is proportional to the magnetic field strength B and the path length l, so that

$$\psi = VBl \qquad (7.25)$$

where V is the *Verdet constant* for the medium. In Table 7.2 values of V are quoted for a specific wavelength; for most substances there is a large variation with wavelength.

A particularly simple explanation of the Faraday effect is available for propagation of a radio wave through a cloud of free electrons (Section 19.5), such as in the ionosphere and in interstellar space, where Faraday rotation is easily demonstrated. The refractive index depends on the amplitude of the oscillation of the electrons in response to the electric field of the wave, which now includes a gyration round the steady magnetic field (see Chapter 19). The amplitude of the oscillation depends on the hand of the circular polarization as compared with the direction of natural gyration round the magnetic field.

Example.   A solenoid 10 cm long consists of a core of flint glass wound with 300 turns of wire and carrying 2.0 amps. If the Verdet constant of the glass is 3.17×10^4 arcmin T^{-1} m^{-1}, find the rotation angle ψ this would induce in plane polarized light.

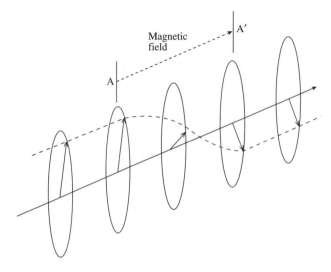

Figure 7.17   The Faraday effect. Between the planes A, A', a longitudinal magnetic field separates the refractive index into different values for the two hands of circular polarization. The relative phases of the two circularly polarized components of the plane polarized wave change, and the plane rotates

Table 7.2   Examples of the Verdet constant V for $\lambda_0 = 589.3$ nm

Substance	Temp. (°C)	V (arcmin $T^{-1}\,m^{-1}$)
Glass (light flint)	18	3.17×10^4
Phosphorus	33	13.3×10^4
Sodium chloride	16	3.59×10^4
Acetone	15	1.11×10^4
Carbon disulphide	20	4.23×10^4
Ethyl alcohol	25	1.11×10^4
Water	20	1.31×10^4

Solution. Inside a solenoid with current $I = 2.0$ A and turns per unit length $n = 3000\,m^{-1}$, the magnetic induction is $B = \mu_0 nI = (4\pi \times 10^{-7}\,T\,m\,A^{-1})(3 \times 10^3\,m^{-1})(2.0\,A) = 7.54 \times 10^{-3}$ T. Hence $\psi = VBl = 3.17 \times 10^4$ arcmin $T^{-1}\,m^{-1} \times 7.54 \times 10^{-3}$ T $\times 0.1$ m $= 24$ arcmin $= 0.40°$.

A device which allows light to travel in one direction but not in the opposite direction, i.e. an *optical isolator*, can be made by placing a Faraday rotating medium between polarizers P_1 and P_2 which are set at 45° to each other. The longitudinal magnetic field in the Faraday rotator is arranged to give a rotation of 45°. Polarized light produced by the first polarizer P_1 is rotated by 45° by the Faraday cell and is transmitted by the second polarizer P_2. Light travelling from the opposite direction through P_2 receives a 45° rotation in the *same* direction, and is rejected by polarizer P_1. These devices find application in eliminating back reflections in optical fibre systems and in high-power laser amplifiers.

Problem 7.1
Verify that equation (7.10) gives the correct value of R (ratio of maximum axis to minimum) for: (a) $\phi = 0$, and (b) $\phi = \pm\pi/2$.

Problem 7.2
Consider $E_x = a\cos(\omega t - kz)$, $E_y = a\cos(\omega t - kz + \phi)$ for the two cases $\phi = \pm\pi/2$, and verify that the upper sign corresponds to right-circular polarization (clockwise rotation) and the lower to left-circular polarization (anticlockwise rotation).

Problem 7.3
Verify that the Jones vectors $\begin{bmatrix} 1 \\ \mp i \end{bmatrix}$

correspond, respectively, to right circularly polarized (upper sign) and left circularly polarized (lower sign) light.

Problem 7.4
Use the Jones calculus to find out what kind of polarization results if (the matrix of) a "right-hand circular (RHC) polarizer" acts upon: (a) waves polarized linearly along the x axis; (b) circularly polarized waves of either hand.

In each case specify the change in irradiance $I \propto |E_{0x}|^2 + |E_{0y}|^2$, if any.

Problem 7.5
Repeat Problem 7.4 for a "polarization-plane rotator".

Problem 7.6
Light is incident from air at angle θ onto a uniaxial crystal face cut perpendicular to its optic axis. Find the angle ϕ at which the e-wave is transmitted.

Problem 7.7
(a) A Glan–air polarizer is cut at an angle θ as shown in Figure 7.11. If light of wavelength 589.3 nm is incident, find the allowed range of θ when the prism is made of quartz. Which wave, o or e, is reflected out of the beam?
(b) Repeat the above for a prism made of calcite.
(c) What would happen to the two waves if either the quartz or calcite prism were cut at $\theta = 30°$ or $\theta = 45°$?

Problem 7.8
A prism in the form of an equilateral triangle is made of calcite and has its optic axis parallel to the edge at its apex. If unpolarized light of wavelength 589.3 nm is incident near the angle which produces minimum deviation, what is the angular spread between the e- and o-waves when they emerge into the air?

Problem 7.9
Calculate the thickness of a calcite quarter-wave plate for sodium D light ($\lambda_0 = 589$ nm), given the refractive indices $n_o = 1.658$ and $n_e = 1.486$ for the two linearly polarized modes.

Problem 7.10
A pair of crossed polarizers, with axes at angles $\theta = 0°$ and 90°, is placed in a beam of unpolarized light with irradiance I_0, so that light emerges from the first with $I_1 = \frac{1}{2}I_0$ and from the second with $I_2 = 0$. A third polarizer is placed between the two at angle $\theta = 45°$. What then is I_2?

If the third polarizer rotates at angular frequency ω show that

$$I_2 = \frac{I_0}{16}(1 - \cos 4\omega t). \tag{7.26}$$

Problem 7.11

A plane polarized wave propagates along the optic axis of quartz as two circularly polarized waves, so that the difference in refractive indices $n_L - n_R$ introduces a phase difference δ between the two. Show that the plane of polarization is rotated by angle $\delta/2$.

Calculate the thickness of quartz plate that will rotate the plane by 90° at wavelength 760 nm, given $|n_L - n_R| = 6 \times 10^{-5}$.

Problem 7.12

A printed page appears double if a doubly refracting crystalline plate is placed upon it. Why is it that a distant scene does not appear double when viewed through the same plate?

Problem 7.13

Why does a thin plate of doubly refracting crystal generally appear faintly coloured when it is placed between two polarizers?

Problem 7.14

Show that an elliptically polarized wave can be regarded as a combination of circularly and linearly polarized waves.

Problem 7.15

An elliptically polarized beam of light is passed through a quarter-wave plate and then through a sheet of polaroid. The quarter-wave plate is rotated to two positions where the polaroid shows the light to be plane polarized, and it is found that the plane of polarization is then at angles of 24° and 80° to the vertical. Describe the original elliptical polarization.

8 Interference

... diversely coloured with all the Colours of the Rainbow; and with the microscope I could perceive, that these Colours were arranged in rings that incompassed the white speck or flaw, and were round or irregular, according to the shape of the spot which they terminated; and the position of Colours, in respect of one another, was the very same as in the Rainbow.

Hooke, 1665, on interference colours in a flake of mica.

A man alike eminent in almost every department of human learning...[who] first established the undulatory theory of light, and first penetrated the obscurity which had veiled for ages the hieroglyphics of Egypt.

Tablet in Westminster Abbey commemorating Thomas Young (1773–1829).

We have seen in Chapter 1, where the idea of Huygens' secondary waves was introduced, that the future position of a wavefront may be derived from a past position by considering every point of the wavefront to be a source of secondary waves. If the wavefront effectively propagates along rays, the geometric optics approach of Chapters 2 and 3 may be the most appropriate description of the progress of a wavefront, taking no account of the physical nature of the wave, including its amplitude and polarization. We now turn to the phenomena of interference and diffraction, where light is treated as a periodic wave, and ray optics provides a totally inadequate description.

Interference effects occur when two or more wavefronts are superposed, giving a resultant wave amplitude which depends on their relative phases.[1]

Diffraction is the spreading of waves from a wavefront limited in extent, occurring either when part of the wavefront is removed by an obstacle, or when all but a part of the wavefront is removed by an *aperture* or *stop*. The general theory which describes diffraction at large distances is due to Fraunhofer,[2] and is referred to as *Fraunhofer diffraction*.

---

[1]In this book, we shall simplify our discussions of interference and diffraction by ignoring the relative polarization of the constituent waves. This amounts to treating the waves as simple additive scalars. It also applies to many basic systems where the combining electric fields are approximately parallel. (Theory predicts and experiment confirms that two EM waves polarized perpendicular to one another will not show interference.)

[2]J. von Fraunhofer (1787–1826), optician in Munich, known mainly in his lifetime for his skill in making telescope lenses and for solar spectroscopy. The dark absorption lines in the solar spectrum were named 'Fraunhofer lines'.

Optics and Photonics: An Introduction, Second Edition   F. Graham Smith, Terry A. King and Dan Wilkins
© 2007 John Wiley & Sons, Ltd

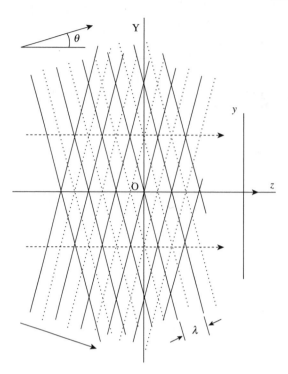

Figure 8.1   Interference between two plane waves. The waves are crossing at angles $\pm\theta$ to the z axis

8.1   Interference

Figure 8.1 shows two monochromatic plane waves, with the same r.m.s. amplitude[3] A and wavelength λ, propagating at angles $\pm\theta$ to the z axis. In the figure at a particular moment of time the solid and broken lines correspond to positive and negative maxima. The two waves combine to give resultant positive and negative maxima of $+2A$ and $-2A$ where two solid and two broken lines intersect. Where a solid line intersects with a broken line the resultant is zero. Along the line OY in the y direction the resultant varies from $2A$ to zero to $-2A$ to zero to $+2A$ and so on. The intensity[4] of the resultant, the square of the amplitude, varies as $4A^2, 0, 4A^2, 0, 4A^2$, etc. The pattern of intensity forms uniformly spaced *interference fringes*. We now find the shape and spacing of these fringes.

We have already noted that in any harmonic wave with a plane wavefront the phase changes linearly with distance along the direction of the wave, changing by 2π in distance λ; the phase is constant across the wavefront. Along the direction OY a distance y has a component $y\sin\theta$ in the direction of the wave, so that the phase change[5] relative to $y = 0$ is

$$\pm\phi/2 = \pm 2\pi \frac{y\sin\theta}{\lambda} \tag{8.1}$$

---

[3]A sinusoid with root mean square (r.m.s.) amplitude A has a peak amplitude $\sqrt{2}A$.

[4]The energy flux of any wave (the power across a unit area perpendicular to the flow) is called *intensity* for arbitrary waves, and *irradiance* for optical waves. In this book we do not use 'radiant' or 'luminous' intensity, which are optical terms with a different meaning from conventional intensity (see Appendix 2).

[5]In this and the following section, ϕ stands for the phase difference of the two waves.

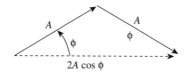

Figure 8.2   Phasor diagram for crossing waves

where the plus and minus signs correspond to the two waves.

The phasor diagram of Figure 8.2 shows the phasors for the two crossing waves at a general point y on the line OY, with a phase difference ϕ given by equation (8.1). Phasor diagrams for $\phi/2 = 0$, $\pi/4$, $\pi/2$, $3\pi/4$ and π are shown in Figure 8.3, with the corresponding intensities. The intensity is given by the square of the resultant amplitude:[6]

$$I = (A_{\text{resultant}})^2 = (2A \cos \phi/2)^2$$
$$= 4I_0 \cos^2 \phi/2 \tag{8.2}$$

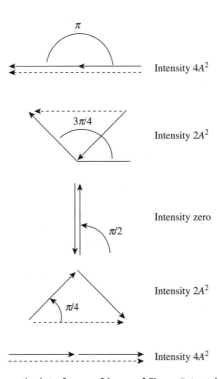

Figure 8.3   Phasor diagrams across the interference fringes of Figure 7.1, at intervals of $\pi/4$ in the phase $\phi/2$

---

[6]The relation $I = A^2$ should be understood as a proportionality $I \propto A^2$, valid for many kinds of waves in physics. The specific constant of proportionality for electromagnetic waves, when A is identified with the amplitude of the electric field, is given in Section 5.5.

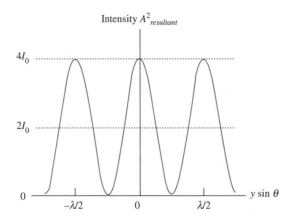

Figure 8.4   The pattern of intensity along the y axis for the crossing waves. Note that each wave alone would give an intensity I_0. The average intensity is $2I_0$, and the peak is $4I_0$

where I_0 is the intensity of each plane wave alone. The variation along the y axis of the irradiance, which for light is the luminance, is the cosine curve shown in Figure 8.4. This is known as a pattern of *cos-squared* fringes.

8.2   Young's Experiment

A simple example of interference between two crossing waves is Young's experiment, which provided the first demonstration of the wave nature of light. Two closely spaced narrow slits, A and B in Figure 8.5, transmit two elements of a light wave from a single source. The two sets of waves spread by diffraction, then overlap and interfere. If they then illuminate a screen, there will be light and dark bands across the illuminated patch; these are the interference fringes. Figure 8.5 shows that the geometry of the fringes becomes simpler at increasing distance from the slits, where the two sets of waves from A and B behave like two sets of nearly plane waves crossing at a small angle, like the plane waves of Figure 8.1. We will consider the effect for monochromatic light.

The light incident on the slits is a plane wave, so that each slit is the source of identical expanding waves. Consider the sum of the two waves at a point P, which is sufficiently far from A and B for the amplitudes of the two waves to be taken as equal. There is a phase difference between the two waves depending on the small difference l between the two light paths, so that the waves add as in Figure 8.6(b). The path difference is $l = d \sin \theta$, giving a phase difference $\phi = 2\pi l/\lambda$, and the intensity I at P varies with the phase difference as in equation (8.2), giving

$$I = 4I_0 \cos^2 \frac{\pi l}{\lambda}, \tag{8.3}$$

where I_0 is the intensity of each wave at P. The phase reference may be taken at O, half-way between the slits. The waves are then advanced and retarded on the reference by $\pi l/\lambda$, as in the phasor diagram of Figure 8.6(b). Thus where constructive interference occurs the intensity is four times that due to

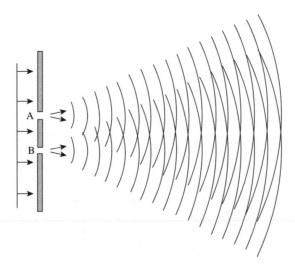

Figure 8.5 Young's experiment. Light from the two pinholes or slits A,B spreads by diffraction, and the two sets of waves overlap and develop an interference pattern

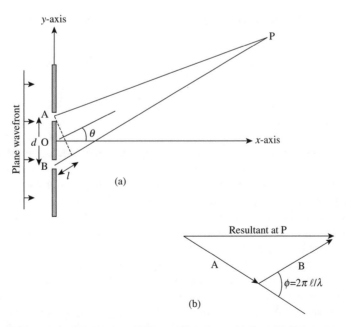

Figure 8.6 Young's experiment. (a) Two wavelets spread out from the pair of slits, and interfere. Constructive or destructive interference occurs at P according to whether the path difference l is $N\lambda$ or $(N + \frac{1}{2})\lambda$. (b) Phasor diagram for the sum at P, at a large distance from the slits. The phase reference (giving a horizontal phasor) is at O and the phase difference between the two waves is $\phi = 2\pi l/\lambda$

one slit, or *twice* the intensity due to two slits if interference did not happen. The conditions for constructive and destructive interference are as follows:

$$\text{constructive } l = d \sin \theta = N \lambda$$

$$\text{destructive } l = d \sin \theta = \left(N + \frac{1}{2} \right) \lambda \tag{8.4}$$

where N is an integer. N is called the *order* of the interference; it is the number of whole wavelengths difference in the paths to points where constructive interference takes place. The bright and dark 'fringes' which appear on a screen placed anywhere to the right of the double slit system can therefore each be labelled according to their order. The highest possible order is the integral part of d/λ. The fringes are spaced uniformly in $\sin \theta$ at intervals λ/d, and at a distance x the linear spacing, in the small-angle approximation, is $x\lambda/d$.

The condition that the distance is sufficient to allow the use of the simple relation $l = d \sin \theta$ is important; it is equivalent to the condition that the phase of any elementary wave from the screen is a linear function of coordinates x and y in the plane of the screen. This is the condition for *Fraunhofer diffraction* (Chapter 10), of which the present example is a special case. Under this condition, the whole of the screen, whatever the pattern of apertures, can be considered as a single diffracting object.

The amplitude of the interference pattern of a pair of slit sources is the function

$$A(\theta) = A(0) \left| \cos \left(\frac{\pi d}{\lambda} \sin \theta \right) \right|. \tag{8.5}$$

$A(0)$ is the amplitude of the diffracted wave when $\theta = 0$. The intensity is the square of $A(\theta)$, giving cos-squared fringes

$$I(\theta) = I(0) \cos^2 \left(\frac{\pi d}{\lambda} \sin \theta \right). \tag{8.6}$$

Notice that the average intensity across several fringes is $I(0)/2 = 2I_0$ (see Figure 8.4), because the peaks of the upper half of the cos-squared fringes just fill the troughs of the lower half. This is an example of an important general principle of all interference and diffraction effects: the energy is *redistributed* in space by these effects, but remains in total the same. This rather obvious remark enables some not so obvious predictions to be made; for example, if light is diffracted *into* the geometrical shadow of an object, the intensity must begin to fall *outside* the geometrical shadow to compensate. We examine this in Chapter 10; it is an example of *Fresnel diffraction*.

A Young's double slit interference can easily be made with two parallel scratches on an overexposed photographic film. Fringes will be seen if a distant street lamp is viewed through the double slit; for slits 0.5 mm apart the angular spacing λ/d will be about $4'$.

Interference between two beams of light usually requires them to be derived from the same source. There are two ways of achieving this. The first, *division of wavefront*, means utilizing spatially separate parts of the wavefront as distinct sources, as in Young's double slit or by using a diffraction grating (Chapter 10). The second, which is the main subject of this chapter, is to divide the amplitude of the wave by partial reflection, obtaining identical wavefronts which can be brought together by different paths. The most familiar example of interference by this *division of amplitude* is the pattern of coloured fringes seen in soap bubbles and thin oil films.

8.3   Newton's Rings

As an introduction to interference by amplitude division we describe Newton's rings. These may be observed with the optical system shown in Figure 8.7. A long-focus lens is placed in contact with a flat glass plate, and illuminated at vertical incidence by the source reflected by the glass plate tilted at 45°. Observations are made with the microscope. The wavefront travelling downwards is partially reflected at each boundary of the lens and plate. The two wavefronts that interfere to cause Newton's rings arise from partial reflection at the lower surface of the lens and the upper surface of the plate. As they are both derived from the same source, the two waves can interfere with each other. Their relative phases are determined by two factors:

1. The reflection coefficients of the two boundaries are of opposite sign (see Equation 5.29).

2. There is an extra path traversed by the wave reflected by the flat surface.

The first factor provides a phase difference of π, so that the centre of the pattern is dark, with the two reflections in antiphase. (No reflection is to be expected anyway from the central area where the glass is in contact and effectively continuous.) Succeeding rings are light and dark as the extra path length is $(N - 1/2)\lambda$ or $N\lambda$, and the two reflections become in and out of phase. If the radius of curvature of the bottom face of the lens is R the condition for brightness at r from the axis is

$$\text{path difference} = 2h = (N - 1/2)\lambda \tag{8.7}$$

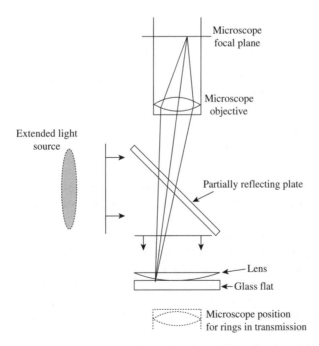

Figure 8.7   Newton's rings. The rings may conveniently be observed in reflection with the system shown. The much lower visibility rings seen in transmission may be seen from below

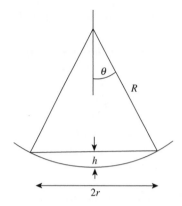

Figure 8.8   The intersecting chord theorem. The sagittal distance *h* is related to the chord 2*r* and radius *R* by $r^2 = R^2 - (R - h)^2 = h(2R - h)$; for $h \ll 2R$ the sagittal distance is $h \approx r^2/2R$

where the value of *h* in terms of *R* and *r* can be determined approximately by using the expansion $\cos\theta = 1 - \theta^2/2 + \ldots$:

$$h = R(1 - \cos\theta) = R\left(\frac{\theta^2}{2} + \ldots\right) \approx \frac{r^2}{2R}. \tag{8.8}$$

This approximation, shown in Figure 8.8, often appears in optics; it is important to be clear under what circumstances it is applicable. The expression

$$h = \frac{r^2}{2R} \tag{8.9}$$

is a *parabola* that approximates to a circle of radius *R* for small *r*. Usually this approximation will be good enough if the deviation between circle and parabola is small compared with a wavelength. If the expansion of equation (8.8) is taken to one more term, this condition becomes

$$R\frac{\theta^4}{4!} \ll \lambda \tag{8.10}$$

so if $R = 100\,\text{cm}$ and $\lambda = 5 \times 10^{-5}\,\text{cm}$, the condition is $\theta \ll 5.8 \times 10^{-2}$ rad or a few degrees. Now at $r = 1\,\text{cm}$, $h = 0.005\,\text{cm}$ so $2h = 0.01\,\text{cm}$. The order *N* of the ring is then $10^{-2}/(5 \times 10^{-5}) = 200$. In this example the approximation is good for many tens of rings.

These ring fringes are localized in the sense that to see them the microscope must be focused on the boundary between the lens and optical flat. When this condition is satisfied the light reflected from a point on the lens surface is brought ultimately to an image point by the same optical path, even though it may have arrived from an extended source. The light reflected from the corresponding point below on the flat is likewise brought to the same focal point, the only difference being that the path for all rays from the flat is longer by approximately 2*h* than that from the lens surface. Combining equations (8.7) and (8.9) gives

$$r = [R\lambda(N - 1/2)]^{1/2} \tag{8.11}$$

as the condition for brightness. The ring system then has a dark centre surrounded by rings getting more and more crowded as r increases.

The *visibility V* of interference fringes, sometimes called fringe contrast, is defined in terms of the maximum and minimum intensities as

$$V = \frac{I_{\max} - I_{\min}}{I_{\max} + I_{\min}}. \tag{8.12}$$

The visibility of Newton's rings is not intrinsically close to unity, as in the case of Young's fringes formed by two equal amplitude waves. It is instead determined by the relative amplitudes of the two waves reflected from the lower surface of the lens and the upper surface of the flat. In practice these are nearly the same, and the ring system is well defined and of high visibility. Quite the opposite is true in the case of Newton's rings seen in transmission rather than reflection; the interference is now between a directly transmitted wave and a much smaller wave that has been twice reflected (see Chapter 5 for the coefficient of reflection). The transmitted fringes are complementary: bright where those reflected are dark, but the visibility is very low. A system for observing them is shown in Figure 8.7. One way to see that they must arise is from a consideration of the conservation of energy. Most of the light incident on the system is transmitted, but due to the interference effects discussed above some areas (the bright rings) reflect some light back, whilst others (the dark rings) reflect hardly at all. Hence, as the energy not reflected back is transmitted, the complementary low-visibility system is seen in transmission. The well-known property of a photographic negative to look like a positive if seen in reflected light from the front is somewhat analogous to this: the high-transmission (transparent) parts look dark compared with the low-transmission (opaque) parts.

If instead of a monochromatic source, a white light source is substituted, a dark spot is still seen in the centre of the reflected pattern. The scale of the pattern increases with wavelength, and the overlapping ring systems in all the different wavelengths get increasingly out of step, giving a white field only a few orders away from the centre. Between the dark and white regions is a system of coloured rings in a sequence called Newton's colours, as the various colours are added or subtracted according to their wavelength and the distance between the surfaces.

8.4   Interference Effects with a Plane-Parallel Plate

The case of Newton's rings is only one example of interference effects observed between two beams derived by the division of amplitude by partial reflections from two surfaces. Consider first a monochromatic point source S illuminating a parallel-sided slab of transparent material of refraction index n, shown in Figure 8.9. Then if the direct path is excluded there are two paths from S to P corresponding to reflections from the front and back of the slab, and P will be light or dark according to whether the optical paths (taking into account the phase change at the upper reflection) differ by an odd or even number of half wavelengths. Clearly there will be circular symmetry about the line SN through the source normal to the slab. Geometrically the system is like Young's double slit, the interference being between light from the two images S_1 and S_2 of S, in the front and back surfaces of the slab. The fringes are non-localized in space, and a photographic plate in any plane parallel to the slab (for example) would record circular fringes centred on SN. The order of the fringes is high if the thickness of the plate is large compared with the wavelength, so that the source must be highly monochromatic if fringes are to be observed. Similarly if the source is not a point, but extended, the fringes from different parts of the source will overlap and the pattern may be lost.

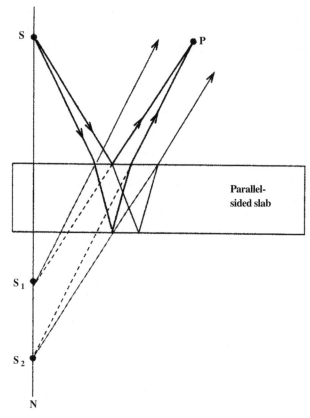

Figure 8.9 A point source S has two images S_1 and S_2 in the top and bottom surfaces of a parallel-sided slab. The interference in the light from these gives non-localized fringes similar to Young's

A surprising change is made in the system by inserting a lens as shown in Figure 8.10, and observing the distribution of brightness in its focal plane. The lens brings together at a point P all the rays that leave the plate at a particular *angle*; the figure shows two of these for each of three points in the source. The source may now be extended, since the path difference for all pairs of rays reflected in the front and back faces of the slab is the same, if they leave the plate at the same angle, regardless of which part of the extended source they come from. Each element of the extended source thus contributes twice to the light wave at P, once by reflection at the back, and once by reflection at the front. These contributions interfere either constructively or destructively, as determined by their different path lengths.

As only the angle of incidence determines the brightness or darkness, these fringes are called *fringes of equal inclination*. It is easy to see from Figure 8.11 that the conditions for bright and dark fringes are related to the angle of refraction r, inside the slab, by

$$2nh\cos r = \left(N + \frac{1}{2}\right)\lambda \quad \text{for bright fringes}$$

$$2nh\cos r = N\lambda \quad \text{for dark fringes.}$$

(8.13)

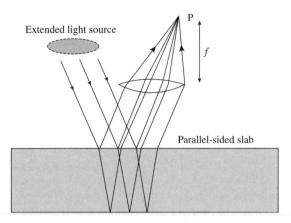

Figure 8.10 The effect of introducing a lens which brings all *parallel* rays to a single point P is to allow fringes of equal *inclination* to be observed with an extended source

The visibility of the fringes is again not necessarily unity. Let A_f and A_b be the amplitudes of light arriving at P from the front and back of the slab. Then the irradiance will be proportional to

$$A^2 = A_f^2 + A_b^2 + 2A_fA_b \cos \phi \tag{8.14}$$

based on a phasor diagram similar to Figure 8.6(b) and where

$$\phi = \frac{4\pi nh \cos r}{\lambda} + \pi. \tag{8.15}$$

From equation (8.14) putting $\cos \phi = 1$ for maximum irradiance and $\cos \phi = -1$ for minimum irradiance,

$$V = \frac{2A_fA_b}{A_f^2 + A_b^2}. \tag{8.16}$$

So with a thick slab of material and a monochromatic extended source, fringes may be observed in the focal plane of the lens in Figure 8.10. The irradiance profile of the fringes follows a squared cosine

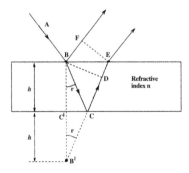

Figure 8.11 The path difference for rays ABF and ABCDE in a film with refractive index n. With point B′ the mirror image of B, right triangles B′C′C and BC′C are congruent Since B′CD is the same length as BCD, and optical path difference between the two rays is 2nh cos r. By Snell's law, there is no additional path difference as wavefront BD in the film becomes FE in the air.

function, as in equation (8.3) for Young's double slit; they are often referred to as cos-squared fringes, but their visibility is less than unity. Notice that the present discussion has excluded multiple reflections, a good approximation unless special measures are taken to increase the reflectivity. This point is returned to in Section 8.7.

8.5 Thin Films

The parallel-sided slab discussed in the previous section can produce fringes of a very high order; that is to say, the path difference between the interfering beams might be many thousands of wavelengths. Thin films in which the thickness is only a few wavelengths display interesting and somewhat different interference phenomena, which do not depend on the film having parallel sides. Interference fringes now appear in the surface of the film; their position is determined mainly by the thickness of the film and very little by the angle at which they are seen.

Consider first the familiar observation of light from the sky reflected in a thin oil film on water (Figure 8.12). Each colour forms a set of interference fringes across the film, the positions of the fringes depending on wavelength and the thickness of the film. Each part of the film reflects two waves, one from the front and one from the back of the oil film; the phase difference of these two waves depends on both angle r and thickness h, but in practice mainly on h so that the interference effects appear to outline the contours of equal thickness in the film.

There are in fact two extreme cases for interference in thin films. If the thickness is completely uniform, then only a variation of the angle r can change the path difference; fringes will therefore be seen outlining directions where r is constant, as in the thick slab already discussed. The fringes will be very broad, as a large change in $\cos r$ corresponds to a small change in path difference when h is small. These are *fringes of constant inclination*. Alternatively, as with the oil film or a soap bubble, the thickness may vary rapidly from place to place while $\cos r$ changes little; *fringes of constant thickness* are then seen. These are known as *Fizeau fringes*.

The fringes of constant thickness seem to be located within the film, as their position is determined by h and not by r; by contrast the fringes of constant inclination are seen in fixed *directions*, and therefore appear to be at infinite distance. In practice, the fringes seen in oil films are also determined in small part by the angle r, so that they are located just behind the film, as can be verified by the observer moving his or her head from side to side, looking for parallactic motion of the fringes across the film.

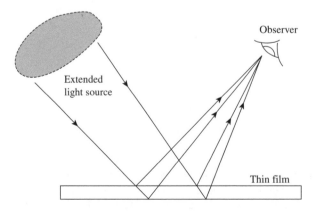

Figure 8.12   Fringes are seen in a thin film by reflection of an extended source of light

Fringes of constant thickness have many practical applications, as they allow measurements of thickness to be made to a fraction of a wavelength.

8.6 Michelson's Spectral Interferometer

An instrument that has been very influential in the development of interference optics, both theoretical and practical, is Michelson's interferometer (not to be confused with his stellar interferometer, Chapter 9). Its simplest form is shown in Figure 8.13, which may seem at first sight to bear little relationship to the optical systems discussed in connection with thin films in the last section. In fact they are closely related.

Light from the extended source S is split in amplitude by a half-silvered mirror D, and the two beams are then reflected by the mirrors M_1 and M_2. A further partial reflection in D sends the two beams out towards the observer at P. An observer can see the source S reflected simultaneously in the two mirrors. The observer can also see an image M_2' of the mirror M_2 close to the surface of M_1. The mirror M_2 is equipped with screws to allow its inclination to be adjusted, so that M_2' can be made parallel with M_1; M_1 itself is mounted on a screw-controlled carriage so that the perpendicular distance between M_1 and M_2' can be adjusted over a considerable range.

When M_1 and M_2' nearly coincide, the view of the source S is exactly the same as the view reflected in a thin film, and fringes will be seen in the surface of M_1. The equivalent thin film is the space between M_1 and M_2', so that the path difference between two rays reaching P from a point on S is determined by the thickness of this space. Exact equality of the optical paths is ensured for the situation when M_1 and M_2' coincide by the insertion of the glass plate C, which is the same thickness

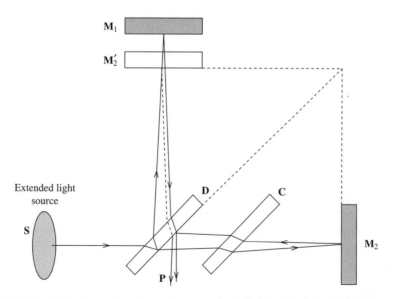

Figure 8.13 Michelson's interferometer. Observations are made from P either directly or with a microscope or camera. An observer at P sees the extended light source S reflected in a 'thin film' consisting of the mirror M_1 and the image M_2' of the mirror M_2. Fringes of constant thickness are generally seen in this thin film, although if M_1 and M_2' are made accurately parallel the fringes take the form of a circular pattern of fringes of constant inclination

as D. Each ray then traverses a glass plate twice, and the paths will remain equal for all wavelengths even if the refractive index of the glass is dispersive.

Suppose that the distance between M_1 and M_2' is small, but that M_1 is not parallel to M_2'. As the mirrors are accurately plane the fringes will then be light and dark lines across the mirrors; these are fringes of constant thickness. The angle of M_2 can then be adjusted, and M_1 and M_2' can be made parallel to a small fraction of a wavelength; the linear fringes of equal thickness are then replaced by fringes of equal inclination. For a bright fringe of order N seen at angle θ to the mirror normal, with an apparent mirror spacing d

$$2d \cos \theta = N\lambda. \tag{8.17}$$

This represents the special case of equation (8.13) with $n = 1$ (air-filled thin film), $h = d, r = \theta$, and without the extra term of $\lambda/2$ due to the phase reversal. Adjustment of the position of M_1 will now make the fringes expand outwards from the centre if the distance $M_1 M_2'$ is decreasing, and shrink inwards towards the centre if it is increasing. Finally a position can be attained when the field is uniformly bright all over. This corresponds to the planes of M_1 and M_2' coinciding. The lightness or darkness of the field in this condition depends on the difference ϕ in phase change at the two reflections in D; if $\phi = \pi$ and the division in amplitude by D has been accurately equal the field will be black. Suppose that M_2 is now tipped about its centre so that the linear fringes of equal inclination are seen. The centre black fringe then corresponds to the zero order. If the monochromatic source S is now replaced by a white light, the central dark fringe will remain, and on either side of it a few fringes will be seen. These will display Newton's colours, merging into white light a few fringes away on each side when the overlapping of the different coloured fringes is complete. So substitution of a white light source is useful as it allows identification of the zeroth-order fringe. The visibility of fringes in monochromatic light is unity even for high-order fringes.

The basic properties then of the Michelson interferometer are the ability:

1. To make both arms equal in optical length to within a fraction of a wavelength.

2. To measure changes of position as measured on a scale (the positions of M_1) in terms of wavelength by counting fringes.

3. To produce interference fringes of a known high *order* (number of wavelengths difference in path lengths).

The last of these is the basis for the use of Michelson's interferometer for measuring the width and shape of spectral lines, as discussed in Chapter 12.

A closely related interferometer is the Mach–Zehnder interferometer described in Chapter 9. Here the two beams are again separated by a mirror system, but in an arrangement which is convenient for inserting optical components into one of the beams to measure differences in optical paths by observing fringe displacements.

8.7   Multiple Beam Interference

In the discussion of interference effects in parallel-sided slabs and in thin films we have so far taken account of only two reflected beams, and ignored any multiple reflections. Unless special measures are taken to increase them, such reflections are very weak, but if appropriate steps are taken so that

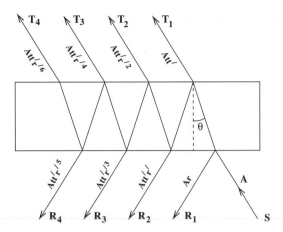

Figure 8.14   Multiple reflections in a parallel-sided slab

they are enhanced, allowing perhaps 10 or more beams to be combined, the fringes change their character and become very much sharper. By suitably coating the faces of a slab with a thin metallic film it is possible to increase the reflection coefficient, so that the front face reflects a large fraction of the light incident upon it. The small amount that does get through to the inside of the slab is reflected back and forth many times, a small proportion emerging at each reflection. It is the interference of these *many* emerging rays, rather than just two, that gives multiple interference its special character.

This simple case is illustrated in Figure 8.14. Most of the incident light S is reflected into the ray R_1, but after that the further rays on the reflection side R_2, R_3,... are all of similar strength, dying away gradually. Let A be the amplitude of the incident ray. Then if r and t are the reflection and transmission coefficients from the surrounding medium to the slab, and r' and t' the corresponding quantities from slab to medium, we can write down the amplitude of the reflected rays

$$Ar, \ Att'r', \ Att'r'^{3}, \ \ldots \tag{8.18}$$

and of the transmitted rays

$$Att', \ Att'r'^{2}, \ Att'r'^{4}, \ \ldots . \tag{8.19}$$

With the exception of the first large reflected ray these amplitudes go in geometric progression, and the closer to unity that r' is made, the more slowly does their size die away. Their *phase* depends on θ, the angle of refraction inside the plate, its thickness h and its refraction index n. The relative phase of the rays on a plane outside the slab perpendicular to the ray depends on θ (see Figure 8.11) as

$$\psi = \frac{4\pi nh \cos \theta}{\lambda}. \tag{8.20}$$

If now, as in the case of two-beam interference, we provide a lens to bring the transmitted rays to a focus we can find the conditions for constructive and destructive interference. For constructive interference, all the slowly declining vectors will be in phase if

$$2nh \cos \theta = N\lambda. \tag{8.21}$$

Consequently with an extended source fringes of equal inclination, that is to say circles, will be seen, each corresponding to a particular value of N. It is when we consider the spaces between these fringes that the special character of multiple beam interference emerges. The angle between *each* of the many phasors is given by ψ in equation (8.20); a change in ψ of 2π takes us to the next maximum, but a change from 2π by only a very small amount causes the long string of nearly equal phasors to curl up into a near circle. For this reason the fringes are very sharp, a plot of the irradiance showing almost no light transmitted except close to the maxima.

If p beams are transmitted the complex amplitude is the sum of the geometric series

$$A(p) = Att'[1 + r'^2 \exp(i\psi) + \ldots + r'^{2(p-1)} \exp(i(p-1)\psi].\tag{8.22}$$

The sum of this geometric series is

$$A(p) = Att' \left(\frac{1 - r'^{2p} \exp(ip\psi)}{1 - r'^2 \exp(i\psi)} \right).\tag{8.23}$$

If the number of beams p becomes large so that r'^{2p} becomes very small, we can ignore that term, and calculate the irradiance I as $A(\infty)A^*(\infty)$, giving

$$I = \frac{(Att')^2}{[1 - r'^2 \exp(i\psi)][1 - r'^2 \exp(-i\psi)]}\tag{8.24}$$

or

$$I = \frac{(Att')^2}{1 + r'^4 - 2r'^2 \cos\psi}.\tag{8.25}$$

In this expression we can recognize A^2 as the irradiance of the incident wave, I_i. The expression for transmitted irradiance I_t is more neatly expressed in terms of $\sin^2(\psi/2)$:

$$\frac{I_t}{I_i} = \frac{(tt')^2}{(1 - r'^2)^2 + 4r'^2 \sin^2(\psi/2)}.\tag{8.26}$$

Now it may be shown from a study of the Fresnel coefficients in Section 5.3 that $tt' = 1 - r'^2$ so that equation (8.26) may be further simplified to

$$\boxed{\frac{I_t}{I_i} = \frac{1}{1 + F \sin^2(\psi/2)}}\tag{8.27}$$

where

$$F = \frac{(2r')^2}{(1 - r'^2)^2}.\tag{8.28}$$

Notice that the parameter F becomes very large as r'^2 approaches unity. For example, if $r'^2 = 0.8$, $F = 80$. The effect of this is to keep the transmitted irradiance (equation (8.27)) always very small, except when ψ is close enough to a multiple of 2π for $\sin^2(\psi/2)$ to become less than $1/F$. The value of

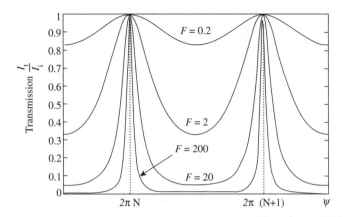

Figure 8.15 Cross-sections of Fabry–Pérot fringes for several values of the parameter F

I_t/I_i then shoots rapidly up to unity when ψ is a multiple of 2π. Figure 8.15 is a plot of I_t/I_i for several values of F. The sharp fringes of multiple beam interference are known as *Fabry–Pérot fringes*.

The sharpness of Fabry–Pérot fringes is often specified as the *finesse* $\mathcal{F}$, which is the ratio of the separation of adjacent maxima to the half-width of a fringe, defined as the width between points of half irradiance. From equation (8.27) the phase shift $\psi_{1/2}$ for the fringe irradiance to be halved is given by

$$1 + F\sin^2\frac{\psi_{1/2}}{2} = 2 \tag{8.29}$$

giving

$$\psi_{1/2} = 2\sin^{-1}(1/\sqrt{F}). \tag{8.30}$$

In practice F is large, and $\sin^{-1}(1/\sqrt{F}) \approx 1/\sqrt{F}$. Since the phase difference between two adjacent fringes is 2π, the finesse $\mathcal{F}$ is the ratio $2\pi/2\psi_{1/2}$, giving

$$\mathcal{F} = \frac{\pi\sqrt{F}}{2}. \tag{8.31}$$

The multiple beams of the Fabry–Pérot act very like those of a diffraction grating (see Chapter 11): the many phasors arising from the multiple reflections give only a small resultant unless the angle θ at which they traverse the slab is just right to put them all in phase. So the slab is a device that will allow light of any fixed wavelength to traverse it only at certain extremely well-defined angles. It can therefore be used as a filter transmitting only selected narrow-wavelength ranges, or as a spectrometer.

8.8 The Fabry–Pérot Interferometer

This instrument, illustrated in Figure 8.16, uses the effect discussed in the previous section to produce circular, sharply defined interference fringes from the light from an extended source. The rings on the

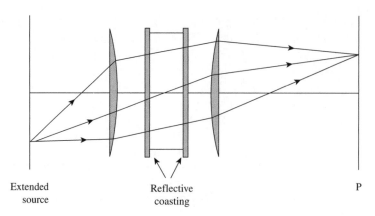

Extended   Reflective   P
source   coasting

Figure 8.16   A simple Fabry–Pérot interferometer. Without the cavity or echelon an image of the broad source is formed on P. When the echelon is in place, only those parts of the image corresponding to allowable angles through the echelon are transmitted, giving the extremely well-defined ring system

plate P are images of those points on the source producing (with the help of the first lens) light going in suitable directions between the lenses. The central cavity is usually made of two glass plates, with their inner surfaces coated with partially transparent films of high reflectivity. These plates are held apart by an optically worked spacer made of invar or silica, to which they are pressed by springs. This device is called a *Fabry–Pérot etalon*.[7]

If the light from the source is not monochromatic but contains two spectral components, the ring system is doubled. It is possible by this means to distinguish optically, or to 'resolve', even the hyperfine structure of spectral lines directly. Since its introduction by Fabry and Pérot in 1899 the instrument has dominated the field of high-resolution spectrometry. We refer again to the Fabry–Pérot interferometer and its use in high-resolution spectrometry in Chapter 12.

8.9   Interference Filters

A parallel-sided glass plate, or a cavity between two parallel glass plates, acts as a spectral filter on light which falls on it at any given angle. Figure 8.15 shows how the irradiance of the transmitted light varies with the phase difference ψ, which is inversely proportional to wavelength. The curve in Figure 8.15 is therefore the *transmission characteristic* of the filter, showing its relative transmission properties over a range of wavelengths.

A combination of two or more such filters, using plates or cavities with different thicknesses, can be arranged to transmit only one narrow spectral band. Filters transmitting a band only 1 nm wide at optical wavelengths can be made in this way. They can even be made tunable by making the thickness variable; a convenient way of doing this is to move one of the reflecting surfaces by attaching a piezoelectric transducer in which an applied electric field induces a small mechanical movement. An alternative is to vary the optical thickness of the cavity by changing the gas pressure within the cavity, which varies the refractive index.

---

[7]From *étalon*, a standard; it can be used as a standard of length calibrated in terms of the wavelength of spectral lines.

A very high reflection coefficient, giving a high finesse, is essential for high resolution in interferometers and in interferometric filters. Since thin metal films absorb rather too much light, it is preferable to use dielectric coatings on glass. The reflection coefficient then depends on the step of refractive index at the interface; it can be increased by using a series of layers of dielectric, with alternate high and low refractive indices. The reflectivity at a given wavelength depends on the thickness of the layers; multiple layers are used for the selective reflection of a defined range of wavelengths. A very high reflection coefficient can be obtained at a single wavelength; mirrors used in the laser interferometers described in Chapter 9 may have reflection coefficients up to 99.999%.

Fabry–Pérot etalons often form an essential component of lasers (Chapter 15), where they are referred to as resonant cavities. A helium–neon gas laser, for example, has a cavity formed by two mirrors enclosing a gas-filled tube; the cavity resonator determines the wavelength of the laser action in the gas. The wavelength of light from semiconductor lasers is similarly determined by a resonant cavity, which is formed by the polished faces of the semiconductor material.

Problem 8.1 Numerical examples

(i) An air-filled wedge between two plane glass plates is illuminated by a diffuse source of light, wavelength 600 nm. Fringes are seen in the light reflected by the air wedge, spaced 5 mm apart. Find the angle α between the glass plates. Assume $\alpha \ll 1$ rad and near-normal incidence.

(ii) Newton's rings are formed by a lens face with radius of curvature 1 m in contact with a plane surface, using sodium light with wavelength 589 nm. Find the radii of the first and second bright rings.

(iii) Newton's rings are formed using the bright sodium spectral line, which is a doublet with wavelengths 589.0 and 589.6 nm. Find the order of rings where the bright ring of one component of the doublet falls on a dark ring of the other.

Problem 8.2
Young's double slit fringes are formed by two side-by-side coherent sources. Consider in contrast the fringe pattern formed by two coherent point sources one in front of the other, and relate it to the pattern from a single source seen reflected in a parallel-sided slab.

Problem 8.3
The colours in a soap bubble often fade just before the bubble bursts, and the film becomes dark. Estimate the film thickness at this stage.

Problem 8.4
When two object-glasses are laid upon one another, so as to make the rings of the colours appear, though with my naked eye I could not discern above eight or nine of those rings, yet by viewing them through a prism I could see a far greater multitude, insomuch that I could number more than forty...But it was on but one side of these rings. Newton, *Opticks*. Explain!

Problem 8.5
The centre of Newton's rings is usually dark when observed in reflected light. Is this the same phenomenon as in the soap film of Problem 8.3? What happens to the size and intensity of the fringe pattern if oil with refractive index close to those of the lens and the flat surface fills the air space between them?

Problem 8.6
Show that, contrary to expectation, a transparent film on a perfectly reflecting surface does not show interference fringes in reflected light. This will require analysis following the lines of Section 8.7, noting the change of phase

at internal reflection at the top face. Based on Fresnel's theory of section 5.3, you can assume that the reflection and transmission coefficients for the bottom face satisfy $r = -r'$ and $tt' = 1 - r'^2$.

Problem 8.7

What happens to the fringe spacing in the air-filled wedge of Problem 8.1 if a liquid with refractive index n fills the wedge? What happens to the brightness and visibility of the fringes if n approaches the refractive index of the glass?

Problem 8.8

The following three problems all relate to the Fabry–Pérot etalon.

Prove that $tt' = 1 - r'^2$. Do this separately for polarization parallel and perpendicular to the plane of incidence.

Problem 8.9

By solving the separate contributions as in Section 8.7, evaluate the reflected irradiance, I_r, of a Fabry–Pérot etalon and verify that $I_r + I_t = I_i$. (Hint: In addition to $tt' = 1 - r'^2$, you will need to apply the correct relation between r and r' based on the Fresnel results of Section 5.3.)

Problem 8.10

Find the highest order fringe visible in a Fabry–Pérot etalon with spacing 1 cm, for light with wavelength 600.000 nm. What is the highest order for a wavelength 600.010 nm? What are the angular radii of the highest order fringes for these wavelengths?

Problem 8.11

The structure of a doublet spectral line is to be examined in a Fabry–Pérot spectrometer. If the separation of the doublet is 0.0043 nm at a wavelength of 475 nm, what is the spacing of the etalon which places the Nth order of one component on top of the $(N + 1)$th order of the other near the centre of the pattern, where $\theta \approx 0$?

Problem 8.12

If the reflectance (see Section 5.3) in a Fabry–Pérot etalon is 60%, find the ratio of the irradiance at maximum to that half-way between maxima.

Problem 8.13

A parallel beam of white light is passed through a Fabry–Pérot etalon at normal incidence and focused on the slit of a spectrograph. Describe the appearance of the spectrum.

   The slit is illuminated also by mercury light. If 200 bands are seen in the spectrum between the blue and green mercury lines (wavelengths 546 and 436 nm), what is the spacing h of the etalon?

Problem 8.14

Consider a Fabry–Pérot etalon with light bouncing within as analogous to a dampened harmonic oscillator. For high reflectivity ($r'^2 \approx 1$), the losses per cycle (two successive reflections, which is considered as 2π radians) are relatively small, and by analogy with a weakly damped oscillator, we can define a "quality factor", Q, in terms of irradiance by $Q =$ (initial I/loss of I per radian); more precisely, if 1,2,3 are three successive rays within the etalon, we define $Q = 2\pi I_1/(I_1 - I_3)$.

   Find Q and show that, for high reflectivity, $F = (2Q/\pi)^2$.

9 Interferometry: Length, Angle and Rotation

Following a method suggested by Fizeau in 1868, Professor Michelson has... produced what is perhaps the most ingenious and sensational instrument in the service of astronomy – the interferometer.

Sir James Jeans, *The Universe Around Us*, Cambridge University Press, 1930.

Optical interferometers, such as the Michelson and the Fabry–Pérot (Chapter 8), can be used to measure distances in terms of the wavelength of light. Most of the interferometers for this purpose are two-beam interferometers, using amplitude division. Their performance may often be improved by using multiple beams, as in the Fabry–Pérot; the principle is the same, but the fringes are sharper.

Interferometers can also be used to measure angular distributions of brightness across sources with small angular diameters. Michelson was again the pioneer, with his stellar interferometer which made the first measurements of the angular diameters of stars.

In this chapter we describe measurements of lengths ranging from a small fraction of a wavelength, in which the purpose may be to test the optical quality of a surface, to some tens or hundreds of metres, where the objective may be to measure the stability of a large structure or to test the theory of relativity. Interferometers measure optical path along a light beam, i.e. the product of geometric path and refractive index; by comparing the optical paths of two light beams very small differences in physical length or refractive index can be measured.

We continue with the measurement of the angular size of light sources, which is achieved by interferometers using elements separated not along a light beam but across a wavefront.

9.1 The Rayleigh Refractometer

We start with the simplest of refractive index measurements. In any two-path interferometer there is a comparison of the optical length of two separate paths. Rayleigh put this to use in measuring the refractive index of a gas. His refractometer (Figure 9.1) was based on Young's double slits, although any other two-beam interferometer could be used. The tubes T_1 and T_2 are in the

Optics and Photonics: An Introduction, Second Edition   F. Graham Smith, Terry A. King and Dan Wilkins
© 2007 John Wiley & Sons, Ltd

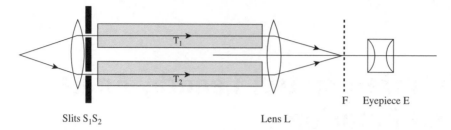

Figure 9.1   The Rayleigh refractometer. S_1, S_2 are illuminated by a common source of light. Interference fringes are formed at the focal plane F of the lens L, and viewed with an eyepiece E. The fringes move across the field of view when the gas pressure is changed in one of the tubes T_1, T_2

separate light paths from the slits S_1 and S_2, illuminated coherently from a single source. When the pressure of gas is changed in one of the tubes, the fringe system, viewed by an eyepiece E at the focus of a long-focus lens, moves across the field of view. A count of the fringes (N) as they move provides a direct measurement of the change in optical path through the tube, and hence the change in refractive index δn as the amount of gas changes. For a tube length l and vacuum wavelength λ_0

$$N = \frac{l\delta n}{\lambda_0}. \tag{9.1}$$

For a dilute gas the refractive index n differs from unity by an amount proportional to density, so that for a fixed temperature $n - 1$ is proportional to pressure. The refractive index obtained from a single measurement can then be used to calculate the value for any other pressure by simple proportion.

Example.   A Rayleigh refractometer is used to measure the refractive index of hydrogen gas. The tubes are $100\,\text{cm}$ long, and a pressure change of $50\,\text{cm}$ mercury gives a count of 145.7 fringes at $\lambda = 589.3\,\text{nm}$. Show that the refractive index n at normal atmospheric pressure ($76\,\text{cm}$ mercury) is given by $n - 1 = 1.305 \times 10^{-4}$.

9.2   Wedge Fringes and End Gauges

Interferometers measuring physical length in terms of light wavelength are used to establish and compare standards of length in the form of *end gauges*. These are metal bars with polished ends that can be used as reflecting mirrors in interferometers.

The simplest comparison that can be made is between two end gauges which are nominally of the same length. They can be placed together on a flat surface, and a partially silvered optically flat glass plate placed on top (Figure 9.2). Thin-film fringes (Fizeau fringes) are then seen in the wedge cavity over each of the gauges, with a spacing depending on the angle between the gauge and the glass plate; the difference in height of the two gauges can easily be found by counting the fringes. Irregularities in the reflecting surfaces are seen as deviations from straight lines in the reflected wedge fringes. The partial silvering of the glass plate gives a multi-beam interference effect, like the transmission fringes in the Fabry–Pérot interferometer. These fringes are so sharp that surface discontinuities down to $0.3\,\text{nm}$ can be detected.

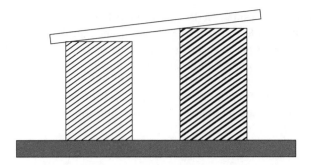

Figure 9.2 Comparison of the lengths of two end gauges. Wedge fringes are formed in the air gap between a glass plate and the ends of the gauges. The spacing of the fringes, which are viewed from above the plate, is a measure of the wedge angle

Interferometers for measuring larger differences in light paths are usually based on the Michelson interferometer. An example of the direct measurement of the length of an end gauge is shown in Figure 9.3. Here the end gauge G_1G_2 is placed firmly in contact with an optically flat reflector M_1, so that M_1 and the end of the gauge G_1 can both reflect light in the field of the interferometer.

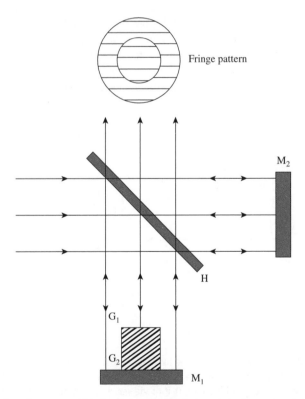

Figure 9.3 End gauge interferometer. This is similar to a Michelson interferometer, but with fixed mirrors. The gauge G_1G_2 is in contact with M_1, so that the centre of the field of view, seen from above, shows fringes from G_1, while the outer part shows fringes from M_1, as shown in the inset diagram

The reflector M_2 is inclined at a small angle, so that the field of view is crossed with parallel fringes, as in Figure 9.3. Part of the field shows fringes from G_1 and part from M_1.

The distance G_1G_2 is measured as a shift of the fringe pattern, in wavelengths of the particular light in use. Only the fractional part can be measured, however. The whole number of fringes can be found by repeating the measurement with other wavelengths of light, e.g. using four different wavelengths of light from a cadmium lamp, each of which is known to about 1 part in 10^8. When the fractional parts are all known, the whole numbers can be found by computation (see Problem 9.1(iii)).

9.3   The Twyman and Green Interferometer

The fringes shown in Figure 9.3 are straight and evenly spaced. This would only be so if the surfaces were precisely flat: any departures from flatness would be seen as distortions in the fringe pattern. Twin beam interferometers can evidently be used to measure the flatness of a reflecting surface. The Twyman and Green interferometer (Figure 9.4(a)) is a twin beam interferometer designed for this purpose. It can also be used to test the transmission properties of transparent optical components.

The Twyman and Green interferometer is a development of the Michelson interferometer in which the source of light is a plane wavefront coherent over the whole field rather than a broad incoherent source. Originally this was achieved by the use of a pinhole source at the focus of a high-quality lens, but now a suitable source is a laser beam. With this arrangement light paths can be compared over the whole field. For example, the arrangement of Figure 9.4(a) may be used to compare the surfaces of two mirrors, while in (b) the optical path through a lens is measured across the whole field. As with the measurement of end gauges in Figure 9.3, the mirror M_1 is slightly tilted, so that a perfect optical system yields a system of parallel fringes across the field of view; imperfections then show as deviations from straightness.

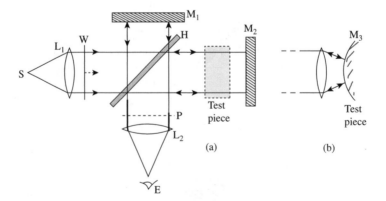

Figure 9.4   Twyman and Green interferometer. The half-silvered plate H and the two mirrors M_1, M_2 are arranged as in a Michelson interferometer, but the source of light is a coherent plane wavefront W, and the fringes are either recorded directly on a photographic plate at P or viewed at the focus of the lens L_2. The field of view is uniformly bright if M_1 and M_2 are exactly aligned. Imperfections in optical path, as for example, (a) through a test piece, show as variations of brightness, or (b) (inset) M_2 can be replaced by a spherical mirror M_3 when a converging lens is to be tested

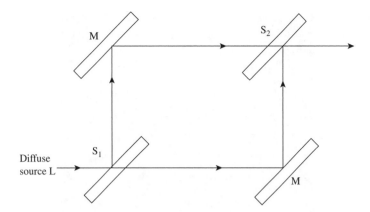

Figure 9.5   Ray paths in the Mach–Zehnder interferometer. The beam splitter S_1 and recombiner S_2 are half-reflecting mirrors. The mirrors M are fully reflecting. An extended light source L may be used

The Mach–Zehnder interferometer, shown in Figure 9.5, is another amplitude-splitting device intended for comparing optical paths in separated beams. The separation of the beams may be large, so that one beam may for example traverse a wind tunnel in the region of a shock wave (Figure 9.6), or it may traverse a plasma cloud in a thermonuclear test reactor. Variations of refractive index across the field are seen as deviations from straightness in a set of plane-parallel fringes.

9.4   The Standard of Length

The internationally defined standard of length is the distance travelled in vacuo by light in unit time.[1] Before 1983 the metre was defined as a number of wavelengths of a narrow spectral line in krypton, but because of the finite width of this line this standard was reproducible only to about 3 parts in 10^8. In contrast, the standard of time can be reproduced with an accuracy of 1 part in 10^{13}; this is achieved by linking a clock such as a quartz crystal oscillator through a series of harmonic generators to the caesium standard.

The velocity of light $c = 299\ 792\ 458\ \text{m s}^{-1}$ is therefore a defined constant which was chosen in 1983 to give agreement, to the best possible accuracy, between the new and old definitions of the unit of length. If the frequency of a narrow-bandwidth laser can be measured in terms of the unit of time, then the wavelength is known in standard units, and it can be used to calibrate a secondary standard such as a metre-long end gauge at the wavelength of a chosen spectral line.

Michelson made the first measurements of the metre in terms of wavelengths in a different era, when the metre was defined by a mechanical standard; he was therefore measuring the wavelength of light rather than measuring the metre. The moving mirror of a Michelson interferometer was mounted on a carriage on accurate sliding guides alongside the standard, and with a microscope attached for

---

[1]The definition (by the Conference Gènèrale des Poids et Mesures, 1983) is: "the metre is the length of the path travelled by light in vacuo during a time interval of 1/299 792 458 of a second". The standard of time, the second, is defined using precision atomic clocks as the duration of 99 162 631 770 periods of radiation between two hyperfine levels of the caesium atom.

Figure 9.6   Interference fringes obtained in a Mach–Zehnder interferometer, showing variations of refractive index around a wind-tunnel model

setting on the fiducial marks of the metre. A direct count of fringes as the carriage traversed the metre would involve some millions of fringes; furthermore, the spectral line sources available to Michelson were not sufficiently narrow to allow the use of path difference as large as a metre. He therefore used an intermediate-sized standard of length called an etalon, and built up the full length by adding a series of etalons.

9.5   The Michelson–Morley Experiment

This classic experiment, which is now regarded as a test of special relativity, was originally devised as a measurement of the velocity of the Earth through space. If light could be regarded as a wave in a medium, called the ether, through which the Earth was moving, the velocity of light as measured on Earth would depend on its direction of travel. A Michelson interferometer, with one light path along this direction and the other at right angles, would show the effect as a fringe shift. The shift could be detected by rotating the interferometer so as to interchange the optical paths (Figure 9.7). To preserve the stability of the interferometer, it was mounted on a stone bed floated in mercury.

In the Michelson–Morley experiment both light paths were increased to 11 metres by folding them between a series of mirrors. The expected effect, according to the ether theory, of the Earth's orbital velocity was nevertheless only 0.4 of a fringe. This was to be detected by rotating the whole apparatus by 90°, so interchanging the arms parallel and perpendicular to the Earth's motion.

The path difference expected from simple "classical" arguments is found from the times of travel t_1 and t_2 along the two beams, assigning a velocity v to the ether drift. Assume for simplicity that the Sun is essentially at rest in the ether. For an observer moving with the velocity v of the Earth in its orbit (and ignoring the much smaller rotational velocity $v_{\mathrm{rot}} \simeq 2\% v_{\mathrm{orb}}$), the ether should then be moving past the interferometer with the same velocity v, so that the velocity of light along and against

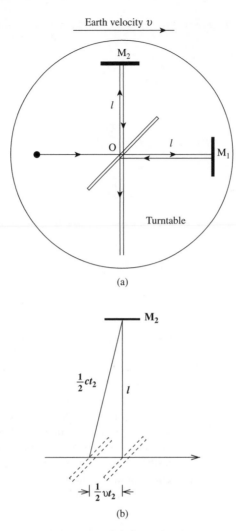

Figure 9.7 (a) The Michelson–Morley experiment. The path to mirror M_1 is initially along the direction of the Earth's orbital motion; the whole interferometer can be rotated to bring the path to M_2 into this direction. (b) The path followed by beam OM_2, according to classical theory, as viewed in the rest frame of the ether. The arrow indicates that the apparatus is moving to the right at speed v

the ether drift would be $c + v$ and $c - v$. For a light path l the time t_1 along and against the ether flow is

$$
\begin{aligned}
t_1 &= \frac{l}{c - v} + \frac{l}{c + v} \\
&= \frac{2l}{c} \gamma^2
\end{aligned}
\tag{9.2}
$$

where $\gamma = (1 - v^2/c^2)^{-1/2}$.

The time t_2 transverse to the flow is increased by the movement of the mirror with velocity v, as shown in Figure 9.7, so that

$$\left(\frac{ct_2}{2}\right)^2 = \left(\frac{vt_2}{2}\right)^2 + l^2 \tag{9.3}$$

giving

$$t_2 = \frac{2l}{c}\gamma \tag{9.4}$$

$$t_1 - t_2 = \frac{2l}{c}(\gamma^2 - \gamma). \tag{9.5}$$

Expanding each term in γ into a series gives a good approximation

$$t_1 - t_2 \approx \frac{l}{c}\left(\frac{v^2}{c^2}\right). \tag{9.6}$$

The expected fringe shift would be $c(t_1 - t_2)/\lambda$, which would reverse when the interferometer was rotated through $90°$. For a path $l = 11$ m, wavelength $\lambda = 550$ nm and an ether velocity $v = 30$ km s^{-1}, which is the Earth's orbital speed, there would be a shift in the fringe pattern by 0.4 fringes on rotation. No such fringe shift was detected. Just in case the Earth happened to be moving at the same velocity as the ether, the experiment was repeated six months later when the Earth's velocity was reversed. Again there was no fringe shift.

We should realize that this result was a great surprise when it was first obtained in 1887, which was 18 years before Einstein published his theory of special relativity. The first reasonable explanation came from Lorentz and from FitzGerald, who independently suggested that moving bodies contract in the direction of motion by the factor γ in our analysis above. Poincaré commented that this was a conspiracy of nature which made it impossible to detect an ether wind by *any* experiment! The null result of the Michelson–Morley experiment is of course now seen to be entirely in accordance with the special theory of relativity, in which the velocity of light is invariant between any pair of frames of reference in uniform relative motion.

9.6   Detecting Gravitational Waves by Interferometry

Although the existence of gravitational waves has been demonstrated by the orbital decay in a binary pulsar system that is consistent with loss of radiated gravitational energy, their direct detection on Earth is extremely difficult. A detectable gravitational wave might be radiated from a catastrophic event such as the coalescence of a binary pair of stars, but the amplitude at the Earth would correspond to a transient or low-frequency change in length scale by a factor of only 10^{-21} or less. Gravitational wave detectors attempt to obtain such a sensitivity by using a laser interferometer shown diagrammatically in Plate 3.[*]

---

[*]Plate 3 is located in the colour plate section, after page 246.

The Michelson interferometer used in the detector compares the length of the two arms, which are typically each about 1 kilometre long. The distant mirrors are mounted on large suspended masses, which will move differentially depending on the direction of the gravitational waves. There are (only!) 3×10^9 wavelengths of visible light (wavelength 600 nm) in a double journey along one beam, so the requirement is to measure a fringe shift of 3×10^{-12} fringes! Obviously the Fabry–Pérot technique must be employed, so obtaining very sharp fringes, and a very stable and high-powered laser must be used to allow the positions of the edges of the fringe profile to be measured with great accuracy. The achievement of a positive detection must be rated as one of the greatest challenges in the measurement of lengths by optical interferometry.

9.7   The Sagnac Ring Interferometer

In 1913 Sagnac[2] constructed an interferometer in which the two beams follow the same optical path but in opposite directions (Figure 9.8a). The ring may have three or four mirrors; as in the interferometers described earlier in this chapter, a small angular misalignment of any of them will produce a parallel fringe pattern. The Sagnac interferometer can be used to detect rotation about an axis perpendicular to the ring, observed as a shift of the fringe pattern. In such a rotation one optical path is effectively shortened in comparison with the other. (Special relativity in no way suggests that rotation cannot be detected, in contrast to the uniform linear velocity which was the objective of the Michelson–Morley experiment.)

As we shall describe in the next section, a modern version of the Sagnac interferometer uses glass fibres to guide light round a circular path, (Fig. 9.8b). Long lengths of fibre can be used, so that light makes many circuits in both directions before being recombined in a single detector. We analyse this circular version, considering a single loop with radius R rotating with angular velocity Ω.

For a non-rotating loop light completes a circuit in time $t = 2\pi R/u$, where $u = c/n$ is the velocity of light in the fibre. The velocity u' observed in the laboratory frame, when the fibre moves with velocity v is found from the relativistic law of velocity addition, which gives

$$u' = \left(\frac{c}{n} + v\right) \Big/ \left(1 + \frac{v}{nc}\right).$$

For $v \ll c$ this reduces to[3]

$$u' = \frac{c}{n} + v\left(1 - \frac{1}{n^2}\right).$$

---

[2]G. Sagnac, *Comptes Rendus*, **157**, 708, 1913; *Journal de Physique*, **4**, ser. 5, 177, 1914.
[3]The factor $(1 - 1/n^2)$ was introduced by Fresnel as a partial dragging of the ether by a moving-material medium, to account for Fizeau's measurements of the velocity of light in water flowing along a tubular path (H. Fizeau, *Comptes Rendus*, **33**, 349, 1851). When special relativity did away with the ether, his explanation became untenable.

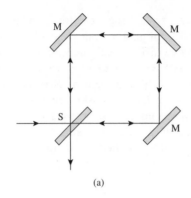

(a)

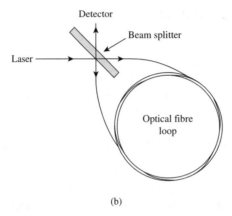

(b)

Figure 9.8   The Sagnac ring interferometer: (a) in the original four-mirror version light traverses the circuit in opposite directions, reflected by the mirrors M, and the two rays recombine at the beam splitter S; (b) the optical fiber version

Setting $v = \pm\Omega R$, the light moving with the rotation travels faster, and that opposite to the rotation slower, according to

$$u_\pm = c/n + v(1 - 1/n^2) = c/n \pm \Omega R(1 - 1/n^2). \qquad (9.7)$$

The ring rotates during the transit times t_+, t_- round the circuit giving an extra path $\Omega R t_\pm$. The two transit times $t_\pm$ then satisfy $uU_\pm t_\pm = 2\pi R \pm \Omega R t_\pm$, which can be solved for the times to give

$$t_\pm = \frac{2\pi R}{(u_\pm \mp \Omega R)} = \frac{2\pi R}{(c/n \mp \Omega R/n^2)}. \qquad (9.8)$$

The difference in arrival times is

$$\Delta t = t_+ - t_- = \frac{4\pi \Omega R^2}{[c^2 - (\Omega R/n)^2]}. \qquad (9.9)$$

In practice, $R^2\Omega^2 \ll c^2$, and

$$\Delta t = \frac{4\pi R^2 \Omega}{c^2} = \frac{4A\Omega}{c^2}, \tag{9.10}$$

where A is the area of the ring. This time difference appears as a shift of ΔN fringes in the interference pattern at the detector, where

$$\Delta N = \frac{c\Delta t}{\lambda} = \frac{4A\Omega}{c\lambda}, \tag{9.11}$$

where, to adequate accuracy, λ is the wavelength inside a fibre at rest.

Note that the fringe shift is directly proportional to the area A of the ring. This is generally true for any shape (see Problem 9.2) including the square in Figure 9.9(a).

The ring interferometer gives a direct measure of the rate of rotation about the axis of the ring, since the displacement of the fringe pattern is proportional to Ω. In 1925 Michelson and Gale[4] used such a system to measure the rate of rotation of the Earth, so demonstrating the contrast between the detectability of linear and rotational motion in special relativity. (See Problem 9.3.)

9.8 Optical Fibres in Interferometers

In the twin beam interferometers which we have described so far, two light beams from a single source are recombined after travelling different paths in air or blocks of glass. We saw in Chapter 6 that light can be guided down long lengths of glass fibre with very little loss, so that some remarkably simple and stable interferometers can be constructed using long lengths of glass fibre. Light from a compact semiconductor laser (Chapter 17) can be split between two fibre paths, and recombined in a simple diode detector (Chapter 20). As in the original Rayleigh interferometer, the sum is dependent on the relative phase of the two beams, allowing measurement of differences in length or refractive index and their dependence on other parameters in the two light paths.

Figure 9.9 shows a basic system in which light from a laser is split between two fibre paths, usually of nearly the same length, and recombined in a detector. The relative phase is detected by incorporating a modulator, which periodically reverses the phase in one arm; the detector output is then observed to be modulated at the switch frequency, and the modulation depth depends on the phase difference between the two beams. The output is selectively amplified at the switch frequency. This is the fibre equivalent of the Mach–Zehnder interferometer. It may be adapted for measurements of any parameter which affects the phase path differently in the two arms, such as temperature differences, or mechanical strain.

The Sagnac ring interferometer has been developed into a useful gyroscope, one form of which uses a fibre optic ring as in Figure 9.8(b). Although a long fibre, making a large number of turns, can be used, the phase differences to be measured are small, as we have seen for the Michelson–Gale experiment.

For example, consider the measurement of the rotation of the Earth ($15° \, \text{h}^{-1}$). For a ring with $R = 0.2 \, \text{m}$, made of 1000 turns of optical fibre and operating with an He–Ne laser at 632.8 nm, the phase shift will be 1.2×10^{-3} radians. (This can be measured by inverting the ring, an expedient which was not available in the Michelson–Gale experiment.)

---

[4]A. A. Michelson and H. G. Gale, *The Astrophysical Journal*, **61**, 140, 1925.

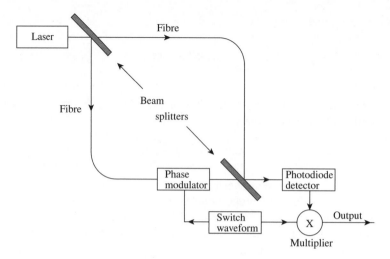

Figure 9.9   Optical fibre interferometer

9.9   The Ring Laser Gyroscope

In another form of the Sagnac gyroscope,[5] the ring laser gyroscope, the laser source is contained within the ring, which acts as a laser cavity (see Chapter 15) producing clockwise and anticlockwise travelling laser waves. The rotation of the ring leads to a difference between the two resonant laser frequencies, which is observable as a beat frequency.

The resonance condition in a ring laser is that the total round-trip distance L must be an integral number of wavelengths, $L = m\lambda$.[6] The optical path in a non-rotating ring is the total perimeter length L, so that the resonant frequency v of the mth mode is $v_m = m\,c/L$. In the rotating ring the optical path for the two directions changes by $\pm \Delta L$, and the resulting beat frequency Δv between the two oscillations is given by

$$\Delta v/v = \Delta L/L = \Delta t/t_R \qquad (9.12)$$

where $\Delta t = 4A\Omega/c^2$, as in equation (9.10), and $t_R = L/c$ is the time for light to travel round the ring. Note that the beat frequency $\Delta v = (4A/\lambda L\Omega)$ is proportional to A/L, indicating that the sensitivity is proportional to the size of the ring.

The measurement of a frequency shift greatly increases the sensitivity of the ring laser gyroscope compared with the passive Sagnac or fibre optic ring, where the light source is external to the ring and rotation is measured as a phase shift.

For an equilateral triangular ring with each side of 20 cm (giving $A = 0.017\,\mathrm{m}^2$) and operating with an He–Ne laser at 632.8 nm, a rotation of $15°\,\mathrm{h}^{-1}$ or $7.3 \times 10^{-5}\,\mathrm{rad\,s}^{-1}$ (equivalent to the Earth's

---

[5]See R. Anderson et al., "Sagnac effect: A century of Earth-rotated interferometers", *American Journal of Physics*, **62**, 975, 1994.

[6]The ring resonator differs from the linear resonator (Chapter 15) in that the laser radiation in the ring resonator is a travelling wave, and the electric field distribution must be reproduced after each transit of the ring. In a linear cavity, length L, the resonance is a standing wave, for which $L_m = m\lambda/2$.

rotation) gives a frequency difference $\Delta v = 8.8\,\text{Hz}$. This frequency difference can readily be measured as a beat frequency using the heterodyne technique described in Section 12.10. The ring laser gyroscope is a highly sensitive instrument; in a large-scale version measurements can be made to an accuracy of $\Delta\Omega/\Omega \sim 10^{-8}$.

For frequency stability the ring laser gyroscope is constructed within a block of low-expansion glass (e.g. the glass ceramic Zerodur, which has near-zero linear expansion coefficient at $300\,\text{K}$). The block is drilled to accommodate the laser medium and laser beams and the mirrors which define the ring. These mirrors represent a notable technical achievement in having not only extremely low scatter (to avoid the clockwise and anticlockwise waves locking in frequency) but also a reflectivity near to 99.9999%.

The ring laser gyroscope has replaced the spinning wheel gyroscope in many applications. It does not require any moving parts and can be directly connected to a vehicle avoiding the need for gimbals. The applications include navigation for aircraft and ships, measurement of the Earth's rotation and its variation, seismic and geophysics monitoring, tidal variation and in large-scale, highly sensitive forms for fundamental tests in general relativity and gravitation.

9.10 Measuring Angular Width

Interference and diffraction theory has in previous chapters considered waves originating from an idealized point source, or from an ideal plane wavefront. In the second part of this chapter we are concerned with the effect on the interference fringes when the source has a finite size; in general this reduces the *visibility* of the fringes. The reduction of visibility enables a measurement to be made of the size of the source; more precisely, the measurement is of the angular spread of plane waves entering the interferometer. The most important example of such a measurement is Michelson's stellar interferometer.

We first consider Young's double source interferometer, which was introduced in Chapter 8. This is an example of an interferometer using *division of wavefront*, in contrast to *division of amplitude*, as discussed in Chapter 8.

Figure 9.10 shows two pinholes or slits S_1, S_2 illuminated by a point source of monochromatic light L. Light from S_1 and S_2 spreads and overlaps in the shaded region: throughout this region there is interference between the two sets of waves, and interference fringes would appear on a screen or photographic plate placed anywhere in the region. (The interference fringes are said to be *non-localized*, in contrast to the *localized* fringes seen in thin films which we discussed in Chapter 8.) We must distinguish between the effects of an extended, rather than a point, source and the effect of slits with a finite width. The same arguments will apply to many related types of interferometer which involve two overlapping light beams from a single small source. Fresnel used

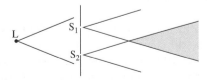

Figure 9.10 Young's double slit. Overlapping beams from two slits, S_1, S_2, illuminated by the same source L. Interference occurs in the whole of the shaded area

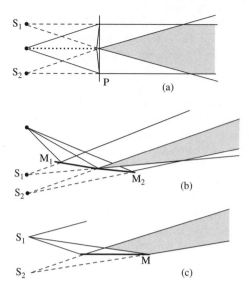

Figure 9.11 (a) Fresnel's biprism; (b) Fresnel's double mirror; (c) Lloyd's mirror. In each arrangement twin beams from the same source overlap in the shaded area, appearing to diverge from the twin sources S_1, S_2

the thin biprism of Figure 9.11(a) and the nearly coplanar mirrors of Figure 9.11(b). A particularly simple arrangement is Lloyd's mirror (Figure 9.11(c); here the two sources are the slit S_1 and its image S_2). An interesting feature of Lloyd's mirror is that the light reflected at grazing incidence to form S_2 suffers a phase reversal at reflection (see Chapter 5), so that the interference fringes are exactly out of step with those of the double slit.

Similar arrangements can be devised for the much longer wavelength radio waves. A classic example is the equivalent of Lloyd's mirror (Figure 9.12(a)) used in early radio astronomical observations in Sydney, Australia. Although the geometry is that of Lloyd's mirror, the radiation moves in the reverse direction and, in place of a slit to emit the waves, there is a receiver to detect them. The radio telescope, mounted on a cliff overlooking the sea, received radio waves of around $1\frac{1}{2}$ metres wavelength (frequency 200 MHz) from the Sun and other celestial radio sources as they rose above the horizon. Both direct and reflected radio waves were received; as the Sun rises the path difference between them changes and produces a set of interference fringes. A typical trace of the interference fringes is shown in Figure 9.12(b); this was recorded at a time of strong and variable radio emission from above a sunspot (see Problem 9.4).

9.11 The Effect of Slit Width

So far each slit in Young's experiment has been supposed to be the source of a uniform cylindrical wave, which for an ideal narrow slit covers 180°. In practice each slit may be many wavelengths wide, and as we will see in Chapter 10 light emerges from each slit only over a restricted angle. The simplest way to think of what happens is to notice in Figure 9.10 that fringes cannot be observed in any particular direction θ unless each slit, acting as a diffracting aperture, sends some light that way. Each slit contributes according to its own diffraction pattern, while the relative phase of the two

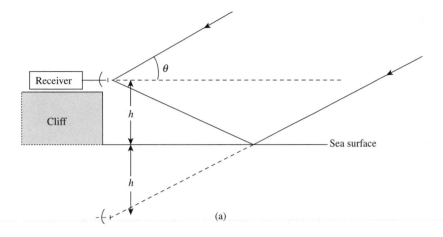

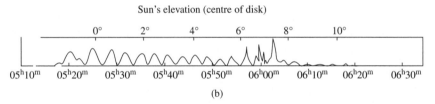

Figure 9.12   Lloyd's mirror in radio astronomy. (a) The radio telescope receives radio waves both directly and indirectly from the sea. (b) The recorded radiation from the Sun as it rises above the horizon. The interference fringes are disturbed by refraction near the horizon, and by solar outbursts (L.L. McCready, J.L. Pawsey and Ruby Payne-Scott. Proc. Royal Soc. A190, 357, 1947)

contributions is $(2\pi d/\lambda)\sin\theta$ where d is the spacing of the slits. Young's fringes are then observed within the intensity envelope of the diffraction pattern of a single slit, as shown in Figure 9.13.

We have seen in Section 8.2 that for two thin slits (width $w \ll \lambda$) separated by distance d, the amplitude for interference between them is proportional to a cosine function, or $A_{\mathrm{I}}(\sin\theta) = \cos(\pi d \sin\theta/\lambda)$, where we normalize to unity on-axis. When the slits are thick, the overall amplitude $A(\sin\theta)$ equals this cosine function modulated by an additional factor due to diffraction by either slit: $A(\sin\theta) = A_{\mathrm{I}}(\sin\theta)A_{\mathrm{D}}(\sin\theta)$. In Chapter 10, we shall show that the diffraction pattern takes the form

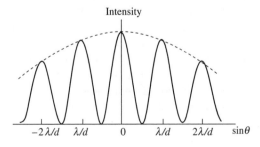

Figure 9.13   Young's fringes with slits of finite width w and separation d. The broken envelope of the fringes is the diffraction pattern of each slit and would reach its first zero at $\pm\lambda/w$

of a sinc function

$$A_D(\sin\theta) = \frac{\sin(\pi w \sin\theta/\lambda)}{(\pi w \sin\theta/\lambda)}. \tag{9.13}$$

Using the notation $l = \sin\theta$, the overall amplitude goes as

$$A(l) = \frac{\sin(\pi w l/\lambda)}{(\pi w l/\lambda)}\cos(\pi d l/\lambda). \tag{9.14}$$

Given that d is the centre-to-centre slit separation, the width w satisfies $w \le d$ because, for $w > d$, the two slits would overlap.

Example. Verify that equation (9.11) makes sense for the limiting values of slit width: (i) $w = 0$, and (ii) $w = d$.

Solution. (i) In the limit $w = 0$, we can show that the sinc function goes to unity because, for small x, $\sin(x)/x = x/x = 1$. So $A(\sin\theta)$ correctly reduces to the cosine function of two thin slits. (ii) For $w = d$, the two slits merge into a single slit of width $2w$. Sure enough, using the identity $\sin(2x) = 2\sin(x)\cos(x)$, the amplitude reduces to

$$A(\sin\theta) = \frac{\sin(2\pi w \sin\theta/\lambda)}{(2\pi w \sin\theta/\lambda)},$$

the diffraction pattern for a slit of width $2w$.

   The same result may be reached via the convolution theorem in Fourier transforms (Chapter 4). In this equivalent approach the twin slits are described as the convolution of a top-hat function, width w, with a pair of delta functions, spaced d apart; the Fourier transform, which is the angular distribution of diffracted plane waves, is the product of the two separate transforms.

9.12   Source Size and Coherence

So far the wavefronts leaving the two apertures have been considered as parts of a single plane monochromatic wave. No wavefront is ideally plane and no wave is perfectly monochromatic, although laser light can approach the ideal very nearly. We now investigate interference between non-ideal wavefronts, such as those obtained from a source of finite size.

   In any ordinary light source, such as an incandescent filament or a sodium lamp, the output of the lamp is made up of the sum of the light waves produced by a very large number of individual atoms. Their phases vary randomly so that, on average, cross-terms vanish and their time-averaged intensities add. Light derived from different parts of the source cannot interfere. Interference is only observable if the two interfering beams are both derived from the same region of the source, when the two beams are said to be *mutually coherent*.[7] Two beams derived from any other part of the

---

[7]The concept of coherence is discussed in more detail in Chapter 13.

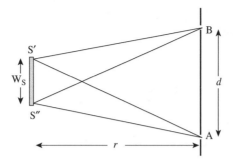

Figure 9.14 A source of finite size illuminating Young's slits. Fringes of high visibility can only be seen if the path difference AS-BS only changes by a small fraction of a wavelength for all positions of S from S' to S''

source may also interfere, but the interference patterns from different parts of the source may not coincide in time and space. But if the interference patterns made by each component of an extended source are identical, then they add to produce the same pattern as that from a point source.

Let us look at the problem in the case of Young's slits. Suppose that the plane wave we have so far considered is replaced by the wavefront from a source S' S'' of finite width w_s at a large distance r from the slits, as in Figure 9.14. Each component of the source produces interference fringes with the same spacing, but the position of the interference fringes depends on the relative phase of the wave as it arrives at the slits. This relative phase depends on the difference in distance between the source and the two slits; if this difference is nearly the same for all components, the fringe patterns will coincide and add to give the normal point source fringe pattern.

The interference patterns will coincide nearly enough to give a fringe pattern with high visibility if the relative path length from each end of the source is the same within a small fraction of a wavelength. Seen from the slits this means that the directions of S' and S'' must be the same within an angle of λ/d. The waves at A and B can then again be regarded as coherent, as they were when the light came from a single point source. This gives us a condition for coherence between the light at A and B:

$$\frac{w_s}{r} \ll \frac{\lambda}{d}. \tag{9.15}$$

A perfect point source would have the property that all pairs of points on its wavefront would be coherent. The important result of equation (9.15) gives the condition under which a finite source may be regarded as a point source as far as a particular diffracting system is concerned. Looking back at the source from the diffracting system, equation (9.15) may be restated as

angle subtended by source $\ll \lambda$ */(linear size of diffracting system).*

9.13 Michelson's Stellar Interferometer

In the previous section it was shown that one condition for fringes to be produced by Young's slits was equation (9.15), a restriction on the angular size of the source seen from the slits. This suggests that such a system might be used to *measure* the angular size of a source by altering the slit separation until fringes could not be seen. This is the principle of *Michelson's stellar interferometer*. Some modification of the simple Young system is needed to make an interferometer capable of

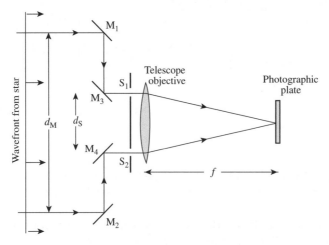

Figure 9.15   Michelson's stellar interferometer. Light from parts of the star's wavefront d_M apart is made to interfere in the focal plane of the telescope. The visibility of the resultant fringes as d_M is varied allows some estimate to be made of the star's angular diameter

producing fringes by the light of a star; this concerns the method of combining the light that comes through the two slits. In Young's system diffraction at the slits is relied upon to make part of the emergent light from each slit reach the same region so that interference can take place. This means that the slits have to be narrow. In Michelson's system (Figure 9.15) the slits are replaced by two large plane mirrors M_1 and M_2 a distance d_M apart. Each reflects light inwards to two further inclined mirrors M_3 and M_4. These reflect two parallel beams of light into a telescope objective (which may be reflecting or refracting). Each beam forms an image of the star in the focal plane F, and fringes are seen crossing the diffraction disc of the combined image. The two apertures S_1 and S_2 in front of the objective, which limit the size of the beams, act like Young's double slits with a spacing d_S.

Suppose first that a star is observed which has so small an angular diameter ϕ_0 that the inequality (9.15) is satisfied, i.e. $\phi_0 \ll \lambda/d_M$. Then the wavefronts falling on M_1 and M_2 are coherent and hence (if the mirrors are perfectly adjusted) S_1 and S_2 are illuminated by coherent wavefronts. In the focal plane F the interference pattern of the slits is observed; it consists of intensity fringes with a spacing $(\lambda/d_S)f$, where f is the focal length of the telescope. Notice that the fringe *spacing* is independent of the separation d_M of the outer mirrors M_1 and M_2. The extent of the area over which fringes can be observed is limited by diffraction at the individual apertures, width w_S, and therefore, recalling the diffraction envelope of Figure 9.13, is of the order of $(\lambda/w_S)f$; this is the approximate width of the central maximum of the diffraction pattern of each aperture.

Now consider the effect of observing a star of finite angular diameter ϕ_0. Suppose it is to be divided up into very narrow strips of width $\delta\phi$ each of which satisfies the condition

$$\delta\phi \ll \frac{\lambda}{d_M}. \tag{9.16}$$

Then each such strip produces a set of fringes, but each set is non-coherent with any other set, originating as it does from a different part of the source. These sets of fringes then add

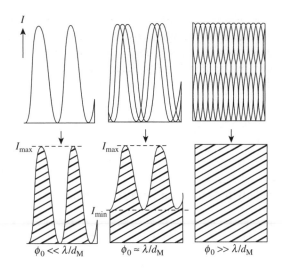

Figure 9.16 Fringes in three cases with the Michelson interferometer. The angular diameter of the observed star is ϕ_0

non-coherently; that is to say, their *intensities* add. Each elementary strip on the star produces its own fringe system which overlaps the others to a greater or lesser degree according to the star's angular diameter. Note that there is no *interference* between fringes from different parts of the star (which are non-coherent), only addition of intensity. Three cases can easily be distinguished, and are illustrated in Figure 9.16, where only the central portion of the pattern is considered, so that the $\sin\psi/\psi$ term in the single slit pattern which gradually modulates the cosine fringes is usually taken as unity. The sets of cosine-squared fringes from different parts of the star will be shifted sideways in y, the total shift in y being $\phi_0 f$, as is easily shown by geometrical considerations. An angular movement ϕ amounting to λ/d_M causes a difference of path of one wavelength for the light going by the two routes. Examining the three cases in turn shows that more or less blurring of the fringes occurs.

1. $\phi_0 \ll \lambda/d_M$. The condition (9.15) is satisfied and the transverse shift of the fringes from different parts of the source is slight. Completely dark minima will be seen.

2. $\phi_0 \simeq \lambda/d_M$. The transverse shift of the fringes from different parts of the source is the same order as their spacing λ/d_S. There will be no completely dark minima, but a sinusoidal variation of the intensity will be seen.

3. $\phi_0 \gg \lambda/d_M$. The overlapping of the sets of fringes from different parts of the source is complete; no variation of the intensity will be seen.

The *fringe visibility* $V(d_M)$, given by

$$V = \frac{I_{max} - I_{min}}{I_{max} + I_{min}}, \tag{9.17}$$

is used in the measurement of source diameter and the brightness distribution across the source. The maximum and minimum intensities used here are illustrated in Figure 9.16. To see how V varies quantitatively with d_M it is necessary to define the brightness distribution of the source as a function of ϕ. Let this be $B(\phi)$, and let $B(\phi)$ be symmetrical about the centre of the source. Then each elementary strip of the source, $d\phi$ wide and at a small angle ϕ from the centre of the source, produces a fringe system with intensity proportional to $B(\phi)$ and displaced by angle ϕ from the centre of the fringe system. The intensity of a fringe maximum now becomes the integral

$$I_{max} = a \int_{source} B(\phi) \cos^2 \frac{\pi d_M \phi}{\lambda} d\phi$$
$$= \frac{a}{2} \int_{source} B(\phi) \left(1 + \cos \frac{2\pi d_M \phi}{\lambda} \right) d\phi \tag{9.18}$$

where a is a constant. Similarly the minimum intensity becomes

$$I_{min} = a \int_{source} B\phi \left(1 - \cos^2 \frac{\pi d_M \phi}{\lambda} \right) d\phi$$
$$= \frac{a}{2} \int_{source} B(\phi) \left(1 - \cos \frac{2\pi d_M \phi}{\lambda} \right) d\phi. \tag{9.19}$$

The fringe visibility is therefore

$$V = \frac{\int B(\phi) \cos(2\pi d_M \phi / \lambda) d\phi}{\int B(\phi) d\phi}. \tag{9.20}$$

For a source of uniform brightness $B(\phi)$ over an angular width ϕ_0

$$V = \frac{\sin(\pi d_M \phi_0 / \lambda)}{\pi d_M \phi_0 / \lambda}. \tag{9.21}$$

This is a sinc function[8] similar to that representing the Fraunhofer diffraction pattern of a single slit. There is in fact a close relation between the fringe visibility function equation (9.20) and the diffraction formula equation (7.20), which can be traced through the fact that the numerator of equation (9.20) is the cosine Fourier transform of the brightness distribution across the source.

Equation (9.21) may be taken to represent the variation of visibility with ϕ_0 at a fixed spacing d_M; alternatively if d_M can be varied the fall of visibility can be used to measure ϕ_0 for a given star.

Michelson in the early 1920s set up such an interferometer at Mount Wilson on the 100 inch diameter telescope, in which the maximum spacing of the outer mirrors was about 6 metres. The angular separation of the interference fringes at this spacing was 10^{-7} radians, or 0.02 seconds of arc. Even this small angle is large compared with the angular diameters of most nearby stars, and thermal instabilities in the air paths of the interferometer did not allow precise measurements of fringe visibility. Nevertheless the diameters of a small number of red giants could be measured. The first of these was Betelgeuse, which was found to have an angular diameter of 0.047 seconds of arc. From

---

[8]According to its definition, equation (9.17), V is always non-negative, but equation (9.21) violates this for certain values of ϕ_0. This reflects reversed identities of I_{min} and I_{max} in equations (9.18) and (9.19). The correction is to replace $\sin(\pi d_M \phi_0 \lambda)$ in equation (9.21) by its absolute value. See Chapter 13 for a further discussion.

this measurement and the distance (known from measurements of parallax) the linear size turned out to be 300 times that of the Sun, and large enough to enclose the Earth's orbit! Only a few such stars could be measured with the 6 m spacing. Michelson attempted to use larger spacings, but thermal instabilities in the separated light paths made the interference fringes fluctuate rapidly and impossible to observe. Larger spacings can, however, be used if a detector with sufficiently rapid response is used; interferometers using pairs of large optical telescopes separated by some tens of metres are now in regular use.

9.14 Very Long Baseline Interferometry

Interferometers with very much longer separations can be used at radio wavelengths, where the effects of random refraction in the atmosphere are negligible. Radio telescopes up to 200 km apart in the MERLIN array centred on Jodrell Bank were connected by radio links in a system closely analogous to Michelson's stellar interferometer. (The same fringe spacing of 0.02 arcseconds used by Michelson is obtained at 200 km spacing using a radio wavelength of 2 centimetres.)

Optical fibres are now used for direct connections between the separate radio telescopes over long baselines, but radio interferometry is routinely used even where direct connection is impossible, using baselines extending over some thousands of kilometres. Instead of directly transmitting the radio signals to a common receiver, they are recorded on magnetic tapes which are subsequently transported and replayed into the common receiver. The relative phase of the signals must be preserved in this operation; this is achieved by using very stable oscillators as phase references at the separated receivers. Very long baseline interferometry (VLBI) with a baseline of 1000 km and a wavelength of 1 centimetre has a fringe spacing of 2 milliarcseconds, giving a very much greater angular resolution than any optical interferometer. Surprisingly, there are some distant and powerful radio sources, the quasars, which demand still longer baselines; this has been achieved by using a radio telescope in an orbiting satellite as one element of a VLBI system, giving baselines up to 6000 kilometres.

9.15 The Intensity Interferometer

The difficulties of achieving stable interference fringes in Michelson's stellar interferometer were overcome in a remarkable way by the intensity interferometer of R. Hanbury Brown and R.Q. Twiss. This was originally conceived by Hanbury Brown at Jodrell Bank for use at radio wavelengths. Longer baselines of radio versions of Michelson's interferometer were required so as to increase their resolving power, but the links which conveyed the radio frequency signals to the central station where they could interfere and produce fringes did not have the necessary phase stability.

Interference between two radio signals can be observed, as in optical interferometers, by detecting the sum of the two signals, producing the product $(A_1 + A_2)^2 = A_1^2 + A_2^2 + 2A_1A_2$; averaged over time the first two terms give the intensities of the separate signals, and the product A_1A_2 is the interference term, which will depend on their relative phase as well as their amplitude. Alternatively in the radio domain it is possible to multiply two electronic signals directly in a *correlator*, to give the product A_1A_2. In either case the interference between the two signals is measured as a *correlation* between them (see Chapter 13).

Hanbury Brown's suggestion was that instead of conveying back the amplitude of the radio waves received at each end from the radio source, the intensity only need be conveyed. This suggestion may

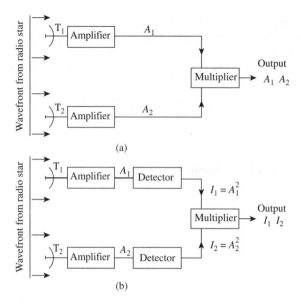

Figure 9.17 Radio interferometers. (a) A conventional form similar to Michelson's. The chief difficulty is in conveying the amplitudes A_1 and A_2 from the radio telescopes T_1 and T_2 which may be intercontinental distances apart. (b) The intensity interferometer where the much more easily handled intensity is brought in, the amplitude being squared (in detection) at each end

seem ridiculous, as the intensity does not depend on phase; however, the signals from radio sources, like other naturally occurring radio signals, are characteristically noisy and have fluctuations in intensity. According to Hanbury Brown, these fluctuations in intensity would correlate if the two stations were close, but as the baseline was increased the correlation would fall off in a way that would allow the angular diameter of the source to be determined. We discuss this concept in more detail in Chapter 13.

These two radio versions of Michelson's interferometer are shown in Figure 9.17. The easing of the problem of conveying the information to the central station this allowed was enormous. The radio frequencies occupying a bandwidth of several megahertz must be transmitted in a way that preserves phase; the intensity fluctuations are instead at the low frequencies normally carried by telephone lines, and can be transmitted without loss over large distances. The system was used first by R. C. Jennison and M. K. das Gupta at Jodrell Bank Radio Observatory in 1951 in a measurement of the angular diameter of the second most powerful radio source in the sky, that in Cygnus. Encouraged by its success at radio wavelengths, Hanbury Brown then proceeded to apply the same technique at optical wavelengths. Here there seemed to be a fundamental question: could an optical photon detector be used to measure correlation between two light beams? An initial laboratory-scale demonstration[9] using a mercury lamp showed that it could, so that the optical stellar interferometer should work.

---

[9]R. Hanbury Brown and R.Q. Twiss, *"Correlations between photons in two coherent beams of light,"* Nature, **127**, 27, 1956.

The equipment for the new intensity interferometer consisted of two searchlight mirrors focussing the light of the observed star onto the cathodes of two photomultipliers. The observed fluctuations in current were then correlated in an electronic multiplying circuit. The equipment was used to observe Sirius on every possible night in the winter of 1955–6, when a total of only 18 hours observing was achieved. As the two ends of the interferometer were moved apart to a final spacing of over 9 m there indeed was a fall-off in correlation, giving an angular diameter for Sirius of 0.0069 ± 0.0004 seconds of arc.

This result and the laboratory experiments in intensity interferometry that had preceded it started a storm of controversy amongst theoretical physicists. The conventional explanation of the intensity fluctuations was that the *numbers* of photons arriving at each photocathode had statistical variations; how could the separate photons caught by searchlight mirrors several metres apart have correlated fluctuations? This is of course a problem that applies to all interferometers, but it was brought into sharp focus by the measurement of intensity at the two telescopes, which was related more obviously to a flux of photons than in the more conventional interferometers.

The work of Hanbury Brown and Twiss epitomizes the dual character of light: a wave nature (which makes interferometers work) and a quantum nature (which is clearly demonstrated in the detection of individual photons). The correct approach to the problem is to apply wave theory to the propagation, diffraction and interference of the light waves, and quantum theory to the interaction of the light waves with the detectors. We consider this problem further in Chapter 13. The wave theory shows what correlation exists between the waves at the detectors, while the quantum theory shows how this correlation may become masked by a statistical 'photon noise'.

Moving to clearer skies than those of Jodrell Bank in Cheshire, England, Hanbury Brown set up at Narrabri in Australia a very large version of his intensity interferometer. The advantage over the Michelson technique was the measurement of correlation over very much larger baselines, giving a crucially important improvement in angular resolution. With the Narrabri interferometer, Hanbury Brown measured the diameters of several hundred of the brightest stars, the first direct measurement of the diameters of any stars smaller than the red giants. This was also the last use of intensity interferometry for measuring stellar diameters; the technique has been overtaken by developments in phase stable Michelson interferometers connecting large conventional optical telescopes, notably at the European Southern Observatory in Chile, and on Hawaii where the two Keck telescopes operate as a Michelson pair.

Problems 9.1 Numerical Examples

(i) In a Michelson interferometer used to measure the wavelength of monochromatic light, 185 fringes crossed the field of view when the mirror was moved by 50 μm. What was the wavelength of the light?

(ii) A simple double slit with separation 1 mm is held immediately in front of the eye, and a distant sodium street lamp is observed. If the lamp is 10 cm across, how far away must it be for clear interference fringes to be observed?

(iii) An etalon used as a length standard known to be near 5 mm, within one or two light wavelengths, was measured in terms of three wavelengths of cadmium light. Only the fractional parts of the fringe numbers were known, as follows:

Wavelength (nm)	Fraction
479.992	0.15
508.582	0.495
643.847	0.80

By trial and error, find the whole numbers of wavelengths and hence the spacing of the etalon.

Problem 9.2

The point of this problem is to show that equations (9.10) and (9.11) for the Sagnac interferometer also apply to any non-circular path. Suppose that: the light rays follow an arbitrary planar curve $r(\phi - \Omega t)$, in polar coordinates, where the argument $\phi - \Omega t$ indicates rigid rotation; ds is an element of arc along the fibre; and the refractive index is n. The angle between the light path and the direction of the rotary velocity is given by $\cos \theta = r d\phi/ds$. Relative to the inertial frame of the laboratory, the two light rays have velocities $u_{\pm} = c/n \pm \Omega r(1 - 1/n^2) \cos \theta$; here we have used the cosine to project the added Fresnel drag speed onto the instantaneous direction of the light path (or fibre). (a) By generalizing the treatment of the circular case, find the time Δt_{+} for the co-rotating ($+$) light ray to travel an arclength Δs. Likewise for the counter-rotating ($-$) light ray. (b) Verify that equations (9.10) and (9.11) are correct, when A is the area enclosed by the light path. (c) In general, the apparatus could be rotated with an angular velocity vector $\boldsymbol{\omega}$ having components both parallel and perpendicular to the light path. Explain why it is that only the perpendicular component, $\Omega = \hat{\boldsymbol{n}} \cdot \boldsymbol{\omega}$, is effective in producing a fringe shift, where $\hat{\boldsymbol{n}}$ is the unit normal to the plane of the light path.

Problem 9.3

Michelson and Gale tested the Sagnac effect by sending two light beams of wavelength 575 nm in opposite directions around an optical path in an evacuated pipe, forming a rectangular loop 610×335 m, and observed interference fringes on the image of a slit source. The experiment was conducted at latitude 42°. Calculate the expected fringe shift.

Problem 9.4

(a) The path difference between the two rays in the "sea interferometer" of Figure 9.13 can be calculated in a manner analogous to that used for a plane-parallel plate in Section 8.4. Find an equation for the bright fringes similar to equation (8.13). (As with the conventional Lloyd's mirror, you may assume a 180° phase reversal for the reflected wave.) Show that for grazing incidence ($\theta \ll 1$), adjacent fringes are separated by an angle $\Delta \theta = \lambda/2h$.

(b) Based on Figure 9.13, and with wavelength $\lambda = 1.5$ m, estimate the height h of the radio receiver above the sea surface.

Problem 9.5

A Fresnel biprism (Figure 9.12(a)) with small wedge angles α and refractive index n_1 is at a large distance from a point source of light with vacuum wavelength λ. It forms fringes on a screen at a distance d from the light source. (a) Show that the spacing of the fringes is $\lambda d/a$, where $a = 2D(n_1 - 1)\alpha$, where D is the distance between the source and the prism. (b) What is the spacing if the whole system is immersed in liquid with refractive index n_2?

Problem 9.6

Hanbury Brown's development of the Michelson interferometer operated at optical wavelengths with spacings up to 50 m between the two mirrors. Calculate the diameter of an object just resolvable by this instrument at a distance equal to the Earth's diameter, $D = 1.28 \times 10^7$m.

Problem 9.7

In an experiment to demonstrate Young's fringes, light from a source slit falls on two narrow slits 1 mm apart and 100 mm from a slit source; the fringes are observed on a screen 1 m away. The source is white light filtered so that only the wavelength band from 480 to 520 nm is used. (a) What is the angular and linear separation of the fringes? (b) Approximately how many fringes will be clearly visible? (c) How wide can the source slit be made without seriously reducing the fringe visibility?

Problem 9.8

In the Rayleigh interferometer of Section 9.1, starting with both tubes at atmospheric pressure, what pressure change will give a minimum fringe visibility due to the pair of sodium D lines at 589.0 and 589.6 nm?

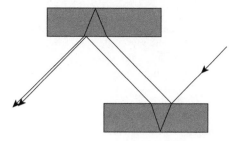

Figure 9.19 The Jamin interferometer

Problem 9.9

The Jamin interferometer shown in Figure 9.19 uses two parallel-sided glass plates to form separated but identical optical paths, like the twin paths of the Rayleigh refractometer. The plates are set at $\theta = 45°$ to the light path, and one is tilted by a small angle $\Delta\theta$ to produce a set of interference fringes. The plates have refractive index n and thickness h. Find the phase difference $\Delta\phi$ between the paths at the centre of the field, using equation (8.15) to show that

$$\frac{\Delta\phi}{\Delta\theta} = \frac{4\pi h}{\lambda} \cos\theta \sin\theta (n^2 - \sin^2\theta)^{-1/2}.$$ (9.22)

10 Diffraction

Augustin Jean Fresnel (1788–1827), . . .unable to read until the age of eight, . . .the first to construct multiple lenses for lighthouses, . . .was enabled in the most conclusive manner to account for the phenomena of interference in accordance with the undulatory theory.

Encyclopaedia Britannica.

Diffraction is the spreading of waves from a wavefront limited in extent, occurring either when part of the wavefront is removed by an obstacle, or when all but a part of the wavefront is removed by an *aperture* or *stop*. The general theory which describes diffraction at large distances is due to Fraunhofer, and is referred to as *Fraunhofer diffraction*.

The Fraunhofer theory of diffraction is concerned with the *angular* spread of light leaving an aperture of arbitrary shape and size; if the light then falls on a screen at a large distance, the pattern of illumination is described adequately by this angular distribution. But if the screen is close to the aperture, so that one might expect to see a sharp shadow at the edges, we see instead diffraction fringes, whose analysis involves a theory introduced by Fresnel. A famous prediction of Fresnel's theory was that the shadow of a circular object should have a central bright spot; the demonstration that this indeed exists was a powerful argument in establishing the wave theory of light.

In this chapter we set out the formal distinction between Fraunhofer and Fresnel diffraction. We start with simple examples of Fraunhofer diffraction which can be understood either intuitively or by using the phasor ideas and constructions of Chapter 2. We shall find that this simple approach leads to a general theory which uses Fourier analysis to analyse diffraction at any aperture. We then show how the diffraction fringes at the edge of a geometric shadow may be analysed using integrals introduced by Fresnel rather than the Fourier integrals appropriate to Fraunhofer diffraction.

10.1 Diffraction at a Single Slit

The simplest case of Fraunhofer diffraction is for a single slit, width w, illuminated by a plane wavefront with uniform amplitude. The slit is perpendicular to the paper in Figure 10.1, so we take as

---

Optics and Photonics: An Introduction, Second Edition   F. Graham Smith, Terry A. King and Dan Wilkins
© 2007 John Wiley & Sons, Ltd

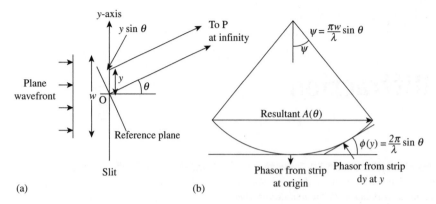

Figure 10.1   Diffraction at a slit. Each elementary strip dy contributes an equal phasor with a phase varying linearly with y. The phasor diagram is then part of a circle, allowing the resultant to be easily calculated

our basic element a strip of width dy, small compared with the wavelength. The diffraction pattern is observed at a large distance from the slit. Then each strip contributes an equal amplitude proportional to dy to the total at P, but the phase of each contribution depends on y as

$$\phi(y) = \frac{2\pi y \sin\theta}{\lambda}. \tag{10.1}$$

The contributions of the elementary strips can be added in the phasor diagram of Figure 10.1. The central strip, at O, is taken as the phase reference origin, appearing as a horizontal phasor in the diagram. Contributions from below O have phases retarded on this reference; these have been placed on the left-hand side of the phasor diagram. Contributions from above O appear on the right-hand side, and the diagram then becomes an arc of a circle. The resultant is a chord, representing the amplitude $A(\theta)$ of the resultant wave in direction θ. When $\theta = 0$ the phasor diagram is a straight line, representing the maximum amplitude $A(0)$. Hence from Figure 10.1 we see that

$$\frac{A(\theta)}{A(0)} = \frac{\sin\psi}{\psi} \tag{10.2}$$

where $\psi = \pi w \sin\theta/\lambda$ is the phase of the component from one edge of the slit. The function $(\sin\psi)/\psi$ is named sinc ψ. The intensity correspondingly varies as

$$\frac{I(\theta)}{I(0)} = \left(\frac{\sin\psi}{\psi}\right)^2. \tag{10.3}$$

Using exponential functions, we can express the sum of the elementary contributions as the integral

$$A(\theta) = \int_{-w/2}^{w/2} a\exp(i\phi)dy. \tag{10.4}$$

which integrates directly to give the result in equation (10.2).

We now see the vital connection between diffraction and the Fourier transform. In Chapter 4 we set out the Fourier transform in terms of time and frequency in equation (4.36), which we now repeat:

$$f(t) = \int_{-\infty}^{\infty} F(v) \exp(2\pi i v t) dv. \tag{10.5}$$

In terms of spatial variables x, u, where u is a spatial periodicity or an inverse length, the Fourier transform is

$$f(x) = \int_{-\infty}^{\infty} F(u) \exp(2\pi i x u) du. \tag{10.6}$$

This is identical to equation (10.4) if $x = \sin\theta/\lambda$ and $u = y$. The integral need not extend to infinity, as there is no contribution from beyond the edges of the slit at $y = +w/2$ and $-w/2$. The integral (10.4) is in fact the Fourier transform of the 'top-hat' function, which was evaluated in Section 4.12 and found to be proportional to the sinc function $(\sin\psi)/\psi$, as also found in equation (10.2). This equality between the Fraunhofer diffraction pattern and the Fourier transform of the aperture function is universal.

The sinc function is important in many branches of physics. By considering the behaviour of the phasor diagram in Figure 10.2 as ψ changes we can easily see its main properties. At $\theta = 0$, $\psi = 0$ and the phasor is a straight line; sinc(0) is indeed unity. As ψ moves away from zero the phasor diagram begins to curve into the arc of a circle; the resultant, the chord, shortens but remains parallel to the contribution from the centre of the slit. Thus the amplitude decreases but the phase remains the same. When $\psi = \pi$, so that $\sin\theta = \lambda/w$, the phasor diagram is a closed circle and the amplitude is zero. As ψ increases beyond π there is again a resultant, but it is in the opposite direction;

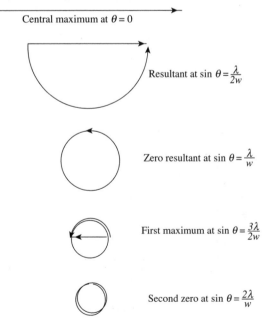

Figure 10.2   As the direction θ of diffraction moves away from zero, the phasor, straight at $\theta = 0$, curls up as shown. Each zero corresponds to the phasor being wrapped round an integral number of times

the amplitude is now negative, or we may say that the phase has changed by π in going through the zero. At approximately $\psi = 3\pi/2$ the resultant reaches its extreme negative value and then begins to shorten as ψ increases. At $\psi = 2\pi$, so that $\sin\theta = 2(\lambda/w)$, the resultant is again zero.

It is easy to see that this oscillatory behaviour continues, giving zeros at exactly $\sin\theta = \pm\lambda/w$, $\pm2(\lambda/w)$, $\pm3(\lambda/w)$, etc., corresponding to $\psi = \pm\pi$, $\pm2\pi$, $\pm3\pi$, etc. The values of ψ to give the maximum and minimum values, where $dA/d\psi = 0$, may be found by differentiation:

$$\frac{dA}{d\psi} = \frac{d}{d\psi}\left(\frac{\sin\psi}{\psi}\right) = \frac{\psi\cos\psi - \sin\psi}{\psi^2} \tag{10.7}$$

so that

$$\frac{dA}{d\psi} = 0 \text{ where } \psi = \tan\psi. \tag{10.8}$$

This intrinsic equation is best solved either graphically or numerically. If n is an integer the extremes for large n are at

$$\psi = \pm\left(n + \frac{1}{2}\right)\pi. \tag{10.9}$$

For small values of n the maxima and minima are somewhat closer in than the values given by equation (10.9). For example, the first extremes come at $\psi = \pm1.43\pi$ rather than at $\pm1.5\pi$. In terms of the phasor diagram these extremes correspond to the same length of phasor being formed into a circle by being wrapped round approximately $1\frac{1}{2}$, $2\frac{1}{2}$, etc., times. Similarly, the zeros are given exactly by the same length of phasor being wrapped round 1, 2, 3, etc., times, and the central

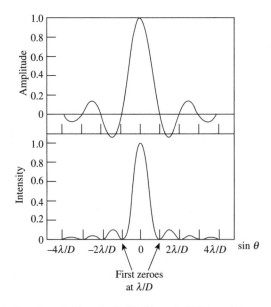

Figure 10.3 The amplitude function $\sin((\pi w\sin\theta)/\lambda)/((\pi w\sin\theta)/\lambda)$ and its square, the intensity function, for diffraction at a slit of width w

maximum by its being straight. Remember that the change in $\sin\theta$ between the zeros on each side of the central maximum is twice that between subsequent zeros. In amplitude the first subsidiary lobe is negative and about 22% of the central lobe.

The eye or a photographic plate is directly sensitive to the intensity which is thus proportional to the amplitude squared. The amplitude and intensity are plotted in Figure 10.3. In intensity the first subsidiary maximum is only about 5% of the main maximum, and has fallen to 0.5% by the fourth.

10.2 The General Aperture

Having looked at a simple but important case of Fraunhofer diffraction, we now go on to make an important generalization. We generalize in two ways: (a) by considering diffraction in two dimensions; (b) by allowing the complex amplitude in the aperture to be non-uniform: that is to say, it can have an arbitrary distribution of amplitude and phase.

Let the aperture be any shape in the x, y plane (Figure 10.4). Then the direction P of interest may be specified by the unit vector $\hat{k} = (l, m, n) = (\hat{k}\cdot\hat{x}, \hat{k}\cdot\hat{y}, \hat{k}\cdot\hat{z})$, where the components l, m, n are called direction cosines.[1] Q is a general point in the aperture with position $\mathbf{r} = (x, y, 0)$. The distance from Q to the distant point P is shorter than that from O by

$$\hat{k}\cdot\mathbf{r} = lx + my. \tag{10.10}$$

Hence on account of path difference the phase of light from Q at P will be advanced by $(2\pi/\lambda)(lx + my)$.

So much for the extension to two dimensions. Now let both the amplitude and phase of the wavefront in the aperture be functions of (x, y). Mathematically we can express this by letting the amplitude be a complex function of position, $F(x, y)$. An element $dxdy$ at (x, y) will then make a

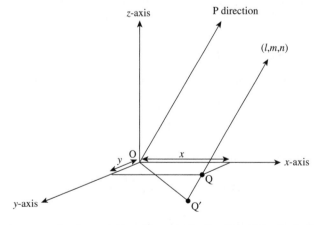

Figure 10.4 A general aperture in the x, y plane. The contribution from Q in the P direction specified by the direction cosines (l, m, n) is advanced in phase compared with those from the origin by $(2\pi/\lambda)QQ' = (2\pi/\lambda)(lx + my)$

[1]The direction cosines are the cosines of the angles between the direction they refer to and the coordinate axes.

contribution to the amplitude at P of $F(x, y)$ dxdy, but rotated in phase by $(2\pi/\lambda)(lx + my)$. Expressing the sum of all such components over the aperture by a double integral,

$$A(l, m) = C' \int \int F(x, y) \exp\left[-\frac{2\pi i}{\lambda}(lx + my)\right] dxdy. \tag{10.11}$$

Be clear what this integral represents. Each element of the aperture dxdy contributes a phasor of length $F(x, y)$ dxdy and phase given by the initial phase of $F(x, y)$ advanced by the phase due to the path difference $\mathbf{k} \cdot \mathbf{r}$ (the minus sign in (10.11) is due to a phase reduction). The complex integral is just a way of arriving at the resultant in the phasor diagram. Equation (10.11) allows the calculation of the complex amplitude at a distant point P in terms of the complex amplitude $F(x, y)$ in the aperture. The dimensions must be the same on each side, and this is taken care of by the constant C' with dimensions $[\text{length}]^{-2}$. For present purposes we are concerned with the form of the diffraction pattern rather than its absolute value.

   Comparison with equation (4.38) shows that equation (10.11) has the form of a Fourier transform in two dimensions, using the pairs of transform variables x/λ and l, y/λ and m. The coordinates x/λ and y/λ in the aperture are lengths measured in wavelengths, while l and m are sines of angles measured from the normal to the aperture. Thus we have the important general result:

   The Fraunhofer diffraction pattern in amplitude of an aperture is the Fourier transform of the complex amplitude distribution across the aperture.

10.3   Rectangular and Circular Apertures

We can now apply the general expression of equation (10.11) to some particularly important examples of diffracting apertures, using Fourier transforms directly. For the first three of the following examples we need only use the one dimensional form of equation (10.11). In the following examples, the amplitude function $F(x, y)$ within the aperture is constant in magnitude and phase. This implies that the aperture is illuminated by a plane wave incident normally.

10.3.1   *Uniformly Illuminated Single Slit*

As in Section 10.1, a uniformly illuminated slit of width w is represented by an amplitude distribution

$$\begin{aligned} F(y) &= F_0, \text{ for } |y| < w/2 \\ &= 0, \text{ for } |y| > w/2. \end{aligned} \tag{10.12}$$

The Fourier transform, i.e. the amplitude function $A(\theta)$, is

$$A(\theta) = A(0) \text{ sinc}\left(\frac{\pi w \sin\theta}{\lambda}\right). \tag{10.13}$$

The intensity distribution is the diffraction pattern shown in Figure 10.3.

10.3.2   *Two Infinitesimally Narrow Slits*

Young's double slit of Section 8.2 is represented by the amplitude distribution

$$F(x) = F_0[\delta(x - d/2) + \delta(x + d/2)] \tag{10.14}$$

where the narrow slits are represented by two delta functions (Section 4.14) located at $(x - d/2)$ and $(x + d/2)$. The Fourier transform, i.e. the double slit interference pattern, is

$$A(\theta) = A(0) \cos\left(\frac{\pi d}{\lambda} \sin\theta\right). \tag{10.15}$$

10.3.3   Two Slits with Finite Width

For slits with width w, separated by d, the results of the previous sections combine to give the amplitude distribution

$$A(\theta) = A(0)\, \text{sinc}\left(\frac{\pi w \sin\theta}{\lambda}\right) \cos\left(\frac{\pi d}{\lambda}\sin\theta\right). \tag{10.16}$$

Note that this result is an example of the convolution theorem (Section 4.13); the double slit pattern is the convolution of a top-hat function with the narrow double slit, and the resultant Fourier transform is the product of the two individual transforms.

10.3.4   Uniformly Illuminated Rectangular Aperture

Here $F(x, y) = F_0$, within the aperture sides extending from $-a/2$ to $+a/2$ and $-b/2$ to $+b/2$. The aperture sides are aligned along the x and y axes. The diffraction pattern in terms of direction cosines l and m is

$$A(l, m) = C'F_0 \int_{-a/2}^{+a/2} \int_{-b/2}^{+b/2} \exp\left[-\frac{2\pi i}{\lambda}(lx + my)\right] dx dy. \tag{10.17}$$

The double integrals in equation (10.17) are separable, so

$$A(l, m) = C'F_0 \int_{-a/2}^{+a/2} \exp\left(-\frac{2\pi i l x}{\lambda}\right) dx \int_{-b/2}^{+b/2} \exp\left(-\frac{2\pi i m y}{\lambda}\right) dy. \tag{10.18}$$

The two integrals both give sinc functions:

$$\int_{-a/2}^{+a/2} \exp\left(-\frac{2\pi i l x}{\lambda}\right) dx = -\frac{\lambda}{2\pi i l}\left[\exp\left(-\frac{2\pi i l x}{\lambda}\right)\right]_{-a/2}^{+a/2}$$
$$= a\, \frac{\sin(\pi l a/\lambda)}{\pi l a/\lambda} \tag{10.19}$$

with a similar expression for the y integral. The amplitude $A(0, 0)$ at the centre of the pattern where l and m are zero is $C'F_0\, ab$, so the amplitude $A(l, m)$ is given in terms of that at $(0, 0)$ by

$$A(l, m) = A(0, 0)\, \frac{\sin(\pi l a/\lambda)}{\pi l a/\lambda}\, \frac{\sin(\pi m b/\lambda)}{\pi m b/\lambda}. \tag{10.20}$$

The intensity is given by the square of $A(l, m)$. Thus the expression for the Fraunhofer diffraction pattern of a uniformly illuminated rectangular aperture is proportional to the product of the

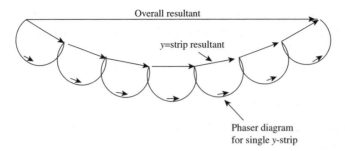

Figure 10.5   Phasor diagram for a rectangular aperture for diffraction in a direction off both l and m axes. The diagram corresponds to a point on the central maximum close to the first minimum in the m direction, but fairly far up the central maximum in the l direction

expressions for the separate diffraction patterns of two crossed slits. Along the l and m axes the subsidiary maxima have the same values as those of each single slit, since one of the product terms of equation (10.20) is unity on each axis. Faint subsidiary maxima exist in the four quadrants. For these both product terms are of the order of a few per cent, and the brightest of them, at approximately $(\pm 1.5\lambda/a, \pm 1.5\lambda/b)$, is $(0.047)^2$ or 2.2×10^{-3} of the intensity at the centre.

Physically we can see why this is so by considering the phasor diagrams shown in Figure 10.5. Along the l axis each strip parallel to the y axis is in the same phase all over. Phasors from each of these strips combine to give the diffraction pattern as in the single slit; the same applies to the m axis. If we consider a general point (l, m), however, the phasor from a strip parallel to the y axis is already bent into the arc of a circle because of the phase along the strip. The total phasor diagram is thus constructed of phasors already bent, and hence comes out very small.

10.3.5   Uniformly Illuminated Circular Aperture

How can we now apply the general theorem (equation (10.11)) to a circular aperture? First we note that the diffraction pattern must be circularly symmetrical. The result is simply written as the Fourier transform

$$A(l, m) = C'F_0 \int \int \exp -\left[\frac{2\pi i}{\lambda}(lx + my)\right] dx\, dy. \tag{10.21}$$

Evaluating this integral is messy (because the limits link x and y). It is simpler to change directly to polar coordinates (h, ψ) in the aperture and (w, ϕ) in the diffraction pattern. Now $h \cos \psi = x$; $h \sin \psi = y$; and an elementary area is $h\,dh\,d\psi$. In the diffraction pattern coordinates $w \cos \phi = l$; $w \sin \phi = m$, so that $w = \sqrt{(l^2 + m^2)} = \sin \theta$; as usual, θ is the angular deviation from the optical axis or normal to the aperture. The amplitude for a circular aperture is now given by

$$A(w, \phi) = C'F_0 \int_0^a \int_0^{2\pi} h\,dh\,d\psi \exp\left[-\frac{2\pi i}{\lambda}hw \cos(\psi - \phi)\right]. \tag{10.22}$$

The integral in equation (10.22) is only soluble analytically in terms of Bessel functions. However, most integrals can only be solved in terms of some sort of tabulated functions; it is merely that

Bessel's are less familiar than, for example, the sines needed to perform the integrals of equation (10.19). This being done, we have

$$A(w, \phi) = A(0,0) \frac{2J_1(2\pi a w/\lambda)}{2\pi a w/\lambda}. \tag{10.23}$$

The square of the right-hand side of equation (10.23) gives the intensity pattern for Fraunhofer diffraction at a circular aperture:

$$I(w, \phi) = I(0,0) \left[\frac{2J_1(2\pi a w/\lambda)}{2\pi a w/\lambda} \right]^2. \tag{10.24}$$

This famous and important result was first derived by George Airy in 1835 at about the time he became Astronomer Royal. It is of especial importance to astronomers as it is the pattern produced in the focal plane of an ideal telescope with a circular lens (or mirror) by a plane wavefront from a distant star. The circular edge of the objective lens of the telescope limits the aperture, and it is the angular width of the diffraction patterns due to two adjacent stars that determines whether or not they can be distinguished.

Airy's pattern in both amplitude and intensity is plotted in Figure 10.6. At first sight it looks like the similar plots for the slit in Figure 10.3, but there are several differences. Most important, it is a *ring* system, so that the plots are radial sections of a pattern possessing circular symmetry. The first zero is at $1.22\lambda/D$ (where $D = 2a$ is the diameter of the aperture), compared with λ/w for a slit of width w. The zeros are not equally spaced but tend to a separation of λ/D for large values of w (see Problem 10.7). The first subsidiary maximum of intensity is lower: 1.75% compared with 4.72% for the slit.

The Airy diffraction pattern is often quoted in relation to the angular resolving power of telescopes and similar optical instruments. If for example a double star is to be seen as two clearly

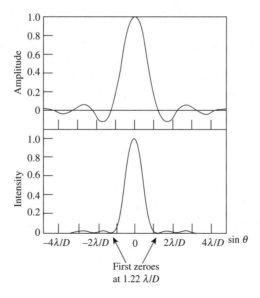

Figure 10.6 The amplitude and intensity functions $2J_1((\pi d \sin\theta)/\lambda)/(\pi d \sin\theta)/\lambda$ and its square for diffraction at a circular aperture of diameter D. The substitutions have been made in the expression in equation (10.23) to allow direct comparison with Figure 10.3 for a single slit

distinguishable images, each image must be smaller than the separation between them. The resolving power of the telescope is therefore ideally $1.22\lambda/D$, where D is the aperture; for example, for $D = 2.4\,\text{m}$ (the Hubble Space Telescope) the angular resolution in visible light is around 0.05 arcseconds. (This is usually unattainable for a comparable terrestrial telescope, because of random refraction effects in the atmosphere.)

10.4   Fraunhofer and Fresnel Diffraction

In any diffraction problem we find the amplitude and phase of the light wave at a point by adding all contributions by every possible path from a source to that point. The simple case of Fraunhofer diffraction is characterized by a *linear* variation of the phase of contributions from elements across an aperture. At a point close to an aperture or an obstacle the phase of these contributions will no longer vary linearly with distance across the aperture, and *quadratic* terms must be introduced. This is typical of Fresnel diffraction; the results are no longer given by Fourier transforms as in Fraunhofer diffraction.

The general problem is illustrated in Figure 10.7. Each element of the wavefront at the aperture is considered as the source of a secondary Huygens wavelet; the resultant amplitude and phase at any point P are determined by summing these wavelets, as in a phasor diagram. The phase of each wavelet is behind that of the wave at Q by an amount depending on the distance PQ and the wavelength. This summation of Huygens' wavelets taking account of their phase is the Huygens–Fresnel diffraction theory; it was Fresnel who contributed the essential idea of the interference of the Huygens secondary wavelets. Figure 10.7 shows how the distance PQ, and accordingly the phases of the wavelets, vary according to the position of the source Q; (b) shows the linear variation typical of Fraunhofer diffraction, and (c) shows the quadratic variation typical of Fresnel diffraction.

The transition from Fresnel to Fraunhofer diffraction is illustrated for slit diffraction in Figure 10.8. To determine the relative phases of contributions across the aperture for a point on any plane P_3 considerably beyond the distance R the linear approximation of Figure 10.7(b) is sufficient. But for a point on the plane P_1 inside the distance R, equal distances lie on a spherical surface rather than a plane, as in Figure 10.7(c). For a point such as P_3 the difference between a sphere and a plane becomes unimportant, since if the maximum deviation between sphere and plane is less than about $\lambda/8$ it has little effect on the phasors which add to give the resultant at P_3. The distance R, dividing the two regimes, is known as the *Rayleigh distance*; for an aperture width d it is given by

$$R = \frac{d^2}{\lambda}.\tag{10.25}$$

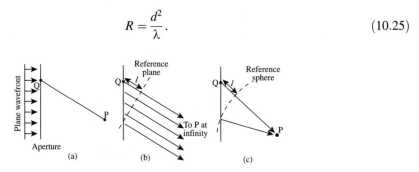

Figure 10.7   (a) The amplitude and phase at P may be considered as the sum of Huygens' wavelets from points such as Q in the aperture. In Fraunhofer diffraction the phase varies linearly across the aperture, as in (b); in Fresnel diffraction it varies quadratically, as in (c)

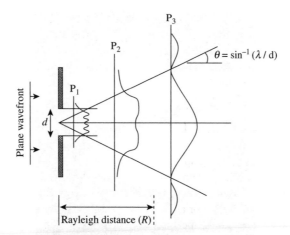

Figure 10.8 Transition from Fresnel to Fraunhofer diffraction. A portion of a plane wave W passes through a slit, width *d*. Intensity distributions across the wave are shown for planes P_1 (close to the slit), P_2 (just inside the Rayleigh distance) and P_3 (beyond the Rayleigh distance)

We return to this definition in Section 10.7 below.

In Section 10.9 we consider two further factors which may complicate some applications of Fresnel diffraction theory:

1. The distances from P of elements of the aperture may vary sufficiently to have a significant effect on wave amplitude as well as phase.

2. The line to P from different parts of the aperture may make considerably different angles with the normal to the surface of the aperture; this *inclination factor* also may have a significant effect on amplitude.

Fortunately there are two cases of particular interest which can be solved without detailed analysis of these factors, and which are well illustrated by graphical means as well as by simplified integrals. These are diffraction at a straight edge and at circular holes or obstacles.

10.5 Shadow Edges – Fresnel Diffraction at a Straight Edge

One of the most interesting predictions of the wave theory of light is that there should be some light within a geometric shadow, and interference fringes just outside it. The effect, seen in the photograph of Figure 10.9 and in the irradiance plot of Figure 10.10, is that at the geometrical edge of a shadow the intensity is already reduced to a quarter of the undisturbed intensity, falling monotonically to zero within the shadow. Outside the shadow the intensity increases to more than its undisturbed value and oscillates with increasing frequency as it approaches a uniform value. These are the 'fringes', a name which has been extended to many other types of diffraction and interference phenomena.

Consider the diffraction of a plane wave incident normally on a straight-edged obstacle. We shall evaluate the contributions of wavelets from strips parallel to the edge of the obstacle to the wave at a point P at distance *s* from the obstacle and on the edge of the geometric shadow. We take as phase

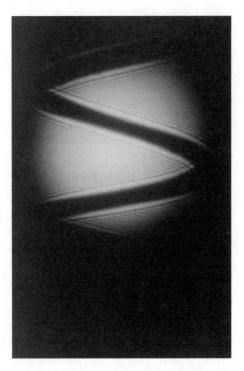

Figure 10.9   Fresnel diffraction at the shadow edges of a spiral spring. (Paul Treadwell, University of Manchester)

reference the phase of a wave from the closest point in the plane of the aperture. The extra path length from a strip distant h from this point gives a phase delay of

$$\phi(h) = \frac{2\pi}{\lambda}\left[(s^2 + h^2)^{1/2} - s\right] \approx \frac{\pi}{\lambda}\frac{h^2}{s}.\tag{10.26}$$

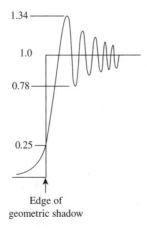

Figure 10.10   Fresnel diffraction; irradiance distribution for a straight edge

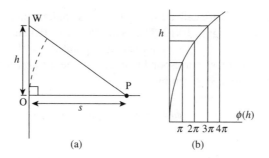

Figure 10.11   (a) Fresnel diffraction. Portions of the wavefront W contribute to the wave at P according to their amplitude (proportional to d*h*) and phase relative to the contribution from 0 (proportional to *h²*). (b) Moving outward from the centre, the half-period zones have successive radii that crowd closer together, though their areas remain constant

The approximation, in which the phase ϕ increases as the square of h, is valid when $h^2 \ll s^2$. If we divide the contributions in equal increments of phase, the corresponding increments of h decrease as h increases. The plot of $\phi(h)$ in Figure 10.11 is marked off at intervals of π in phase, showing the decreasing width of zones across which the phase reverses. Zones marked off in this way at intervals of π in phase are known as 'half-period zones'.

We can now construct a phasor diagram made up of the contribution of infinitesimal strips to the resultant at P. The contribution of an infinitesimal strip of width dh at h has a phase $\phi(h)$ given by equation (10.26) and an amplitude proportional to dh. The phasors from each contribution may be added geometrically by adding their components along two axes x and y. Taking the x axis as the phase reference, the phasor contributed by each strip is $dx + idy = dh[\exp(i\phi(h))]$, so the separate components are

$$dx = dh \cos \frac{\pi h^2}{\lambda s} \quad \text{and} \quad dy = dh \sin \frac{\pi h^2}{\lambda s}. \tag{10.27}$$

The x and y components of the phasor at P resulting from contributions from the origin up to any value of h are now given by the integrals of the expressions of equation (10.27). As h increases, the tip of the phasor traces a spiral, with the property that the angle $\phi(h)$ that its tangent makes with the x axis is proportional to the square of the distance along it from the origin. (It is instructive to notice here that if $\phi(h)$ were simply proportional to the distance along it the spiral would become a circle through the origin with its diameter along the y axis. This is why the corresponding Fraunhofer phasor diagrams in Figures 10.1 and 10.2 are circular!)

It is usual when plotting this spiral to do so in terms of a dimensionless variable v, which is a distance along the spiral. It is related in the present case to the variable h by $dv \propto \sqrt{dx^2 + dy^2} = dh$, or with a convenient choice of proportionality constant equals

$$v = h \left(\frac{2}{\lambda s} \right)^{1/2}. \tag{10.28}$$

Henceforth in this chapter, h and v are taken as signed variables (positive or negative) with the point of observation, P, defining the fixed value $h = 0$. Then the coordinates x, y of any point a distance v along the spiral are

$$x = \int_0^v \cos \frac{\pi v'^2}{2} dv', \quad y = \int_0^v \sin \frac{\pi v'^2}{2} dv'. \tag{10.29}$$

The integrals of equations (10.29) are called *Fresnel integrals*, and the plot of their value as v varies is called *Cornu's spiral*, shown in Figure 10.12. The oscillations of irradiance in the fringes correspond to the turns in the spiral, which for large v contracts to a point $Z(x = \frac{1}{2}, y = \frac{1}{2})$.

The edges of the half-period zones correspond to points on the spiral where its tangent is parallel to the x axis. The Mth such position is given by

$$\phi(h) = M\pi = \frac{\pi}{2} v^2 \tag{10.30}$$

or

$$v = \sqrt{2M}. \tag{10.31}$$

The phasor representing the wave at P, on the edge of the geometric shadow, is the resultant of the whole spiral from the origin to Z. Now imagine removing the obstacle and opening the other half-plane. Then the half of the plane wavefront that was covered by the obstacle can clearly be treated similarly, and contributes another branch of the Cornu spiral in the third quadrant. The whole curve is shown in Figure 10.12. The resultant Z'Z when there is no screen at all is clearly double the resultant OZ with the screen in place. This explains why the irradiance at the position of the geometric shadow is a quarter of that of the undisturbed wave.

Now, starting with the undisturbed wave, consider how the amplitude and intensity at P vary as a half-plane is moved from infinity across the plane wave. Starting at Z', a growing proportion of the spiral is deleted (Figure 10.13). The resultant instead of being Z'Z is DZ. D moves round the spiral, so the amplitude begins to show oscillations above and below its undisturbed value. Each extreme represents the deletion of one half-turn of spiral, corresponding to a movement in by one half-period zone. If w is the coordinate of the edge of the plane, the rate of oscillation increases as w^2; as the edge moves in the spiral gets bigger and the oscillations become larger and less rapid. The last minimum

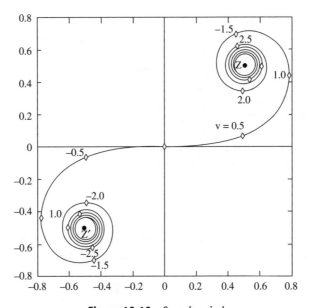

Figure 10.12   Cornu's spiral

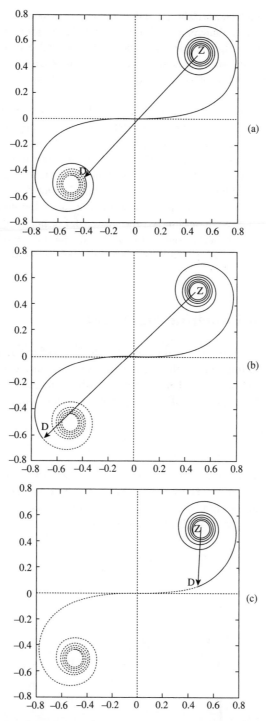

Figure 10.13   Edge diffraction. Phasor diagrams for successive positions of a shadow edge: (a) a diffraction minimum; (b) the first diffraction maximum; (c) inside the geometric shadow

and the last maximum are 0.88 and 1.16 of the undisturbed wave amplitude, giving irradiances of 0.78 and 1.34 of the unobstructed irradiance, as shown in Figure 10.10. From here on the resultant moves smoothly on, arriving at the origin at just half the amplitude and a quarter the irradiance. As w becomes positive, the resultant becomes a short vector joining Z to a point on the spiral around Z which rotates, shortening smoothly in length and reducing rapidly in size as P gets deeper into the geometric shadow.

It is quite easy to observe the first bright fringe around the edge of a shadow in white light, though of course the further ones get progressively out of step due to the large range of wavelengths. For example, the shadow cast by the back of a chair placed half-way across a room, illuminated with a car headlamp bulb at one side of the room, shows the bright fringe quite convincingly around its shadow on the opposite wall. If one looks back from a position in the shadow area towards the obstacle, the edge of the obstacle appears bright. This is the light which is diffracted into the shadow; it appears to originate at the edge itself, and it is sometimes referred to as an 'edge wave'.

An interesting point to notice is that while the scale of Fresnel fringes is determined by the wavelength and the distance s from the edge to the plane on which the shadow is observed, the ratio of the oscillations to the undisturbed irradiance is always the same. So these effects still occur even at short X-ray wavelengths.

10.6   Diffraction of Cylindrical Wavefronts

In the previous section we analysed the diffraction of a plane wavefront at an edge. It is very easy to extend this analysis to the diffraction of cylindrical wavefronts such as the wavefront emerging from a slit. If the source of the wavefront is a distance r from the diffracting screen as shown in Figure 10.14, the extra phase in the path that passes a distance h from the centre line is given by

$$\phi(h) = \frac{2\pi}{\lambda} h^2 \left(\frac{1}{2s} + \frac{1}{2r} \right). \tag{10.32}$$

This is similar to equation (10.26) with the addition of the $h^2/2r$ term to take account of the curvature of the wavefront before diffraction. The same argument can be followed through with this slightly more complicated expression. It turns out that the Cornu spiral can be applied as before if instead of the change of variable in equation (10.28) the substitution

$$v^2 = \frac{2h^2}{\lambda} \frac{(r+s)}{rs} \tag{10.33}$$

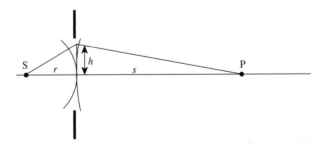

Figure 10.14   Geometry for the diffraction of cylindrical wavefronts

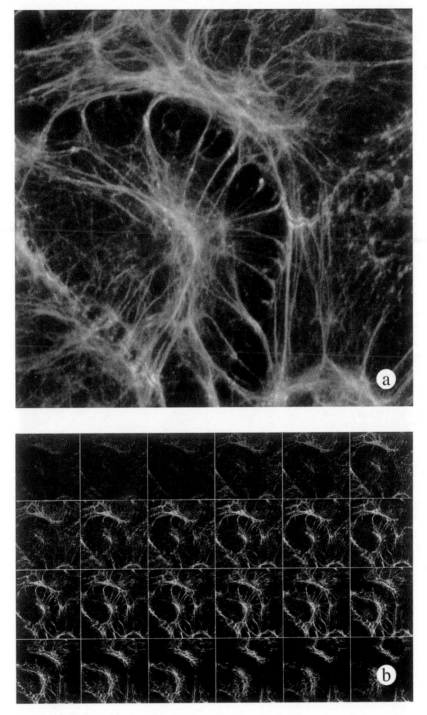

Plate 1 (a) Microphotography, using a confocal scanning microscope to scan and separate images at different depths. This three-dimensional network of cell-to-cell connections in cultured canine epithelia was stained with fluorescent dyes and scanned at 24 different focal depths. (b) Shown separately below are the individual sections, which can be superimposed to show the network from any aspect (S. Bagley, Paterson Institute, Manchester)

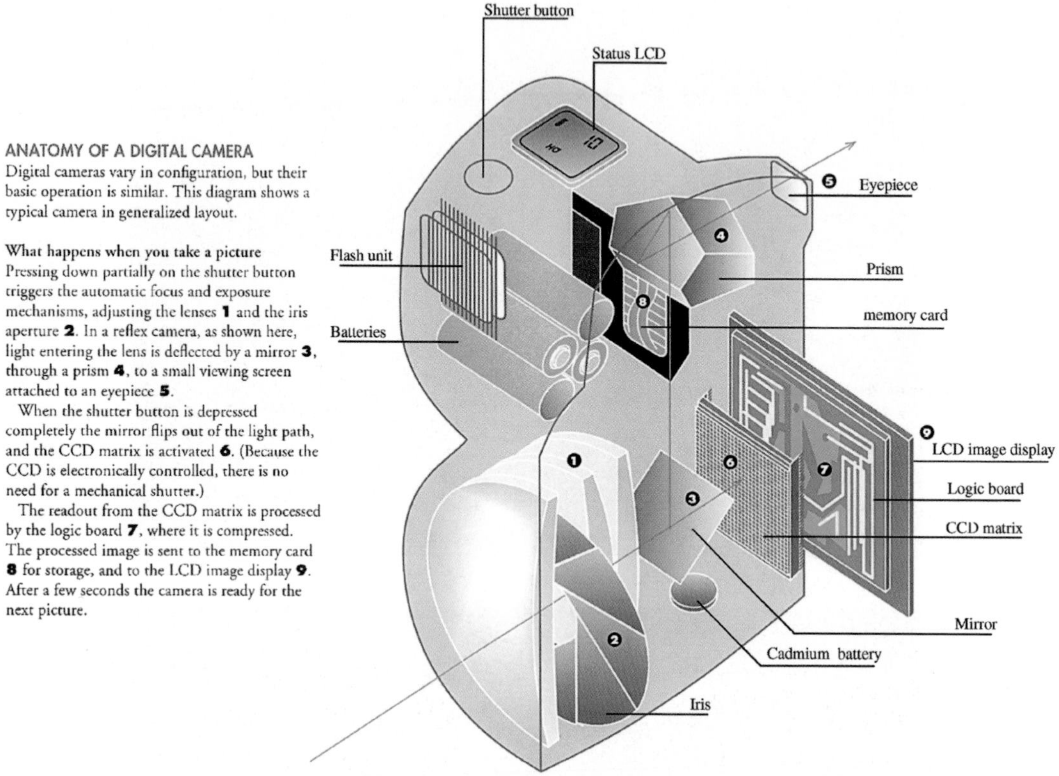

ANATOMY OF A DIGITAL CAMERA

Digital cameras vary in configuration, but their basic operation is similar. This diagram shows a typical camera in generalized layout.

What happens when you take a picture
Pressing down partially on the shutter button triggers the automatic focus and exposure mechanisms, adjusting the lenses **1** and the iris aperture **2**. In a reflex camera, as shown here, light entering the lens is deflected by a mirror **3**, through a prism **4**, to a small viewing screen attached to an eyepiece **5**.

When the shutter button is depressed completely the mirror flips out of the light path, and the CCD matrix is activated **6**. (Because the CCD is electronically controlled, there is no need for a mechanical shutter.)

The readout from the CCD matrix is processed by the logic board **7**, where it is compressed. The processed image is sent to the memory card **8** for storage, and to the LCD image display **9**. After a few seconds the camera is ready for the next picture.

Shutter button

Status LCD

Flash unit

Batteries

5 Eyepiece

Prism

memory card

9 LCD image display

Logic board

CCD matrix

Mirror

Cadmium battery

Iris

Plate 2 Cutaway section of a modern compact digital camera. (Reproduced with permission from J. Odam, "Start with a Digital Camera", © Copyright 1999, Pearson Education Inc.)

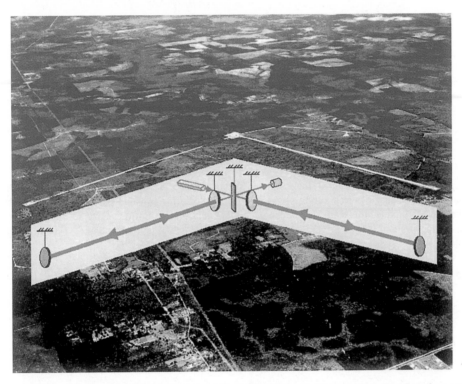

Plate 3 LIGO, a gravitational wave detector in Louisiana, USA. This very large-scale optical interferometer (the arms are 4 km long) is based on the Michelson principle, but using multiple beam reflections as in the Fabry–Pérot interferometer (Bernard Schutz)

M82
Merlin and VLA

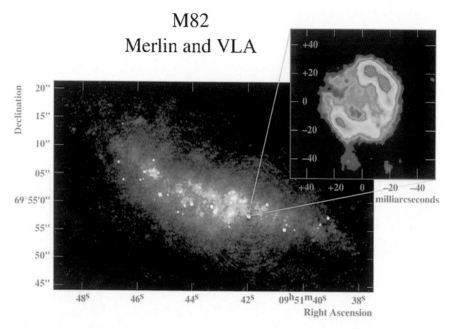

Plate 4 Aperture synthesis map of the radio emission at 20 cm wavelength from the galaxy M82. Radio interferometers with spacings from ten to several thousand kilometres provided the Fourier components, which were combined to produce this map. The prominent bright radio sources in this galaxy are supernova remnants; the ring-shaped remnant (shown inset) is mapped with a resolution of 1 milliarcsecond (T. Muxlow, Jodrell Bank Observatory)

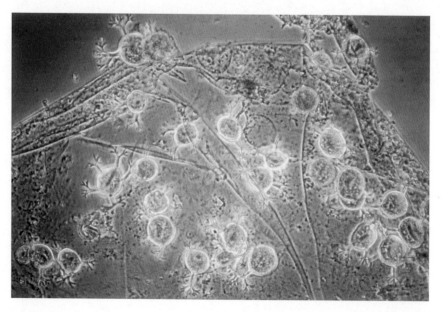

Plate 5 Dark-field and phase-contrast microscopy. The objects in this photo are almost completely transparent, but are made visible by refractive index differences that modify the phase of the light waves travelling through them. The circular objects (diameter 0.3 mm) are protozoa on a gill plate of the fresh water shrimp *Gammamus pulex* (T. Allen, Paterson Institute, Manchester)

Plate 6 A high power fibre laser glowing with visible light. The core of the silica fibre, which is several metres long and has resonator mirrors at each end, is doped with thulium ions which lase in the infra-red at 2 μm. The pump is a semiconductor diode laser array, wavelength 790 nm, focused on one end. The laser power is 5 watts; the visible light emitted from the side of the fibre is from a wavelength up-conversion process. (Stuart Jackson, University of Manchester)

Plate 7 Double rainbow over Flagstaff, Arizona, photographed by W. Livingston (*Color and Light in Nature*, Cambridge University Press, 1995). Note the darker sky outside the primary bow, and the fainter supernumerary bow inside the primary

Plate 8 Aurora borealis photographed by Paul Neiman (*Color and Light in Nature*, Cambridge University Press, 1995)

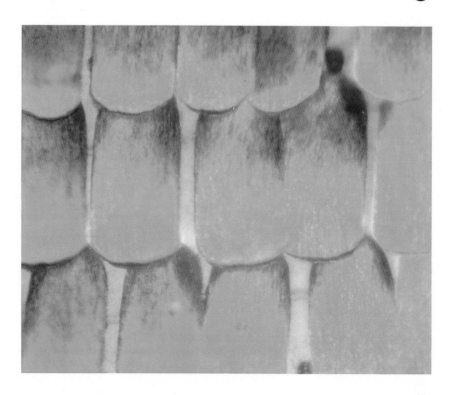

c

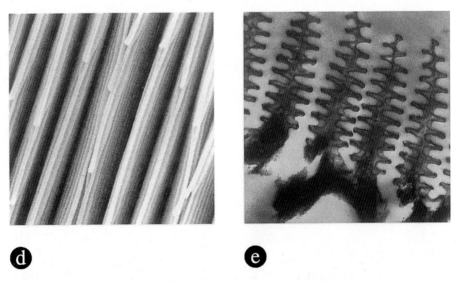

d   e

Plate 9 Iridescent colour in a butterfly wing: (a) The Morpho butterfly; (b) Wing scales ×300. The colour is due to diffraction and selective reflection in the structure of the wing scales, shown by scanning electron microscopy (SEM) and transmission electron microscopy (TEM); (c) the surface has a diffracting array of slots (SEM × 650); (d) the slots are deep and contain serrations (SEM × 10000); (e) a cross-section of the slots shows a pattern of grooves at quarter-wave intervals, which act in the same way as multilayer films (TEM × 15000) (Peter Vukusic, University of Exeter with G. Wakely for TEM))

is made. The same spiral can then be used with just this change of scale factor. For example, if $r = s$ the diffraction pattern would be scaled in h by $1/\sqrt{2}$ compared with the plane wave ($r = \infty$) case. For simplicity we shall go on to discuss the Fresnel diffraction of plane waves by slits, but all the results are easily adapted to cylindrical waves by the change of scale given by equation (10.33).

10.7   Fresnel Diffraction by Slits and Strip Obstacles

The Cornu spiral will now be seen as the phasor diagram obtained by adding the contributions at some point of infinitesimal strips across the whole of a plane or cylindrical wavefront. If an obstacle or a slit deletes some of the spiral, the remainder allows us easily to obtain the amplitude and phase. From this point of view it is natural to work in terms of v as variable, always remembering that v is related to h, the actual dimensional coordinate perpendicular to the strips, by equation (10.28) for a plane wave or equation (10.33) for a cylindrical wave.

In the case of slits it is again conventional to think of the slit being moved past P as was done in the case of the half-plane; the contribution of the uncovered portion of the wavefront is then represented by a segment of the Cornu spiral with a fixed length v_s. Moving the slit relative to P, the point of observation, moves the segment along the spiral. To illustrate this Figure 10.15 shows successive positions of a segment with length $v_s = 1.2$ moved along the spiral in unit steps. This corresponds to a slit too narrow for the undisturbed brightness ever to be attained. As the free length of phasor moves out from the centre – here it is almost straight – to the spiral portions, its resultant decreases monotonically until the spiral is of small enough diameter for it to be wrapped once round between the endpoints of the phasor. From then the resultant *increases* until the spiral is wrapped round $1\frac{1}{2}$ times, after which it again decreases, repeating the process in a series of fringes getting smaller and smaller. This process is highly reminiscent of the *Fraunhofer* diffraction at a single slit, as indeed it should be! As a glance at equation (10.28) shows, to make v_s small at a given λ, we must make s, the distance from the slit to the observation point, large, which is just the condition for Fraunhofer diffraction.

We can now see the basis of the Rayleigh criterion for the minimum distance from the slit for Fraunhofer diffraction to apply. If the slit width covers a range $v_s = 2$ the phasor diagram is just beginning to be seriously bent in the centre: this is evident by inspection of the Cornu spiral. Equation (10.28) then may be used to give the distance s in terms of slit width Δh and wavelength λ. Then

$$2 = \Delta h \left(\frac{2}{\lambda s}\right)^{1/2} \quad \text{or} \quad s = \frac{1}{2}\frac{(\Delta h)^2}{\lambda} \tag{10.34}$$

which is half the Rayleigh distance (Section 10.4) and Fresnel effects should still be appreciable. Similarly, if $v_s = 1$, the bending of the phasor at the centre of the spiral is negligible, its resultant being 99.4% of its unbent length, and a similar calculation shows we are at twice the Rayleigh distance.

As v_s increases the irradiance in the centre goes up, reaching the undisturbed irradiance at $v_s \approx 1.4$, and rising to 1.8 times the undisturbed irradiance at $v_s \approx 2.4$. This great increase in irradiance can be thought of as the coherent superposition of the bright fringes near each edge diffracting separately. The general character of the diffraction from wider and wider slits will now be clear. As the fixed length of phasor slides along the Cornu spiral there are two conditions:

1. *In the geometrically bright area.* The two ends of the phasor are in opposite parts of the spiral. The irradiance is of the same order as the undisturbed irradiance but because the resultant joins the

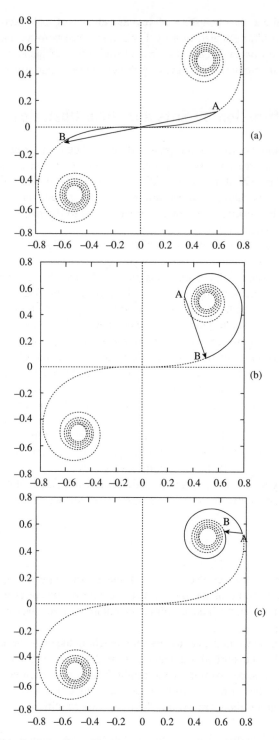

Figure 10.15   Fresnel diffraction by a slit. A fixed length $v_s = 1.2$ of the Cornu spiral forms phasor diagrams (a) at the centre of the diffraction pattern, (b) and (c) at increasing distance off centre

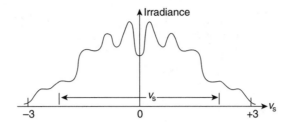

Figure 10.16 A slit diffraction pattern in the Fresnel region. The pattern may be traced on the Cornu spiral, by moving a segment length $v_s = 4$ by distance Δv

ends of the phasor which are independently going round different spirals, complicated beating effects may be seen, as shown in Figure 10.16, in which $v_s = 4$.

2. *In the geometrical shadow.* The two ends of the phasor are on the same part of the spiral. The irradiance is low but, because the ends of the phasor are on the same spiral, fringes are produced having maxima if the two ends are opposite, and minima if they are close.

Between these conditions is the rapid transition through the edge of the geometric shadow, where for a change of position of about one unit of v the edge of the phasor sweeps from one arm of the spiral to the other.

The Cornu spiral can be used in a similar way to analyse the effect of strip obstacles. Here a limited portion of the spiral is *removed*, so that if this is in the centre the two coils Z and Z′ move closer together. In the centre of the shadow of a strip obstacle there is always some light, although it rapidly becomes less as the strip is made wider. Similarly, as we see in the next section, the centre of the shadow of a perfectly circular object contains a narrow spot of light; but this spot has the *same* irradiance as the unobstructed light.

10.8 Spherical Waves and Circular Apertures: Half-Period Zones

This section deals with Fresnel diffraction in axially symmetrical systems, as for example along the axis of a circular diffracting disc or hole. In the limit a large enough circular hole offers no obstacle at all, so the case of free space propagation is also covered.

In Figure 10.17 the wave amplitude at a point P due to a point source P_0 is to be calculated by integrating all contributions originating from a spherical surface surrounding P_0. The limit of the integral will depend on the size of the diffracting aperture, whose circular edge lies on the sphere. At distance h from the axis, the deviation from the planar wavefront of the Fraunhofer case is measured by the distance $\epsilon = r_0 - \sqrt{r_0^2 - h^2} \approx h^2/2r_0$. (When $\epsilon \ll \lambda$, we recover the Fraunhofer limit.) A reference sphere radius b, centred on P, touches the surface on the axis; the phase of a contribution from an annulus at a distance h from the axis is then given to a first approximation by

$$\phi = \frac{\pi h^2}{\lambda}\left(\frac{1}{r_0 - \epsilon} + \frac{1}{b + \epsilon}\right)$$

$$\approx \frac{\pi h^2}{\lambda}\left(\frac{1}{r_0} + \frac{1}{b}\right)$$

(10.35)

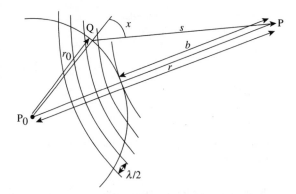

Figure 10.17   Fresnel's half-period zone construction

where we assume that $\epsilon \ll r_0, b$. The area of an annulus between h and $h + dh$ is $ds \approx 2\pi h dh$ (a planar approximation to the area on a sphere). Differentiating equation (10.35) gives

$$d\phi = \frac{2\pi}{\lambda}\left(\frac{1}{r_0} + \frac{1}{b}\right)h dh. \tag{10.36}$$

The element of area ds is therefore proportional to $d\phi$, so that the integral over the surface is conveniently carried out in terms of ϕ. The wave amplitude at P is then given by an integral of the form

$$A(P) = C \int_{\text{wavefront}} \exp(-i\phi)d\phi \tag{10.37}$$

where C is a constant depending on the amplitude of the source, and on r_0, b, λ.

For an aperture which is a circular hole with radius ρ the integral giving the wave amplitude on-axis is between 0 and

$$\phi_0 = \frac{\pi}{\lambda}\rho^2\left(\frac{1}{r_0} + \frac{1}{b}\right) \tag{10.38}$$

giving

$$A_{\text{hole}}(P) = iC\{\exp(-i\phi_0) - 1\}. \tag{10.39}$$

The wave amplitude at P is therefore proportional to $\sin(\phi_0/2)$, and the irradiance is proportional to $\sin^2(\phi_0/2)$. This means that if the radius of the hole is increased progressively from zero, the irradiance at P increases until ϕ_0 reaches π, and decreases and increases cyclically thereafter. The successive annuli opened up between these turning points are known as the Fresnel half-period zones.

The contributions making up the integral (10.37) are shown in the phasor diagram, Figure 10.18. This is shown as an open spiral, but it should be more nearly a circle; indeed it should be exactly a circle given the approximation we have made in neglecting distance and inclination effects (see Problem 10.7). Ultimately these effects shrink the circle to a point at its centre, giving a resultant amplitude for free space which is half the amplitude obtained when a single zone is exposed.

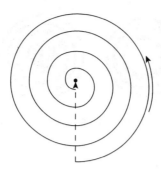

Figure 10.18 The spiral phasor diagram for a spherical wavefront. (The spiral form is exaggerated in this diagram: the phasor diagram is nearly circular

This discovery that opening an aperture wider can make the irradiance decrease comes as a surprise. But our usual intuition is based on experience with incoherent light sources where a bigger hole does indeed admit more light.

A circular disc, acting as an obstacle, requires the integral (10.37) to be carried out from a limit ϕ_0 out to infinity. The contribution from large values of ϕ tends to zero thanks to the aforementioned effects of inclination and distance. We model this gradual damping effect by inserting an extra factor $\exp(-\delta\phi)$, where δ is a small positive number. At the end of the calculation, we will let δ approach zero. Applied to a circular disc of radius ϕ_0, Equation (10.37) becomes

$$A_{\text{disc}}(P, \delta) = C \int_{\phi_0}^{\infty} \exp[-i(1 - i\delta)\phi]\,d\phi = \frac{iC}{(1 - i\delta)} \exp[-i(1 - i\delta)\phi]_{\phi_0}^{\infty}$$

$$= \frac{-iC}{(1 - i\delta)} \exp[-i(1 - i\delta)\phi_0].$$

(10.40)

Letting δ vanish, we obtain the on-axis amplitude behind the disc:

$$A_{\text{disc}}(P) = -iC \exp[-i\phi_0]$$

(10.41)

and the integral becomes

$$A(P) = -iC \exp(-i\phi_0).$$

(10.42)

Surprisingly, the modulus of $A(P)$ is independent of ϕ_0; the irradiance at a point on the axis behind any circular obstacle is the same as the unobstructed irradiance.

The prediction that there should be a bright spot at the centre of the shadow of a circular disc was first made by Poisson[2] on reading a dissertation by Fresnel on diffraction, submitted to the French

---

[2]Siméon Denis Poisson (1781–1840), celebrated French mathematician. His fame was predicted by his teacher M. Billy in a couplet due to Lafontaine:
Petit Poisson deviendra grand
Pourvu que Dieu lui prête vie.
(The little Fish will become great, while God gives him life.)

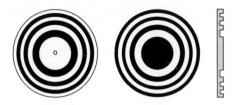

Figure 10.19 Zone plates. Alternate half-period zones are either blacked out as shown above or reversed in phase (as seen in the cross-section on right)

Academy of Sciences in 1818. When the test was made, and the bright spot was found, the wave theory of light was firmly and finally established.

A circular aperture in which alternate half-period zones are blacked out, called a *zone plate*, is shown in Figure 10.19. Alternate semicircles are now removed from the phasor diagram, and a large concentration of light appears at P. Figure 10.20 shows how the phasors from the half-period zones add in phase. The zone plate is acting like a lens; for any given wavelength the relation between object and image distance r_0 and b conforms to a simple lens formula. An improved zone plate can be made by reversing the phase of alternate zones instead of blacking them out; this is done by a change in thickness of a transparent plate. As shown in Figure 10.20(c), the amplitude at the focal point is then doubled.

The zone plate used as a lens is particularly useful at X-ray wavelengths, where there is no transparent refracting material which can be used to make conventional lenses. It has also been used on a minute scale in electron optics, to produce an electron lens only 0.7 μm in diameter and with a focal length of 1 mm.[3] Each transparent zone in this lens consisted of an array of holes only a few

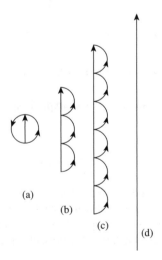

Figure 10.20 Phasor diagrams for (a) a circular aperture containing an odd number of half-period zones, (b) a zone plate with three clear zones, (c) a zone plate of the same size as (b) but with phase reversal instead of obscured zones, (d) a perfect lens

[3]Y. Ito, A. L. Blelock and L. M. Brown, *Nature*, **394**, 49, 1998.

nanometres in diameter, drilled through a thin inorganic film; there were 4000 holes altogether in the complete lens. This astonishing achievement has a practical application: the same lens pattern can be reproduced many times, allowing multiple beams of electrons or X-rays to be used in the fabrication of electronic circuits on silicon chips.

10.9 Fresnel–Kirchhoff Diffraction theory

In both Fresnel and Fraunhofer theory we have assumed that a diffracted wave amplitude can be calculated from the sum of secondary waves originating at an aperture. We have assumed also that the wave can be represented by a scalar, so that polarization can be neglected; and we have assumed that the amplitude distribution across an aperture is that of the undisturbed wavefront. These latter assumptions may be improved in specific cases; for example, we know that at the edge of a slit in a metal sheet the electric field must be perpendicular to the conducting surface, so that the parallel component is zero near the edge of the slit. The effective width of the slit will therefore be affected by the direction of polarization, and this will influence the angular spread of the diffraction pattern. Such cases can be dealt with by the application of boundary conditions in determining the amplitude distribution across the aperture. Some fundamental questions still remain, however, which were clarified by Kirchhoff and added to the Fresnel theory.

One of the problems of the Huygens–Fresnel principle was to assign to each wavelet an *inclination factor*, which would give it unit amplitude in the forward direction and zero backwards. Fresnel assumed, incorrectly, that it was also zero at 90° to the forward direction. The inclination factor is obtained explicitly in Kirchhoff's analysis, which involves not only the amplitude and phase on a diffracting surface but also their differentials *along* the wave normal.

A harmonic wave from a point source in a homogeneous and isotropic medium travels at the same speed in all directions, but with an amplitude decreasing inversely with distance. At a diffracting aperture, distance r_0 from the source, the wave amplitude of this spherical wave can be written as $(A_0/r_0) \exp(ikr_0)$. Figure 10.21 shows a small element of the aperture at Q with area da, which is the origin of a wave reaching a field point P at a further distance r, giving a contribution at P with the form

$$dA = A_0 da \frac{1}{r_0} \exp(ikr_0) \frac{1}{r} \exp(ikr). \tag{10.43}$$

The Fresnel–Kirchhoff analysis adds two further factors, an inclination factor and a change in phase, giving the diffracted wave amplitude $A(P)$ at P as the integral over the diffracting surface S

$$A(P) = -\frac{ik}{4\pi} \int_s A_0 \frac{\exp[ik(r + r_0)]}{rr_0} (\cos \chi_0 + \cos \chi) da. \tag{10.44}$$

Inside the integral sign the exponential term determines the phase of each component from an area da, while the amplitude is proportional to $1/r$, the distance of P from the area da. The factor $(\cos \chi_0 + \cos \chi)/2$ is the inclination factor, where χ_0 is the angle to the normal of the incident wave at the diffracting surface, and χ is the angle to the normal at P. Outside the integral the factor $-ik/4\pi$ normalizes the amplitude and phase of $A(P)$; the factor $-i = \exp(-i\pi/2)$ accounts for a 90° phase shift of the diffracted wave relative to the incident wave.

In most of the diffraction problems encountered in this chapter the surface S may be made to coincide with a wavefront, so that the incidence angle χ_0 is zero; the inclination factor

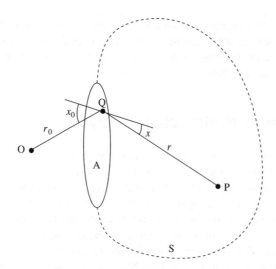

Figure 10.21 Fresnel–Kirchhoff theory. An aperture forms part of the surface S enclosing P

$(\cos \chi_0 + \cos \chi)/2$ then becomes $(1 + \cos \chi)/2$. The propagation of Huygens' wavelets forwards but not backwards is now clear, as the inclination factor becomes zero for $\chi = 180°$. The correct factor for $\chi = 90°$ is not zero, but one-half; we should point out, however, that diffraction through such a large angle is very dependent on the boundary conditions at the edge of the aperture.

This integral may look formidably complicated, and indeed it can be so for an arbitrary shape of diffracting aperture or obstacle. As we have seen, however, the evaluation of equation (10.44) can be greatly simplified in many practical situations.

10.10 Babinet's Principle

A consequence of the Kirchhoff theory, due to Babinet, concerns complementary diffracting screens. Consider a surface S_1 with some open and some opaque areas, and a complementary surface S_2 in which all the apertures are made opaque, and all the opaque regions are made open. With neither screen in place the complex amplitude at a point beyond the screen can be regarded as $A_1 + A_2$, the sum of the two diffracted amplitudes from S_1 and S_2. If P is outside the unobstructed light beam, so that $A_1 + A_2 = 0$, it follows that $A_1 = -A_2$. If either screen diffracts light so that it reaches P, then the complementary screen also diffracts to give exactly the same irradiance at P.

Example. When they are the same size, the circular hole and disc of Section 10.8 are complementary apertures. Check whether they fulfil Babinet's principle.

Solution. From equations (10.39) and (10.41) we find

$$A_{\text{hole}}(P) + A_{\text{disc}}(P) = iC[\exp(-i\phi_0) - 1] - iC \exp(-i\phi_0) = -iC. \tag{10.45}$$

This does not vanish because P, the point on-axis, lies within the beam of the incident plane wave. The sum correctly equals the amplitude for unobstructed free space, or, equivalently, for a vanishingly small disc.

Babinet's principle applies to any situation where light is diffracted by an obstacle or aperture into an otherwise dark region. For example, if a small obstruction is placed in a large parallel light beam, the light diffracted out of the beam is the same as that which would be diffracted out of the beam by an aperture of the same shape and size. Astronomical photographs often show this effect as a cross-like diffraction pattern extending from images of bright stars: this is due to a support structure for a secondary mirror, forming an obstructing cross in the telescope aperture. The diffraction pattern is the same as would be obtained from crossed slits of the same dimensions in an otherwise totally obscured telescope aperture.

10.11   The Field at the Edge of an Aperture

In the diffraction theory of this chapter, and indeed in most of the later diffraction theory, we have assumed that the wave can be described by a scalar variable, and made no mention of polarization. It is not usually necessary to calculate diffraction separately for each component of the polarization of the vector wave, but we can easily see one situation where this is necessary. It concerns the assumption, made in Section 10.3, that the diffraction of a plane wave at a slit may be calculated as if the amplitude of the wave were uniform over the whole slit.

Suppose the diffracting slit is made of a perfectly conducting metal sheet. Then the electric field must be zero in the sheet; immediately outside the sheet the component parallel to the slit edge must also be zero. Only at distances greater than about one wavelength from the edge can the field reach its full value. The wavefront passing through the slit is therefore narrower for polarization parallel to the edge than it is for polarization perpendicular to the edge, and the width of the diffraction patterns will correspondingly be somewhat different. This effect is only important if the scale of the diffracting object or slit is not large compared with one wavelength.

Evidently there can be considerable complications introduced by the behaviour of the wavefront close to a diffracting object. The full solution of such problems involves a detailed description of the wavefront, which must accord with the boundary conditions at the edge of the object. When the wave is described, then the diffraction pattern can be calculated either by the simple theory of this chapter, or in more difficult cases by the full wave theory due to Kirchhoff, which we have discussed briefly in Section 10.9. Fortunately it is often possible to proceed without the full rigour of the Kirchhoff theory.

Problem 10.1   Numerical examples

(i)   A Young's slit experiment has two very narrow slits separated by 0.1 mm. At what angles are the first- and second-order fringes for red and blue light (700 nm and 450 nm respectively)?

(ii)   If the slits in the previous problem are each of width 0.01 mm, how many red fringes might one see easily?

(iii)   A simple demonstration of diffraction and interference can be made by scratching lines through the emulsion of an undeveloped photographic plate, and looking through the lines at a distant bright light with the plate held close to the eye. Find the angular breadth of the pattern given by a sodium lamp ($\lambda = 589$ nm) with a slit width of 0.1 mm. What will be the effect for two such slits 1 mm apart?

(iv)   What is the limiting angular resolution of the astronomical telescope with objective diameter $D_1 = 40$ mm described in Problem 3.1? Assume light of wavelength 600 nm.

Problem 10.2

A single slit width D is made into a double slit by obscuring its centre with a progressively wider opaque strip, leaving two slits each with width a. Draw phasor diagrams for the single slit diffraction pattern at that angle which, prior to removal of the central strip, would have been at the edge of the main maximum and at the first zero, and show how these are changed as the opaque strip is widened.

Sketch the diffraction and interference patterns for the single slit and double slit with opaque strip $D/2$ wide, on the same scales of irradiance and angle.

Problem 10.3

Two pinholes 0.1 mm in diameter and 0.5 mm apart are illuminated from behind by a parallel beam of monochromatic light, wavelength 500 nm. A convex lens of diameter 1 cm and focal length 1 m is placed 110 cm from the holes. Describe the pattern formed on a screen placed (a) 1 m, (b) 11 m, from the lens.

Problem 10.4

Show that the single slit interference pattern in Figure 10.3 can be observed using a narrow line source of light which is extended along a line parallel to the slit. What is observed when the line source is rotated in a plane parallel to the screen containing the diffracting slit?

Problem 10.5

Estimate the smallest possible angular beam width of (i) a paraboloid radio telescope, 80 m in diameter, used at a wavelength of 20 cm, (ii) a laser operating at a wavelength of 600 nm, with an aperture of 1 cm.

Problem 10.6

An aperture in the form of an equilateral triangle diffracts a plane monochromatic wave. The side of the triangle is 20 wavelengths long. Find the directions of the zeros of the diffraction pattern closest to the normal.

Problem 10.7

We have seen that the irradiance pattern for a circular disc is $I(\theta) = 4I_0[J_1(\sigma)/\sigma]^2$, with $\sigma = 2\pi a \sin\theta/\lambda$. Standard mathematical tables show that the first four zeros of $J_1(\sigma)$ are $\sigma_j = 0$, 3.8317, 7.0156, 10.1735 where $j = 1,2,3,4$. Find out what features these zeros correspond to, and give expressions for their $\sin\theta$ values. (Note that the limit as σ tends to zero of $J_1(\sigma)/\sigma$ is $\frac{1}{2}$.)

Problem 10.8

From the asymptotic expansion for large z

$$J_1(z) = \frac{\sin z - \cos z}{(\pi z)^{1/2}}$$

show that the angular distance between diffraction minima far from the axis of a circular aperture, diameter d, when θ is not small, is approximately $(\lambda/d)(\cos\theta)^{-1}$.

Problem 10.9

The altitude of aircraft approaching land is controlled by a 'glide path' in which a radio transmitter of wavelength 90 cm forms interference fringes. The fringes are formed from a transmitter at height h above a conducting ground plane. Find the height h for a maximum signal to be received along a path at 3° elevation from the airfield. (This is similar to Lloyd's mirror in Figure 9.13. Assume that the signal is strongest in the first interference maximum.)

Problem 10.10

An image of a narrow slit, illuminated from behind by light of wavelength 500 nm, is formed on a screen by a convex lens of focal length 100 cm. The slit is 200 cm from the lens.

A second slit, parallel to the first, now limits the beam of light to a width of 0.5 mm. This slit is placed successively (a) 100 cm from the screen; (b) in contact with the lens; (c) 100 cm from the first slit. What is the width between the first zeros of the diffraction pattern in each case?

Problem 10.11

Calculate approximate values for the theoretical angular resolution of: (i) A 100 m radio telescope working at $\lambda = 5$ cm. (ii) The unaided human eye, aperture 4 mm, at $\lambda = 500$ nm. (iii) An 8 m diameter optical telescope at $\lambda = 1$ μm. (iv) An optical interferometer with 100 m baseline at $\lambda = 500$ nm. (v) A radio interferometer, working at $\lambda = 10$ cm, with baseline 6000 km.

Problem 10.12

The beam shape of a 15 m diameter millimetre-wave telescope is to be measured from beyond the Rayleigh distance. Calculate this distance for a wavelength of 0.5 mm.

Problem 10.13

The Cornu spiral (Figure 10.12) represents a phasor diagram giving the amplitude and phase of contributions at a point P from strips of a plane wave at a distance s from the nearest component. When $s = 10000\lambda$, how large is h (in wavelengths) for the phase of the contribution to be 5π behind that of the component at $h = 0$? Where on the Cornu spiral is this contribution, and what is the value of v?

Problem 10.14

Equation (10.44) gives the inclination factor of Fresnel–Kirchhoff theory. What is the fractional decrease in this factor for the contributions in Problem 10.13 from $h = 0$ to $h = 224\lambda$?

Problem 10.15

Consider the possibility of observing optical Fresnel diffraction, as in Figure 10.10, when a star is occulted by the Moon (given the small size of the star relative to the Moon, you can ignore the curvature of the Moon and regard it as a straight edge). Calculate for wavelength 600 nm (i) the width of the first half-wave zone at the Moon, (ii) the angular width of a star just filling this zone. Compare these with the size of irregularities on the Moon's surface, and the actual angular width of bright stars. (Moon's distance $= 3.76 \times 10^5$ km.)

Problem 10.16

A distant point source of light is viewed through a glass plate dusted with opaque particles. The light now appears to have a diffuse halo about $1°$ across. Use Babinet's principle to explain this and estimate the diameter of the particles.

Problem 10.17

Compare the intensities of light focused from a point source by a zone plate and by a lens of the same diameter and focal length. What change is made by reversing the phase of alternate zones rather than blacking them out? Where has the remaining energy gone?

Problem 10.18

An infinite screen is made of polaroid, and divided by a straight line into two areas in which the polaroid is oriented parallel and perpendicular to the division. Describe the diffraction pattern due to the edge when unpolarized light is incident normally on the screen.

Problem 10.19

A shadow edge for demonstrating Fresnel diffraction is made by depositing a metallic film on glass. What will be the effect of using a film that transmits one-quarter of the light irradiance?

Problem 10.20

We can construct spiral phasor diagrams that are more tractable analytically than the Cornu spiral. Consider the differential phasor elements

(a) $dA(\phi) = \exp[\phi(\alpha + i)]d\phi$ where α is any real number

(b) $dA(\phi) = \phi \exp(i\phi)d\phi$.

$$(10.46)$$

(The parameter ϕ is evidently the angle each differential phasor makes with the real axis.) In each case, integrate over an integral number of turns of the phasor, i.e. from $\phi = 0$ to $2\pi N$, to find the resultant $A(2\pi N)$. Then evaluate what we call the "winding factor",

$$W = |A(2\pi N)| \bigg/ \int_0^{2\pi N} |dA(\phi)|,$$

$$(10.47)$$

which measures the amount by which the phasor is shortened relative to the total length of all its elementary constituents.

11 The Diffraction Grating and its Applications

In 1912 Laue (1879–1960) had the inspiration to think of using a crystal as a grating.
S.G. Lipson and H. Lipson, *Optical Physics*, Cambridge University Press, 1969.

In Chapter 10 the Fraunhofer diffraction pattern in intensity from two slits illuminated by a plane wavefront was shown to be a set of equally spaced cos-squared fringes. We now discuss the more general problem of how such a system performs when it has a large number of slits instead of just two. This provides a description of an important optical element, the diffraction grating. The general solution for any grating is to evaluate the Fourier transform of the aperture function. However, much physical insight can be gained by using a phasor approach.

In this chapter we use phasor diagrams to illustrate the mathematics of diffraction by gratings, and develop the relation between the grating function and its transform, which is the relation between the properties of the grating and its diffraction pattern. We give examples of diffraction theory applied to radio antenna theory and to X-ray crystal diffraction.

11.1 The Diffraction Grating

Consider first the simple case of five slits illustrated in Figure 11.1, each separated by d, centre-to-centre distance, from its neighbours. With two slits maxima were produced by beams from *both* slits being in phase, which occurred when $d \sin \theta = \pm m\lambda$. Clearly we can have a situation when beams from *all five* slits are in phase, and this will again be when

$$d \sin \theta = m\lambda. \tag{11.1}$$

This condition means that the path difference between adjacent slits is in all cases an integral multiple of the wavelength. The phasors lie on a straight line and add up to the same value of amplitude as when there is no path difference between the slits at $\sin \theta = 0$. Note that the condition in equation (11.1) applies for a grating with any number of slits: it is worth remembering as the simplest form of

Optics and Photonics: An Introduction, Second Edition   F. Graham Smith, Terry A. King and Dan Wilkins
© 2007 John Wiley & Sons, Ltd

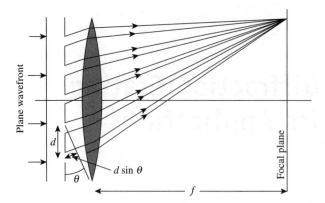

Figure 11.1   A diffraction grating with five slits. All the light diffracted into the direction θ is brought to one point in the focal plane of the lens where the Fraunhofer diffraction pattern may be seen

the *grating equation*. The successive maxima in the diffraction pattern of a grating are called its *orders*, first, second, third, etc., according to the value of m. The *central* or *zeroth* order is the one for $m = 0$.

The phasor amplitude pattern for the five-slit grating as $\sin \theta$ moves away from zero is illustrated in Figure 11.2; for comparison the amplitude pattern for a pair of slits at spacing d is also shown. The remarkable thing is that in the five-slit case the light has been diffracted mainly into the strong maxima at the several orders, with only weak maxima coming between them. The first zeros on each side of an order are $\pm\lambda/5d$ apart in θ ($\approx \sin \theta$, for $\lambda \ll d$). The orders then are approximately the angular width we should expect for the whole diffraction pattern of a slit as wide as the whole grating; that is to say, $5d$ in this case. Also we can narrow the intensity distribution of the orders by adding more slits at the same spacing of adjacent slits.

Diffraction gratings in use at optical wavelengths often have many thousand slits per centimetre, or *lines* as they are more usually called. In practice gratings are often used in *reflection* rather than *transmission*. The discussion will be continued in terms of transmission, which is probably easier to follow, but reference will be made to reflection gratings where necessary. Figure 11.3, showing the

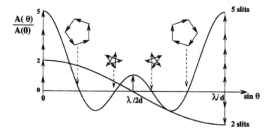

Figure 11.2   Amplitude diffraction patterns for five slits d apart and two slits d apart. Phase is in each case referred to the centre of the slit pattern. Notice that in the five-slit case the phasor contributed by the central slit remains unchanged throughout

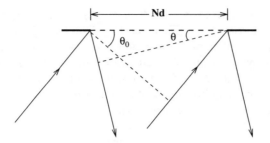

Figure 11.3 Geometry for a reflection grating

geometric relations of a reflection grating, has been arranged so that the discussion that follows can easily be related to it. In general a grating is not illuminated with a wavefront arriving exactly parallel to its plane, but with one at an angle θ_0 as measured in a plane perpendicular to the grating's lines. We shall continue in this analysis to assume the lines are narrow enough for each to be considered as the source of a single cylindrical wavelet. Take the first line as the phase reference and let there be N lines, so that the optical path difference across the width of the plane wavefront incident to the reference plane for emergence at θ in Figure 11.4 is

$$Nd(\sin\theta - \sin\theta_0).\tag{11.2}$$

It will be convenient to write the phase difference between light paths that pass through two adjacent slits as $\psi = 2\pi d(\sin\theta - \sin\theta_0)/\lambda$. Then the phase of light that has gone through the nth line is (with respect to the first) $(n-1)\psi$. The complex amplitude obtained as the sum of all the light leaving the grating in the direction θ is then

$$A(\theta,\theta_0) = A_0[1 + \exp(i\psi) + \exp(2i\psi) + \ldots + \exp((N-1)i\psi)].\tag{11.3}$$

This geometrical series may be summed in the usual way,[1] giving

$$\begin{aligned}A(\theta,\theta_0) &= A_0\left[\frac{1 - \exp(iN\psi)}{1 - \exp(i\psi)}\right]\\ &= A_0\exp\left[\frac{i(N-1)\psi}{2}\right]\frac{\sin(N\psi/2)}{\sin(\psi/2)}.\end{aligned}\tag{11.4}$$

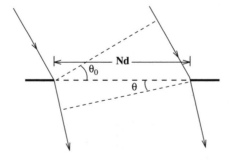

Figure 11.4 Geometry for a transmission grating

---

[1] $\sum_{j=0}^{N-1} a^j = (1 - a^N)/(1 - a)$.

The exponential term gives the phase of the resultant $A(\theta, \theta_0)$ relative to the zero of phase, which was taken to be the first line. If instead we take the centre of the grating (be it line or non-line) as our phase reference, the exponential term becomes unity and we see once again that the amplitude remains real, although the phase can be reversed (see Figures 8.2 and 10.2).

If we are interested in the pattern as a function of θ for a fixed θ_0, it is convenient to regard the inclination of the illuminating wavefront as putting a linear phase shift across the grating. This may be conveniently designated by δ radians per line. Then

$$\delta = \frac{2\pi d}{\lambda} \sin \theta_0. \tag{11.5}$$

Rewriting equation (11.4) in terms of θ and θ_0 and with the grating centre as phase reference gives

$$A(\theta, \delta) = A_0 \frac{\sin((N\pi d/\lambda) \sin \theta - N\delta/2)}{\sin((\pi d/\lambda) \sin \theta - \delta/2)}. \tag{11.6}$$

This is an important general expression for the diffraction pattern from N narrow lines d apart.

11.2 Diffraction Pattern of the Grating

For most purposes it may be sufficient to remember the basic equation for the diffraction maxima at normal incidence

$$d \sin \theta = m\lambda \tag{11.7}$$

and for incidence at angle θ_0

$$d(\sin \theta - \sin \theta_0) = m\lambda. \tag{11.8}$$

We may also need to know the width and shape of the diffraction maxima, for which we need the general diffraction pattern of equation (11.6). This is illustrated in Figure 11.5 where the intensity is plotted for a grating with six lines. It has several important properties:

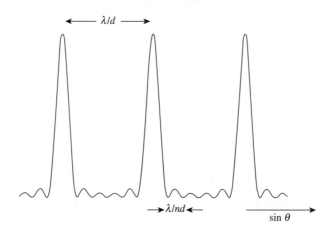

Figure 11.5 General irradiance pattern for a grating with six very narrow slits

1. The major maxima are equally spaced in $\sin \theta$, and occur whenever the phase difference between adjacent lines is an integral multiple of λ, as in equation (11.8). When this occurs, $\psi = 2\pi m$, so that the numerator and denominator of equation (11.6) both tend to zero together. In other words

$$\frac{\pi d}{\lambda} \sin \theta - \frac{\delta}{2} = m\pi \qquad (11.9)$$

or

$$\sin \theta = \frac{m\lambda}{d} + \frac{\delta\lambda}{2\pi d}. \qquad (11.10)$$

Here m is the diffraction grating order number. So δ, the phase shift per line caused by the angle of the illumination, determines the position of the orders. Hence changing δ by altering θ_0 shifts the pattern so that $(\sin \theta - \sin \theta_0)$ remains constant.

2. On the other hand, the separation of the major maxima in $\sin \theta$ is independent of δ. Thus

$$(\sin \theta)_m - (\sin \theta)_{m-1} = \frac{\lambda}{d} \qquad (11.11)$$

and the orders are equally spaced in $\sin \theta$, this spacing depending only on the separation of the lines, d, and the wavelength.

3. Zeros are given by the numerator of equation (11.6) being zero when the denominator is not. That is to say, when

$$\frac{N\pi d}{\lambda} \sin \theta - \frac{N\delta}{2} = p\pi \quad (p \text{ is integral, } p/N \text{ is non-integral}) \qquad (11.12)$$

or, with the same restriction,

$$\sin \theta = \frac{p\lambda}{Nd} + \frac{\delta\lambda}{2\pi d}. \qquad (11.13)$$

This is similar to the expression in equation (11.10), i.e. the condition for orders, except for the restriction that p/N is not integral. So to sum up: from equations (11.10) and (11.13) we see that the N phasors produced by the N lines of the grating form a closed polygon at each zero; zeros thus appear whenever the phase shift across the whole grating is a multiple of 2π. However, in the rare cases where this condition is satisfied, but also the phase shift between *each pair* of lines is a multiple of 2π, the phasor diagram is not a polygon at all, but a straight line, and instead of a zero a major maximum is produced.

11.3 The Effect of Slit Width and Shape

So far the diffraction grating has been taken to consist of N slits so narrow that the phase change across each could be neglected. This condition is very restrictive as lines of actual gratings may be several wavelengths wide. To analyse the situation where this restriction is not valid, consider first the

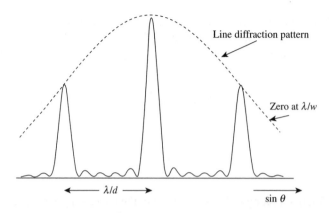

Figure 11.6   Intensity pattern for a grating with six slits of width comparable with λ. The modulating envelope (broken line) is the squared amplitude function shown more completely in the lower part of Figure 10.3

case of a grating which is opaque except for lines of width w spaced, as before, d apart. Then the diffraction at a single slit follows the analysis of Section 10.3.1, restricting the light to a range of angles depending on the width of the slit. The amplitude contributed by an individual line to the light transmitted in any direction by the whole grating is governed by the diffraction pattern of that line itself. Hence the resultant diffraction pattern from the grating is the product of the intensity pattern of a single line with the intensity pattern of Figure 11.5 for the ideal grating. This is illustrated in Figure 11.6.

11.4   Fourier Transforms in Grating Theory

We pointed out in Chapter 10 that the Fraunhofer diffraction pattern of an aperture is the Fourier transform of the amplitude distribution across the aperture. An idealized grating has the aperture distribution shown in Figure 11.7(a); this infinite, equidistant series of lines, or delta functions, is known as the *Dirac comb*, or grating function. Its Fourier transform is also a grating function, as shown in Figure 11.7(b). Following Section 10.3, the transform applies to the diffraction grating if the scales are in terms of wavelengths (for the aperture distribution) and in terms of direction cosines (for the angular distribution). As expected, the closer the lines of the grating, the wider apart in angle are the diffracted beams of successive orders.[2]

The grating function represents an idealized grating, an infinitely wide grating with infinitely narrow lines. Practical gratings, with finite overall width and with lines of finite width, are also very conveniently analysed by Fourier theory; the theory becomes essential for the more complex case of three-dimensional diffraction encountered in X-ray crystallography.

---

[2]The Dirac comb is the limit as $N \to \infty$ of a grating with N infinitely narrow slits. Its Fourier transform, as shown in Figure 11.7(b), displays only the primary maxima. What has become of the smaller secondary maxima seen, for example, in Figure 11.5? The answer is that they have disappeared in the limit of $N \to \infty$.

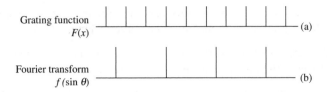

Figure 11.7 The Fourier transform of a Dirac comb is another Dirac comb. The two scales are inversely proportional

We now develop the Fourier transform approach. As we shall see, this easily extends to cover the general case of arbitrary slit structure and of arbitrary distribution of illumination over the grating. Mathematically, the aperture distributions are constructed as products and convolutions (see Section 4.13) of various functions with the grating function. These cases develop as follows:

1. The grating function: an infinite array of delta functions, spacing d, with the form

$$\psi(x) = \sum_{n=-\infty}^{n=+\infty} \delta(x - nd). \tag{11.14}$$

2. Repeated line structure, i.e. an infinite array of elements each with the form $f(x)$. The overall aperture distribution $F(x)$ is then a convolution of $\psi(x)$ with $f(x)$:

$$F(x) = f(x) \star \psi(x). \tag{11.15}$$

3. Finite grating length, i.e. a grating with narrow (delta function) lines, and limited in extent by a function $H(x)$, where $H(x) = 1$ for $|x| < L$ and $H(x) = 0$ for $|x| > L$. Then $F(x)$ is the product

$$F(x) = \psi(x) \cdot H(x). \tag{11.16}$$

4. Finite array of structured lines, i.e. a combination of cases 2 and 3 above. Then

$$F(x) = [f(x) \star \psi(x)]H(x). \tag{11.17}$$

Recalling from Section 4.13 that

the Fourier transform of the convolution of two functions is the product of their individual Fourier transforms,

and correspondingly

the Fourier transform of the product of two functions is the convolution of their individual Fourier transforms,

we find the required transforms, i.e. the diffraction patterns, of the four cases as follows:

1. The ideal grating $\psi(x)$ transforms into an ideal angular distribution $A(l)$, consisting of diffracted beams at equal intervals of the direction cosine l.

2. The grating with finite linewidth transforms into the product of the transforms of the grating function and the line structure, as in the envelope curve of Figure 11.6.

3. The grating with finite length transforms into the convolution of the transforms of the grating function and the overall illumination function of the grating, as in Figure 11.5.

4. The general case is a combination of the above, as in Figure 11.6. In this example the individual line structure throughout the grating is a 'top-hat' function, which transforms into a wide sinc function forming the overall envelope; the uniform illumination of the grating is also a top-hat function, which transforms into a narrow sinc function which is convolved with the grating function to produce diffracted beams with finite width.

Notice also the ideal case of a sinusoidal grating, such as may be produced holographically (Chapter 14), in which $F(x) = \sin 2\pi(x/a)$. Here the diffraction pattern consists simply of single sharp (delta-function-like) diffraction maxima at $\pm\theta = \sin^{-1}(\lambda/a)$. It is often valuable to consider complicated diffraction problems, such as those encountered in determining crystal structures (Section 11.11) or in holography (Chapter 14), as the sum of diffraction effects by many simple components which each give single diffraction maxima.

11.5   Missing Orders and Blazed Gratings

The combination of the individual line pattern and the grating pattern, which modulates it, can produce the effect of *missing orders*. Suppose the linewidth w is commensurate with the slit spacing d; that is to say, d/w is a rational fraction. Then a zero of the individual line diffraction pattern will fall on a major maximum of the grating pattern, so that this order will not appear. Algebraically, the orders of the slit pattern satisfy $d \sin\theta = m\lambda$ while the zeros of the modulating diffraction envelope satisfy $w \sin\theta = n\lambda$, with m, n integers, $n \neq 0$. If a zero and an order coincide in the same direction θ, then $d/w = m/n$. Given that one order is missing, all integer multiples of it will also be suppressed for the same reason.

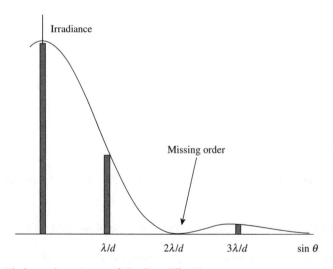

Figure 11.8   Missing orders. A zero of the line diffraction pattern can suppress an order entirely

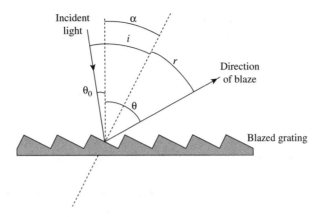

Figure 11.9   A blazed grating

This effect is shown in Figure 11.8 for $w = d/2$. So it is possible to remove orders by the skilful choice of the diffraction pattern of the lines. More important, particularly in reflection gratings, it is possible to arrange the grating so that most of the light goes into one particular order. Such a grating, illustrated in Figure 11.9, is called a *blazed* grating. The lines are ruled so that each reflects specularly in the direction of the desired order. Another way of looking at this is to observe that the angle of illumination and the tilt of the reflecting lines is such that each line has across it the appropriate phase shift to make it (as a single slit) diffract in the direction of the required order.

Example. It is amusing to notice that a plane mirror can be regarded as a limiting case of a diffraction grating, in which the separation d is equal to the linewidth w. Consider diffraction of a normally incident wave by a grating with N wide lines:

(a) For arbitrary d and w, write an expression for the diffracted amplitude by combining the amplitude for N narrow lines with the diffraction pattern of a single wide line.

(b) When all lines merge into one because $w = D$, show that zeros in the line diffraction pattern eliminate all orders except $m = 0$, which corresponds to specular reflection.

(c) Show that the resulting diffraction pattern for $w = d$ is just what you would expect for a single line (or mirror) of width equal to Nd.

Solution

(a) From equations (11.6) and (10.2),

$$A(\theta) = A_0 \frac{\sin(N\pi d \sin\theta/\lambda)}{\sin(\pi d \sin\theta/\lambda)} \frac{\sin(\pi w \sin\theta/\lambda)}{\pi w \sin\theta/\lambda} \tag{11.18}$$

where the first ratio gives the interference between the lines, and the second ratio gives the diffraction pattern of each line.

(b) The mth order satisfies the grating equation $d \sin \theta = m\lambda$. Putting $w = d$, the numerator of the line pattern becomes $\sin(\pi d \sin \theta / \lambda)$, which will vanish for the mth order. (The reader can check that, according to L'Hôpital's rule, the preceding factor from interference alone has the finite, non-zero value of $\pm N$.)

(c) When $w = d$, cancelling out the common factor in equation (11.18) yields $NA_0 \times \operatorname{sinc}(\pi Nd \sin \theta / \lambda)$, which is the diffraction pattern of a single line of width Nd.

11.6   Making Gratings

The main grating effects are easy to observe, and do not require very fine gratings. If a handkerchief is held in front of the eye and one looks through it at a well-defined edge (such as a distant roof against the sky) it is found that as well as the actual edge at least two more can be seen. These, which are displaced by equal increments of angle, correspond to the first and second orders of the grating formed by the threads of the handkerchief and are spaced at λ/d. The human eye can resolve angles down to about 1 minute of arc, or 0.0003 rad. With $\lambda = 500$ nm, this means that grating effects can just be detected if a grating of spacing 1 mm is held in front of the eye. The millimetre graduations on a transparent ruler will serve, but only just. A better grating to look through may be made by ruling a 10 cm $\times$ 10 cm square with 100 lines 1 mm apart. Photographic reduction to produce a negative 5 mm $\times$ 5 mm then gives a grating with a line spacing of 0.005 cm, in which the order separation is 0.01 rad or about 34 minutes of arc. The Sun viewed through this shows a spectacular and colourful series of orders, overlapping more and more as higher orders are reached. (Safety warning: be careful not to look directly at the Sun!)

Fraunhofer made his first grating in 1819 by winding fine wire between two screws. Later he made them by ruling with the help of a ruling machine, in which a ruling point was advanced between lines by means of a screw. The ruling was either of a gold film deposited on glass, or directly onto glass with a diamond point. Later in the nineteenth century, Henry Rowland improved the design of ruling machines and was able to rule 14 000 lines to the inch on gratings as much as 6 inches (15 cm) wide. He also invented the *concave* grating (Section 11.7).

Excellent gratings can be made by exposing a photographic plate to the interference pattern made by two crossing plane waves, as in Figure 8.1. The two waves must be essentially monochromatic, so that high-order interference fringes still have full visibility; this means in practice that they are both derived from a single laser source. The process is an elementary form of *holography* (see Chapter 14). Holographic gratings are used in most modern optical spectrometers (Chapter 12).

If a grating is ruled by a machine which is not perfect, confusing effects are observed which make its use for spectroscopy difficult. Each single spectral line is seen with several equally spaced and dimmer lines on each side of it. These are called *ghosts* and in a complicated spectrum of many lines may be difficult to distinguish from genuine lines. They arise from *periodic errors* in the ruling. Suppose that the machine's error was such that the depth of the ruling varied so as to go through a cycle of deep, shallow, deep, every m lines, and that the transmission of the lines was proportional to their depth. Then the grating would be like a perfect grating with another perfect grating m times as coarse in front of it. When illuminated by monochromatic light, each *order* of the perfect grating would be further split into orders separated by λ/md caused by the coarse grating. It is these satellite orders that are the ghosts. In fact any type of imperfection with a periodicity every m lines causes such ghosts spaced at λ/md, whether it be of amplitude or *phase*. The case we considered was of amplitude, in which the ability to transmit light was periodically variable. In a phase variation the *spacing* of the lines, whilst remaining on the average d, is periodically variable. In this case the lines are first too close, then too far apart, repeating this cyclically. This is like phase modulation of a carrier wave in radio engineering. The spectrum produced has numerous ghosts spaced at multiples of λ/md and of various intensities.

11.7 Concave Gratings

In an ordinary spectrograph a grating is usually illuminated by a plane wave, requiring a collimator lens or mirror with the light source at the focus. The diffracted spectrum is then focused onto a detector, so that two lenses or mirrors are needed; these may introduce losses and aberrations, especially for infrared and ultraviolet light. The difficulty may be avoided by using an arrangement due to Rowland in which a concave grating is itself used for focusing. In this grating the lines are ruled on the surface of a concave mirror. An interesting piece of geometry shows that if the slit is located on a circle tangential to the mirror and with the circle's *diameter* equal to the mirror's radius of curvature, then the several orders of diffraction are also in focus along this circle.[3]

In Figure 11.10, S is the slit and C the centre of curvature of the grating. Then all rays from S that are reflected from the grating at R have the same angle of incidence $\alpha = $ SRC because CR is normal to the grating. The directly reflected ray SQP crosses the circle again at P. Other rays such as SR are very nearly also focussed on the same point P; if R were on the circle the angle SRP $= 2\alpha$ would be independent of the position of R, so that all rays from S would be reflected through P. In fact, if the size of the mirror is small compared with the diameter of the circle, this is true enough. Now, if one wavelength in one of the orders is diffracted by an angle β more than the direct reflection, so that it

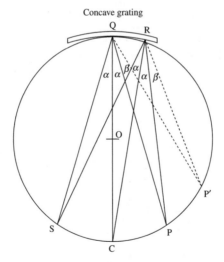

Figure 11.10 The geometry of the Rowland circle

---

[3]A theorem from plane geometry states that an arc of a circle subtends the same angle from any point of the circle outside the arc. For example, from any point on a circle, a semicircle (or the diameter across it) subtends an angle of $\pi/2$ radians. Any triangle inscribed in a semicircle, with the diameter as one side, is thus a right angle. This theorem applies to Rowland's circle, because the grating's departure from the circle is assumed small enough to be ignored.

cuts the circle at P′, the same argument applies to angle SRP′ which is constant at $2\alpha + \beta$. Hence if the slit, the grating and the photographic plate are all located on this *Rowland circle*, sharp spectra may be recorded without the intervention of further optics.

11.8   Blazed, Echellette, Echelle and Echelon Gratings

When a grating is used in a spectrometer, its usefulness in distinguishing between adjacent features of a spectrum is measured by its *resolving power*. We discuss this in detail in Chapter 12, where we show for a grating that the resolving power is mN, the product of the number of lines and the order of the diffraction. For a given number of lines, this may be increased by concentrating the diffracted light into a high-order m. A *blazed* grating (Figure 11.11) is a reflection grating with tilted reflection faces, so that light is reflected predominantly in the direction of one of the higher orders, giving the advantage of greater resolution at a high light level. The angle α between the normal to the grating and the normal to the grooves is called the *blaze angle*. The diffracted light satisfies the grating equation $d \sin \theta = m\lambda$ and the major peak in the diffracted light is at $\theta = 2\alpha$.

   To obtain still higher resolution from gratings it is easier to use fewer lines but increase the order of the diffraction. By setting $\theta = \theta_0 = 90°$ in equation (11.8) it can readily be seen that for a conventional plane grating the order cannot be higher than $2d/\lambda$, twice the number of wavelengths in the space between lines, so that close ruling does not permit the use of high orders. For example, if a grating with 5000 lines/cm was to be used at high resolution at wavelength 500 nm, the highest order it could possibly be used in would be

$$m_{\text{max}} = \frac{2d}{\lambda} = \frac{2}{5 \times 10^{-5}\,\text{cm}}\frac{1\,\text{cm}}{5 \times 10^3} = 8 \tag{11.19}$$

and to realize this extreme case the light would be at grazing incidence (at $\pi/2$, parallel with the surface) and be diffracted back through π. High-order diffraction is achieved in practice by the use of blazed gratings, and the *echelette, echelle* and *echelon* gratings. The idea of all these basically similar systems is to separate the fixed relationship between the line spacing and the order by making the

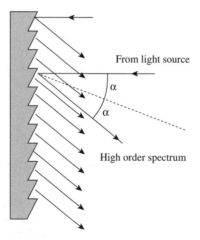

Figure 11.11 A blazed reflection grating, with blaze angle α and illuminated at normal incidence. The diffracted light is concentrated in the direction θ

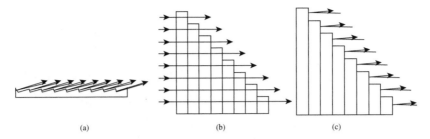

Figure 11.12 Gratings for high-order spectrometers: (a) echellette (reflection); (b) echelle (transmission); (c) echelon (reflection)

grating not flat but rather like a flight of stairs viewed from a distance. The riser of the stairs corresponds to the line, and the tread to a displacement backwards of each line. The lines have thus become reflecting surfaces, each one displaced backwards from the previous one to give a high order of interference. The angle of these reflecting surfaces can now be adjusted to reflect light into the direction in which it is desired to observe spectra.

An echelle grating about 25 cm across with 10^4 steps or grooves can be used in the 1000th order for visible light, giving the product $mN = 10^7$. The echelle grating is often used as a tuning element in lasers, since it gives high angular dispersion and high efficiency. For use at longer wavelengths, into the far infrared, a small number of grooves may be ruled directly onto metal: these are called echellettes, meaning 'little ladders'. Similar systems due to Michelson called echelons consist of a pile of glass plates arranged like a flight of stairs, which may be used in either transmission or reflection at orders as high as 20 000. The difficulties of realizing high resolution in this system become very great, and Michelson never in fact perfected the reflection echelon, though he made transmission echelons (Figure 11.12) successfully with some tens of plates.

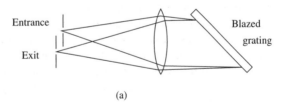

(a)

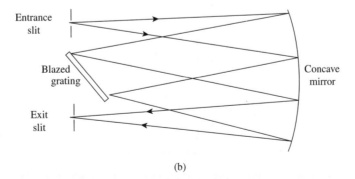

(b)

Figure 11.13 Mountings for blazed gratings: (a) Littrow; (b) Ebert

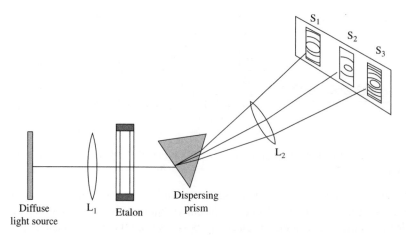

Figure 11.14 Cross-dispersion with low-resolution prism and high-resolution echelle spectrometers. The spectral lines S_1, S_2, S_3 are images of the source, dispersed by the prism. The echelle produces high dispersion spectrum within each of the spectral lines

Blazed gratings are often used in an arrangement due to O. Littrow (Figure 11.13(a)) in which the diffracted light returns almost along the incident path. This allows the same lens to be used as a collimator and for focussing. A similar arrangement due to H. Ebert is also shown in Figure 11.13(b); here the collimator and focussing elements are combined in a single concave mirror, avoiding the losses inherent in lens systems.

With the very high orders of interference obtained in these devices the problem of overlapping orders becomes extreme. Overlapping orders may, however, be dealt with by crossing any high-resolution spectrometer with a low-resolution spectrometer, such as a prism, whose resolution is in a perpendicular direction. The various orders are then separated in a two-dimensional format. An example is shown in Figure 11.14, where a prism is used as a cross-disperser for a Fabry-Perot spectrometer.

The combination of a grating and a prism, often called a 'grism', has another advantage when it is used in reflection (Figure 11.15). If the grating is bonded to, or etched into, the glass of the prism, the wavelength of the incident light is reduced by the refractive index of the glass, giving a larger angular dispersion at the grating; furthermore the resolving power may be increased in proportion by using a grating with a smaller line spacing.

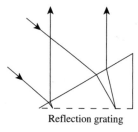

Reflection grating

Figure 11.15 A combination of a blazed grating and a prism, used in reflection. This 'grism' has a higher resolution than a grating in air with the same geometry

11.9   Radio Antenna Arrays

From metre wavelengths to centimetre wavelengths it is often convenient to construct large antennas or aerials from many similar radiating or receiving elements, arranged on a one- or two-dimensional grid. Such an arrangement is called an array. The radiating elements do not here concern us: they may for example be half-wave dipoles. In the present discussion they are considered to be identical so that they have all have the same polar diagram. The *power* polar diagram used in radio engineering is simply the angular pattern of intensity produced at a large distance from the antenna, often expressed as a fraction of the maximum of intensity. It is a Fraunhofer diffraction pattern. Similarly the less familiar *voltage* polar diagram is the complex amplitude. Further nomenclature that is usual in antenna work is that the main maximum or maxima of a polar diagram are called the *main beam* or beams. Subsidiary maxima are referred to as *sidelobes*.

Consider first a one-dimensional array of elements uniformly spaced d apart. The elements can all be excited separately, using suitable lengths of transmission line from the transmitter. New possibilities now arise as compared with the optical diffraction grating, since the phase of each element can be separately controlled. The main beam can be directed at any angle to the line of elements, as in the following examples, illustrated in Figure 11.16.

11.9.1   End-Fire Array Shooting Equally in Both Directions

An end-fire array means an array with a polar diagram having equal main beams directed each way along its length. To achieve this it must be arranged that one order appears at $\sin \theta = \pm 1$, with no order in between. To make the distance in $\sin \theta$ between orders equal to 2,

$$\frac{\lambda}{d} = 2 \quad \text{so} \quad d = \frac{\lambda}{2}. \tag{11.20}$$

To put a main beam at $\sin \theta = -1$ it can be seen from equation (11.10) that

$$\frac{\lambda \delta}{2\pi d} = -1 \quad \text{so} \quad \delta = -\pi. \tag{11.21}$$

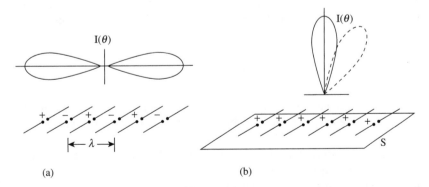

(a)   (b)

Figure 11.16   Directive radio antennas made from separately excited dipole elements spaced $\lambda/2$ apart. The polar diagrams show the main beam. In the end-fire array (a) the phase is reversed in alternate elements. The broadside array (b) is constructed a short distance (typically $\lambda/8$) above a reflecting sheet S. A progressive phase difference between elements changes the direction of the main beam, as shown in the broken line polar diagram

So an array of spacing $\lambda/2$ between (say) dipole elements phased alternately positive and negative would have the required property. It is easy to see that in either direction along the array the contributions from each dipole would be in phase, the delay in space from dipole to dipole being matched by the shift $\delta = \pi$ between the dipoles. On the other hand, in the direction at right angles to the array (for example) the contributions from alternate dipoles would cancel each other.

11.9.2   End-Fire Array Shooting in only One Direction

Here we must put an order at $\sin \theta = -1$, but not another anywhere. To ensure the last condition

$$\frac{\lambda}{d} > 2 \quad \text{or} \quad d < \frac{\lambda}{2} \tag{11.22}$$

and to put an order at $\sin \theta = -1$

$$\frac{\lambda\delta}{2\pi d} = -1 \quad \text{so} \quad \delta = -\frac{2\pi d}{\lambda}. \tag{11.23}$$

A reasonable arrangement now might be to make $d = \lambda/4$ and hence $\delta = \pi/2$. Seen from one end of the array, the delay in space would now be compensated for by the phase shift of $\pi/2$ between elements. Seen from the other end, the dipoles' contributions would cancel in pairs.

The technical problem of obtaining the required phase shift is not here our concern. In practice, because of the mutual coupling between elements, these sorts of arrays are difficult to realize. The familiar television Yagi aerial is a realization of the end-fire array in which a single 'driven' element couples with a reflector and several 'director' elements to approximate to the correct phase conditions.

11.9.3   The Broadside Array

In the broadside array a single main beam is required, emerging perpendicular to the array at $\sin \theta = 0$. A reflecting metal sheet spaced a quarter wavelength behind the array reinforces the forward beam (note the phase reversal on reflection). The single main beam with low sidelobe level elsewhere is ensured if

$$\frac{\lambda}{d} = 2 \quad \text{or} \quad d = \lambda/2. \tag{11.24}$$

Equation (11.13) then becomes $\sin \theta = 2m + \delta\lambda/2\pi d$. If the magnitude of the δ term is much less than 1, we can see that the only permitted order number is $m = 0$, and corresponds to values of $\sin \theta$ close to zero.

The sidelobes along the array at $\sin \theta = \pm 1$ will then be those midway between two orders of the general grating pattern and therefore they will be low.

The main beam will emerge perpendicular to the array if

$$-\frac{\lambda\delta}{2\pi d} = 0 \quad \text{or} \quad \delta = 0. \tag{11.25}$$

So in a broadside array all the elements must be fed in phase. Suppose we do not make $\delta = 0$. Then the beam will emerge at an angle determined by

$$\sin \theta = -\frac{\lambda \delta}{2\pi d}. \tag{11.26}$$

Evidently with a fixed broadside array the direction of the main beam can be steered simply by altering the phase shift from element to element at the rate given by equation (11.26). This principle is used in many large fixed arrays for beam swinging.

11.9.4   Two-Dimensional Broadside Arrays

In the discussions above of one-dimensional arrays we have considered only the plane of the elements. Considering now all three dimensions the one-dimensional broadside array above would have a main beam rather like a pancake with its plane perpendicular to the line of the array. In almost all applications a beam narrow in both dimensions is required, and this may be achieved by making the array two dimensional. As in the rectangular aperture treated in Section 10.3.4, the two-dimensional polar diagram is now given by the product of the two grating patterns appropriate to each dimension, and the resultant pencil beam may be steered independently in each direction by applying suitable phase shifts.

11.10   X-ray Diffraction with a Ruled Grating

Diffraction gratings are expected to work well when the line spacing is a few wavelengths. If the spacing is very many wavelengths, the diffraction angles will be small and hard to measure. Diffraction of X-rays by a ruled grating is therefore difficult; furthermore X-rays are not easily reflected or absorbed, so that no ordinary grating can be used.

A.H. Compton solved these difficulties very neatly by using a metal grating at nearly glancing incidence, when X-rays are reflected quite well. For X-rays the refractive index of all substances is less than unity by a few parts in 10^6 and hence the phenomenon of total internal reflection is observed when X-rays attempt to pass from free space into a medium rather than the other way round. As the angle of incidence of a beam of X-rays on a highly polished surface approaches $90°$ there is a critical *glancing* angle θ_R $(= 90° - i)$ at which reflection is observed; since the refractive index n is near unity we write

$$\cos \theta_R = n = 1 - \delta, \quad \sin \theta_R = \theta_R = \sqrt{1 - \cos^2 \theta_R} = \sqrt{2\delta}. \tag{11.27}$$

A typical example is copper, for which $\theta_R = 20'24''$ at a wavelength of $0.1537\,\text{nm}$, giving $\delta = 17.6 \times 10^{-6}$. The observation of θ_R is not a very accurate method of determining δ because absorption and other unwanted effects make the transition from non-reflection to reflection gradual rather than sharp. But the ability to reflect X-rays at once leads to the possibility of measuring their wavelength with a grating. Compton and Doan in 1925 successfully achieved this measurement. Using a grating of speculum metal with 50 lines per mm they found the wavelength of the K_α line of molybdenum to be $0.0707 \pm 0.0003\,\text{nm}$.

In Figure 11.17 the relationships of the zeroth and first order are shown. For the *m*th order

$$BD - AC = d(\cos \alpha - \cos(\alpha + \beta)) = m\lambda. \tag{11.28}$$

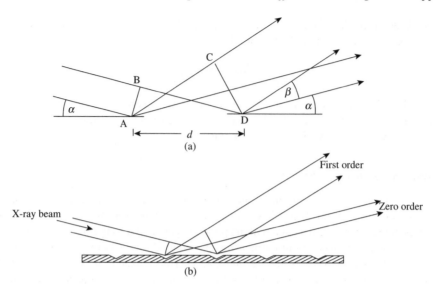

Figure 11.17   Diffraction of X-rays at glancing incidence on a ruled grating

Since α and β are small, and since $\cos\theta \approx 1 - \theta^2/2$ for small angles, this may be simplified to

$$d\left(\alpha\beta + \frac{\beta^2}{2}\right) = m\lambda. \tag{11.29}$$

Notice that there is no difficulty in knowing which order the observed diffraction is at; the orders starting at zero simply come out in series as β increases.

Spectrometers using grazing incidence gratings have been used successfully in satellite X-ray telescopes to resolve spectral lines at around 1 nm wavelength.

11.11   Diffraction by a Crystal Lattice

The short wavelengths of X-rays are generally not well suited for diffraction by ruled gratings. They are, however, conveniently close to the spacings of atoms in crystal lattices, which therefore provide excellent three-dimensional diffraction gratings for X-rays. The diffraction pattern is intimately connected with the arrangement and spacing of atoms within a crystal, so that X-rays can be used for determining the lattice structure.

It was Max von Laue who first suggested that a crystal might behave towards a beam of X-rays rather as does a ruled diffraction grating to ordinary light. At the time it was not certain either that crystals really were such regular arrangements, or that X-rays were short-wavelength electromagnetic radiation. In 1912 Friedrich, Knipping and von Laue performed the experiment illustrated in Figure 11.18. X-rays from a metallic target bombarded by an electron beam were collimated by passing through holes in screens S_1 and S_2 and fell on a single crystal C of zinc blende, passing through to a photographic plate. When the plate was developed, as well as the central spot at O where the beam struck the plate there were present many fainter discrete spots also. This showed that there were a few directions into which the three-dimensional array of atoms in the crystal selectively diffracted the X-rays. These depended on the orientation of the single crystal used. How could the

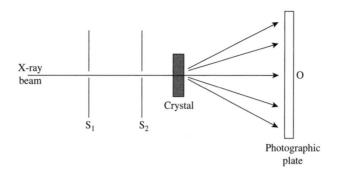

Figure 11.18   The experiment of Friedrich, Knipping and von Laue

significance of these directions be understood, when the three-dimensional and doubtless very complicated crystal structure was not known?

The answer was given by W.L. Bragg in 1912 and is so simple that (in hindsight) it may seem completely obvious. Bragg pointed out that whatever the structure of a crystal, so long as it is *repetitive* in three dimensions, it is possible to draw sets of parallel planes on each of which the arrangements of atoms will be the same. Such planes are called *Bragg planes* and their separations *Bragg spacings*. In any crystal structure, it is possible to draw many such sets of planes, but the numbers of atoms on each will vary very much. If plane monochromatic X-rays fall upon the atoms of a Bragg plane, each atom acts as a scatterer and a secondary wavelet spreads out in all directions. If we consider a *single* Bragg plane these wavelets will combine in phase in the undeviated direction, which is uninteresting, but also in a direction corresponding to ordinary reflection just like a mirror. Now add all the other Bragg planes parallel to the first. The specularly reflected waves from the various planes will in general be out of phase and will interfere destructively. They will all combine in phase, however, if a rather simple condition involving the glancing angle θ is satisfied. In Figure 11.19 it is easy to see that the paths $R_1 A$ and $R_1 C$ are identical, so that the extra path in reflection number 2 is $CR_2 + R_2 B$. The conditions for reinforcement are therefore the simple law of reflection, $\theta = \theta'$, plus[4]

$$2d \sin \theta = m\lambda. \tag{11.30}$$

These two statements together form Bragg's law for X-ray reflection from a crystal. (Note that observationally θ is one-half the deflection angle of the radiation.)

The spots found in the original experiment of Figure 11.18 represented reflection at Bragg planes. The X-ray source contained a wide range of wavelengths, so that the Bragg law was obeyed for each spot by a wavelength within the range. Modern X-ray diffraction techniques use monochromatic sources; since λ is fixed, a spot can then only be found by rotating the crystal until a suitable value of θ occurs. An alternative to rotating a single crystal is to use a powder containing many small crystals at all possible random orientations.

The importance of the many methods of investigation of crystal and molecular structures which sprang from these fundamental discoveries of von Laue and Bragg is hard to over-emphasize. The full analysis of an X-ray diffraction pattern requires measurement not only of spot positions but also of

---

[4]The path difference of equation (11.30) also follows directly from the construction of Figure 8.11.

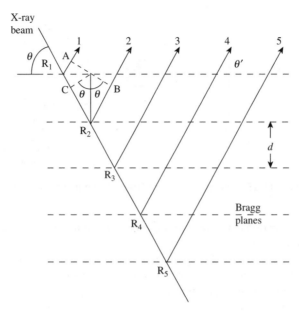

Figure 11.19   Diffraction of X-rays at the Bragg planes in a crystal. The extra path of ray 2 is the distance $CR_2B = 2d \sin \theta$

their relative intensities, since these reveal the internal structure of the repeating elements in the crystal lattice. This may not, however, be sufficient to determine the structure of large and complex biochemical molecules, even when they can be prepared and analysed in crystalline form; here a Fourier transformation of a diffraction pattern requires a determination of *phase* as well as intensity. The solution to this problem is found by labelling a particular site in the molecule by adding chemically an atom of a heavy metal. This atom produces a diffraction pattern of comparable intensity with the rest of the molecule, adding to or subtracting from the intensity of the elements of the diffraction pattern according to the relative phase. The heavy molecule acts as a phase reference, allowing a full Fourier transform to be made. The structure of very complex molecules can be completely determined, even if they contain some tens or hundreds of thousands of atoms, as in a protein molecule.

11.12   The Talbot Effect

Diffraction gratings are normally used at a distance where light from the whole of the grating contributes equally; in contrast, if a microscope is focussed on the surface, the lines of the grating are seen individually. If the microscope is racked away from the surface the lines at first become blurred and disappear, but remarkably they reappear when the microscope is racked out further and is focussed on another plane above the surface. In monochromatic light the image at this plane is an exact copy of the grating. Figure 11.20 shows how at this distance z the light waves diffracted from adjacent grating lines combine in phase to produce an element of the phantom grating. The path difference between direct and adjacent rays is $2d^2/\lambda$, where d is the line spacing in the grating (note the similar calculations of sagittal distance in Figure 8.8 and the Rayleigh distance in Chapter 10), so that the so-called Talbot distance z_T where the image appears is given by $z_T = 2d^2/\lambda$. This effect,

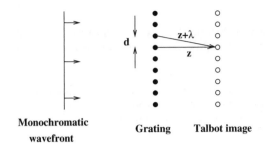

Monochromatic   Grating   Talbot image
wavefront

Figure 11.20   The Talbot effect. An image of a grating is observable by a microscope focussed on a plane at some distance above it

discovered by Talbot[5] in 1836, has recently been used in the reproduction of gratings without the use of lens systems.

Problem 11.1 Numerical examples
(i) Sodium chloride, NaCl, has a face-centred cubic crystal lattice with Na and Cl atoms alternately, spaced equidistant along perpendicular x, y and z directions. The spacing of adjacent atoms is 2.8×10^{-10} m. At what angles would you expect X-rays of wavelength 1.54×10^{-10} m to be reflected by (a) planes of atoms perpendicular to the z direction and (b) planes of atoms parallel to the z axis but at $45°$ to the x and y axes?
(ii) A diffraction grating consisting of just 10 narrow lines separated by a spacing of 0.001 cm is used at normal incidence. At what angles are the first and second orders for light of wavelength 500 nm? What is the angular half-width of the main peaks of the diffraction pattern for these two orders? What wavelengths might therefore be separated in a spectrograph using this grating?

Problem 11.2
Light with wavelength small relative to the line spacing, $\lambda \ll d$, strikes a grating at a glancing angle $\theta_0 \approx \pi/2$. As viewed by the incident beam, the line spacing has a projected value d_{proj}, much smaller than d. The mth order, where m is a small integer, diffracts off at angle θ. Show that the angular dispersion $\Delta\theta$ between adjacent orders is essentially that for normal incidence on a grating with the projected line spacing.

Problem 11.3
Consider a sinusoidal aperture function $F(x) = (1/2)[1 + \cos(2\pi x/d)]$, which is similar to that mentioned at the end of Section 11.4 but everywhere non-negative. Using the well-known integral expansion of a Dirac delta, $\delta(x) = (2\pi)^{-1} \int_{-\infty}^{+\infty} \exp(ixy)dy$, verify that all the light is diffracted into only a few orders near $m = 0$.

Problem 11.4
Sketch the diffraction pattern (in intensity) of a grating with N slits, in which the slit width is w, and the spacing is d, when the grating is illuminated normally by a plane monochromatic wave.

   Indicate how this pattern would be modified if: (a) the illumination is reduced at the edges of the grating; (b) alternate lines are blacked out; (c) the amplitude in alternate lines is reduced by a grey filter to a fraction α.

---

[5]Henry Fox Talbot (1800–77), inventor of photography. Like Young (see Chapter 8), he was also an expert in ancient languages, specializing in Syrian and Chaldean inscriptions.

Problem 11.5
How is the pattern of the grating in Problem 10.2 modified if transparent material which retards the wave by half a wavelength covers (a) alternate slits, (b) every third slit, (c) one half of the grating?

Problem 11.6
A plane diffraction grating consists of alternate transparent and opaque lines with width a, b respectively. Babinet's principle shows that the Fraunhofer diffraction pattern is the same for a grating with widths a and b interchanged, apart from the central maximum. Confirm this by means of Fourier analysis.

Problem 11.7
Derive the Fraunhofer diffraction pattern of a set of three equally spaced, equal-width slits by adding the transform of the outer pair of slits to that of the single central slit. Repeat the exercise for a four-slit grating, taking the inner and outer pairs separately.

Problem 11.8
The inverse of the line spacing d in a badly ruled grating increases linearly with distance x across the grating so that

$$d^{-1} = a + bx. \tag{11.31}$$

Show that the nth-order diffraction maximum at angle θ, formed by plane-incident light, does not emerge from the grating as a plane wave but converges on a focus at a distance $\cos^2 \theta/(bn\lambda)$ from the grating. (Hint: The grating equation implies that nth-order rays from adjacent parts of the grating with different spacings $d(x)$ will not emerge parallel, but instead they will intersect at a finite distance.)

Problem 11.9
The transparency of the lines in a badly ruled grating varies periodically across the grating, so that the pth line has transparency proportional to $1 + a\sin(2\pi p/q)$ where a is small and the pattern of transparency repeats every q lines. Use the convolution theorem (Section 4.13) to find the diffraction pattern of the grating.

12 Spectra and Spectrometry

One important object of this original spectroscopic investigation of the light of the stars and other celestial bodies, namely to discover whether the same chemical elements as those of our Earth are present throughout the universe, was most satisfactorily settled in the affirmative; a common chemistry, it was shown, exists throughout the universe.

The scientific papers of Sir William Huggins (1824–1910).

Modern improvements in optical methods lend additional interest to an examination of the causes which interfere with the absolute homogeneity of spectral lines.

Lord Rayleigh, 1915.

Spectroscopy, the study of spectra, scored a spectacular success at the solar eclipse of 1868, when Lockyer[1] noted that a bright yellow–orange line emitted by solar prominences corresponded to no known element. He named it helium; it was isolated from the terrestrial atmosphere by W. Ramsay in 1895. Spectroscopy is still the general term for the study of both absorption and emission spectra, but we may distinguish between a *spectroscope* for direct visual use, a *spectrograph* providing simultaneous measurement over a wide spectral range and a *spectrometer*,[2] which measures and records photoelectrically details of spectral lines at any part of the electromagnetic spectrum.

The choice of an appropriate spectrometer depends primarily on its *resolving power*, i.e. the ability to distinguish between light of closely adjacent wavelengths. In this chapter we briefly describe the rich variety of spectral lines and bands in the optical domain, and the factors determining their width. We then analyse the resolving power of simple prism and grating spectrometers, and show how the higher resolving powers which are needed for complex spectra with narrow lines demand interferometric techniques such as the Fabry–Pérot or Michelson interferometers, and finally we show how Fourier analysis techniques are used in spectral analysis.

The spectral analysis of light by filtering techniques such as the diffraction grating and Michelson and Fabry–Pérot interferometers measures the irradiance as the filter frequency is changed. There is a

---

[1]Sir Norman Lockyer (1836–1920), founded and edited for 50 years the journal *Nature*.

[2]The term spectrometer was originally used for an instrument to measure refractive index!

Optics and Photonics: An Introduction, Second Edition   F. Graham Smith, Terry A. King and Dan Wilkins
© 2007 John Wiley & Sons, Ltd

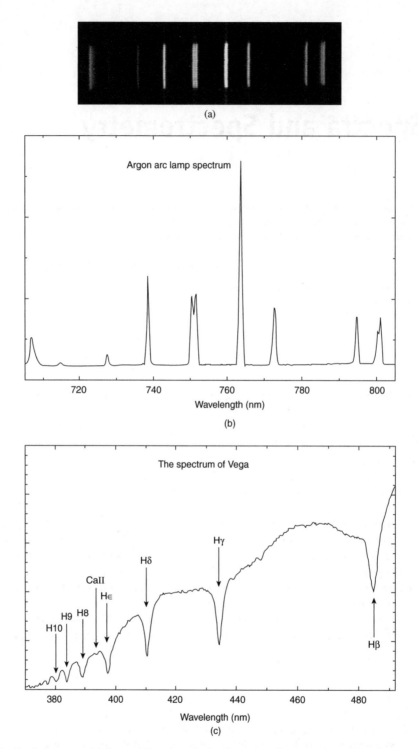

Figure 12.1 Typical spectra. (a) Spectral lines in emission from argon in an arc lamp (Jodrell Bank Observatory). (b) The argon spectrum seen by a spectrometer. (c) Part of the spectrum of light from the bright star Vega, showing the Balmer series of hydrogen lines in absorption within a continuum (Jodrell Bank Observatory)

lower limit on the bandwidth of those methods. We describe how irradiance fluctuation spectroscopic techniques acting on the detected signal can be used to measure much smaller spectral features.

12.1   Spectral Lines

Every elementary particle, whether atom or molecule, exists in one of a series of discrete quantized energy states. The energy released or absorbed in transitions between these states appears as discrete features in spectra radiated or absorbed by the particles; these are the spectral lines which are characteristic of each atom or molecule, and which are the principal subject of spectroscopy. Transitions giving spectral lines in the optical domain have energies of order 1 eV (electronvolt); these are usually transitions between electronic states in isolated atoms. Lower energy transitions, of order 0.1 eV and below, related to infrared, millimetric and radio radiation, are associated with molecules. We discuss in Chapter 18 the nature of the various types of transition, and the theoretical basis of these quantized states.

The series of quantum levels in each regime each allow a series of transitions, either to a lower energy level in the radiation of a photon or to a higher level in the absorption of a photon. Even the simplest atom, hydrogen (Figure 12.1(c)), therefore has several series of spectral lines. Most atomic species, however, including atoms at various states of ionization, can be identified by some particularly prominent and characteristic spectral lines (Figure 12.1(c)). Molecules, even such simple molecules as O_2, CO_2 and H_2O, which have multiple rotational and vibrational energy levels, have very complex spectra. The band structure of polyatomic molecules often appears as a distribution of continuous irradiance rather than a set of resolved lines. Despite the complexity, each species of atom or molecule has its own typical spectrum, and spectral analysis can identify and quantify constituents of light sources such as a discharge tube, or of ionized atomic species in the solar corona, or absorbing gases such as the molecular constituents of the terrestrial atmosphere.

Spectral lines may contain fine structure which only appears if the line is intrinsically narrow. For example, a spectral line may be split by the effect of a magnetic field into two or more components with different polarizations (see Chapter 18); this may be observed throughout the electromagnetic spectrum from radio to X-rays. Again, electronic transitions at optical wavelengths may reveal a split in quantized energy levels due to interactions between the spin of the electron and its orbital angular momentum, and between the spins of the electron and the nucleus; these appear in spectra as *fine structure* and *hyperfine structure* respectively (see Chapter 18).

12.2   Linewidth and Lineshape

The simplest spectral line arises from the spontaneous radiative decay of atoms between two well-defined states with energies E_2 and E_1; the spectral line then has wavelength $\lambda = c/v$, where frequency $v = (E_2 - E_1)/h$. The energies of both states have a finite width, so that light from a set of atoms is emitted over a range of wavelengths. Every spectral line therefore has a natural width, shown as $\Delta\lambda$ in Figure 12.2. This natural width depends on the probability (the inverse of the lifetime) of the quantum transition within an isolated atom or molecule; the longer the lifetime, the narrower the linewidth.

Often the observed width, particularly of intrinsically narrow lines, is considerably increased by external factors; in a gas this broadening may be due to thermal motion and collisions. The narrowest

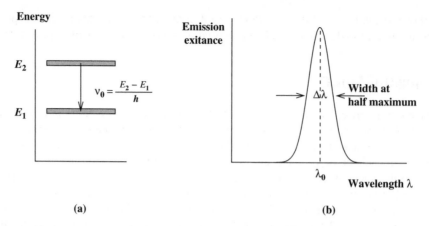

Figure 12.2   Spontaneous radiative decay between two energy levels, and the resulting emission lineshape and width

lines are found in isolated heavy atoms at low temperatures, such as caesium atoms whose resonant spectral lines are exploited in atomic clocks. In contrast, broad lines occur in the ultraviolet and X-ray spectra of highly ionized species, notably in very hot gas such as in the solar corona.

The shape of the broadened line is dependent on the mechanism of broadening. Thermal motion in a gas causes a spread in velocity and therefore Doppler shift among individuals of the radiating species; the resulting spread in wavelength reflects the Gaussian distribution of thermal velocities in the gas. In contrast, collisions between atoms or molecules in a gas act to reduce the lifetime or interrupt the phase of a transition, which has the effect of broadening the emitted line; a set of emitting particles will have a spread of widths all centred on the same wavelength, and the lineshape will reflect the statistical distribution of times between collisions.

The lineshape due to the collisional broadening processes, in which the central wavelengths are unchanged, is termed *homogeneous broadening*; in this case the lineshape is *Lorentzian* (Figure 12.3). The Lorentzian frequency profile for the radiative transition probability per unit frequency interval in a homogeneous transition with central frequency v_0 is of the form

$$g(v, v_0) \propto \frac{1}{1 + [2(v - v_0)/\Delta v_{\mathrm{H}}]^2}. \tag{12.1}$$

Here Δv_{H} is the linewidth (full width at half maximum, often abbreviated to FWHM), which is related to the collision frequency.

For the Doppler broadening mechanism, termed *inhomogeneous broadening*, in which the radiating frequencies are spread by thermal motion, the lineshape is *Gaussian*:

$$g(v, v_0) \propto \exp\left\{ -\left[\frac{4 \ln 2}{\Delta v_{\mathrm{D}}^2} (v - v_0)^2 \right] \right\}. \tag{12.2}$$

Here Δv_{D} is linewidth (FWHM), which is related to temperature. For atoms with mass m at Kelvin temperature T

$$\frac{\Delta v_{\mathrm{D}}}{v} = 2[2\, kT \ln 2/mc^2]^{1/2} \tag{12.3}$$

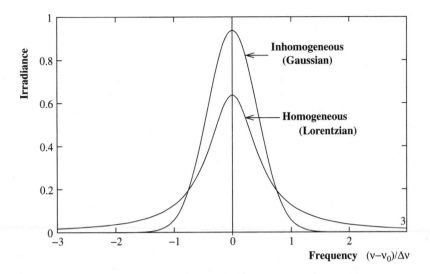

Figure 12.3 Spectral emission lineshape for homogeneously broadened (Lorentzian lineshape) and inhomogeneously broadened (Gaussian lineshape) profiles. The lineshapes are shown for the same centre emission frequency and $\Delta \nu_H = \Delta \nu_D$

where k is Boltzmann's constant. For atomic or molecular weight A

$$\frac{\Delta \nu_D}{\nu} = 7.16 \times 10^{-7} \left(\frac{T}{A}\right)^{1/2}. \tag{12.4}$$

The derivation of these lineshapes can be found in Appendix 4. The two basic lineshapes are illustrated in Figure 12.3 for equal integrated areas of emission. Distinguishing such lineshapes is often an important objective of high-resolution spectroscopy.

12.3 The Prism Spectrometer

The simplest way of examining a spectrum is to use a prism to spread a light beam in angle, following the example of Newton, who placed a prism in a beam of sunlight and showed how the light was split into a spectrum of colours. The geometry is best understood from the thin prism, as presented in Chapter 2, where the angles are small. The prism deviates a wavefront through an angle θ given by

$$\theta \approx (n-1)\alpha \tag{12.5}$$

where α is the apex angle of the prism and n its refractive index. Since the refractive index n varies with wavelength λ, different colours emerge as wavefronts with different angles when a plane wavefront is incident on the prism. The angular separation $\delta\theta$ of two components of the wavefront with wavelengths separated by a small amount $\delta\lambda$ depends on $dn/d\lambda$, and differentiation of equation

(12.5) leads to

$$\delta\theta \approx \alpha(\mathrm{d}n/\mathrm{d}\lambda)\,\delta\lambda. \tag{12.6}$$

The *angular dispersive power* of the prism $\mathrm{d}\theta/\mathrm{d}\lambda$ is therefore directly proportional to the angle α and to $\mathrm{d}n/\mathrm{d}\lambda$, the rate of variation of refractive index with wavelength in the prism. As we have seen when discussing chromatic aberration in a lens (Chapter 2), the *dispersive power* of a particular type of glass is often quoted in terms of its refractive indices at specific wavelengths; for example, a typical flint glass has refractive indices $n_F = 1.632$ at $\lambda = 486\,\text{nm}$, $n_D = 1.620$ at $\lambda = 588\,\text{nm}$ and $n_C = 1.616$ at $\lambda = 656\,\text{nm}$. The conventional dispersive power, Δ, is then $(n_F - n_C)/(n_D - 1) = (1.632 - 1.616)/(1.62 - 1) = 1/39$. The differential $\mathrm{d}n/\mathrm{d}\lambda$ is approximately $-10^{-4}\,\text{nm}^{-1}$, varying slowly over the wavelength range.

The design of a practical spectrometer requires the angular spread of the spectrum to be large enough for an adequate separation of components with different wavelengths; for the prism this requires glass with a large dispersive power and a large prism angle. Light with a single wavelength will, however, spread over a certain range of angles, so that light with two closely spaced wavelengths may overlap and not be clearly *resolved*. This limit on the possible resolution of closely spaced wavelengths is due to diffraction at the whole aperture. As we see later in this chapter, the best possible *spectral resolving power* of a prism depends mainly on its overall dimensions rather than on its apex angle or refractive index.

Equation (12.6) may be restated for any shape of prism, thick or thin, given a symmetrical configuration of the ingoing and outgoing rays, as in Figure 12.4. Let B be the base length of the prism, l the length of its vertical sides and w the width of the wavefront. From the geometry of that figure, $w = l\cos[(\theta + \alpha)/2]$ and $B = 2l\sin(\alpha/2)$. Equation (2.5) gives $\sin[(\theta + \alpha)/2] = n\sin(\alpha/2)$.

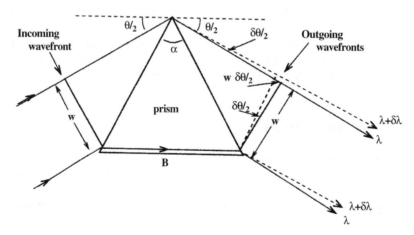

Figure 12.4 The angular dispersion of a prism. An incoming wavefront, width w, containing two nearby wavelengths λ and $\lambda + \delta\lambda$, is refracted at or near minimum deviation. The wavefronts can be regarded as defining constant optical path along the rays. For the wavelength change $\delta\lambda$, the change of optical path is therefore the same for the rays travelling through the base of the prism as for the rays travelling through the apex; this gives $B\delta n = w\delta\theta$. Equation (12.7) for $\delta\theta$ follows. (Note that the angular deviation is independent of angle of incidence at minimum deviation.)

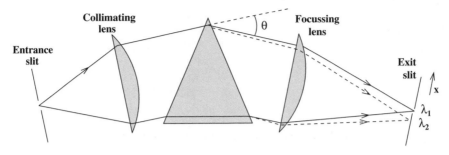

Figure 12.5 Optical system of a prism spectrometer. The entrance slit is at the focus of the collimating lens; the exit slit with its focussing lens forms a telescope which is moved to collect plane wavefronts over a range of angles

Differentiating, we find $(\delta\theta/2)\cos[(\theta+\alpha)/2] = \delta n \sin(\alpha/2)$. Multiplying by l gives

$$\delta\theta = \frac{B}{w}\frac{\mathrm{d}n}{\mathrm{d}\lambda}\,\delta\lambda. \tag{12.7}$$

A practical prism spectrometer requires in essence an incident plane wavefront, and a means of recording separately the plane wavefronts which emerge in different directions. The incident plane wave may be obtained from a source at a great distance, or from a point source of diverging waves which are made plane by a lens with a positive power, such as a simple biconvex lens. The emergent wavefronts may be examined by eye, since the eye is designed to sort out waves travelling in different directions, or by the eye with the help of a telescope, or by a form of camera, as shown in Figure 12.5. A narrow source of light, usually an *entrance slit*, is needed. A large part of the spectrum may be seen or recorded simultaneously on a photographic plate or a photodiode array detector such as a CCD (see Chapter 20); the instrument may then be called a *spectrograph*. A spectrograph using a linear array of diode detectors may be termed an optical spectrum analyser (OSA) or an optical multichannel analyser (OMA). Alternatively, a narrow range may be selected by an exit slit as a source of monochromatic light; this is then a *monochromator*. If the selected wavelength range is recorded in a photoelectric detector, this becomes a *spectrometer*. (Although this is the formal definition, we shall follow common practice and use the term spectrometer in the more general sense of any device used to produce, view, or measure and record a spectrum, and thus include the spectroscope and spectrograph as special cases.)

It will be seen that for each wavelength the exit slit accepts a monochromatic image of the original point source, formed by the two lenses, referred to as the *collimating* and the *focussing* lenses. If the source is made a line, such as a slit in front of a flame or a discharge tube, then a line will appear for each colour of the spectrum. The width of the line depends on the width of the slit; a wide slit admits more light, but produces wider final images and thus gives poorer resolution. The line from a narrow slit will be broadened by diffraction, which limits the resolving power of the spectrometer.

As shown in Figure 12.5, the focussing lens images the entrance slit onto the plane of the exit slit. The exit slit may be scanned across the spectrum; if the focal length of the focussing lens is f, the distance between the images for λ and $\lambda + \Delta\lambda$ is

$$\Delta x = f\Delta\theta = f\frac{\mathrm{d}\theta}{\mathrm{d}\lambda}\Delta\lambda = \frac{\mathrm{d}x}{\mathrm{d}\lambda}\Delta\lambda. \tag{12.8}$$

The quantity $\mathrm{d}x/\mathrm{d}\lambda$ is the *linear dispersion* of the spectrometer when used at wavelength λ.

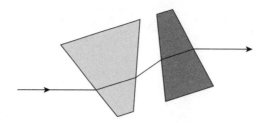

Figure 12.6   Direct vision prism for a central wavelength λ

A small hand-held prism spectrometer, which is useful for detection of elements in a flame or discharge tube, would probably use the *direct vision prism* of Figure 12.6. Here there are two prisms made of glasses with different dispersive power, deviating in opposite directions. The prism angles are chosen to give zero deviation at a central wavelength.

12.4   The Grating Spectrometer

As we have seen in Chapter 11, a diffraction grating can act like a prism in deviating a wavefront through an angle which depends on wavelength; a diffraction grating could therefore be substituted for the prism in Figure 12.5. The diffraction grating is, however, usually used in reflection, since it is often advantageous to avoid transmission through glass which loses light by either absorption or partial reflection. The grating spectrometer offers improved throughput, dispersion and resolution compared with the prism spectrometer. If collimation and focussing are achieved with concave mirrors instead of lenses, the spectrometer can work not only at visible wavelengths but also at ultraviolet wavelengths. The Czerny–Turner spectrometer of Figure 12.7 is a common arrangement; here the scanning in wavelength is accomplished by rotating the plane grating. For use at wavelengths λ < 200 nm the spectrometer requires evacuation to avoid the ultraviolet absorption by air.

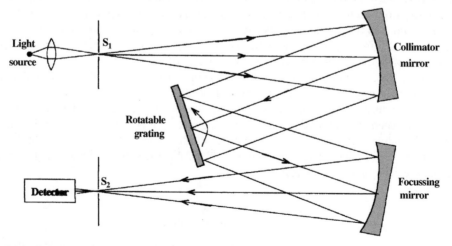

Figure 12.7   Czerny–Turner spectrometer. Only reflecting optical components are used, allowing operation at ultraviolet wavelengths

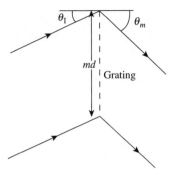

Figure 12.8   A diffraction grating showing the angles of incident and emerging light. The total number of lines in the grating is N

The essential geometry of diffraction at a grating, which applies equally in transmission and reflection, is shown in Figure 12.8. Here θ_I is the angle of incidence, and θ_m is the angle of emergence for wavelength λ in the mth order. The grating equation, from Chapter 11, is

$$d(\sin \theta_m + \sin \theta_I) = m\lambda. \tag{12.9}$$

where d is the line spacing and m is the order of the diffraction.

The *angular dispersive power* of the grating used at order m is the rate of change of angle of emergence against wavelength. For a given angle of incidence, we obtain from equation (12.9)

$$\frac{d\theta_m}{d\lambda} = \frac{m}{d \cos \theta_m}. \tag{12.10}$$

Thus for any given order a grating may be used just like a prism, except that it is possible to get a much greater angular dispersion with a grating. Let us substantiate this for visible light. Based on the table in Section 2.12, and on equation (12.7), a wide prism ($p/w \simeq 1$) of crown or flint glass has $d\theta/d\lambda_{\text{prism}} \simeq dn/d\lambda \simeq -0.1\,\mu\text{m}^{-1}$. For comparison, from equation (12.10) and the grating equation, $d\theta/d\lambda_{\text{grating}} \simeq m/d = \sin \theta_m/\lambda \simeq 2\,\mu\text{m}^{-1}$, which is an order of magnitude greater.

A new problem arises, however, when a wide range of wavelengths is being observed: the spectra in different orders may overlap. If a range from λ_1 to λ_2 is observed in the mth and $(m+1)$th order, there is an overlap if

$$m\lambda_2 > (m+1)\lambda_1. \tag{12.11}$$

The wavelength range between the overlapping orders is the *free spectral range*. If the spectrometer is set for operation at wavelength λ_1 in order m, it will pass $m\lambda_1$ in first order, $m\lambda_1/2$ in second order and so on. For a grating used at normal incidence, and with the diffracted beam in the mth order at angle θ, the overlap occurs when

$$d \sin \theta = m\lambda' = (m+1)\lambda_1. \tag{12.12}$$

The free spectral range $\delta\lambda_{\text{FSR}} = (\lambda' - \lambda_1)$ for order m is then

$$\delta\lambda_{\text{FSR}} = \frac{\lambda_1}{m}. \tag{12.13}$$

The confusion this may cause when observing a spectral range greater than $\delta\lambda_{FSR}$ may be avoided by using a filter to restrict the wavelength range of the incident light, or by adding a *cross-dispersing* device such as a second grating or prism which spreads the spectrum in an orthogonal direction. Note that a prism used alone concentrates light into a single spectrum, with no overlapping orders; for this reason astronomical telescopes may use a large thin prism in front of the objective lens or mirror, to display a small spectrum for each of many stars observed over a large angular field.

12.5   Resolution and Resolving Power

The purpose of a spectrometer is to distinguish between light waves separated by a small wavelength difference $\delta\lambda$. The prism and the grating spectrometers change the *wavelength* difference $\delta\lambda$ into a difference of *emergence angle* $\delta\theta$ in the wavefronts at the two wavelengths. The relation between $\delta\lambda$ and $\delta\theta$ is determined for a prism by the geometry of the prism and the dispersive power of its material, and for the grating by the line spacing and the order of diffraction. Light from a single wavelength will, however, emerge over a spread of angles, so that there is a limit to the possibility of distinguishing two spectral lines closely spaced in wavelength.

The resolution of a spectrometer is a measure of its ability to distinguish two adjacent spectral lines, such as the two sodium D-lines at 589.0 and 589.6 nm. Even if the entrance slit of the spectrometer is made very narrow, the exit slit will be scanning across two diffraction images whose width is determined by the characteristics of the grating, as in Figure 12.9(a). If these are well separated, the spectral lines are resolved. If they are so close as to merge into a single image, as in Figure 12.9(b), they are unresolved. In Figure 12.9(c) the separation is such that the first diffraction zero of one image falls on the maximum of the other, giving an obviously double line. This is known as *Rayleigh's criterion* for the limit of resolution of the spectrometer. When the line profile is a sinc-squared function, as in Figure 12.9, the dip is 20% below the maxima; for different profiles, including those without a clear minimum, it is useful to define resolving power in terms of the FWHM of a

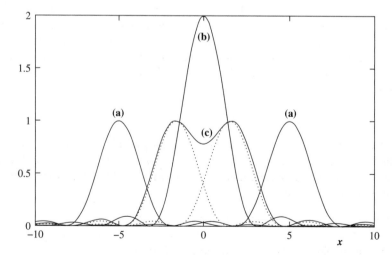

Figure 12.9   Diffraction images of adjacent spectral lines in a spectrograph. In (a) the lines are clearly resolved, while in (b) they merge and are unresolved. The separation in (c) illustrates Rayleigh's criterion for resolution. Each image is represented as $[\sin(x - x_0)/(x - x_0)]^2$, where x is in radians and x_0 locates the peak

single line. (A check that this gives a similar result may be usefully made on the sinc-squared function of Figure 12.9.)

The quantity $\lambda/\delta\lambda$ is obviously a useful measure of the power of any device to distinguish different wavelengths and is called the *chromatic resolving power R* of the spectrometer:

$$R = \frac{\lambda}{\delta\lambda} = \frac{v}{\delta v}. \tag{12.14}$$

For example, a resolving power greater than 1000 is needed to resolve the two sodium D-lines.

We may conveniently distinguish three ranges of wavelength resolution in spectrometers, from the point of view both of technique and of application. The simplest, using prisms and gratings, are useful for distinguishing the various spectral lines of a complex source and deducing its atomic or molecular content. The higher resolution $R > 5 \times 10^5$ demanded for measuring the detailed shape of spectral lines usually demands an interferometric technique, such as the Fabry–Pérot interferometer described in Chapter 8. Finally, the very narrow bands in scattered laser light may be resolved by a totally different technique in which fluctuations of irradiance, measured through optical mixing spectroscopy, are related to the spectrum (Section 12.10).

12.6   Resolving Power: The Prism Spectrometer

Following a similar argument to the discussion of diffraction in Chapter 10, we note that the minimum width (at a given wavelength) of the image at the exit slit in Figure 12.5 is due to diffraction in the limited width of the wavefront emerging from the prism. The angular spread is determined by the ratio of the wavelength to the width w of the wavefront. This angular width $\delta\theta$ is λ/w, measured from the line centre to the first minimum. Using the thin prism approximation, equation (12.6) this is related to the dispersion in the prism by

$$\delta\theta = \frac{\lambda}{w} \leq \alpha\frac{dn}{d\lambda}\delta\lambda \tag{12.15}$$

giving the criterion for resolving the two spectral lines

$$\frac{\lambda}{\delta\lambda} \leq w\alpha\left(\frac{dn}{d\lambda}\right). \tag{12.16}$$

Instead of extending the simple geometry applicable to equation (12.16) from *thin* to the geometrically more complicated case of *thick* prisms, we choose to derive an interesting simple expression for the chromatic resolving power of any prism spectrometer by a direct consideration of optical paths and Rayleigh's criterion (Section 12.5 above). In Figure 12.10 a thick prism is shown at the position of minimum deviation showing the approximate paths for plane waves of light with wavelengths λ and $\lambda + \delta\lambda$. Now for the diffraction maximum of one emerging wavefront to lie on the minimum of the other there must be one wavelength difference between them at the top of the wavefront emerging from the prism (see for example the way the phasors curl up in Figure 10.2). So for light of wavelength λ, equating optical path lengths for the extreme rays in air and in the prism,

$$2a = nB \tag{12.17}$$

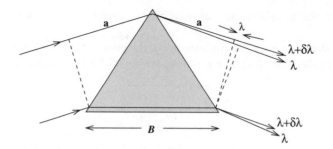

Figure 12.10 Geometry for the chromatic resolving power of a prism

where n is the refractive index of the prism at wavelength λ and B is its base length. For wavelength $\lambda + \delta\lambda$ the refractive index is $n + \delta n$. The plane waves for λ and at the resolved wavelength $\lambda + \delta\lambda$ are separated by a small angle because of the extra optical path in the prism, so that

$$2a - \lambda = (n + \delta n)B \qquad (12.18)$$

giving

$$\lambda = -\delta n\, B \qquad (12.19)$$

which may be written as

$$\boxed{R = \frac{\lambda}{\delta\lambda} = -B\left(\frac{dn}{d\lambda}\right).} \qquad (12.20)$$

At minimum deviation the resolving power of the prism spectrometer depends on the base length and the spectral dispersion of the material of the prism. Equation (12.20) for the chromatic resolving power of a thick prism shows that the *angle* of a prism is unimportant; what matters is the distance B traversed in the prism by the extreme ray, and the value of $dn/d\lambda$ for the material of the prism. For a heavy flint glass $dn/d\lambda$ can be about $-10^{-4}\,\text{nm}^{-1}$ so that for a wavelength of 500 nm and a large prism with $B = 10\,\text{cm}(= 10^8\,\text{nm})$ we have

$$\frac{\lambda}{\delta\lambda} = 10^4 \qquad \text{and} \qquad \delta\lambda = \frac{500\,\text{nm}}{10^4} = 0.05\,\text{nm} \qquad (12.21)$$

which is adequate for the resolution of the two sodium D-lines but insufficient for detailed measurement of the structure of each line.[3] The practical limit of resolution of the prism spectrometer is often set by aberrations in the imaging optics.

---

[3]The concept of spectral *lines* is very deep in the language: we talk of atoms having emission lines, and of the 21 cm hydrogen line in the radio spectrum, and so on. But of course the atoms do not have lines; they emit or absorb at certain wavelengths. It is the spectrograph that displays the different wavelengths in the light presented to it as a series of lines, each of which is an image of the slit, each at a different wavelength. So that when we speak of lines in an X-ray spectrum, or of the emission lines of molecules in millimetre-wave astronomy, we are using a word which is an interesting fossil originating in the simplest and oldest technique of spectral analysis, the prism spectrometer.

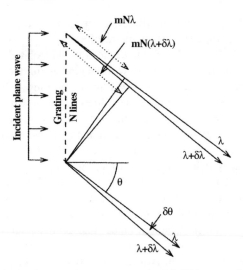

Figure 12.11 Chromatic resolution in a grating spectrograph with *N* lines, used in the *m*th order, showing the separation of two components of a plane wavefront

12.7 Resolving Power: Grating Spectrometers

We have seen that the chromatic resolving power of a prism is related to its overall size. The same arguments applied to the grating spectrometer give a similar result: the resolving power is again related to its overall size. Figure 12.11 shows diffracted wavefronts for two wavelengths λ and $\lambda + \delta\lambda$ emerging from a grating. The angular distribution of irradiance in these two spectral components is shown for a separation $\delta\lambda$ where they are just distinguishable. Again following Rayleigh's criterion, the maximum of one diffraction image falls on the first zero of the other.

The diffraction angle θ for wavelength λ at normal incidence is given by the grating equation

$$d \sin \theta = m\lambda \tag{12.22}$$

where d is the line spacing of the grating and m is the order of diffraction. For a grating of width W and a total number of lines N we can write $W \sin \theta = mN\lambda$. Across the emerging wavefront there is a difference in path $mN\lambda$. Now concentrate on the irradiance in this one direction as the wavelength is changed by a small amount $\delta\lambda$. Light of wavelength $\lambda + \delta\lambda$ has its principal maximum at the same angle as the first minimum for light of wavelength λ (see Section 11.2). If the extra path $W \sin \theta$ changes by one wavelength the irradiance will fall to zero. So for two adjacent spectral lines to be distinguished the criterion is

$$mN\lambda + \lambda = mN(\lambda + \delta\lambda) \tag{12.23}$$

or

$$R = \frac{\lambda}{\delta\lambda} = mN. \tag{12.24}$$

That is to say, the chromatic resolving power of a grating is the product of the order in which it is used and the total number of lines across it. The order acts like a kind of gearing: in the third order a given change of wavelength $\delta\lambda$ changes the path difference between adjacent lines by three times as much as it does in the first order, giving three times the resolution.

Note from equation (12.10) that the dispersion of the grating is related to the line spacing, while the resolving power is related to the number of wavelengths m in the extra path labelled $m\lambda$ in Figure 12.11. To obtain the same resolving power with a grating in the second order as for the large prism in Section 12.6 above would need 5000 lines across it. So a grating on the scale of the prism which was 10 cm across would need only 500 lines per cm. Fraunhofer, Rowland and Michelson all improved techniques for ruling conventional gratings, Michelson eventually producing gratings more than 15 cm across giving resolving powers of 4×10^5. Modern gratings are produced by a simple form of holography (Chapter 14), in which two crossing beams of monochromatic laser light form an interference pattern on a photographic plate. The plate surface is a film of photoresist which is subsequently etched to leave lines of clear glass on which a metallic coating is deposited. Holographic gratings can be made with up to 6000 lines per millimetre; furthermore, they are very uniform, avoiding the periodic errors which produce 'ghosts' (Chapter 11).

A further development of holographic gratings uses volume phase holography (VPH), which gives gratings in which none of the light is lost at a partially reflecting surface. The crossed laser beams used for making holographic gratings on a surface will make a three-dimensional pattern in a thicker film of gelatin; this pattern can be preserved as a three-dimensional pattern of changed refractive index in a completely transparent film. The refractive index changes are produced by a hardening process in the gelatin, in which the collagen molecules become cross-linked when exposed to blue light. The result is a grating which behaves like a crystal in X-ray diffraction (Chapter 11). It can be used in reflection or transmission, and has the normal dispersive power of a plane grating. The Bragg wavelength, which is the centre of the envelope of efficiently reflected wavelengths, can be tuned by tilting the grating. The width of the envelope is related to the thickness of the gelatin film.

12.8   The Fabry–Pérot Spectrometer

The Fabry–Pérot interferometer described in Chapter 8 forms the basis of a spectrometer of very high resolution over the spectral range from the ultra-violet to the near infrared. There can be large optical path differences between the multiple beams emerging from a Fabry–Pérot etalon, so that the interferometer behaves like a grating used at very high order. An equivalent grating would have a number of lines approximately equal to the finesse $\mathcal{F}$ (see Section 8.8).

The interferometer may be either a solid glass or quartz disc with parallel sides or two plane-parallel discs of glass or quartz separated by a small gap (Figure 12.12). When the transmitted interference fringes are focussed onto a screen, different wavelengths produce rings of different radius, so that there is radial dispersion. The centre of the ring system can be isolated by setting a circular aperture in the screen which enables a small wavelength band to be selected and then photoelectrically detected (Figure 12.13). That aperture is equivalent to the exit slit used in the grating spectrometer. The spectrum may then be scanned by changing the effective spacing nh of the interferometer, where using equation (8.22), $2nh\cos\theta = m\lambda$, the irradiance can be recorded as a function of h. The change in nh may be achieved by changing n, through a change in the pressure in an air-spaced interferometer, or by changing h using a piezoelectric drive on one of the interferometer plates. At the centre of the ring pattern, and for normal incidence, $m\lambda = 2nh$. With pressure scanning and a 1 cm gap, a change of one order requires a change in refractive index $\Delta n = 2 \times 10^{-5}$ or a change in pressure of about 0.1 bar (recall the discussion of the Rayleigh interferometer in

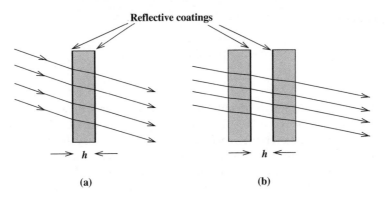

Figure 12.12   Fabry-Pérot interferometer etalons: (a) solid, (b) air spaced

Section 9.1, and that for air under standard conditions $n - 1 = 3 \times 10^{-4}$). The central aperture should isolate only a small band of wavelength or equivalently only a small fraction of an order (see Problem 12.7). The reflecting surfaces of the solid etalon and air-spaced interferometers have coatings whose reflectivities determine the finesse.

When used at high order the Fabry–Pérot interferometer suffers from the problem of overlapping orders, described in Section 12.4 for the grating spectrometer. Following the same argument, the free spectral range $\delta\lambda_{\mathrm{FSR}}$ is the wavelength spacing of two lines whose interference maxima coincide at orders m and $m + 1$. For an etalon spacing h and $n \simeq 1$,

$$\delta\lambda_{\mathrm{FSR}} = \lambda/m = \lambda^2/2h. \tag{12.25}$$

Since the free spectral range is very small, there may often be overlapping of orders, and the interferometer is normally used in conjunction with a grating spectrometer. The effective resolution of the grating spectrometer may be set to the free spectral range of the interferometer $\delta\lambda_{\mathrm{FSR}}$, while the resolution of the combined spectrometer and interferometer is the resolution $\delta\lambda$ of the interferometer.

From equation (8.28), the resolution of the Fabry–Pérot spectrometer is related to the finesse $\mathcal{F}$ and the free spectral range which are given by

$$\mathcal{F} = \frac{\delta\lambda_{\mathrm{FSR}}}{\delta\lambda} = \frac{\pi}{2}\left(\frac{2r'}{1 - r'^2}\right). \tag{12.26}$$

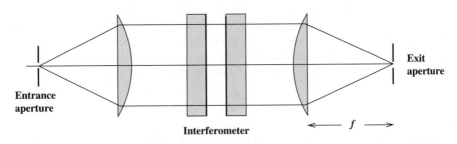

Figure 12.13   Fabry–Pérot spectrometer

The resolving power is

$$R = \frac{\lambda}{\delta\lambda} = \frac{\lambda\mathcal{F}}{\delta\lambda_{\text{FSR}}} = \frac{\pi r'}{1 - r'^2}m, \qquad (12.27)$$

from which $R = m\mathcal{F}$.

The Fabry–Pérot interferometer is usually used in high order. For an interferometer spacing $h = 1\,\text{cm}$ and $\lambda = 500\,\text{nm}$, $m = 4 \times 10^4$. Then for a typical finesse of 25 a resolving power of $R = 10^6$ is obtained. For an interferometer with a larger spacing of say $10\,\text{cm}$, the resolving power becomes $R = 10^7$. We see that the resolving power of the Fabry–Pérot spectrometer can be at least an order of magnitude greater than for the diffraction grating. In addition to limits on the practical finesse attainable set by the reflection coefficient of the Fabry–Pérot surfaces there are limitations from imperfections in the flatness of the plates. A further advantage of the Fabry–Pérot spectrometer is that the amount of light which passes through the spectrometer from the source to the detector, termed in general for a spectrometer the *étendue*,[4] is about two orders of magnitude greater than for the grating spectrometer. The *étendue* is defined as $L = A\Omega$, where A is the area of the exit slit and Ω is the solid angle subtended at the exit slit by the final focussing lens; it may be interpreted as the limiting aperture, e.g. the size of the grating or interferometer plate. The solid angle is that subtended by the slit at the collimating lens in a grating spectrometer, or that subtended by the aperture at the focussing lens in a Fabry–Pérot spectrometer. The quantity L is a constant through the spectrometer if there are no losses, such as from absorption or scattering.

A type of interferometer offering extremely high resolution is the confocal Fabry–Pérot interferometer, constructed with two spherical reflecting surfaces of radius r and separated by a distance $d = r$. This provides very high finesse up to 1000, resolving powers greater than 10^9 and with high luminosity. In this case the input light needs to be mode matched to the interferometer. The confocal Fabry–Pérot interferometer is useful in measuring narrow linewidth sources such as the mode structure and linewidth of laser sources, isotope shifts and atomic beams. For mirror reflectances $r'^2 = 0.99$ a finesse of $\mathcal{F} = 300$ can be obtained, so that for $r = d = 3\,\text{cm}$ two spectral lines only $6 \times 10^{-7}\,\text{nm}$ apart at $\lambda \sim 500\,\text{nm}$ could be resolved; in frequency terms they would be separated by only $0.72\,\text{MHz}$.

12.9   Twin Beam Spectrometry; Fourier Transform Spectrometry

We now turn to another interferometric method of spectrometry, which also extends the resolution by many orders of magnitude beyond that available from grating spectrometers. We have so far described the performance of a twin beam interferometer in terms of an ideally narrow spectral line. The next step is to consider its action with a single spectral line with finite width or structure, and then to generalize for a wide spectral range and complex spectrum. The twin beam interferometer, in its many and varied forms and in many ranges of wavelength, will then be seen to be a very powerful spectrometer, capable of resolving and measuring the width and shape of narrow line profiles.

Suppose that sodium light provides the illumination in a twin beam interferometer, such as the Michelson seen in outline in Figure 12.14. As is known from examination in any optical spectroscope that can resolve wavelengths separated by a fraction of a nanometre, the prominent yellow light

---

[4]With alternative terms luminosity (which is not to be confused with the photometric term luminance), light-gathering power or throughput.

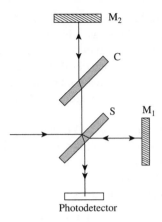

Figure 12.14 Outline of a Michelson interferometer

from sodium is made up of the two D-lines, at approximately 589.0 and 589.6 nm, or in Ångstrom units[5] $\lambda = 5890\,\text{Å}$ and $5896\,\text{Å}$, of almost equal irradiance. These wavelengths differ by just over 0.1%. As a first approximation consider them as very narrow compared to their separation. Suppose that the interferometer is first set up with one mirror M_1 and the image M_2 of the other in coincidence at the centre and at a slight angle so that vertical fringes of near-zero order are seen. Then as M_1 is moved further away the fringes move sideways, and as each crosses the centre of the field of view it indicates a change of one wavelength in the optical paths between the two arms; that is to say, a movement of $\lambda/2$ in the position of M_1. As M_1 moves further and the order of the interference increases, the fringes become less and less visible. This is because the two sets of fringes from the two wavelength components get progressively out of phase until a point is reached when the maxima of one set coincide with the minima of the other, giving a nearly uniform irradiance. The condition for this is that the mirror M_1 moves a distance d_1, and N_1 fringes have crossed the field, where

$$2d_1 = N_1\lambda_1 = \left(N_1 - \frac{1}{2}\right)\lambda_2.\tag{12.28}$$

We must be clear that there is no interference between the two sets of fringes, as light on different frequencies cannot be coherent. It is simply that the addition in *irradiance* of two nearly equal but antiphase sine waves gives a more or less uniform irradiance. Increasing d still further, the visibility improves and the fringes become sharp again when

$$2d_2 = N_2\lambda_1 = (N_2 - 1)\lambda_2.\tag{12.29}$$

Changing d by about 3 cm allows about 100 such cycles of visibility variation to be counted, each with about 1000 fringes between them. Such an observation allows the separation of the two lines to be accurately determined.

In general when light from a spectral line with any structure is examined in the twin beam interferometer it is found to give high-visibility fringes at zero path difference d, which decrease in

---

[5]The wavelengths of the sodium D-lines and other familiar lines are often quoted in Ångstrom units $(1\,\text{Å} = 10^{-10}\,\text{m})$.

visibility as d is increased, and finally disappear. (See Sections 8.3 and 13.2 for the definition of fringe visibility.) A recording of the fringe visibility as d is varied is an *interferogram*. In the example of sodium light above it is easy to see that the form of the interferogram implies the spectrum of the light causing it. Michelson realized this and pointed out that the two quantities are related as a Fourier transform pair. He was then able to use a twin beam interferometer to find the shape and structure of a single spectral line, and discovered the hyperfine structure of many spectral lines previously regarded as monochromatic.

The Fourier relationship is analysed in Chapter 13 using the Wiener–Khintchine theorem of Section 4.15, but the concept is easily demonstrated as follows. Consider first a single spectral component with wave number k (where $k = 2\pi/\lambda$). Two waves of equal irradiance $I(k)$ arrive at the detector with phase difference kx resulting from a path difference x (this is the path difference d in the Michelson interferometer). The measured irradiance then varies with x in the familiar pattern of cosine fringes:

$$I(x) = I(k)(1 + \cos kx). \tag{12.30}$$

Each component of an extended spectrum $I(k)$ incident on the splitter produces a pattern of cosine fringes with amplitude $I(k) \cos kx$, which adds as the integral

$$I(x) = \frac{1}{2} \int_0^\infty I(k)[1 + \cos(kx)]\mathrm{d}k = \frac{1}{2}I(0) + \frac{1}{2} \int_0^\infty I(k) \cos kx \, \mathrm{d}k. \tag{12.31}$$

With $\int_0^\infty I(k)\mathrm{d}k = I(0)$, the quantity $[2I(x) - I(0)]$ is the cosine Fourier transform of the spectrum. Leaving aside the more general formulation via the Wiener–Khintchine theorem, this result shows that in an interferometer *the measurement of fringe visibility as a function of order of interference gives the profile of a spectral line via a Fourier transform*. Furthermore, the resolving power of the interferometer is equal to the order of interference reached in the measurement.

Another way of specifying the order of interference in this relationship is in terms of the difference in travel time for the two light beams; for a path difference x this is simply $\tau = x/c$, and the order of interference is $m = x/\lambda = v\tau$, where $v = c/\lambda$ is the frequency of the light.

As an example of Fourier transform spectrometry we show in Figure 12.15 how the fringe visibility $V(\tau)$ varies with τ (and therefore with order m) for two different line profiles: these are the Lorentzian

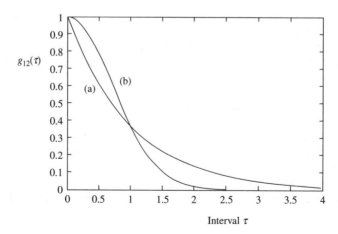

Figure 12.15 The fringe visibility as a function of delay τ between two beams for (a) Lorentzian and (b) Gaussian line profiles

and Gaussian profiles, shown in Figure 12.3, which result from two different processes of line broadening. The fringe visibility is shown as a function of τ, the difference in travel time for the two beams of the interferometer. For both profiles, at small path differences, $\tau \rightarrow 0$ and $V(\tau) \rightarrow 1$, while for large path differences, i.e. as $\tau \rightarrow \infty$, $V(\tau) \rightarrow 0$. The shapes of the two visibility functions are considerably different: the Gaussian profile transforms[6] into a Gaussian visibility function, while the transform of the Lorentzian extends to larger values of τ. Comparison with Figure 12.3 shows that this extension is due to the sharp peak at the centre of the Lorentzian.

A comparison of a conventional spectrometer such as a prism or grating spectrometer with a twin beam interferometric spectrometer such as the Michelson of Figure 12.14 shows that the interferometer has practical advantages in sensitivity as well as in resolving power. First, an extended source can be used, instead of a narrow slit; second, a single detector can be used to record light from the whole spectrum simultaneously while the interferometer is scanned by varying the delay τ, in contrast to a detector scanning a narrow part of a dispersed spectrum. The efficient use of a single detector, called multiplex advantage, is vital for efficient measurements in the far infrared where multiple element detector arrays are not available.

The only loss of light in the interferometer occurs at the arrangement for splitting the beam, but even this can be avoided by systems such as those of Figure 12.16. In (a) double mirrors are used in a Michelson interferometer to allow both beams to be detected at D_1 and D_2, while in (b), due to J. Strong, an ingenious interleaved mirror reflects all the light into a single detector.

Starting in the 1950s, the speed and sensitivity of Fourier transform spectroscopy revolutionized infrared astronomy; for example, observations of the spectra of planetary atmospheres could be made in a single night, which previously would have required many years to complete.

12.10 Irradiance Fluctuation, or Photon-Counting Spectrometry

When the width of a spectral line is so small that the required resolution exceeds that available from the Fabry–Pérot interferometer, and the path difference in a twin beam interferometer with sufficient resolving power becomes impracticably long, a different technique becomes available for measuring spectral lineshapes. As we have seen in Section 12.9, the twin beam interferometer is measuring the correlation between the *amplitudes* of light in two light beams one of which is delayed by time τ, which is the same as the correlation between two points in a single beam separated by a path difference $x = c\tau$. Instead of sampling a light beam at two separated points, the technique of irradiance fluctuation spectrometry (also termed photon-counting spectroscopy) is concerned with fluctuations of *irradiance* at a single point; the differences between the wave at two separate points are then converted into fluctuations as the wave passes a single point. These fluctuations are usually on a very short time scale, and are averaged out in most photometric and interferometric measurements. The irradiance of light from a spectral line with width δv fluctuates only on a time scale of order $1/\delta v$, which is usually so small that it is unresolvable, and the fluctuations are unnoticed. But if they can be resolved, using techniques with a time resolution better than $1/\delta v$, the frequency spectrum of the irradiance fluctuations can be related to the

---

[6]The astute reader will note that the Fourier transform of a spectral lineshape is a complex function (see Chapter 4), while we have treated fringe visibility V simply as a real quantity. This will be dealt with in Chapter 13, where we discuss visibility in terms of an autocorrelation function. We note, however, that for symmetrical lineshapes such as those considered here the phase of the transform is constant and may be set to zero.

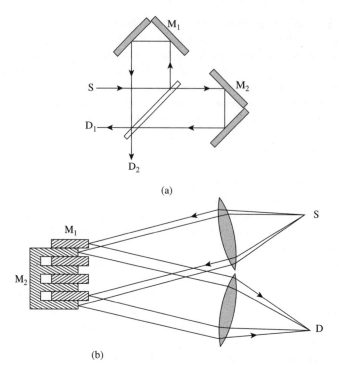

(a)

(b)

Figure 12.16 Examples of efficient twin beam interferometers. (a) The double mirrors in a Michelson interferometer allow the returning beam to be detected at D_1 in addition to the beam at D_2. (b) All the light from the source S reaches a single detector in this arrangement due to Strong. The path difference is changed by moving the multiple mirror M_2, which interleaves with a fixed mirror M_1

spectral lineshape and width through a Fourier transformation similar to that of Fourier transform spectrometry.

We now consider the amplitude and irradiance of light from a number of atoms radiating independently, so that their phases are randomly distributed. (This an example of *chaotic light*, as contrasted with laser light; see Chapters 13 and 16.) The radiation from each atom is coherent for a time τ; then the phase changes discontinuously by a random amount, as might occur at a collision in a gas. We add the contributions of a large number n of atoms to the observed instantaneous irradiance. Assuming the contributions all have equal amplitudes, the sum contains the resultant of the individual phases as an amplitude factor $a(t)$ and the irradiance averaged over a long time is proportional to

$$\bar{I} = \overline{|a(t)|^2} = \overline{|\exp(i\phi_1 t) + \exp(i\phi_2 t) + \ldots + \exp(i\phi_n)|^2}. \tag{12.32}$$

Since the cross-terms between the phase factors for different radiating atoms give a zero average contribution, the average irradiance is, as expected, simply n times the irradiance from an individual atom.

Instantaneously, however, the irradiance may be very different. The sum of many amplitude contributions with random phase is shown in Figure 12.17. After a time greater than τ the phases change and the sum will change unpredictably. The *probability distribution $P(I)$* of the irradiance I at

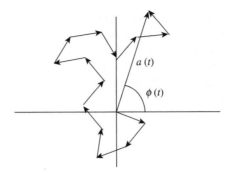

Figure 12.17   The sum of many unit vectors with random phases, as in a random walk. The amplitude and phase of the sum are shown as $a(t)$ and $\phi(t)$. This is a phasor diagram for chaotic light

an instant of time t follows a statistical law familiar in the theory of the random walk:

$$P(I)\mathrm{d}I = \bar{I}^{-1} \exp\left(-\frac{I}{\bar{I}}\right)\mathrm{d}I. \tag{12.33}$$

The average amplitude of the irradiance fluctuations given by the difference ΔI between the instantaneous irradiance and the mean is

$$\left[\overline{(\Delta I)^2}\right]^{1/2} = \left[\overline{I^2} - \bar{I}^2\right]^{1/2} = \bar{I} \tag{12.34}$$

so that the r.m.s. fluctuations equal the mean irradiance itself.

Figure 12.18 shows an example of the form of fluctuations in the irradiance of chaotic light, on a time scale comparable with the coherence time τ. The rate of fluctuation is inversely proportional to the coherence time, and in more detail the spectrum of the fluctuations in irradiance is related to the shape and width of the spectral line by a Fourier transform; however, the information about the line is not as comprehensive as in normal Fourier transform spectrometry.

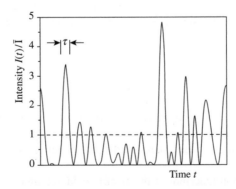

Figure 12.18   An example of the fluctuations in irradiance for a collision-broadened chaotic light source. $\bar{I}$ is the mean irradiance averaged over a long time compared with the mean time τ between collisions

Irradiance fluctuation spectroscopy requires a time resolution better than the coherence time: this is achieved by the high time resolution of detectors such as the photomultiplier, which can reach 10^{-9} s. Narrow linewidths are often expressed as a frequency bandwidth; the irradiance fluctuation technique therefore applies to bandwidths up to about 10^8 Hz. In the same terms diffraction grating spectroscopy, which is a filter technique, is applicable to bandwidths of 10^{10} Hz and higher, while Fabry–Pérot interferometry methods are applicable in the range 10^6 to 10^{12} Hz, overlapping with irradiance fluctuation and diffraction grating methods. Irradiance fluctuations were first observed for a low-pressure mercury lamp in a famous experiment by Hanbury Brown and Twiss which we describe in Chapters 9 and 13.

The fluctuations in irradiance are measured by an optical mixing technique in which the light is incident on a photodetector and the resulting post-detection signal analysed. The photodetector responds to the irradiance $I(t)$, or square of the light electric field, with photocurrent $i(t) \propto I(t) \propto |E(t)|^2$, and hence is termed a 'square-law detector'. The incident light $I(\omega)$ has oscillating frequencies $\sim 10^{14-15}$ Hz; the waves at these frequencies interfere and produce beat frequencies in the detected photocurrent at all the difference frequencies $(\omega_a - \omega_b)$, or beat frequencies, within the linewidth $\Delta\omega$. Examples of this are the light scattered from moving particles, or the difference frequencies between laser modes, in which cases frequencies may be produced in the range up to 10^8 Hz.

The frequency spectrum of the incident light $I(\omega)$ is related to the frequency spectrum $P(\omega)$ of the photodetector output, as follows. The beat frequency content of the photocurrent $P(\omega)$ may be related to the time autocorrelation function $C(\tau)$ of the photocurrent by a Fourier transform. In the autocorrelation function the photocurrent at time t is compared with delayed versions at $(t + \tau)$ for the range of delay times τ.

$$C(\tau) = \frac{\langle i(t).i(t + \tau)\rangle}{\langle i \rangle^2}. \tag{12.35}$$

Then

$$P(\omega) = \int_0^\infty C(\tau) \exp(i\omega\tau)\mathrm{d}\tau. \tag{12.36}$$

A light signal with a distribution of frequencies $\Delta\omega$ implies that it has a fluctuating irradiance. The photocurrent $i(t)$ is derived directly from the irradiance of the incident light, so that the time autocorrelation function of the photocurrent is directly related to the correlation function of the light irradiance

$$C(\tau) \propto \frac{\langle I(t)I(t + \tau)\rangle}{\langle I(t)\rangle^2}. \tag{12.37}$$

This in turn can be related to the correlation function of the electric field. (The proportional sign is used in equation (12.37) since the measured autocorrelation function of the photocurrent depends on the optical mixing process, particularly the scattered light coherence and the detector area.

The nature of the correlation functions of the electric field amplitudes and intensities in describing the coherence properties of the light are discussed in Chapter 13. The spectral distribution of the light incident on the photodetector can be obtained from the measurement of the time dependence of the

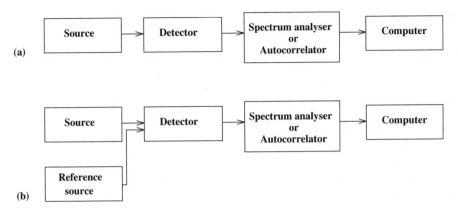

Figure 12.19 Arrangements for irradiance fluctuation spectroscopy. (a) Homodyne spectroscopy. (b) Heterodyne spectroscopy, optical mixing with a reference source

photocurrent $i(t)$, followed by the Fourier transform to give the frequency spectrum $I(\omega)$ of the incident light.

There are two main forms of irradiance fluctuation spectroscopy, illustrated in Figure 12.19. In *homodyne spectroscopy*, also referred to as *self-beat spectroscopy*, the incident light only is detected. The frequency spectrum of the detected signal can be measured by an electronic spectrum analyser or, more commonly, by determining the time autocorrelation function $C(\tau)$ of the photocurrent.

Alternatively in *heterodyne spectroscopy* the light is mixed with a reference beam on the photodetector, i.e. a local oscillator. For example, in a light scattering arrangement the reference beam is split off from the incident laser beam.[7] With a reference signal at angular frequency ω_0, the heterodyne beat frequency contains terms in the difference frequencies $(\omega - \omega_0)$ which contain the spectral information on $I(\omega)$.

Coherent mixing of the light at the detector is necessary to maintain the interference condition, and optical mixing is ensured by the use of an aperture before the detector to select one coherence area. A source of wavelength λ having a diameter d_1 and spaced a distance D from the aperture will be spatially coherent at the aperture for an aperture diameter $d_2 \leq D\lambda/d_1$. As an example, for the conditions $\lambda = 500$ nm, $d_1 = 1$ mm and $D = 0.5$ m, a detector aperture $d_2 = 0.25$ mm is required. Further considerations of coherence area are discussed in Chapter 13.

The time autocorrelation $C(\tau)$ of the photocurrent is determined electronically by either a digital or analogue autocorrelator. In the digital mode the autocorrelation is performed on the arrival of the stream of photons $n(t)$, which are converted into current pulses. The time scale is divided into equal time channels and the number of photons detected in any one channel equal to or above a set number is counted as a '1' or, if below, as a '0'. The autocorrelation can be performed digitally and rapidly. This method of digital correlation is known as *photon correlation spectroscopy* and is particularly appropriate to low light levels.

---

[7]An alternative nomenclature that is sometimes used is to term this arrangement homodyne and to use heterodyne to refer to optical mixing with a reference signal which is shifted in frequency.

12.11   Scattered Laser Light

The technique of measuring the width and shape of very narrow spectral lines through irradiance fluctuations finds its most useful application in the examination of the scattering of laser light in substances such as colloids, polymers and biopolymers.

An example of the application of homodyne and heterodyne spectroscopy is the measurement of the size distribution of microparticles dispersed in a liquid and undergoing Brownian motion. The diffusion coefficient D of spherical particles of mean radius R in a fluid of viscosity η and temperature T is described by the Einstein diffusion equation $D = kT/6\pi\eta R$. Scattering of a laser beam by the particles confers a linewidth Δv on the scattered light which is dependent on the diffusion coefficient and the scattering angle θ: light scattered at angle θ in a medium of refractive index n has linewidth

$$\Delta v = D\left(\frac{4\pi n}{\lambda}\right)^2 \sin^2(\theta/2). \tag{12.38}$$

This is measured by a homodyne spectrometer, or in a heterodyne spectrometer by combining the scattered light with a reference beam direct from the laser (Figure 12.20). Fluctuations in irradiance are measured by photon counting in time intervals shorter than $1/\Delta v$. (The laser light itself contains effectively no fluctuations, and the dominant term in the fluctuations is the second-order correlation of the scattered light, as discussed in Chapter 13.) The magnitude of the fluctuations, and the ease of

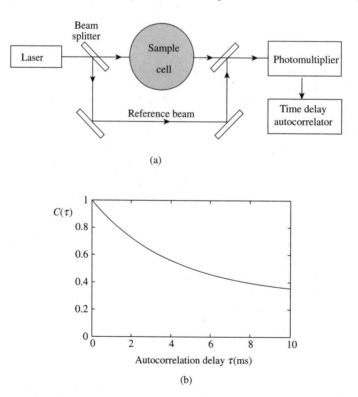

(a)

(b)

Figure 12.20   Photon correlation spectrometry applied to laser light scattered from a colloidal solution

measurement, are greatly enhanced if the irradiance of the reference beam is much larger than that of the scattered light.

As an example, measurement of the mean size of microparticles in a water dispersion by a typical heterodyne spectrometer might use an He–Ne laser operating at a wavelength of 632.8 nm, detecting scattered light at angle $\theta = 90°$. The linewidth of the scattered light would be measured by the photon correlation technique; if this gave a decay constant $\tau_c = 5 \times 10^{-3}$ s, the linewidth $1/\tau_c = 200$ Hz. From equation (12.38), the mean diffusion coefficient of the microparticles is 5.73×10^{-13} m^2 s^{-1}. The mean radius of the microparticles determined by this measurement is $R = kT/6\pi\eta D = 3.8 \times 10^{-6}$ m, for $T = 300$ K and $\eta = 10^{-3}$ N s m^{-2}.

Problem 12.1

A spectrometer uses a prism with base width 5 cm and apex angle 11.5°, i.e. 0.2 radians, made of glass with refractive index $n = 1.70$ at $\lambda = 650$ nm and 1.72 at $\lambda = 590$ nm. Calculate the resolving power, using equation (12.15), and the angular separation of the two sodium lines at 589.0 nm and 589.6 nm. Will this spectrometer resolve the hydrogen doublet at $\lambda = 656.272$, 656.285 nm?

Problem 12.2

In a high-resolution spectrograph three prisms are arranged with their bases on a semicircle with diameter 20 cm as in Figure 12.21, so as to deflect light through 180°. Show that for refractive index 1.5 the prism angle must be approximately 82°. Find the resolving power if

$$\frac{dn}{d\lambda} = 5 \times 10^4 \, \text{m}^{-1}.$$

Problem 12.3

Calculate the spectral resolving power for wavelengths near 500 nm of the following spectrometers: (i) A glass prism, base length 4 cm, with refractive index varying linearly between $n = 1.5477$ at $\lambda = 546$ nm and $n = 1.5537$ at $\lambda = 486$ nm. (ii) A grating 4 cm across with 1500 lines per cm, used in the third order. (iii) A Fabry–Pérot interferometer in which $F = 40$, and with a spacing 4 cm between the plates.

Problem 12.4

Light falls normally on a reflection echelon grating in which the step height is h and the step width is w (the light falls vertically, not horizontally as in Figure 11.12(c)). Show that the path difference between light beams reflected in direction θ to the normal from corresponding points on adjacent step faces is

$$h(1 + \cos \theta) - w \sin \theta. \tag{12.39}$$

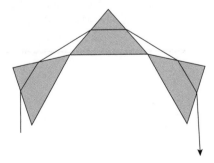

Figure 12.21 A prism spectrometer with increased dispersion and resolving power

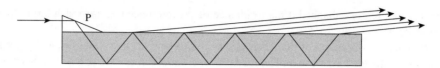

Figure 12.22   The Lummer plate

For small θ the mth order then emerges at

$$\theta = \frac{2h - m\lambda}{w}.$$ (12.40)

For an echelle with $h = 1\,\text{cm}$ and $w = 0.1\,\text{cm}$, and with 40 such steps, find for wavelengths near 500 nm:

(i)   the order m for θ near zero

(ii) the angular separation of orders

(iii) the resolving power.

Problem 12.5
The resolving power $\lambda/\delta\lambda$ of a grating spectrograph is the difference between extreme optical paths measured in wavelengths. Show that the resolvable frequency difference δv is related to the difference τ in light travel times in the extreme paths by

$$\delta v = \frac{1}{\tau}.$$ (12.41)

Problem 12.6
For a Lummer plate (Figure 12.22), which produces a fringe pattern with high resolution, show that the resolving power at grazing emergence angle is approximately

$$\frac{\lambda}{\delta\lambda} = \frac{L}{\lambda}(n^2 - 1).$$ (12.42)

Note that the plate is used with the emergent beams at a very small angle to the surface of the plate. The light beam enters the plate via prism P; λ is the vacuum wavelength, the refractive index is n and the length is L.

Problem 12.7
To use the Fabry–Pérot spectrometer shown in Figure 12.13 as a filter it should transmit only one order of interference fringe. Show that for order m this requires a restricted cone of light passing through the etalon. Show that if f is the focal length of the focussing lens, this requires the diameter d of the aperture to be less than $2f\sqrt{2/m}$.

13 Coherence and Correlation

All nature is but art unknown to thee,/ All chance, direction which thou canst not see;/ All discord, harmony not understood.

Alexander Pope, *An Essay on Man.*

How can a particle go through both slits? Nobody knows, and it's best if you try not to think about it.

Richard Feynman.

In much of the discussion of diffraction and interference phenomena in previous chapters we have been concerned with monochromatic light produced by a point source. No actual source is either a point or strictly monochromatic, so that no light has a perfect sinusoidal waveform extending indefinitely in space or in time. In practice there is a loss of *coherence* both in space and in time, whose consequences have already been encountered in the two basic types of interferometer, of which Michelson's stellar interferometer and spectral interferometer are examples. The stellar interferometer investigates the waves from a source which is nearly, but not quite, a point, finding that the loss of coherence *across* the wavefront is a measure of the angular diameter of the source. The spectral interferometer investigates the waves from a narrow spectral line by exploring the loss of coherence between two points separated *along* the path of the wave, which is a measure of the coherence in time. In this chapter we define coherence more precisely, and apply the concepts of coherence and correlation to the practical issues of spatial and temporal coherence, and to angular and spectral resolution in optical instruments. We also discuss the concept of spatial filtering, in which the Fourier components of an object are modified in instruments such as the phase contrast microscope.

As in previous chapters, there is barely any need to introduce the concept of a photon into these discussions. Inevitably, however, the question addressed (or avoided) by Feynman (see the epigraph above) will be asked, together with the related question about interference involving material particles. We address these briefly at the end of this chapter.

13.1 Temporal and Spatial Coherence

The loss of coherence along the path of a wave from a source which is nearly, but not quite, monochromatic can be understood by supposing that the wave is made up of a large number of individual

Optics and Photonics: An Introduction, Second Edition   F. Graham Smith, Terry A. King and Dan Wilkins
© 2007 John Wiley & Sons, Ltd

wavetrains of finite length, each produced by a single atom or other emitter, and that a large number of such wavetrains pass a point in the time taken to make an observation of irradiance. Light from two points closer together than the length of an individual wavetrain will be coherent and will interfere as for a monochromatic source. Light from two points along the wavetrain separated by more than the length of the wavetrain is incoherent, and cannot show interference effects. (Instantaneously the two samples will add according to their phase relation, but this will change randomly during the observation, since the relative phase of different wavetrains is randomly distributed.) There is a typical *coherence length* in the light beam, which is the length of an elementary wavetrain. There is also a typical *coherence time*, which is the time for the elementary wavetrain to pass any point.

Coherence time is fundamentally related to spectral width, as may be seen from the Fourier analysis of Chapter 4. The precise relation depends on the shape of the spectral line, but it is useful to remember that for a coherence time Δt, and an oscillation with angular frequency bandwidth $\Delta \omega$, there is a general relation, known as the *bandwidth theorem*[1]

$$\Delta t \, \Delta \omega \sim 1. \tag{13.1}$$

The corresponding coherence length l_C can be estimated by

$$l_C = c\Delta t = \frac{c}{\Delta \omega}. \tag{13.2}$$

An important example is a wave consisting of a randomly phased assembly of Gaussian wave groups. We show later (Section 13.3) that the coherence time of the assembly is that of a single group, and analyse here a single group. For a Gaussian group with amplitude E_0 and central wavelength ω_0, $E(t) = E_0 \exp(-at^2 + i\omega_0 t)$, i.e. a cosine wave modulated by a Gaussian envelope (Figure 13.1), the spectral width (full width at half maximum, FWHM) is $\Delta \nu = \Delta \omega / 2\pi$. The duration t_G is the coherence time where, from Fourier analysis (again using the FWHM for t_G),

$$t_G = 2\ln 2/\pi\Delta \nu = 0.441/\Delta \nu. \tag{13.3}$$

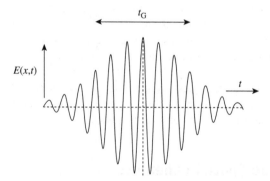

Figure 13.1   A wave group with Gaussian profile. The spectral width of the group $\Delta \nu$ and the coherence time t_G are related by $t_G = 2\ln 2(\pi\Delta \nu)^{-1}$

---

[1]This is analogous to the Heisenberg uncertainty relationships in quantum mechanics between the momentum p, position x and energy E of a particle $\Delta p \, \Delta x \geq \hbar/2$ and $\Delta E \, \Delta t \geq \hbar/2$.

The minimum time–bandwidth product for a Gaussian signal, using FWHM values, is $\Delta v\, t_{\mathrm{G}} = 0.441$. (A general Fourier analysis theorem states that uncertainties in time and bandwidth, using r.m.s. values, are related as $\Delta v\, \Delta t \geq 1/2$; this is a precise equality for a Gaussian signal.)

Correspondingly, for a wave velocity c the coherence length is

$$l_{\mathrm{C}} = ct_{\mathrm{G}} = \frac{2\ln 2\, c}{\Delta\omega} = \frac{2\ln 2\, \lambda^2}{\pi\Delta\lambda} = 0.441 \frac{\lambda^2}{\Delta\lambda}. \tag{13.4}$$

Typical coherence lengths for light are readily estimated from equation (13.4). A colour filter on a white light might isolate a band 50 nm wide at a wavelength of 500 nm; the coherence length is then about 2 µm; a Fabry-Perot filter with a bandpass width of 1 nm increases this to 100 µm. A narrow spectral line from a sodium or mercury lamp can have a coherence length of 1 cm. Light from a carefully constructed laser can have a coherence length of 10 km or more, although very short wavetrains only a few microns long can be also be made by lasers specially designed to operate in a pulsed mode (Chapter 16).

In the discussions of the stellar interferometer (Chapter 9) the condition for obtaining interference between light derived from mirrors transversely separated across the wavefront was expressed as the inequality $\phi_0 \ll \lambda/d_{\mathrm{M}}$, where ϕ_0 is the angle subtended by the source and d_{M} is the separation of the mirrors. Put in a way more suitable for the present discussion, the maximum distance apart of the mirrors for interference fringes to be observed is of order λ/ϕ_0; this is a measure of the *transverse coherence distance*. The pair of mirrors can be thought of as exploring the degree of coherence across the light wave; full coherence can only be found if the pair are close together, while if they are separated by more than the transverse coherence distance the light at the two mirrors becomes incoherent and no interference can be observed.

The concept of transverse coherence concerns the phase relation between waves at different points in a wavefront perpendicular to the direction of propagation. In accordance with the most common usage, we will term this spatial coherence; it is also sometimes referred to as lateral coherence. For a wave with perfect spatial coherence, any two points of a wavefront are in phase and remain in phase. The degree of spatial coherence of a thermal light source is dependent on the size and distance of the source.

Typical transverse coherence widths d_{M} can be estimated from $d_{\mathrm{M}} \leq \lambda/\phi_0$. Light from a source 1 arcsecond across has a coherence width of about 10 cm: light from the nearest large-diameter stars (angular diameter ~ 0.01 arcseconds) has a coherence width of 10 m, and light from the Sun, which subtends an angle of 30 arcminutes at the Earth, has a coherence width of only 50 µm.

We are thus led to the idea that around any point in the light field produced by a real source there is a *region of coherence*, with a transverse size governed by the angular diameter of the source, and a longitudinal size governed by the bandwidth of the radiation from the source. Any interferometer that is to produce fringes from the light of the source must derive its two beams from points within this volume. The two sorts of Michelson interferometer we have discussed are the archetypes, the stellar (Chapter 9) using transverse separation and the spectral (Chapter 12) using longitudinal separation. We now define coherence more precisely and quantify these relationships.

Light sources may be divided broadly into two types with different coherence properties: chaotic and laser sources. Chaotic sources include gas discharge lamps, filament lamps and other thermal sources in which radiation is produced by independently emitting atoms. Lasers, described in Chapter 15, produce radiation by an entirely different mechanism of stimulated emission.

13.2   Correlation as a Measure of Coherence

The previous section provided a qualitative description of the coherence of a light beam. It is now useful to put the nature of coherence on a more quantitative basis and introduce the concepts of degree of coherence and partial coherence.

Let $E_1(t)$ and $E_2(t)$ be the amplitudes at points P_1 and P_2 in a light field *in vacuo*. The irradiances at P_1 and P_2 are then $E_1(t)E_1^*(t)$ and $E_2(t)E_2^*(t)$ where the asterisk indicates the complex conjugate.[2] An interferometer, of any type, combining the light from these two points adds the two amplitudes with, in general, a time delay and measures the square of the sum. The interferometer measures an irradiance $I(\tau)$ as a function of τ, the relative delay, given by

$$I(\tau) = \langle \{E_1(t+\tau) + E_2(t)\}\{E_1^*(t+\tau) + E_2^*(t)\}\rangle. \tag{13.5}$$

The brackets $\langle\rangle$ denote an average over time t, extending over many oscillation periods. If this expression is multiplied out it gives

$$I(\tau) = \langle E_1(t+\tau)E_1^*(t+\tau)\rangle + \langle E_2(t)E_2^*(t)\rangle + \langle E_1(t+\tau)E_2^*(t)\rangle + \langle E_1^*(t+\tau)E_2(t)\rangle. \tag{13.6}$$

The first two terms are simply the average irradiances at P_1 and P_2, $I_1 = \langle E_1(t+\tau)E_1^*(t+\tau)\rangle$ and $I_2 = \langle E_2(t)E_2^*(t)\rangle$. The second two terms give the interference fringes (note that they are each other's complex conjugate, so that their real parts are equal and their imaginary parts cancel). Suppose the fields at P_1 and P_2 are from a monochromatic point source of period T. Then when $\tau = NT$ (N is an integer) all four terms are equal and the irradiance is four times that at P_1 or P_2. On the other hand, when $\tau = (N + \frac{1}{2}T)$, the second pair of terms are negative (each being the average of the product of cosines in antiphase) and they exactly cancel the first pair of terms. Thus fringes of 100% visibility are observed.

Evidently it is the second pair of terms that are of interest, expressing the relationship between the complex amplitudes at the two points. As they are each other's complex conjugate, each has the same information as the other and conventionally the first is taken. Mathematically this is the *cross-correlation* of $E_1(t)$ and $E_2(t)$, regarding these as complex functions of time; in optics it is the *mutual coherence* $\Gamma_{12}(\tau)$. Thus

$$\boxed{\Gamma_{12}(\tau) = \langle E_1(t+\tau)E_2^*(t)\rangle.} \tag{13.7}$$

Notice that when P_1 and P_2 coincide and $\tau = 0$, the mutual coherence reduces to $\langle E_1(t)E_1^*(t)\rangle$, which is simply the irradiance.

The correlation of the field at the same point but at different times is described by the *first-order correlation function*

$$\begin{aligned}
\Gamma_{11}(\tau) &= \langle E_1(t+\tau)E_1^*(t)\rangle \\
\Gamma_{22}(\tau) &= \langle E_2(t+\tau)E_2^*(t)\rangle.
\end{aligned} \tag{13.8}$$

---

[2]As shown in Section 5.6, irradiance is properly related to a peak field E_0 as $I = \frac{1}{2}\epsilon_0 c E_0^2$. For clarity in the following discussion we omit the constant factor $\frac{1}{2}\epsilon_0 c$.

For zero delay time, $\tau = 0$, the self-coherence functions are proportional to the irradiances:

$$\Gamma_{11}(0) = \langle E_1(t)E_1^*(t) \rangle = I_1$$
$$\Gamma_{22}(0) = \langle E_2(t)E_2^*(t) \rangle = I_2. \tag{13.9}$$

More generally, $\Gamma_{12}(\tau)$ may be normalized to give a *complex degree of mutual coherence* where $\Gamma_{11} = I_1$ etc.

$$\boxed{\gamma_{12}(\tau) = \frac{\Gamma_{12}(\tau)}{\sqrt{I_1 I_2}} = \frac{\Gamma_{12}(\tau)}{\{\Gamma_{11}(0)\Gamma_{22}(0)\}^{1/2}}.} \tag{13.10}$$

In terms of $\gamma_{12}(\tau)$ the interferometer output measures an irradiance

$$I(\tau) = I_1 + I_2 + 2\sqrt{I_1 I_2} \, \text{Re} \, (\gamma_{12}(\tau)). \tag{13.11}$$

The complex quantity $\gamma_{12}(\tau)$ may be expressed as $|\gamma_{12}| \exp[i\phi_{12}(\tau)]$. Two quasi-monochromatic waves of frequency ω_0, $E_{1,2} = E_0 \exp(i\omega_0 t - ikr_{1,2})$, having an optical path difference $(r_1 - r_2)$ and hence a phase difference $\phi_{12}(\tau) = k(r_1 - r_2)$, give a resultant irradiance

$$I = I_1 + I_2 + 2\sqrt{I_1 I_2}|\gamma_{12}| \cos \phi_{12}(\tau). \tag{13.12}$$

For equal irradiances $I_1 = I_2$

$$I = 2I_1[1 + |\gamma_{12}| \cos \phi_{12}(\tau)]. \tag{13.13}$$

The *visibility* of a set of interferometer fringes, as in Newton's rings or Young's double slit fringes of Chapter 8, is defined as

$$V = \frac{I_{\text{max}} - I_{\text{min}}}{I_{\text{max}} + I_{\text{min}}}. \tag{13.14}$$

From equation (13.12)

$$I_{\text{max}} = I_1 + I_2 + 2\sqrt{I_1 I_2}|\gamma_{12}(\tau)|$$
$$I_{\text{min}} = I_1 + I_2 - 2\sqrt{I_1 I_2}|\gamma_{12}(\tau)|. \tag{13.15}$$

Then the visibility is

$$V = \frac{2\sqrt{I_1 I_2}}{I_1 + I_2} |\gamma_{12}(\tau)|. \tag{13.16}$$

When $I_1 = I_2$ the visibility of the fringes is equal to the modulus of the complex degree of mutual coherence.

The function $\gamma_{12}(\tau)$ makes precise the conceptual ideas of the previous section. If we regard P_1 as fixed and P_2 as exploring the space around it, there is in general a complex number $\gamma_{12}(\tau)$ for each position of P_2 and value of τ. The degree of correlation, i.e. the magnitude of γ_{12}, varies between 0 and 1.

The degree of *first-order temporal coherence* $\gamma^{(1)}(\tau)$ is given by the normalized coherence function

$$\gamma^{(1)}(\tau) = \frac{\langle E(t+\tau)E^*(t)\rangle}{\langle E(t)E^*(t)\rangle}. \qquad (13.17)$$

Two waves have complete temporal coherence when $\gamma^{(1)}(\tau) = 1$, and complete incoherence for $\gamma^{(1)}(\tau) = 0$. For $0 < |\gamma^{(1)}(\tau)| < 1$ there is partial coherence. The ability to form interference fringes is determined by the value of $\gamma^{(1)}(\tau)$.

The coherence function $\gamma^{(1)}(\tau)$ may be generalized to include the spatial dependence of the fields at space and time points (r_1t_1) and (r_2t_2):

$$\gamma^{(1)}(r_1t_1, r_2t_2) = \frac{\langle E(r_1t_1)E^*(r_2t_2)\rangle}{[\langle|E(r_1t_1)|^2\rangle\langle|E(r_2t_2)|^2\rangle]^{1/2}}. \qquad (13.18)$$

Most of the interference phenomena we have discussed will be seen to be interpretable in terms of the complex degree of correlation. For example, in the case of Young's double slit, the delay τ varies across the plane where the fringes are seen. If the light illuminating the slits is monochromatic and the slits are effectively point sources, the light will be completely coherent and $\gamma_{12}(\tau)$ will be unity everywhere. Fringes of unit visibility result. If a wide source is used the light at the two slits is only partially correlated, $\gamma_{12}(\tau)$ is less than unity and so is the visibility. Similarly, if a broad spectrum source is used, the fringe on-axis, where $\tau = 0$, will be visible, but those off-axis where $\tau \neq 0$ rapidly decline in visibility. The explanation in Michelson's stellar interferometer (Section 9.12) in terms of overlapping of fringes from different *parts* of the source, and in the case of twin beam spectrometry (Section 12.9) from different *wavelengths*, is now seen to be more elegantly expressed in terms of coherence.

13.3   Temporal Coherence of a Wavetrain

As an example we calculate the first-order temporal coherence for the elementary case of a wavetrain consisting of a large number of Gaussian wave packets uniformly spaced in time, at interval T, and having random, uncorrelated phases ϕ_j, so that the field amplitude is

$$E(t) = \sum_{j=1}^{n} \exp[i(\omega t + \phi_j)] \exp\left[-\frac{1}{2}a^2(t - jT)^2\right]. \qquad (13.19)$$

We can view the separate terms as a crude model for the quasi-monochromatic, but mutually incoherent, flashes of radiation emitted by a collection of excited atoms. Each term in equation (13.19) has a temporal width given by the standard deviation $\sigma = 1/a$. The uniform spacing in time, and the Gaussian profile, are adopted for simplicity.

From equation (13.17)

$$\gamma^{(1)}(\tau) = \frac{\int_{-\infty}^{\infty} E(t+\tau)E^*(t)\mathrm{d}t}{\int_{-\infty}^{\infty} |E(t)|^2\mathrm{d}t}. \qquad (13.20)$$

The numerator is

$$\int_{-\infty}^{\infty} E(t+\tau)E^*(t)\mathrm{d}t = \sum_{j,k} \exp[i(\omega\tau + \phi_j - \phi_k)] \exp\left\{-\frac{1}{2}a^2[(t+\tau-jT)^2 + (t-kT)^2]\right\} \qquad (13.21)$$

Because the phases are uncorrelated, the factors $\exp[i(\phi_j - \phi_k)]$ will fluctuate randomly in sign and tend to suppress the contribution from all terms except those with $j = k$. With the help of the integral[3]

$$\int_{-\infty}^{\infty} \exp(Ax^2 + Bx)\mathrm{d}x = \sqrt{\frac{\pi}{-A}} \exp(-B^2/4A) \tag{13.22}$$

equation (13.21) can be evaluated:

$$\int_{-\infty}^{\infty} E(t + \tau)E^*(t)\mathrm{d}t = \sum_j \exp(i\omega t) \int_{-\infty}^{\infty} \exp\{-1/2a^2[(t + \tau - jT)^2 + (t - jT)^2]\}\mathrm{d}t$$

$$= n\sqrt{\pi}a^{-1} \exp\left(i\omega\tau - \frac{1}{4}a^2\tau^2\right). \tag{13.23}$$

Setting $\tau = 0$ gives the denominator of equation (13.20), and we find

$$\gamma^{(1)}(\tau) = \exp\left(i\omega\tau - \frac{1}{4}a^2\tau^2\right) = \exp[i\omega\tau - \frac{1}{4}(\tau/\sigma)^2]. \tag{13.24}$$

We saw in equation (13.16) that for equal irradiances of the two fields, the fringe visibility is equal to the modulus of this function

$$V(\tau) = |\gamma^{(1)}(\tau)| = \exp\left[-\frac{1}{4}(\tau/\sigma)^2\right]. \tag{13.25}$$

This is the Gaussian function which is plotted in Figure 12.15(b).

Notice that the coherence falls off rapidly with the time difference τ. We can identify the coherence time with the temporal width of each wave packet: $t_C = \sigma$. The coherence length is then $l_C = c\sigma$.

It should be noted that even though the entire wavetrain may be unlimited, coherence disappears beyond the time scale of a single wave packet. This correctly reflects the assumed lack of phase correlation between pairs of wave packets.

13.4 Fluctuations in Irradiance

The light from a chaotic light source, such as a gas discharge lamp, contains fluctuations in phase and irradiance due to the random nature of the light emission. The fluctuations in irradiance may be quantified in a similar manner to the first-order electric field correlation function. The light is sampled with measurements of the irradiance I separated by a time interval τ. Each measurement of I is an average over one cycle, and the fluctuations are recorded as differences from the mean irradiance[4] $\bar{I}$. The product of the differences is averaged over a time longer than the coherence time, as indicated by the angle brackets in

$$\langle(I(t) - \bar{I})(I(t + \tau) - \bar{I})\rangle = \langle I(t)I(t + \tau)\rangle - \bar{I}^2 \tag{13.26}$$

---

[3]R.P. Feynman and A.R. Hibbs, *Quantum Mechanics and Path Integrals*, McGraw-Hill, 1965, p. 357.

[4]In Section 5.5 we defined irradiance, for rapid harmonic oscillations, as a time average of the energy flux S: $I = \bar{S}(t)$. To allow for more complex time variations of irradiance, in this chapter we define its instantaneous value by $I(t) = \bar{S}(t)$, with the bar standing for a time average over a short time, preferably the response time of the detector.

since

$$\langle I(t)\rangle = \langle I(t+\tau)\rangle = \bar{I}. \tag{13.27}$$

The second term of equation (13.26) is a *second-order correlation function*. Expanding in terms of the electric fields this is

$$\langle |E(t)|^2 |E(t+\tau)|^2\rangle = \langle E^*(t)E^*(t+\tau)E(t)E(t+\tau)\rangle. \tag{13.28}$$

Expanding each term as $E\exp(i\omega t)$ or $E\exp[i\omega(t+\tau)]$, and averaging over times large compared with $1/\omega$, we find

$$\langle |E(t)|^2 |E(t+\tau)|^2\rangle = |\langle E^*(t)E^*(t+\tau)\rangle|^2 + \bar{I}^2. \tag{13.29}$$

The second-order (intensity) correlation function is therefore determined by the magnitude of the first-order (field amplitude) correlation function (equation (13.8).

These ideas can be formalised in the form of a normalized second-order degree of temporal coherence, defined as[5]

$$\boxed{\gamma^{(2)}(\tau) = \frac{\langle I(t)I(t+\tau)\rangle}{\bar{I}^2}.} \tag{13.30}$$

The normalized $\gamma^{(2)}(\tau)$ may be expressed in terms of the electric fields as

$$\gamma^{(2)}(\tau) = \frac{\langle E(t)E(t+\tau)E^*(t)E^*(t+\tau)\rangle}{\langle E(t)E^*(t)\rangle^2}. \tag{13.31}$$

For chaotic light the range of $\gamma^{(2)}(\tau)$, in contrast to $\gamma^{(1)}(\tau)$, is $1 \le \gamma^{(2)}(\tau) \le 2$. Figure 13.2 illustrates the dependencies of $\gamma^{(1)}(\tau)$ and $\gamma^{(2)}(\tau)$ as a function of the delay time; the figure shows these for both Gaussian and Lorentzian spectral lineshapes. A connection between the second-order and first-order correlation functions may be derived for chaotic light (but not for laser light) from equation (13.29). Dividing each side by $\langle E(t)E^*(t)\rangle^2$ we obtain

$$\boxed{\gamma^{(2)}(\tau) = 1 + |\gamma^{(1)}(\tau)|^2.} \tag{13.32}$$

For chaotic light and zero delay time, $\gamma^{(2)}(0) = 2$, so that for zero delay the detection rate is twice that for long delay times. This indicates that photons arrive in pairs at zero time delay and independently at long time delays. This is known as the photon bunching effect for thermal (chaotic) light sources.

13.5 The van Cittert–Zernike Theorem

The van Cittert–Zernike theorem provides a useful connection between the complex degree of spatial coherence and diffraction theory; it enables the forms of calculated diffraction patterns to be used in

---

[5]The normalized first- and second-order correlation functions in the quantum electrodynamics description of the light field are usually designated $g^{(1)}$ and $g^{(2)}$.

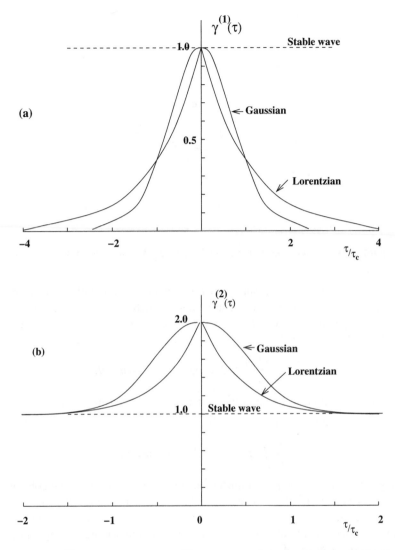

Figure 13.2   First-order $\gamma^{(1)}(\tau)$ and second-order $\gamma^{(2)}(\tau)$ coherence functions for chaotic light having Gaussian or Lorentzian frequency distributions

connection with coherence theory, provided the functional terms are interpreted correctly. This relationship can be illustrated by a simplified one-dimensional analysis: the construction is shown in Figure 13.3. A quasi-monochromatic incoherent source illuminates a distant screen at which the spatial coherence is to be determined. The amplitude of the source at point S is $S(\theta)$ and the field amplitude at some fixed point of reference $P_1(x = 0)$, derived from amplitude $S(\theta)$ over the source, is

$$A(0) = \frac{1}{d} \int_{\text{source}} S(\theta) \exp(ikr_0) d\theta \qquad (13.33)$$

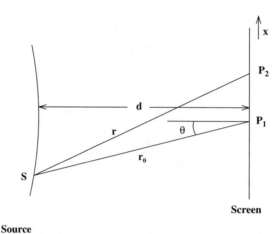

Figure 13.3 The geometry of the van Cittert–Zernike theorem, in one dimension

where k is the wave vector (we have assumed that the screen is sufficiently far away from the source that radial distances from source to screen can be approximated by the perpendicular distance, d).

By dropping a perpendicular from P_1 to line SP_2, we find $r \simeq r_0 + x \sin \theta$ provided $|x| \ll r_0$. Consequently at point P_2 on the screen

$$A(x) = \frac{1}{d} \int S(\theta) \exp[ik(r_0 + x \sin \theta)] d\theta. \tag{13.34}$$

The correlation $C(x)$ between the amplitudes at $x = 0$ and at position x is

$$C(x) = \langle A(0) A^*(x) \rangle \tag{13.35}$$

$$= \frac{1}{d^2} \int \int S(\theta) S^*(\theta') \exp(-ikx \sin \theta') d\theta d\theta'. \tag{13.36}$$

The time average of $C(x)$ only contains contributions from $S(\theta).S^*(\theta')$. For an incoherent source, $S(\theta)$ and $S(\theta')$ are uncorrelated and the product $S(\theta).S^*(\theta')$ has a time average only for $\theta = \theta'$, when $\langle |S(\theta)|^2 \rangle = I(\theta)$, the irradiance of the source. The angle brackets denote the time average.

The complex degree of spatial coherence $\gamma(x)$ in this one-dimensional example may be expressed as the normalized time average of $C(x)$:

$$\gamma(x) = \frac{\langle C(x) \rangle}{\langle A(0) A^*(0) \rangle}. \tag{13.37}$$

Substituting for $C(x)$ we find

$$\gamma(x) = \frac{\int I(\theta) \exp(-ikx \sin \theta) d\theta}{\int I(\theta) d\theta}. \tag{13.38}$$

When the source is small compared with the distance of observation d, so that $\sin \theta \simeq \theta$, the complex degree of spatial coherence $\gamma(r)$ is equal to the normalized Fourier transform of the irradiance distribution $I(\theta)$ within the source. The degree of spatial coherence is equal to the amplitude produced at P_2 by a spherical wave passing through an aperture of the same size and shape as the extended source and converging to P_1.

In the description of diffraction contained in Section 10.2 it was shown that the Fourier transform of the complex amplitude distribution across the aperture represented the Fraunhofer diffraction pattern. It is seen that equation (13.38) is a normalized one-dimensional representation of the Fourier transform of the irradiance at the source. The integral has the same form as the diffraction integral (equation (10.11)) with the quantity $I(\theta)$ interpreted in the diffraction equation as equivalent not to the irradiance, but to the field amplitude distribution at the source, when acting as an aperture.

The analogy may readily be extended to two dimensions. Then the general statement of the van Cittert–Zernike theorem is:

the complex degree of spatial coherence $\gamma(r_1 r_2)$ between a fixed point and a variable point in a plane illuminated by an extended source is equal to the normalized Fourier transform of the irradiance distribution $I(\theta_x, \theta_y)$.

The analogy with the diffraction theory developed in Chapter 10 can be taken further. For a slit source of uniform irradiance, $\gamma(r_1 r_2)$ is a sinc function; similarly for a uniform circular source, e.g. from a star, $\gamma(r_1 r_2)$ is a Bessel function. Similarly, the transverse coherence diameter measured by a Michelson interferometer is seen to be related to the separation of two points in the observing screen at which $\gamma(r_1 r_2) = 0$.

13.6 Autocorrelation and Coherence

In Section 4.15 we considered autocorrelation of a time-varying quantity $A(t)$. The autocorrelation function is defined as the time average

$$\Gamma(\tau) = \langle A(t + \tau)A^*(t) \rangle. \tag{13.39}$$

This was shown to be the Fourier transform of the power spectrum of $A(t)$. Comparison with equation (13.7) shows that the longitudinal coherence function for a plane wave, where $E_1(t) = E_2(t)$, is the autocorrelation function, which is the Fourier transform of the power spectrum. The transverse autocorrelation $\Gamma(x)$ is similarly the Fourier transform of the angular distribution of radiance across the source. Any interferometer which measures coherence along a wavetrain can find $\Gamma(\tau)$, and hence the spectrum of the wavetrain; any interferometer which measures coherence along an axis x transverse to a wavefront can find $\Gamma(x)$, and hence the radiance distribution across the source.

Fourier transform spectrometry, as described in Chapter 12, is therefore a process of measuring the autocorrelation *along* a wavetrain, using an interferometer such as Michelson's spectral interferometer over a range of path differences. The fringe amplitude is measured as a function of path difference, and a Fourier transformation gives the spectrum. The spectrum can be measured with a resolution which depends only on the maximum delay τ_{max} between the two beams; the frequency resolution is approximately $1/\tau_{max}$.

The extent of the coherence *across* a wavefront, as measured in the stellar interferometer, depends on the angular width of the source of light; as we have seen in the previous section, there is a Fourier transform relation between angular distribution of radiance across the source and the decrease of coherence across the wavefront. Autocorrelation in space, i.e. transverse to the wave, is related to the angular distribution of the source; autocorrelation in time, i.e. along the wave, is related to its spectrum.

The concept of coherence is also useful in communications, where a narrow-bandwidth electrical signal is analogous to an optical spectral line. Any modulation of the signal will give a finite width to the spectrum, and very broad-bandwidth electrical noise is analogous to white light. In radio

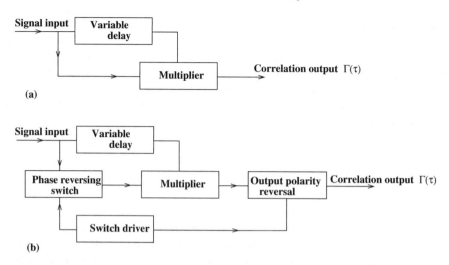

Figure 13.4   Measuring the spectrum of an electrical signal by autocorrelation. (a) The detector measures the product of the signal with the same signal delayed by a variable amount. (b) Phase switching: the correlated component of the direct and delayed signals reverses in sign when the phase reverses

astronomy the spectral lines of interstellar gas, such as hydrogen at 21 cm wavelength and carbon monoxide at 2.7 mm wavelength, have a width which is due to a combination of thermal broadening and Doppler shifts within an interstellar cloud. A typical linewidth might be $\Delta v = 1\,\mathrm{MHz}$; then according to equations (13.3) and (13.4), the coherence time would be about 1 µs and the coherence length about 100 m. The coherence length and the whole autocorrelation function can be measured by an autocorrelation technique shown in Figure 13.4. Here the electrical signal passes through a circuit containing a variable delay (ranging up to about 1 µs in the example above), and the direct and delayed signals are recombined in a detector. The detector multiplies the sum, giving an average product which measures their correlation. The correlation function is obtained by measurements over a range of delays. The spectrum of the signal is then found by a Fourier transform of the autocorrelation function, using the Wiener–Khintchine theorem set out in Section 4.15.

The output of the detector in Figure 13.4(a) also contains an unwanted constant component, proportional to the intensity of the input signal, as in equation (13.11). This can be removed by the switching system of Figure 13.4(b), where a phase-reversing switch has been included in the direct signal path. When this operates the sign of the correlation reverses, while the intensity component is unchanged. The output of the detector is the square of the input; if the signal amplitude is $A + a$, where a is a correlated component which is small compared with the uncorrelated component A, the detector output switches between $A^2 + 2aA + a^2$ and $A^2 - 2aA + a^2$, giving a difference signal $4aA$ which is proportional to a. The phase switch is operated periodically by a driver which also reverses the output of the detector. The intensity component then averages to zero, leaving only the correlated signal. This technique of *phase switching* has many other applications in electronics and optics.

13.7 Two-Dimensional Angular Resolution

We have seen in Chapter 9 that coherence across a wavefront is related to the angular distribution of the source of the radiation, so that a measurement of the coherence as a function of distance across a wavefront gives the width and shape of the source. This applies in two dimensions; we now show that a two-dimensional mutual coherence across a wavefront is directly related to the two-dimensional angular distribution of radiance across the source. Exploring the coherence of the wavefront allows a map to be drawn of the angular distribution of radiance across the source of the wavefront.

Suppose that the amplitude of the wavefront at a point in the x, y plane is $A(x, y)$, and that at another at a distance X, Y is $A(x + X, y + Y)$. There is no time delay between these samples of the wavefront, so that the two-dimensional mutual coherence is

$$\Gamma(X, Y) = \langle A(x + X, y + Y)A^*(x, y)\rangle. \tag{13.40}$$

From Section 13.5, the two-dimensional Fourier transform of this turns out to be the two-dimensional distribution of radiance with *angle*. Put in simpler terms, this is the distribution of radiance giving rise to the sampled interferometer outputs. The coherence $\Gamma(X, Y)$ is complex; in circumstances where phase as well as amplitude of the correlation can be measured the Fourier transform will give the radiance distribution across the source without any assumptions about symmetry. The resolution in angle is of the order of λ/X and λ/Y in the x and y directions.

This relationship is the basis of *aperture synthesis* in radio astronomy. At radio wavelengths, typically of order 10 cm, it is difficult to obtain sufficient angular resolution by using a single radio telescope. It is, however, straightforward to measure the correlation between radio waves received by two or more telescopes separated by large distances (Figure 13.5), even up to some thousands of kilometres (see very long baseline interferometry, Chapter 9). Through a succession of measurements of the two-dimensional coherence using pairs of telescopes at various spacings and orientations a

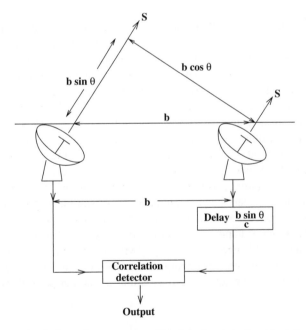

Figure 13.5 Aperture synthesis in radio astronomy. The interferometer's output is the complex degree of coherence between the two radio telescopes, spaced a distance b apart, which together sample the transverse coherence function. The source under observation may be at an angle θ to the normal to the baseline, so that the correct correlation requires one signal to be delayed by $b\sin\theta$, and the effective baseline length is $b\cos\theta$. Observations at many baselines are combined and transformed to produce maps such as that of the radio galaxy M82

sufficient map of $\Gamma(X, Y)$ can be obtained. The phase of the correlation can be found by comparison between coherence between different pairs. A map of complex correlation is constructed which, when Fourier transformed, gives a map of the angular distribution of radio radiance (brightness) across the source.

Radio interferometers using aperture synthesis may require a network of 10 or more radio telescopes operating simultaneously, so that sufficient baselines are available for measuring the distribution of the complex correlation. They can, however, be operated with baselines up to some thousands of kilometres, using wavelengths of a few centimetres. Since the angular resolution of the resulting source map is of order λ/D, where D is the largest available baseline, it is possible by this method to construct maps of radio brightness with resolutions down to 10^{-3} arcseconds. It is interesting to note that this is several orders of magnitude better than the resolution of the largest optical telescopes, even though the radio wavelength is four orders of magnitude larger than the optical wavelength. An example is the map of the radio emission from the galaxy M82 in Plate 4.[*]

13.8   Irradiance Fluctuations: The Intensity Interferometer

In Chapter 9 we described the intensity[6] interferometer used by Hanbury Brown and Twiss for measuring very small angular diameters of stars, and in Chapter 12 we summarized its application to the measurement of very narrow linewidths, namely $\Delta v \sim 1\,\text{kHz}$ in photon correlation spectroscopy. No explanation was given there as to why the intensity fluctuations observed at separated positions should correlate but we can now see why it works in terms of the correlation analysis in Sections 13.2 and 13.4.

The first intensity interferometer in 1950 was set up by Hanbury Brown and Twiss using two 2.4 m diameter radio telescopes and was used to measure the diameter of the Sun and the angular diameters of the Cassiopeia A and Cygnus A radio sources. Hanbury Brown and Twiss then set out to demonstrate that interference could be detected in the intensity (irradiance) of light, as already demonstrated for the intensity of radio waves, despite the fact that light was detected as a stream of photons. Their initial optical experiment, shown in Figure 13.6, used a mercury lamp as a source and measured the correlation of detected photons at two photomultiplier detectors. It was demonstrated that intensity (irradiance) correlations could be measured by detecting individual photons, and that this measurement could be used to determine coherence area or time for chaotic light.

The relation between Michelson's stellar interferometer and the Hanbury Brown interferometer (Chapter 9) may be seen qualitatively as follows. Each atomic emitter in a source gives rise to a finite wavetrain of random phase. We can imagine a multiplicity of spherical waves spreading out from the source. At any point P_1 in space the amplitude at a particular time depends on how many wavetrains are present and how their phases happen to be arranged. Sometimes favourable interference will take place and the amplitude – and hence the intensity – will go up; sometimes destructive interference will make it go down. In these rather oversimplified terms one can see that irradiance fluctuations should exist. Now let us consider whether the fluctuations at another point P_2 will be correlated with

---

[6]This interferometer was first developed by Hanbury Brown for radio astronomy, where the term intensity is used for the radio equivalent of the radiometric optical quantity irradiance (or the photometric quantity illuminance), and he continued this usage when he transferred into the optical domain. We follow this traditional usage.

[*]Plate 4 is located in the colour plate section, after page 246.

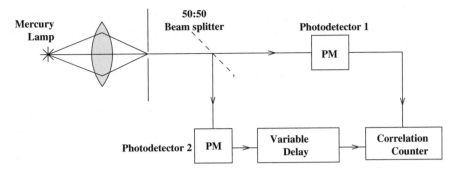

Figure 13.6 The original Hanbury Brown and Twiss experiment, using two photomultiplier detectors to measure the correlation between two photon streams

those at P_1. The same wavetrains reach P_2 as reach P_1: the only difference is in their relative phases caused by the different paths they have travelled.

The condition for identical fluctuations at P_1 and P_2 is the same as that for interference at P_1 and P_2: the relative phases of the wavetrains must be the same, which is to say that the waves are coherent at P_1 and P_2. The phase condition has already been found in Section 9.11; it is

$$\phi_0 \ll \frac{\lambda}{d} \tag{13.41}$$

where ϕ_0 is the angular width of the source and d is the separation between P_1 and P_2. The discussion can be put on a quantitative basis in terms of $\gamma^{(1)}(\tau)$ and $\gamma^{(2)}(\tau)$, the first- and second-order degrees of coherence.

Interference effects such as those occurring in the Young's double slits arrangement and involving two interfering electric field *amplitudes* may be quantified by $\gamma^{(1)}(\tau)$, the normalized first-order degree of coherence, and are used to describe temporal and spatial coherence. Correlations in *irradiance* (intensity) were defined in equations (13.27) and (13.28). As indicated in equations (13.29) and (13.32) for chaotic light sources, the second-order degree of coherence $\gamma^{(2)}(\tau)$ is related to $|\gamma^{(1)}(\tau)|^2$. These equations may be generalized to include either (or both) time and spatial coherence:

$$\gamma^{(2)}(\tau, r) = 1 + |\gamma^{(1)}(\tau, r)|^2. \tag{13.42}$$

Thus temporal or spatial coherence properties can also be measured by determination of $\gamma^{(2)}(\tau, r)$.

The Hanbury Brown and Twiss effect is concerned with the fluctuations in intensity and their correlations $\langle I(t)I(t+\tau)\rangle/\bar{I}^2$. The effect has both a classical explanation arising from irradiance fluctuations and a quantum theoretical explanation arising from fluctuations in photon count. It is an illustration of the correspondence between the classical and quantum theories of light. However, while here we will describe the classical approach, the quantum theory provides a more extensive description and an explanation of other phenomena, such as photon anti-bunching, which cannot be explained classically.

The fluctuations in the light beams are measured by a photodetector, usually a photomultiplier giving a photoelectron current, and for low light levels this is carried out by the counting of detected photons. The average rate of emission of photoelectrons is proportional to the instantaneous irradiance. The origin of the fluctuations in the irradiance arises from two sources: from the light beam itself and from the detection process. We look first at fluctuations in the light beam itself. The irradiance fluctuations of a thermal (chaotic) light source may be simulated by the superposition of the independent radiation from many atoms. For emitters with amplitude a_n and phase ϕ_n which are each independent from the others, the combined electric field for N emitters is then

$$E = \sum_N a_n \exp[i(\omega_n t + \phi_n)] = E_0 \exp(i\omega t).$$ (13.43)

The fluctuations in the electric fields of the chaotic waves lead to fluctuations in the irradiance $I(t) = |E_0(t)|^2$. The statistical fluctuation of light wave irradiance I has a probability distribution

$$P(I)dI = (1/\bar{I}) \exp(I/\bar{I})dI.$$ (13.44)

For a mean irradiance $\bar{I}$ recorded in a time interval δt, the mean number of emitted electrons = $\bar{n} = \bar{I}\eta\delta t/\hbar\omega$. Here the quantum efficiency η gives the probability of a photoelectron being emitted following the detection of a photon.

The mean value $\bar{n}$ depends on the mean irradiance $\bar{I}$, which is also fluctuating, and the time over which the averaging is carried out. We can relate these times to the coherence time t_G discussed in Section 13.1. Averaging may be carried out over times long compared with the coherence time, $t_1 \gg t_G$, and over times short compared with the coherence time, $t_2 \ll t_G$. The mean irradiance corresponding to the short time will be different from the mean irradiance over the long time. There is a further characteristic time involved which is the response time t_p, of the detectors; this is required to be more rapid than the fluctuations that are to be detected, otherwise the fluctuations are smoothed out.

The mean square fluctuation is then

$$[\langle I(t)\rangle_{t_2} - \langle I(t)\rangle_{t_1}]^2 = \overline{[(\Delta I)^2]} = (\overline{I^2} - \bar{I}^2)$$ (13.45)

with $\overline{I^2} = \int I^2 P(I)dI = 2\bar{I}^2$. We obtain that the mean square fluctuation is $\overline{(\Delta I)^2} = \bar{I}^2$.

It is seen that in the emitted beam the short-term mean fluctuations occur about the long-term mean and that these can be large, being equal to the mean irradiance.

In terms of photon counting the mean square fluctuation in photocounts is

$$\overline{(\Delta n)^2} = \bar{n}^2.$$ (13.46)

A second source of fluctuation arises from the photon nature of the beam and the photoelectric detection process. In the photoelectric detection a constant irradiance onto the detector gives photoelectron emission pulses having a Poissonian statistical distribution. The probability of n electrons being emitted in a certain time interval and with a mean number $\bar{n}$ is

$$p(n) = (\bar{n}^n/n!) \exp(-\bar{n}).$$ (13.47)

The variance or mean square fluctuation for the Poisson distribution is equal to its mean

$$\langle\overline{(\Delta n)^2}\rangle_{t_2} = \langle n^2\rangle - (\bar{n})^2 = \bar{n}.$$ (13.48)

The two sources of fluctuation, equations (13.46) and (13.48), can be combined by taking the sum of their variances. The total variance in photoemission in a time $t < t_G$ for a chaotic light source is then

$$\overline{(\Delta n)^2} = \langle(n - \bar{n})^2\rangle_{t_1} = \bar{n} + \bar{n}^2.$$ (13.49)

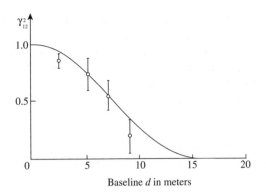

Figure 13.7 Hanbury Brown and Twiss's results for the variation of the normalized intensity (irradiance) fluctuations with baseline for the star Sirius. The solid curve is the variation of Γ_{12}^2 with d, calculated from an assumed angular diameter of 0.0069 arcseconds. The optical system consisted of two searchlight mirrors 1.56 m in diameter and 0.65 m focal length, capable of focussing the light from the star into an area 8 mm in diameter. A photomultiplier was mounted at the focus of each mirror and the anode currents were multiplied and integrated to give Γ_{12}^2

The positive term in $\bar{n}^2$ for a chaotic source for times $t < t_G$ shows that the photoemissions have an excess fluctuation or are correlated. The term in $\bar{n}^2$ has an interpretation as a bunching of photons in the beam.

The intensity interferometer takes the time average of the product of $I(t)$ and $I(t + \tau)$ which, as seen in equation (13.32), is related to the square of the *modulus* of the mutual coherence between P_1 and P_2. A variation of the spacing of P_1 and P_2 thus allows $\gamma_{12}^{(2)}(\tau)$ to be measured over the lateral coherence area of the source. As in the aperture synthesis discussed in the previous section, the source radiance distribution must then be obtained by Fourier transformation of $|\Gamma_{12}|$. This cannot be achieved unambiguously without knowledge of the phase of Γ_{12}, but it may be allowable to assume that the source is symmetrical, giving a constant phase at all interferometer spacings. This is at least a reasonable assumption when the diameter of a star is first measured. The optical intensity interferometer was first used in 1956 at Jodrell Bank on the bright star Sirius. Figure 13.7 shows the measured fall-off of γ_{12}^2 with baseline increasing up to 9 metres. The angular diameter of Sirius is known to be 7×10^{-3} seconds of arc; a circular disc of this size would give the theoretical variation of γ_{12}^2 shown in the figure, agreeing well with the observations. Notice that this account of the intensity interferometer is a purely *wave* explanation. We presented in Chapter 12 a similar discussion of the relation between the shape of a spectral line and intensity fluctuations on a time scale related to the width of the line. In both cases consideration should be given to the photon nature of light. There is no need for this if the flux of photons is large enough for a large number to arrive within a single measurement time, but if the flux is small the random variations in photon count become important. These appear as statistical fluctuations in intensity, and the correlator output becomes noisy. This does not change the coherence and correlation, but it reduces the accuracy of the measurements. Ultimately, when on average fewer than one photon is detected at each measurement the intensity interferometer becomes practically impossible to operate.

Although two detectors were used in the original demonstration by Hanbury Brown and Twiss the effect applies also to a single detector observing a single source. This is the arrangement used in the photon correlation spectroscopy technique described in Chapter 12.

After the initial observations on radio astronomical sources Hanbury Brown and Twiss conducted the optical experiments confirming the effect at visible frequencies. The experiments raised some controversy at the time that they were reported. It was questioned that if the photons emitted by the thermal source were emitted at random, how could the signals received at the two detectors be correlated? That is, how could the detection of a photon at one photomultiplier be correlated with the detection of a different photon at the other photomultiplier? As we have remarked earlier, the correlation in the two signals suggested that the photons arrive in pairs at the detectors, i.e. they are bunched. This observation is supported by an explanation considering statistical fluctuations in a system of bosons. Shortly after these observations a detailed quantum theory of coherence was developed by R.J. Glauber for which he received the 2005 Nobel Prize for Physics, which gave a firm explanation for the effect. In this regard the Hanbury Brown and Twiss effect has also made an important contribution to the development of the subject of quantum optics.

13.9   Spatial Filtering

We have seen that the resolving power of instruments such as the telescope and microscope, and of the Fourier transform spectrometer, is best understood by considering the range of Fourier components which contribute to the output, whether it is an optical image or the shape of a spectral line. This concept can be extended further to consider what happens if instead of a direct reconstruction of the original light source we modify some of the Fourier components before reconstruction. Such a process is familiar in communication engineering, where a signal may be modified by a filter; for example, an unwanted oscillation might be removed by a narrow-band filter. The directly analogous process in optics is spectral filtering in the Fourier transform spectrometer (Chapter 12), where the output can be modified by adjusting the amplitude and phase of the measured longitudinal correlation function. In this section we consider the modification of measured transverse correlation functions, and its effects on an optical image. This is the process of *spatial filtering*, shown schematically in Figure 13.8.

The first application of spatial filtering (although not then described as such) was to the microscope, in Abbe's theory of image formation. Consider the formation by a microscope objective lens of an image of a grating-like object, illuminated by fully coherent light. In Figure 13.9 the objective, shown as a single lens, collects light leaving the object over a wide range of angles. If we consider this light as an angular spectrum of plane waves, we find that these components are focussed on the focal plane F of the lens, which therefore contains the angular spectrum. The plane waves continue beyond this focal plane, forming an image further away which is then examined by an eyepiece; in this discussion, however, we are only concerned with the angular spectrum in the focal plane F of the objective. How can this be modified, and with what effect on the final image?

As in all Fourier analysis, the finer detail is contained in the highest order components; if these are lost, the resolution is reduced. An object which is a periodic grating will produce a series of components $S_1, S_2, \ldots, S_n, S_{-1}, S_{-2}, \ldots, S_{-n}$ as shown in Figure 13.9. A purely sinusoidal grating would produce only the two first-order components S_1, S_{-1}; a grating with sharp narrow lines will

Figure 13.8   Spatial filtering. Light from the object O is Fourier transformed into the spectrum S. The spectral components are modified by filtering, producing the spectrum S′, and an inverse Fourier transform produces the reconstructed object O′

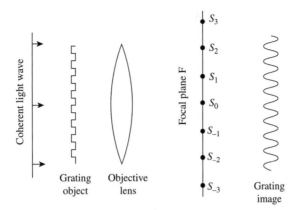

Figure 13.9   The Abbe theory of microscopy. The object is a grating, coherently illuminated. The diffraction pattern in the focal plane F of the microscope objective is the Fourier transform of the complex amplitude across the object; as the object is periodic, the resulting diffraction pattern comprises discrete components S_0, S_1, $S_2, \ldots, S_n, S_{-1}, S_{-2}, \ldots, S_{-n}$

produce a series of high-order components. If a mask is placed in the focal plane so that only the first-order components are admitted to the rest of the microscope, a grating with any lineshape will be seen in the image plane simply as a sinusoidal grating. The reason for the resolution limit is clear: the objective must accept plane waves leaving the object over a sufficiently wide range of angles. If this range is $\pm\theta$, and the space between the grating and the objective has refractive index n, the finest detail of the source that can appear in the final image has a size $d = \lambda/n \sin\theta$. The spectrum in the focal plane has a zero-order component at S_0 in Figure 13.9, which may be very intense for a nearly transparent object. This provides the first example of spatial filtering: the central zero-order component can be removed by placing a simple mask in the focal plane. This provides *dark-field microscopy*; an example is shown in Plate 5.[*]

A more subtle effect is obtained by changing the phase of the central component, by using a phase filter, i.e. a transparent mask in which a central zone is thicker. Transparent objects then become visible because of the pattern of phase changes which they impose on the light passing through them. This is often important in biological specimens, which otherwise would have to be stained if they are to be made visible in an ordinary microscope. *Phase contrast microscopy* was introduced by Zernike, and is often named after him. Consider again a simple grating in the object plane (Figure 13.9), but a phase-changing grating instead of an amplitude grating. The grating has no effect on the modulus of an incident plane wave, but introduces a small phase shift which varies periodically across the object plane. The diffraction pattern in the focal plane F then contains components S_1, $S_2, \ldots, S_n, S_{-1}, S_{-2}, \ldots, S_{-n}$ as before, except that the components are in quadrature with the light at S_0, as may be seen by describing the complex amplitude in the object plane as

$$A(x) = A_0 \exp\left(i\phi_0 \cos\frac{2\pi x}{d} \right). \tag{13.50}$$

---

[*]Plate 5 is located in the colour plate section, after page 246.

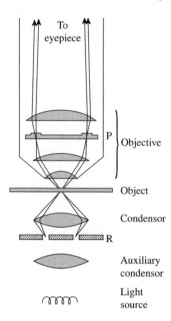

Figure 13.10 Practical arrangement of a phase contrast microscope. The ring source of light R is focussed on the phase contrast plate P, within the objective lens system. The undeviated light is retarded by the plate, while light diffracted by the object passes through the thinner part of the plate and appears in quadrature in the final image

If ϕ_0 is small we may write this as

$$A(x) = A_0 \left(1 + i\phi_0 \cos \frac{2\pi x}{d} \right) \qquad (13.51)$$

and the wave has two components, one with the unchanging amplitude, the other in phase quadrature (because of the i) and with an amplitude varying periodically across the aperture. These produce respectively the central zero-order component S_0 and the diffracted components S_1, S_{-1}, etc.

The idea of phase contrast microscopy is to retard the phase of the large undeviated component S_0 by a quarter wavelength, so as to reproduce the diffraction pattern of an amplitude grating. This may be achieved by inserting in the focal plane F a glass plate with an extra thickness in the central region, so that the light at S_0 is retarded by $\lambda/4$ or $3\lambda/4$. In the first case regions of the object having greater optical thickness will appear brighter, and in the second darker. These are called respectively bright and dark, or positive and negative, phase contrast.[7]

The undeviated light at S_0 forms an image of the light source. Instead of using a point source it is convenient to use a ring source, as shown in the practical arrangement of Figure 13.10. The phase-changing plate is then also in the form of a ring, which covers the image of the light source. This arrangement allows a larger light source to be used, giving greater illumination in the image.

---

[7]"Bright phase contrast" should not be confused with conventional "bright-field microscopy".

Problem 13.1

Calculate the transverse coherence length for sunlight and starlight at the surface of the Earth, given that the Sun subtends an angle of $\frac{1}{2}°$, while atmospheric scintillation spreads light from a star typically over $\frac{1}{2}$ arcsecond.

Problem 13.2

Calculate the longitudinal coherence length for laser light with a bandwidth of 60 MHz. What bandwidth Δv and linewidth $\Delta \lambda$ would be required in a laser to produce a coherence length of 10 km?

Problem 13.3

An argon laser beam at a wavelength of 515 nm is repetitively chopped by a shutter to produce pulses of 10^{-10} s duration. Calculate for the pulses: (a) the frequency bandwidth, (b) the wavelength spread, (c) the coherence length.

Problem 13.4

The coherence length of a light source may be measured by determining its time autocorrelation function. A correlator was used to measure the magnitude squared of the time autocorrelation function of a light wave having a wavelength of 532 nm, and gave an exponential decay with a time constant of 60 ns. Estimate (a) the coherence length of the source, (b) the power spectral density of the light.

Problem 13.5

A 'cross' type of radio telescope consists of two perpendicular strips of receiving area each with length D and width d. (They may for example be large arrays of dipoles or parabolic reflectors.) The radio signals from these two are multiplied together in a receiver which records only their product. What is the angular resolution of the system?

Problem 13.6

The wavelength of a beam of particles, mass m and velocity v, is given by $\lambda = h/mv$, where h is Planck's constant. (i) Show that the wavelength for electrons accelerated by a field of V volts is approximately $1.23 V^{-1/2}$ nm. (ii) Calculate the best possible resolving power of an electron microscope with numerical aperture 0.1 and accelerating field 30 000 volts.

Problem 13.7

A spectrograph used in radio astronomy is required to resolve the structure of the hydrogen spectral line at 1420 MHz, as observed when a radio telescope is receiving radiation from several hydrogen gas clouds moving with different speeds in the line of sight. The spectrograph divides the radio signal into two paths, inserting a variable digital delay into one path, and then recombines them in a multiplier. The smallest increment of delay is τ and the largest total delay is $N\tau$. If gas clouds with velocity differences between $10 \, \mathrm{km \, s^{-1}}$ and $1000 \, \mathrm{km \, s^{-1}}$ are to be distinguished, what values of τ and N are required?

14 Holography

But soft! what light through yonder window breaks?

William Shakespeare, *Romeo and Juliet.*

If a scene is viewed through a window or any aperture large compared with the wavelength it is seen as three dimensional and is completely lifelike. The scene changes as we alter our viewpoint: we approach a window and look up through it to see an object in the sky. As the viewpoint alters, objects in the scene show *parallactic* displacements relative to each other; if we move from left to right nearby objects seem to move from right to left compared with more distant ones. Another effect is that of being able to focus the eye on a particular object at a specific distance.

How different is this view through an aperture from a photograph of the same view! The photograph may give an impression of depth, but it is only two dimensional. No parallactic displacements of objects within it may be seen by a shift of viewpoint. The eye must be focused on the plane of the photograph to see it, and no eye focusing can make sharp any part of the photograph not originally brought into focus by the camera.

What is the information that has been lost in the photograph? According to the diffraction theory which we have used in Chapter 10, the amplitude and phase of the light reaching any point on the viewer's side of the window can be deduced from the amplitude and phase of the light in the plane of the aperture. The photograph, however, only records intensity, which is the square of the amplitude, and not the phase. If we are to replace the window with a record of the wavefront, we must record its phase as well as the amplitude. It is the *complex amplitude* in the aperture which must be recorded if we wish to reconstruct, completely lifelike and indistinguishable from reality, the view through the window.

In this chapter we show how holography enables us to record the complex amplitude over an aperture, and so store all the information necessary to construct a three-dimensional image of the original scene behind the aperture. Hence the term *holography*, from the Greek word *holos* meaning whole.

Holography is achieved by combining the required object wavefront with a reference wave, forming an interference pattern on a photographic plate or film. At any point, the recorded irradiance depends on the relative phase of the object and reference waves. When the developed image, or *hologram*, on the photographic plate is illuminated by the same reference wavefront, the light leaving the hologram contains the original object wavefront, with both amplitude and phase. (It also contains

Optics and Photonics: An Introduction, Second Edition   F. Graham Smith, Terry A. King and Dan Wilkins
© 2007 John Wiley & Sons, Ltd

other diffracted components, which we shall discuss later.) The original object cannot be seen in the hologram itself, but information on the wavefronts which came from the object is coded within the interference pattern. The object wavefronts can be reconstructed by re-illuminating the hologram.

14.1   Reconstructing a Plane Wave

We first explore the holographic process for the simple example of an object wave which is a single plane wave. Any wavefront passing through the plane of any aperture can be regarded as an assembly of elementary plane waves at various angles and with various amplitudes. We show how one of these plane waves may be recorded by combining it with a reference wave, and how it may be reconstructed. The object wave is incident at angle α, as shown in Figure 14.1(a). For simplicity we choose a reference wave which is a plane wave at the complementary angle α. These two waves are coherent, which requires them to be derived from the same source. The upper beam is the object beam, which is to be recorded in the hologram, and the lower beam is a reference beam. These crossing plane waves form an interference pattern in the aperture plane, which is recorded on the photographic plate. This recording is a pattern of lines forming a diffraction grating; as in Figure 4.19, the line spacing d is given by the familiar grating formula[1]

$$d = \lambda/2 \sin \alpha. \tag{14.1}$$

The developed plate contains a grating which ideally has a sinusoidal distribution of transparency; it is then referred to as a *sinusoidal diffraction grating* (see Chapter 11, Problem 11.3). It is also a simple hologram corresponding to an object with no structure. The amplitude transmittance of the developed plate is proportional to the irradiance distribution on the plate. When the developed grating is placed in its original position, illuminated by only one of the plane waves, which is the reference beam, three diffracted beams are generated. Figure 14.1(b) shows the diffracted beams emerging from the grating, labelled according to the order m of diffraction at the grating; the undeviated beam is at $m = 0$. The beam at $m = -1$ is now a reconstruction of the object beam, travelling in the original direction. As a check on the direction, following Section 11.2, note that for a beam to emerge at angle θ

$$\sin \theta - \sin \alpha = m\lambda/d. \tag{14.2}$$

Since $d = \lambda/2 \sin \alpha$ the beam must be at angle $-\alpha = \theta$ as shown. The second beam, at $m = 1$, gives rise to an unwanted second image.

Reconstruction of the original object beam occurs for any angle of incidence of the reference wave, provided that exactly the same reference wave is used in the reconstruction. Having reconstructed an elementary plane wave, we now see that any plane wave will produce a grating pattern on the photographic plate; the whole wavefront can therefore be recorded simultaneously and reproduced by illuminating the grating with the reference beam. The reference beam itself may take almost any form, provided that exactly the same beam is used in the reproduction; it need not even be a plane wave, as we shall show in a more general analysis.

---

[1]More generally, crossing waves with equal amplitude and with wavevectors $\mathbf{k}_1$ and $\mathbf{k}_2$ give a pattern of field irradiance $I = 2I_0[1 + \cos(\Delta\mathbf{k} \cdot \mathbf{r})]$ where $\Delta\mathbf{k} = \mathbf{k}_1 - \mathbf{k}_2$.

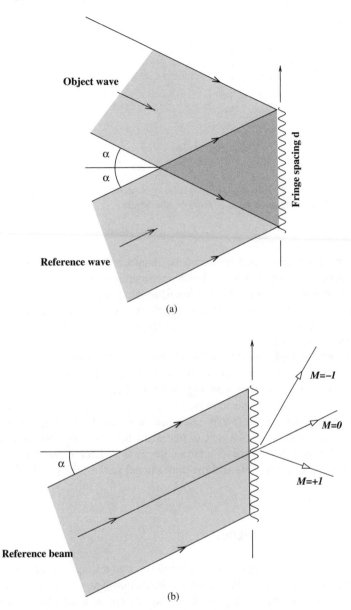

Figure 14.1 Holographic reconstruction of a plane wave: (a) two crossing plane waves form fringes on a photographic plate; (b) diffraction at the sinusoidal fringe pattern illuminated by one of the beams. The first-order diffracted beam is the holographic image of the other beam

The holographic process is illustrated in another simple form in Figure 14.2, in which the object is a single point. Here the photographic plate receives a coherent plane wave, which is the reference beam, and light from the same wave scattered from the point object O. The reference and scattered waves interfere to produce a circular holographic pattern on the plate, similar to a zone plate

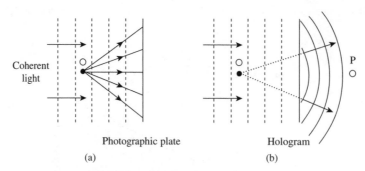

Figure 14.2 Holography using plane wave illumination of a point object. The wave diverging from the point object O interferes with the reference wave to form the hologram

(Figure 10.17). When the developed hologram is illuminated by the same plane wave, diffraction creates a diverging primary spherical wave which appears to originate in a point source at the position of O. As in most holograms, there is a second beam, which in this case converges towards the point P. This is a conjugate beam and forms a real pseudoscopic image.[2] The two images O and P are each at a distance from the hologram equivalent to the focal length of the Fresnel zone plate which has been formed by the interference pattern.

The light used for holography must be coherent over a large volume, which includes the object and the photographic plate. This requires laser light both for illuminating the object and for the reference beam. A typical arrangement is shown in Figure 14.3. Here a beam of laser light illuminates the object through a beam splitter; one beam provides light which is reflected or scattered back from the object to interfere with the other off-axis reference beam at the photographic plate. After development, the plate is a hologram; when it is placed in the same position and illuminated by the reference beam in the same way, it shows a virtual image from behind the plate, as though the object is being seen through a window. This is three-dimensional photography, achieved without a camera lens!

14.2 Gabor's Original Method

Dennis Gabor, the inventor of holography, was led to the idea through the problem of interpreting an X-ray diffraction pattern from a crystal. Since the diffraction pattern was a Fourier transform of the crystal structure, he was attempting to use the X-ray pattern as a diffraction grating in an optical system, so producing the reverse Fourier transform which would be a visible image of the crystal structure. As we noted in Section 11.11, the problem was that phase information had been lost in the original diffraction process. Gabor's idea was to add a reference beam, which would be very difficult to achieve in X-rays, so he set out first to demonstrate the principle using light.

---

[2]A pseudoscopic image is one that has its relief reversed (depth inversion), so that points of the object further from the viewer appear closer, and vice versa.

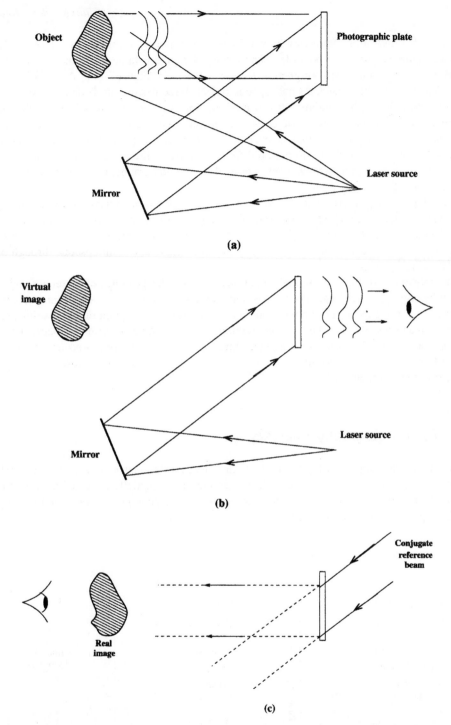

Figure 14.3 Recording and reproducing a holographic image. (a) The object and the photographic plate are illuminated by the same source of light. Scattered light from the object combines with the reference beam to form the hologram on the photographic plate. (b) The developed hologram is illuminated from behind by the same laser beam, and a virtual image of the object is seen through the hologram exactly as it would appear through an open aperture. (c) Projection of a real image. The hologram is illuminated by a reference beam which is conjugate to the original

Gabor's first demonstration of holography was in 1948, before the invention of the laser. Only a very small-scale demonstration was possible because the coherence volume of ordinary monochromatic light sources is so small. His original system is shown in Figure 14.4. A pinhole source of monochromatic light illuminates a small opaque object (three narrow lines) on a transparent screen close to the pinhole. A photographic plate behind the object records the irradiance of the diffracted light. From the viewpoint of geometric optics the object would cast a shadow on the photographic plate; what actually happens is that at each point of the plate interference occurs between the undisturbed light wave and the transmission diffraction wave of the object. The undisturbed wave is then the reference beam; this arrangement is termed in-line holography, because the reference beam and object wave lie along one line.

The developed hologram (in which a positive transparency has been made from the photographic negative) is illuminated through the same pinhole, using a lens or microscope to see the tiny object. A primary beam diverges from the position of the original scattering object. The true three-dimensionality of the image was demonstrated by racking the focal plane of the microscope through different layers of the holographic image.

Further movement of the microscope's focal plane reveals the existence of a pseudoscopic second image of the original object located behind the pinhole. This is inverted and has the property that each point on it is the same distance from the pinhole as the corresponding point on the first image. This is the second image discussed in the previous section. The reconstructed primary image in the Gabor demonstration was degraded by the conjugate image which was superimposed on it, also by light scattered from the directly transmitted beam. Off-axis holography, which we discuss later, provided an answer to these difficulties.

14.3   Basic Holography Analysis

In recording a hologram, as in Figure 14.3(a), an object is illuminated by a laser beam, which on reflection or scattering creates object wavefronts, and these are partially collected by a photographic emulsion. Part of the laser beam is also used to illuminate the photographic emulsion directly, often with the help of one or two mirrors, to create a reference wave.

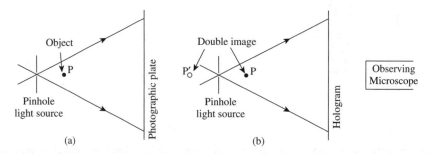

Figure 14.4 Gabor's original system of making a hologram and reconstructing the image. (a) The hologram is recorded on a photographic plate. (b) The object is removed and the hologram is illuminated via the original pinhole. The image and its pseudoscopic partner can be seen by looking through the hologram

Let $E_O(x, y, t)$ and $E_R(x, y, t)$ be the complex amplitudes of the object wave and the reference wave in the plane of the photographic plate ($z = 0$):

$$E_O(x, y, t) = A(x, y) \exp[i(\phi(x, y) - \omega t)]$$
$$E_R(x, y, t) = A_R(x, y) \exp[i(\rho(x, y) - \omega t)] = A_R \exp[i(k_1 x - \omega t)] \tag{14.3}$$

To illustrate the argument, we take the reference wave simply as a plane wave incident on the plate at angle θ in the x, y plane. Hence the spatial part of its phase reduces for $z = 0$ to $k_1 x$, where $k_1 = 2\pi \sin \theta / \lambda$, but its amplitude is constant. The object wave, however, may vary in a complicated way.

Leaving aside an uninteresting constant of proportionality, the resultant irradiance at the plate is

$$I(x, y) = |E_R(x, y) + E_O(x, y)|^2$$
$$= |E_R|^2 + |E_O|^2 + E_R E_O^* + E_O E_R^* \tag{14.4}$$
$$= A_R^2 + A^2 + 2AA_R \cos[\phi(x, y) - k_1 x].$$

Assuming a linear relation between the transmittance T of the hologram and the integrated irradiance, the developed negative darkens according to

$$T = T_0 - KI. \tag{14.5}$$

Here T_0 is the transmittance of the unexposed plate and the constant K is proportional to the exposure time. (The photographic process and the response of the emulsion are described in detail in Chapter 20.) The transmittance of the hologram is therefore

$$T(x, y) = T_0 - K\{A_R^2 + A^2 + AA_R \exp[i(\phi(x, y) - k_1 x)] + AA_R \exp[i(k_1 x - \phi(x, y))]\}. \tag{14.6}$$

A^2 and A_R^2 represent the irradiances of the two waves. Only the cross-terms, from their interference, carry information about the phase of the object wave.

In the holographic reconstruction, or "readout", the developed plate is illuminated with a wave identical to the reference wave. Leaving out the time dependence $\exp(-i\omega t)$, the complex amplitude of the transmitted wave is

$$E_{\text{read}}(x, y) = T(x, y) E_R(x, y)$$
$$= [T_0 - KA_R^2 - KA(x, y)^2] E_R(x, y) - KA_R^2 E_O(x, y) - KE_R(x, y)^2 E_O(x, y)^*. \tag{14.7}$$

The three terms in square brackets correspond to the beam directly transmitted by the plate, and a halo surrounding it. The fourth term is the one desired: aside from a constant multiplying factor, it is identical to the original object wave and reconstructs a virtual image of the object. The last term, with its complex-conjugate object wave, corresponds to a real image, which is usually unwanted. The extra phase factor of $\exp[i(2k_1 x)] = \exp[i(4\pi \sin \theta x / \lambda)]$ multiplying the conjugate wave determines that it will be separated from the object wave by roughly twice the incident angle of the reference beam. Hence when the reference beam is off-axis by a large enough angle, the virtual image can be separated from the conjugate image as well as from the direct beam.

If a real image of the object is required, it can be reconstructed by illuminating the hologram with a wave which is the conjugate of the reference beam. The conjugate wave is such that its complex amplitude is the complex conjugate of the original wave. With illumination by the conjugate wave E_R^*, the transmitted amplitude is the same as in equation (14.7) but E_R and E_O are replaced by their complex conjugates. The fourth term, which contains $A_R^2 E_O^*$, is proportional to the complex-conjugate wave. This wave converges to a real image of the object; because of its reversed spatial phase, the image produced is pseudoscopic, with inverted depth and modified parallax.

The essence of the process is that the recording of the hologram enables both the amplitude and the phase of the object wavefront to be stored, even though the photographic plate only responds to irradiance.

14.4   Holographic Recording: Off-axis Holography

A difficulty with the original Gabor on-line holographic method was that the virtual and real images overlapped, leading to poor quality of the virtual image. The off-axis technique, which was developed starting in the 1960s by E. Leith and J. Upatnieks, overcomes that problem by separating the images.

In equation (14.7) the final term corresponding to the conjugate real image has depth which is inverted, so that the real image is *pseudoscopic* (whereas the virtual image is *orthoscopic*). The primary and conjugate images are separated from each other and from the directly transmitted beam, ensuring no overlap between the beams.

Changes in phase difference between the reference and object beams, e.g. arising from mechanical or acoustic disturbances, need to be minimized during the exposure. The arrangement would normally be mounted on an anti-vibration table.

The two forms of holography illustrated in Figures 14.3 and 14.2 are related to the two categories of diffraction, Fraunhofer and Fresnel, which we distinguished in Chapter 10. When the recording photographic plate is in the near field a Fresnel hologram is formed with the wavefronts from the object being closely spherical. The real and virtual images on the illuminated Fresnel hologram are positioned on either side of the hologram. A Fraunhofer hologram is formed when the distance between the object and the plate is large, in which case the object wavefronts are nearly planar.

14.5   Aspect Effects

When the reconstructed object is viewed through a hologram the edges of the hologram act rather like a window frame. Within the limits set by the frame movement of viewpoint changes the aspect of the reconstructed scene, so that if it is three dimensional one can see more of the image to the right by moving the head to the left and vice versa. Similarly, parallactic displacements of different elements of the scene may be observed.

A consideration of these aspect effects brings out another interesting property of a hologram. From a given direction of observation light reaching the eye from the image only comes from a small portion of the hologram determined by the position of the eye and the angle subtended by the object as shown in Figure 14.5. Evidently if all the rest were removed leaving only this piece the object could still be seen, but only from that aspect. Thus if a hologram is broken into fragments, the reconstructed object can still be seen through each fragment, as seen from the appropriate aspect. This

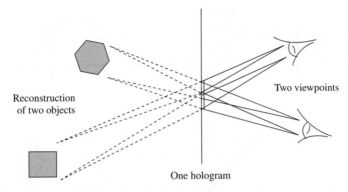

Figure 14.5   To see a reconstructed object from one aspect, only a small portion of a hologram is used. The same portion viewed from another angle allows a different reconstruction (or a different part of the same one) to be seen

is rather like looking through a window that is completely obscured except for a small hole. The view is still to be seen, but only from the viewing position allowed by the hole.

It is inherent in the holographic process that the many different elements of a scene can all be recorded on the same small area of a hologram. Figure 14.5 shows this effect for two separate objects, which can be seen separately from two different aspects. A remarkable and very important extension of this property is the superposition of two or more separate holograms on the same photographic plate, using different reference beams for each object or scene. Any individual object can then be reconstructed and seen by using the appropriate reference in the reconstruction.

A limitation of normal holograms is that they can only be viewed over a limited range of angles. Wide angle or full 360° viewing can be made by extending the range of angles that the photographic film subtends, e.g. the object can be surrounded by a photographic film and illuminated from above or below. A very large amount of information can evidently be stored on a hologram. Extension of the holographic principle to three dimensions expands the possibilities even further; we discuss in Section 14.12 the use of three-dimensional holographic memories for data storage and computing.

14.6 Types of Hologram

The holograms discussed so far are recorded as developed photographic negatives which are then used as transmission gratings. They have an inherent disadvantage of low efficiency in the brightness of the reconstructed image, since light must be lost in the grating. The amplitude hologram may, however, be converted into a phase hologram, in which the hologram grating operates by changing the phase of the light wave instead of its amplitude. This is achieved by storing the interference pattern as a corresponding distribution of refractive index changes within the recording film, and bleaching out the developed amplitude hologram. For silver halide photographic plates the deposited silver metal can be converted into a transparent silver compound, with a refractive index which is different from the gelatin base of the emulsion. In the wave reconstruction, the phase of the wave is altered in proportion to the exposure energy forming the interference pattern.

Holographic reconstruction in which a change of phase is induced can equally well be achieved by reflection, as in Figure 14.6. This is particularly important for phase holography, since a modulation of phase can be achieved by a corrugation of the reflecting surface, using a *photoresist* material. These organic materials are sensitive to light intensity, and after development a photoresist film yields

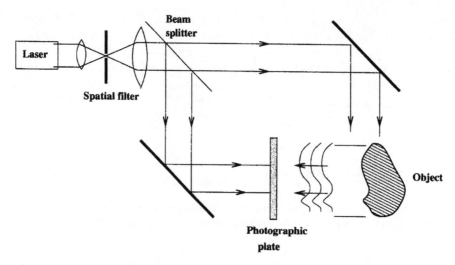

Figure 14.6   Arrangement for recording a reflection hologram

a relief surface whose corrugations provide the phase changes in a reflected wavefront. A major advantage is the ease of replication, since the surface can easily be replicated in a press, using thermoplastic materials. The replication process begins by the making of a stamper in which the relief image recorded on the photoresist is overplated with a layer of nickel by electrodeposition. The nickel layer is separated from the master hologram and put on a metal backing plate. The surface relief of the stamper is transferred in a heated embossing press onto a thermoplastic film. A reflection layer of aluminium is vacuum deposited on the film for subsequent illumination. This is the basis of the familiar holographic logos and icons impressed into bank cards and the like. (But we have yet to explain how these are usefully viewed in white light; see Sections 14.7 and 14.9 below.)

We have so far introduced four categories of holograms (amplitude and phase transmission, and amplitude and phase reflection), as two-dimensional recordings on a surface. There is a further distinction for amplitude and phase holograms depending on whether the recording medium is thin or thick. Photographic films many wavelengths thick can store a three-dimensional interference pattern; as we see in the following sections this presents additional possibilities in colour holography and in high-density data storage. Holograms can also be distinguished by the angle between the object beam and the reference beam (Figure 14.7). For a thin hologram where the angle is small (a few degrees),

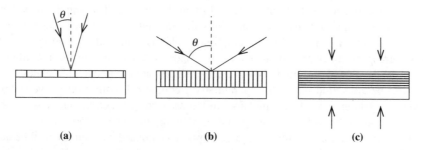

Figure 14.7   Variation of fringe spacing with illumination angle. (a) Thin hologram, small θ, fringe spacing large compared with emulsion thickness. (b) Thick hologram, intermediate θ, fringe spacing small compared with emulsion thickness. (c) Reflection hologram; the fringes are nearly parallel to the surface of the emulsion

the fringe spacing is about the same size as the emulsion thickness (typically 5–15 μm). Diffraction by a thin hologram is described by the diffraction equation. For a larger angle between the object and reference beam, the fringe spacing is small compared with the emulsion thickness.

A Fourier hologram can be formed by interference between the Fourier transforms of the complex amplitude of the object and reference waves (Figure 14.8). The reconstructed image of the Fourier

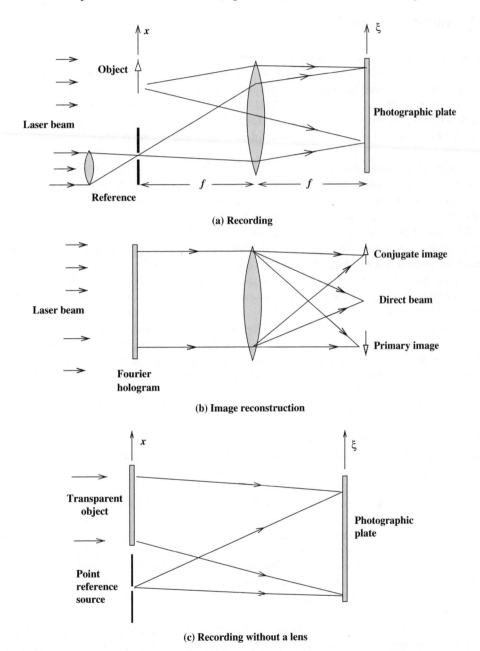

Figure 14.8 Fourier hologram: (a) recording; (b) image reconstruction; (c) recording without a lens

hologram does not move when the hologram is translated sideways. Arrangements to record a Fourier hologram and reconstruct from it are shown in Figure 14.8(a) and (b), together with an alternative lensless arrangement (c). The Fourier hologram technique is used in spatial filters and in pattern recognition.

14.7   Holography in Colour

A disadvantage of the methods described above is that both the recording and reconstruction processes demand monochromatic light, usually from a laser, both for recording and for reconstruction. Reconstructions seen through such holograms are in bright red, or whatever monochromatic laser light is used; the three-dimensional and aspect effects have been gained at the expense of unreality of colour. If white light is used to illuminate such a hologram no reconstruction is seen at all, as an infinite number of overlapping and different-sized images are produced by each wavelength present.

One approach to colour holography is to record three holograms simultaneously, e.g. as volume reflection holograms, using three differently coloured lasers, each with its own reference beams. Reconstruction then needs the same three reference beams; each produces its own set of unwanted beams as well as the required image, but in practice the system is too complicated for common use.

A three-colour hologram may also be illuminated with white light, out of which wavelength bands at the three reference wavelengths are selected by Bragg reflection for reproduction. This technique is based on the historic work of Lippmann in 1891 (Section 4.6) and of Bragg in 1912 (Section 11.11). It will be recalled that Lippmann demonstrated the existence of standing waves close to a reflecting surface by showing that a thick photographic emulsion on top of a mirror was darkened in layers corresponding to maxima of the interference pattern between the direct and reflected waves. Such a plate serves as a selective reflector of light of the wavelength in which it was made, for only at that wavelength do the reflections from the different layers in depth add constructively. Similarly, Bragg's work on crystals shows how a three-dimensional structure can single out a particular direction or directions and reflect a monochromatic beam to it selectively. This effect was the basis of Lippmann's process of colour photography.

The holograms we have considered so far have been essentially flat two- dimensional patterns. To make a three-dimensional hologram with sufficient structure in depth two main changes are necessary. First, a thick emulsion (up to a few millimetres, much thicker than the fringe spacing) is used for making the hologram; throughout its depth the film is transparent except where it has been blackened by an interference maximum during exposure. Second, the angle between the reference beam and the scattered light from the object is made large (Figure 14.7(c)). In the most extreme case of this the angle is made almost $180°$, so that the scattered light and the reference beam arrive at the photographic plate from opposite sides. In equation (14.1) the angle α becomes $90°$ and an interference structure of the order of $\lambda/2$ in depth is produced, with interference fringes parallel to the emulsion surface. The hologram is reconstructed in reflection rather than transmission.

Such holograms when illuminated by diffused white light from say a tungsten filament or quartz halogen lamp will only transmit light of the right colour which is going in the appropriate direction. To produce exact full-colour holograms, it is necessary to illuminate the object with red, green and blue light from three separate lasers, three corresponding reference beams being used. However, when illuminated with an ordinary white light source this hologram produces a realistic three-dimensional coloured reconstruction. This is called a reflection or white light hologram. The planes of the interference fringes act like Bragg planes in X-ray crystal diffraction and select the reflected wavelength. A practical technique to record the hologram is to use part of the beam transmitted by the photographic emulsion to illuminate the object.

14.8   The Rainbow Hologram

The familiar holograms impressed on plastic surfaces work passably well in white light, although they are only two dimensional and cannot use the Bragg reflection principle. The rainbow hologram is a transmission hologram which reconstructs a bright, sharp monochromatic image when illuminated with white light. This is achieved at the cost of some loss of function. When looking at such a hologram, a sideways movement of the viewpoint shows the normal parallax effect of a three-dimensional image, but a movement up and down does not; instead the colour of the image changes, as though the eye is exploring across the colours of a rainbow. One dimension of geometric reality has been sacrificed and replaced by a colour dispersion.

Although reproducing such holograms is a simple matter of impressing a pattern on a plastic surface, the initial construction is complex and involves two stages of holography. Figure 14.9 shows the two processes. In the first stage a normal hologram is made, as in (a), and then illuminated by the reference beam from the opposite side, as in (b), so producing a real image.

A screen with a narrow horizontal slit about 1 cm long is then placed over the hologram as in (c), so that the vertical extent of the hologram is insufficient to give a parallax effect in the vertical direction. The second stage of recording (d) is made with a photographic plate located close to the real image, and illuminated by a new reference beam which is inclined in the vertical plane. This second reference beam is shown in (d) converging on a focus which will be the position of the white light source in the reconstruction. (These two steps may be combined into one in a more complex system.) Finally, the photographic amplitude hologram must be converted into a surface phase hologram suitable for bulk reproduction.

When the hologram is viewed with illumination by a monochromatic source, as in Figure 14.10, the two steps produce two images, one of the object and the other of the slit. The vertical position of the slit image is wavelength dependent, with the effect that a vertical movement of the eye traverses a "rainbow" spectrum, so that the colour of the object depends on the eye position. The images formed by the rainbow hologram are bright since all the light falling on the hologram is used to form the image.

14.9   Holography of Moving Objects

Here we come to another interesting technical challenge in holography. The process of forming the hologram depends on the phase differences between the reference beam and the scattered light from the object remaining constant within a few degrees during the exposure. Clearly the object must not move more than a fraction of a wavelength. Any larger movement will not just cause blurring of the reconstruction; there will be nothing to reconstruct. A consideration of a simple case is helpful. Suppose we were making a diffraction grating by allowing two coherent beams of light to meet at an angle and interfere at a photographic plate. Then obviously a movement of a wavelength or so of the source of one of the beams would move the interference fringes on the photographic plate so we would get no grating at all. The solution is to use a very short exposure time.

Happily lasers, which are universal source of light for holographic recording, can produce astonishingly short pulses. It is instructive to calculate how short an exposure is needed. If we

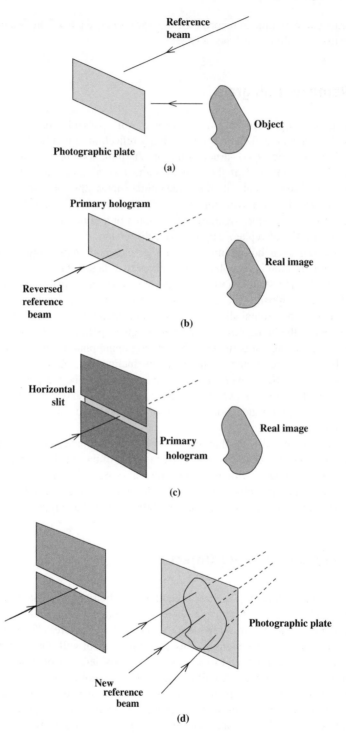

Figure 14.9 Steps in the production of a rainbow hologram. (a) Recording the primary hologram. (b) Projecting the real image. (c) Real image with no vertical parallax. (d) Recording the final hologram

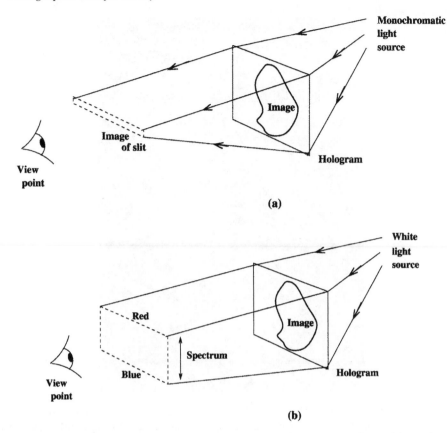

Figure 14.10 Rainbow hologram: image reconstruction. (a) Reconstruction with a laser source. (b) Reconstruction with a white light source

take it that for human scenes we need to record objects moving at up to $10\,\mathrm{m\,s^{-1}}$ the exposure must be so short that movement of only $\lambda/10$ (say) happens in that time. If $\lambda = 5 \times 10^{-7}\mathrm{m}$ the exposure must last only for $5 \times 10^{-9}\mathrm{s}$; 5 nanoseconds is a short time for conventional photography: light itself only moves 1.5 m in that time. A laser pulse can, however, be much shorter than 1 nanosecond (the shortest is less than a femtosecond), and repetitive pulses are easily obtained. A series of separate holograms can often resolve fine details of an object's motion, but in the next section we turn to a more powerful method of detecting movement.

14.10 Holographic Interferometry

The sensitivity of holography to small movements can be turned to advantage in measuring small physical displacements within an object, due for example to vibration, thermal expansion, distortion or stress. In the reconstruction of a holographic image the object is normally removed. If instead it is replaced in the same position it will appear superposed on its image, so that light from any point will originate from the laser and reach the eye by two routes, directly and via the hologram. These will interfere, so revealing any movement of the object between the recording and the reconstruction. In

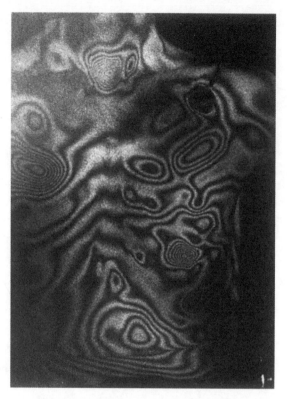

Figure 14.11   Holographic interferometry of a human torso, showing surface movement due to the action of the beating heart. The movement in 70 ms is recorded by superposing two holographic exposures. (Hans Bjelkhagen, De Montfort University)

this way holographic interferometry can measure displacements or distortions of objects within a small fraction of a wavelength. Alternatively, in double exposure holographic interferometry two holograms are recorded on the same photographic plate, e.g. without stress and then under stress. The superposition of the two holograms will create fringes if the dimensions or position of the object have changed between the two exposures. An example is shown in Figure 14.11.

The conversion of the phase difference between the two light waves into visible interference fringes can be followed using the notation of Section 14.3. In the first and second exposures on the photographic plate the irradiances are

$$I_1(x,y) = |E_0 + E_R|^2$$
$$I_2(x,y) = |E_0' + E_R|^2. \qquad (14.8)$$

The amplitude transmittance of the hologram is

$$T(x,y) = T_0 - K(I_1 + I_2). \qquad (14.9)$$

When the hologram is illuminated with the same reference beam, the transmitted amplitude of the hologram is

$$E_{\text{read}}(x,y) = E_R(x,y)T(x,y). \qquad (14.10)$$

Retaining only the term which corresponds to the superimposed primary images, this has a complex amplitude

$$E_{\text{read}}(x,y) = KTE_{\text{R}}^2|E_0(x,y)|[\exp(-i\phi) + \exp(-i\phi')]. \tag{14.11}$$

The resultant irradiance is

$$I(x,y) \propto |E_0(x,y)|^2\{1 + \cos[\phi(x,y) - \phi'(x,y)]\}. \tag{14.12}$$

The movement of the object between the two exposures has been recorded as the phase change $\Delta\phi(x,y) = \phi(x,y) - \phi'(x,y)$. Then a bright fringe is observed whenever $\Delta\phi(x,y) = p.2\pi$, where p is an integer. The two interfering waves are reconstructed in exact register with each other, so that the positioning of the doubly exposed hologram is not critical. Since the two waves have the same amplitude the fringes have high visibility.

Dynamic effects, such as small but rapid vibrations of mechanical components, can be followed by an electronic TV camera rather than a photographic plate. The object beam is imaged onto the camera detector, together with a reference beam from the same laser source. These combine to form a speckle pattern (see Chapter 16) which can be scanned and recorded at 25 frames per second or even faster. Any small movement of the object is immediately obvious as a movement of the speckles. This technique is known as *electronic speckle pattern interferometry*.

14.11   Holographic Optical Elements

Holographic optical elements (HOE) are optical components produced using holographic techniques. Diffraction gratings made by the holographic technique of interfering two laser beams in a photographic emulsion may be a simple amplitude grating in a thin film of emulsion, or they may be three dimensional, using a thick emulsion; they may also be phase-changing rather than amplitude gratings. More generally, a hologram may be regarded as an optical component which will modify a light wavefront in ways which are usually associated with conventional components such as lenses, spatial filters, beam splitters and optical connections used in microelectronic systems.

The three-dimensional grating made by interfering two plane waves behaves like a crystal in X-ray diffraction. The interference pattern in the film is a regular lattice; transmission through or reflection from the hologram follows Bragg's law. If instead one of the beams is diverging, the resulting grating will behave as a lens, since the reconstructed beam is a copy of the original beam. A plane laser beam will be focussed to a spot. Movement of the hologram causes the spot to be scanned; this is the basis of the holographic scanner. The barcode scanner used in shops uses a mosaic of such holograms with different orientations formed on a circular disc, providing a multiple scan pattern when the disc is rotated and with each scan line focussed to a different position in space.

Holographic optical elements have several valuable advantages over bulk optical components. They can be made with large aperture on thin, light substrates and several elements can be made on the same hologram. Synthetic computer-generated holograms are able to be produced which can produce wavefronts with any required amplitude and phase distribution. In analogy to the off-axis holographic recording, the complex amplitudes of an object wave and a reference wave are computed, superimposed and the resultant square modulus calculated. This is then used to produce a transparency to act as a hologram. Holographic video imaging is being developed in which computer-generated holograms are able to produce real-time holographic three-dimensional displays.

14.12 Holographic Data Storage

A holographic image stored in a thick recording medium (a volume image) may be reconstructed by a laser beam at the same angle as the reference beam used in the recording. At this angle the condition for Bragg reflection is satisfied, but at other angles no reconstruction takes place. This allows many holograms to be superposed in the same volume of recording medium, each able to be accessed by its own particular reference beam angle, or by its own wavelength at a particular angle. A very large amount of information can be stored in this way, which is the basis of holographic data storage and holographic memories.

A high-capacity holographic memory must be transparent through many wavelengths' thickness of recording medium. This makes an amplitude grating unsuitable, and phase grating techniques must be used. Phase grating volume holograms are based on photorefractive crystals or polymers, in which the refractive index is altered by a pattern of space charge formed by photoexcited electrons. The process is reversible; the grating may be erased by illuminating the grating uniformly, so that the same material can be used as an optically rewritable memory. The potential performance is phenomenal: data may be stored at a density of 10^{11} cm^{-3} (100 Gbit cm^{-3}) and may be accessed in less than 100 microseconds, or transferred at a rate of 10^9 bit s^{-1}. Further developments have been made to make volume holograms of the display type which produce large-scale ($1\,\text{m} \times 1\,\text{m}$) images, in full colour and with full parallax.

Problem 14.1

The plane wave beam from an He–Ne laser ($\lambda = 633$ nm) is split into two beams which symmetrically illuminate a photographic plate. A hologram is recorded when the object and reference beams make angles of $+30°$ and $-30°$ with the normal to the photographic plate. Calculate the spatial frequency (i.e. the inverse of the spatial wavelength) of the fringes.

Problem 14.2

The depth of modulation (or visibility) of holographic fringes is defined in terms of the maximum and minimum irradiances I_{max} and I_{min} on the photographic plate as $(I_{max} - I_{min})/(I_{max} + I_{min})$. For reference and object wave irradiances I_R and I_O, determine the depth of modulation for the cases: (a) $I_R = 2I_O$, (b) $I_R = 4I_O$ and (c) $I_R = 10I_O$. Comment on the suitable values for I_R/I_O.

Problem 14.3

The resolution of a photographic plate or film R_p is a measure of the finest fringes that can be recorded, e.g. in units of lines per mm. Deduce in off-axis holography the largest angle that can be recorded by a film with resolution R_p.

Problem 14.4

The photographic emulsion is not linear in its response under conditions of low and high irradiances, as indicated by the characteristic response curve (the HD curve; see Section 20.7). How may the effects of this factor be minimized in holographic recording?

Problem 14.5

In a thick hologram the interference fringes exist throughout the thickness of the emulsion. In reconstruction a diffracted wave may then interact with more than one fringe. Consider a holographic arrangement in which the object and reference waves are incident at equal angles onto the holographic plate. Show that a condition for the plate to act as a thick plate is when the emulsion thickness l is such that $l > 2nd^2/\lambda$, where n is the emulsion refractive index and d is the spacing of fringes formed between the object and reference waves.

Problem 14.6

In a holographic arrangement an object is centred perpendicular to the recording film, at which it subtends an angle θ. The film receives light from all points in the object, which then interfere at the film. (a) What is the smallest fringe spacing (highest spatial frequency) for this arrangement? (b) The film is illuminated by a reference beam at an angle of ϕ. In reconstruction with the original reference beam, what value of ϕ in recording is required to avoid overlap of the object beam and the reconstruction waves?

Problem 14.7

In holography the reconstructed real image is pseudoscopic; for example, the structure in an object will appear with depth inverted. Explain this effect, and suggest a method to enable the real image to be observed as an orthoscopic (i.e. non-pseudoscopic) image.

15 Lasers

...there are certain situations in which the peculiarities of quantum mechanics can come out in a special way on a large scale.

The Feynman Lectures on Physics, vol. III, p. 21–1.

Lasers, the outcome of elegant physical theory and extensive experimentation, have become a vitally important tool in contemporary research in physics and chemistry, and indeed in all branches of science. Lasers are also used extensively in everyday life, from reading barcodes to playing CD recordings, and in technology, where they have many diverse uses such as optical communication and processing materials, and for many types of measurement.

In this chapter we set out the fundamentals of laser action.[1] The laser produces light in a significantly different way from normal light sources. The essential process of *stimulated emission* is considered along with absorption and spontaneous emission. This leads to the Einstein relations between the rate coefficients for these processes. The creation of *population inversion* is seen to produce *optical gain*. In most lasers the laser medium is inside an *optical resonator* to enhance gain by providing a long path length. We look at some of the properties of these resonators and their influence on the laser radiation they emit.

We describe some of the main types of laser, leaving the all-important semiconductor lasers to a separate chapter. The many types of laser have similar operating principles of population inversion, gain, feedback and threshold, but they differ greatly in their characteristics. Many lasers are table-top size, others are the size of a large room or a building, while the semiconductor laser has submillimetre dimensions. The special characteristics of laser light, such as monochromaticity and directionality, which depend on its high degree of coherence, will be described in Chapter 16, which also deals with the tuning of lasers, the conversion of the wavelength of laser light by non-linear optical techniques and the generation of laser pulses of ultrashort duration.

15.1 Stimulated Emission

Before the invention of the laser, the available sources of light were essentially either thermal, such as from a tungsten filament lamp, or spontaneous emission from atoms and molecules, as in a gas

---

[1]The acronym laser stands for Light Amplification by Stimulated Emission of Radiation.

Optics and Photonics: An Introduction, Second Edition   F. Graham Smith, Terry A. King and Dan Wilkins
© 2007 John Wiley & Sons, Ltd

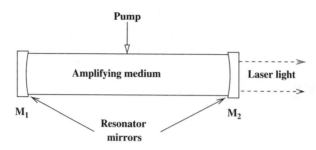

Figure 15.1 The basic elements of a laser: amplifying medium, pumping energy source and resonator $M_1 M_2$

discharge; in either case, their brightness was limited by the temperature of the emitter. The broadband white light of solar radiation, for example, is limited in brightness by the temperature of the photosphere, while the brightness of the solar spectral lines is limited by the ambient gas temperature in the chromosphere or corona. Light from a thermal source is *incoherent*; it is the chaotic sum of a disorderly outpouring of photons from individual atoms, radiating at random without any relation to one another. In a laser, however, the emission from individual atoms is synchronized, giving *coherent* radiation with very much higher brightness. *Stimulated emission*, a concept introduced by Einstein in 1916, is the source of the synchronization.

The first use of stimulated emission to achieve a high brightness was in the microwave spectrum; historically the *maser* was developed several years before the laser. In 1953 Gordon, Zeiger and Townes[2] demonstrated stimulated emission between the two lowest levels of the ammonia molecule, giving a very narrow emission line at a wavelength of 12.6 millimetres. For this achievement, Townes shared the 1964 Nobel Prize in Physics with N. Basov and A. Prokhorov of the USSR. The first laser, originally called the optical maser, followed in 1960, when T. H. Maiman produced red light at wavelength 694.3 nm from the chromium ions in a ruby crystal. Stimulated emission in the maser and laser is the essential effect causing emission from excited atoms of coherent radiation that adds precisely in phase and with the same direction and polarization. Three components are needed to achieve this in a laser (Figure 15.1): an active medium with suitable energy levels, the injection of energy so as to provide an excess of atoms in an excited state, and (in most cases) a resonator system in which multiple reflections allow the build-up of the coherent laser light.

We have already introduced in Section 1.7 the three elemental quantum processes of light–matter interaction: absorption, spontaneous emission and stimulated emission. All three play a role in the laser. Consider, for example, the three-level laser shown in Figure 15.2. Here the atoms in the active medium have three energy levels involved in the laser action. Absorption raises the energy from level 1 to level 3 (this process is called *pumping*), spontaneous emission (or a non-radiative transition) reduces the energy to level 2, which is a metastable state, and stimulated emission occurs between levels 2 and 1. The accumulation of excited atoms in the metastable state results in an overpopulation, or *population inversion*, in relation to the ground state. Stimulated emission leads to the rapid release of this accumulated energy; one photon arrives at the excited atom, and two leave, with the same energy, travelling together and in phase. The stimulated photon has the same momentum as the incident photon, and hence travels in the same direction. Both photons can then repeat the process at other excited atoms, and the resulting chain reaction causes the light wave to grow exponentially.

[2]J.P. Gordon, H.J. Zeiger and C.H. Townes, *Physical Review*, **95**, 282, 1954.

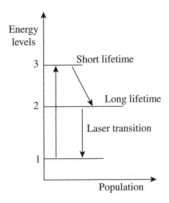

Figure 15.2 Energy levels and the level populations in a three-level laser

The ruby laser is an example of a three-level laser in which the active species is the Cr^{3+} ion rather than a neutral atom.

One further element is needed to make such an amplifier into a self-excited oscillator; the light must be fed back into the laser material. This is achieved by enclosing the lasing material between mirrors, forming a resonant cavity. Emission from the device is obtained by arranging that one of the resonator mirrors has a non-zero *transmittance*.

15.2 Pumping: The Energy Source

As shown in Figure 15.2 the energy which is converted into laser light is injected, or *pumped*, into the laser at a higher photon energy $h\nu_{31}$ than the laser output photons with energy $h\nu_{21}$. The excited atoms (or ions) then lose energy $h\nu_{32}$, falling into the intermediate level 2 which has a longer lifetime. Atoms accumulate in this metastable state, and are available for the stimulated emission process.

The original ruby laser was pumped by an intense flash of white light, which is selectively absorbed by chromium ions dispersed through the aluminium oxide crystal. Only a small part of the energy in the white light is at the right wavelength to be absorbed and produce the population inversion; this is inefficient, which is the reason for the use of an intense source of light. Other types of laser use more finely tuned pumping systems; the very common He–Ne gas laser provides a good example.

The He–Ne laser contains a mixture of the two gases in an electrical discharge tube. Both gases are excited and ionized in the discharge. The amplifying medium is neon, which is pumped into a state of population inversion by collision with excited helium atoms; these in turn have been energized by electron collisions in the discharge. The energy transfer between the two species of gas atoms is very efficient because of a close coincidence between energy levels in the excited helium and the upper levels suited for the laser action in neon. Figure 15.3 shows the outline of the He–Ne laser, and the energy levels involved. The coincidence is between the two metastable levels 2^1S_0 and 2^3S_1 in helium and the two metastable levels 5s and 4s in neon.[3] Stimulated emission from the 5s and 4s levels can be

---

[3]The He levels are described by Russell–Saunders coupling, while for Ne the levels are designated by their electron configuration as $(1s^2 2s^2 2p^5)3s$, ( )4s, ( )5s, etc.; note that in the older Paschen notation the ( )3s configuration is designated 1s and ( )4s is designated 2s, and so on.

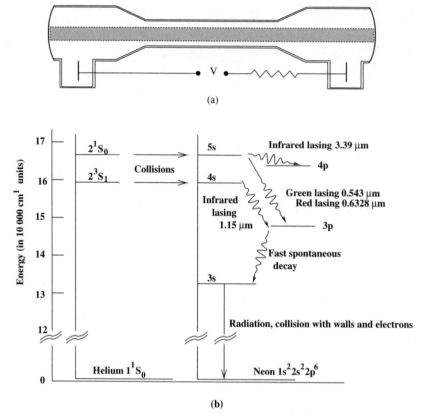

(a)

(b)

Figure 15.3 The helium–neon laser. (a) Laser excited by a d.c. electrical discharge, with potential *V*. (b) Simplified energy level diagram

through transitions to several different energy levels, allowing laser action at 3.39 μm, 1.15 μm, 632.8 nm and 543.5 nm. The familiar red beam of the He–Ne laser is operating on the 632.8 nm transition. Figure 15.3 shows the mirrors which enclose the laser, forming a resonator. As will appear later, a particular laser wavelength can then be selected by a choice of resonator system.

15.3   Absorption and Emission of Radiation

We now review the basic theory of the three processes involved in the interaction of radiation and matter, which we introduced briefly in Section 1.7. The processes of absorption, spontaneous emission and stimulated emission are sketched in Figure 15.4. We suppose that the two states, with energies E_1 and E_2, are populated with number densities n_1 and n_2. Absorption occurs when radiation of frequency $v = (E_2 - E_1)/h$ is incident on the medium, with excitation from the ground state to the excited state. The rate of absorption in which atoms are raised from level 1 to level 2 is

$$\left(\frac{dn_1}{dt}\right)_{ab} = -B_{12}n_1 u(v) \tag{15.1}$$

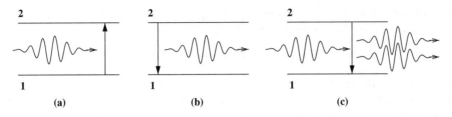

Figure 15.4   Absorption, spontaneous emission and stimulated emission

where n_1 is the population per unit volume in level 1 and $u(v)$ is the energy density of the incident field (units of energy per unit volume per unit frequency interval, $\mathrm{J\,m^{-3}\,Hz^{-1}}$). $u(v)$ is a function of the frequency v of the radiation field. B_{12} is the *Einstein absorption coefficient*, which is a constant characteristic of the pair of energy levels in the particular type of atom.

Spontaneous emission of a photon occurs with transition of the electron from the excited level 2 to the ground level 1 with the emitted photon energy $hv = E_2 - E_1$. The rate of decrease of the population n_2 by spontaneous emission is

$$-\left(\frac{\mathrm{d}n_2}{\mathrm{d}t}\right)_{\mathrm{spon}} = A_{21}n_2. \tag{15.2}$$

The constant A_{21} (unit: $\mathrm{s^{-1}}$) is related to the spontaneous radiative lifetime τ of the excited state as

$$A_{21} = \frac{1}{\tau}. \tag{15.3}$$

In stimulated emission atoms in level 2 are stimulated to make a transition to level 1 by the radiation field itself. The rate at which the transition occurs is proportional to the number of atoms in level 2 and the energy density of the radiation field:

$$\left(\frac{\mathrm{d}n_2}{\mathrm{d}t}\right)_{\mathrm{stim}} = -B_{21}n_2u(v). \tag{15.4}$$

The constant B_{21} is the Einstein coefficient for stimulated emission from energy level 2 to level 1. Note that the rate of stimulated emission is proportional to the energy density at the resonant frequency $v = (E_1 - E_2)/h$, so that for high levels of radiation energy density stimulated emission dominates spontaneous emission.

The rate of change of population in level 2 is the sum of the effects of spontaneous and stimulated transitions given by equations (15.1), (15.2) and (15.4), which yields the rate equation

$$\frac{\mathrm{d}n_2}{\mathrm{d}t} = -B_{21}u(v)n_2 + B_{12}u(v)n_1 - A_{21}n_2. \tag{15.5}$$

Conservation of atoms implies that the ground-state population density obeys $\mathrm{d}n_1/\mathrm{d}t = -\mathrm{d}n_2/\mathrm{d}t$.

The relation between the three Einstein coefficients is found by considering an equilibrium situation, where a collection of atoms within a cavity is in thermal equilibrium with a radiation field. Then the populations of the two levels n_1 and n_2 are constant

$$\frac{\mathrm{d}n_2}{\mathrm{d}t} = \frac{\mathrm{d}n_1}{\mathrm{d}t} = 0. \tag{15.6}$$

In thermal equilibrium, when there is detailed balancing between the processes acting to populate and depopulate the energy levels, setting $dn_2/dt = 0$ we obtain

$$A_{21}n_2 + B_{21}u(v)n_2 = B_{12}u(v)n_1 \tag{15.7}$$

giving the relation between the values of n_1, n_2 and $u(v)$ at thermal equilibrium.

We now use two fundamental laws relating $u(v)$ and the relative populations n_1, n_2 to the temperature T. These are *Planck's radiation law* for cavity radiation (Section 5.7)

$$u(v) = \frac{8\pi h v^3}{c^3} \frac{1}{\exp(hv/kT) - 1} \tag{15.8}$$

and the *Boltzmann distribution* of atoms between the two energy levels:

$$\frac{n_2}{n_1} = \frac{g_2}{g_1}\exp\left(-\frac{E_2 - E_1}{kT}\right) = \frac{g_2}{g_1}\exp\left(-\frac{hv}{kT}\right). \tag{15.9}$$

Here we have allowed for the possibility that either level is degenerate, i.e. that for the jth level, there are $g_j(= 1, 2, 3...)$ quantum states with the same energy ($g_j = 1$ is the non-degenerate case). From equations (15.7) and (15.9)

$$u(v) = \frac{A_{21}}{B_{12}(n_1/n_2) - B_{21}} \tag{15.10}$$

or

$$u(v) = \frac{A_{21}}{B_{12}(g_1/g_2)\exp(hv/kT) - B_{21}}. \tag{15.11}$$

This equation may be combined with equation (15.8) to give

$$\frac{A_{21}}{(g_1/g_2)B_{12}\exp(hv/kT) - B_{21}} = \frac{8\pi h v^3}{c^3}\left(\frac{1}{\exp(hv/kT) - 1}\right). \tag{15.12}$$

Equation (15.12) is satisfied when

$$\boxed{\begin{aligned} \frac{A_{21}}{B_{21}} &= \frac{8\pi h v^3}{c^3} \\ g_1 B_{12} &= g_2 B_{21}. \end{aligned}} \tag{15.13} \\ \tag{15.14}$$

These are the required relations between the three Einstein coefficients (see also Problem 15.3).

These equations and the concept of transition probability are fundamental to the theory of exchange of energy between matter and radiation.

The crucial factor for lasers is the ratio between the rates of stimulated and spontaneous emission:

$$\frac{\text{rate of stimulated emissions}}{\text{rate of spontaneous emissions}} = \frac{B_{21}u(v)}{A_{21}} = \frac{1}{(\exp(hv/kT) - 1)}. \tag{15.15}$$

For lasing to be feasible, this ratio should be much greater than 1. In that case, stimulated emission dominates spontaneous emission, and the latter is less able to erode away a population inversion before lasing can occur.

Example. Assuming thermal equilibrium at room temperature ($T = 300$ K), evaluate the ratio of equation (15.15) for $\lambda = 600$ nm (visible) and $\lambda = 1$ cm (microwave).

Solution. In general, if $T = 300$ K, and we measure λ in μm, $[\exp(h\nu/kT) - 1]^{-1} = [\exp(48/\lambda_{\mu m}) - 1]^{-1}$. Hence

$$\frac{\text{stimulated rate}}{\text{spontaneous rate}} \approx \exp(-80) \simeq 10^{-35} \quad (\lambda = 0.6\,\mu m) \tag{15.16}$$

and 200 at $\lambda = 1$ cm.

The factor $\exp(h\nu/kT)$ shows us that in thermal equilibrium stimulated emission is very unlikely at optical frequencies, and explains why the first successful device was the maser, operating at a much lower radio frequency. It is not surprising then to learn that all lasers developed until now operate with radiation that is far from thermal equilibrium.

We also note that since the stimulated emission rate is $n_2 B_{21} u(\nu)$ we may increase the rate by increasing $u(\nu)$, which is achieved in a resonant cavity, and by increasing n_2 in the population inversion resulting from pumping.

15.4 Laser Gain

We can now consider the growth of a light wave as it passes through an active laser medium, and find the conditions for the wave to grow by stimulated emission. The resulting fractional rate of growth, in equation (15.25) below, depends on four factors: the population inversion, the spectral lineshape, the frequency and the transition probability A_{21}. We start by finding the emission and absorption in a small element dz of the path through the laser medium, and then integrating over the whole path, which may involve many to-and-fro reflections in a resonator.

First we look at the attenuation of an *absorbing* medium in which a plane wave of monochromatic radiation is travelling as illustrated in Figure 15.5. The reduction in irradiance (power flow across unit area) as the wave travels from position z to $z + dz$ for a uniform medium is proportional to the magnitude of the irradiance and the distance travelled:

$$dI(z) = I(z + dz) - I(z) = -\alpha I(z)dz. \tag{15.17}$$

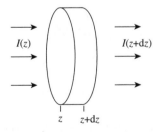

Figure 15.5 Attenuation of a wave in a slab of material

Here α is the *absorption coefficient*. Hence

$$\frac{\mathrm{d}I(z)}{\mathrm{d}z} = -\alpha I. \tag{15.18}$$

On integration

$$I(z) = I_0 \exp(-\alpha z), \tag{15.19}$$

where I_0 is the irradiance of the incident beam. This represents exponential attenuation.

If the number of stimulated emissions exceeds the number of absorptions, rather than being attenuated the wave will grow. The number of stimulated emissions depends on the energy density $u(v)$. The irradiance is the product of the energy density and the velocity, so that in free space or a thin gas

$$u(v) = \frac{I}{c}. \tag{15.20}$$

The change in irradiance $\mathrm{d}I$ of the wave in travelling a distance $\mathrm{d}z$ is now proportional to the difference between the numbers of stimulated emissions and absorptions:

$$\mathrm{d}I = \left[n_2 B_{21} g(v)\frac{I}{c} - n_1 B_{12} g(v)\frac{I}{c} \right] hv \mathrm{d}z. \tag{15.21}$$

Here we have introduced the normalized spectral function, or lineshape $g(v)$ for the transition, which describes the frequency spectrum of the spontaneously emitted radiation. The lineshape is dependent on the mechanism determining the broadening of the transition, as described in Chapter 12 and Appendix 4. In gas lasers inhomogeneous broadening usually dominates due to the thermal motion of the atoms o⁺ ions. Inhomogeneous broadening also applies to transitions in doped glasses where variations in the sites of the doped ions lead to a distribution of centre frequencies. A typical inhomogeneously broadened lineshape is shown in Figure 15.6.

The normalization of the function $g(v)$ is such that

$$\int_0^\infty g(v)\mathrm{d}v = 1. \tag{15.22}$$

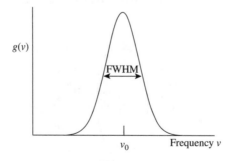

Figure 15.6   A typical inhomogeneously broadened Gaussian lineshape function $g(v)$ showing the full width at half maximum (FWHM)

From equation (15.21), and the Einstein relations (15.13) and (15.14),

$$\frac{dI}{dz} = \left(n_2 - \frac{g_2}{g_1} n_1 \right) \frac{c^2 A_{21}}{8\pi v^2} g(v) I. \tag{15.23}$$

Integrating gives an exponential dependence on distance z

$$I = I_0 \exp(\gamma(v) z) \tag{15.24}$$

where I_0 is the irradiance at $z = 0$, and $\gamma(v)$ is the *gain coefficient*:

$$\gamma(v) = \left(n_2 - \frac{g_2}{g_1} n_1 \right) \frac{c^2 A_{21}}{8\pi v^2} g(v). \tag{15.25}$$

If $n_2 > (g_2/g_1)n_1$, representing *population inversion*, then $\gamma(v) > 0$, and the irradiance grows exponentially with distance in the medium. The gain coefficient depends, as expected, on the transition probability A_{21} and on the lineshape. Note, however, that the frequency dependence (v^{-2}) indicates it is more difficult to make lasers for ultraviolet light than for infrared.

In comparing the suitability of different laser media it is convenient to specify a *stimulated emission cross-section parameter* $\sigma(v)$, which is related to the gain coefficient $\gamma(v)$ by

$$\gamma(v) = \left(n_2 - \frac{g_2}{g_1} n_1 \right) \sigma(v). \tag{15.26}$$

From equation (15.25)

$$\sigma(v) = \frac{c^2 A_{21} g(v)}{8\pi v^2}. \tag{15.27}$$

Since the lineshape $g(v)$ is normalized (equation (15.22)), the central height of the line $g(v_0)$ is inversely proportional to the linewidth,[4] and to a useful approximation $g(v_0) \sim 1/\Delta v$. For the lineshape of homogeneous broadening (see Section 12.2 and Appendix 4)

$$g(v_0) = \frac{2}{\pi \Delta v}. \tag{15.28}$$

Then the cross-section parameter at the peak frequency becomes

$$\sigma_0 = \sigma(v_0) = \frac{c^2 A_{21}}{4\pi^2 v_0^2 \Delta v}. \tag{15.29}$$

This shows that the stimulated emission cross-section for a homogeneously broadened transition is proportional to the ratio $A_{21}/\Delta v$, the spontaneous transition rate over the linewidth. (In liquid and solid state lasers the higher refractive index n of the medium compared with a gas means that the light speed c should be replaced by c/n.)

---

[4]The linewidth here is the full width at half maximum, or FWHM.

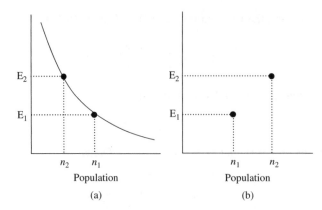

Figure 15.7 Population inversion. The normal Boltzmann distribution (a) of population in two energy levels is shown inverted in (b). (Here we assume $g_2/g_1 = 1$)

15.5 Population Inversion

The population inversion condition $n_2 > (g_2/g_1)n_1$ derived in Section 15.4 is a necessary condition for the gain coefficient to be positive. The two cases of thermal equilibrium and population inversion are shown in Figure 15.7.

To create population inversion, energy is required to be put selectively into the laser medium such that the population of level 2 is increased over level 1 to form a non-equilibrium distribution. Excitation of the laser medium by pumping may be achieved in several ways. In gases at normal pressures, the absorption lines have a narrow bandwidth, which limits their ability to absorb light, and pumping is usually by electron collisions in an electrical discharge. Solid state crystals and glasses doped with an active ion have broader absorption lines than gases and are usually excited optically by absorption of energy from a lamp or from another laser. In semiconductor lasers (Chapter 17), the relevant energy levels correspond to the conduction and valence bands, which are comparatively very broad. Here pumping is achieved by applying an electric field across the semiconductor junction.

Lasers may conveniently be divided into three- and four-level systems depending on the number of levels active in their operation. This is illustrated in Figure 15.8 which shows the inverted population at the laser transition.

15.6 Threshold Gain Coefficient

Laser oscillation is initiated in a system with population inversion by the spontaneous emission of a photon along the axis of the laser. For the laser to sustain oscillation the gain in the laser medium must be greater than the losses in the cavity. The losses arise from transmission at the cavity mirrors (in order to provide the laser output, a typical transmission is 5% for continuous laser operation). Other losses arise from absorption and scattering by the mirrors and in the laser medium, and diffraction out of the sides of the cavity. The threshold for laser oscillation will occur when the gain is equal to the losses. To calculate this threshold gain we combine all the sources of loss into one

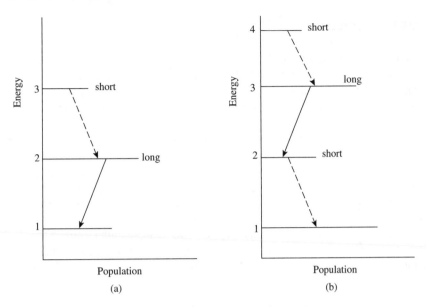

Figure 15.8   Three- and four-level laser schemes

lumped loss coefficient k. At threshold the irradiance neither decreases nor increases; it stays constant.

Consider a cavity made up of mirrors M_1 and M_2 with reflectances R_1 and R_2 and spaced by a distance L. A beam of irradiance I_0 starting at M_1 on reaching M_2 has become $I_1 = I_0 \exp[(\gamma - k)L]$, where γ and k are the gain and loss coefficients. On reflection from M_2 and travelling in return through the medium and undergoing reflection at M_1, the irradiance becomes $I_2 = I_0 R_1 R_2 \exp[2(\gamma - k)L]$. The round-trip gain, G, is defined as I_2/I_0. Then

$$G = I_2/I_0 = R_1 R_2 \exp[2(\gamma - k)L]. \tag{15.30}$$

The threshold condition for laser oscillation is $G = 1$, giving

$$R_1 R_2 \exp[2(\gamma_{th} - k)L] = 1 \tag{15.31}$$

where γ_{th} is the threshold gain coefficient, at which the laser will begin to oscillate.

From equation (15.31) we find

$$\gamma_{th} = k + \frac{1}{2L} \ln\left(\frac{1}{R_1 R_2}\right). \tag{15.32}$$

The first term is the loss within the cavity, and the second term is the loss due to the mirror transmission (or absorption), i.e. including that leading to the useful laser output.

Continuously operating lasers are called CW lasers, standing for continuous wave. Once a CW laser is operating in a steady state, the gain stabilizes at the threshold value, since if the gain were greater or less than unity the irradiance would increase or decrease. The level at which the irradiance stabilizes depends on the pump power.

15.7   Laser Resonators

Most lasers require a long path through the active medium to obtain sufficient overall gain. This is achieved by multiple reflections in an optical resonator, often referred to as a *resonant cavity.*

An optical resonator both increases laser action and defines the frequency at which it occurs. Optical feedback is provided by the optical resonator which retains photons inside the cavity, reflecting them back and forth through the laser medium. The simplest basic optical resonator is a pair of shaped mirrors at each end of the laser medium, as in a Fabry–Pérot interferometer. There are various configurations using plane and curved mirrors used in optical Fabry–Pérot resonators; some of these are shown in Figure 15.9. Not all configurations of mirror curvatures and spacings will give stable operation. Usually one of the mirrors is arranged to be practically 100% reflecting at the laser wavelength; the other mirror (the output mirror) has a finite transmission, so that light will be transmitted out of the optical cavity to provide the laser output.

The optical Fabry–Pérot resonator made up of two plane-parallel mirrors is similar to the Fabry–Pérot etalon or interferometer described in Chapter 8. The resonance condition for waves at normal incidence, along the axis of a cavity with optical length L, as for standing waves, is

$$m\frac{\lambda}{2} = L \tag{15.33}$$

where m is an integer. Then the resonant frequency v_m for each longitudinal mode of the cavity is

$$\boxed{v_m = m\frac{c}{2L}.} \tag{15.34}$$

This equation is important in defining the resonant frequencies at which the laser will oscillate, as it will if they fall within the gain profile of the laser transition, as illustrated in Figure 15.10. The possible oscillating frequencies are termed the *longitudinal* modes of the laser; they are spaced by

$$\Delta v = \frac{c}{2L}. \tag{15.35}$$

Each of these frequencies may, however, be broken into a more narrowly spaced set; these are due to *transverse* modes, in which the field pattern may have different structures transverse to the beam

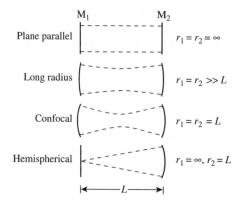

Figure 15.9   Common laser resonator configurations. r_1 and r_2 are the radii of curvature of mirrors M_1 and M_2

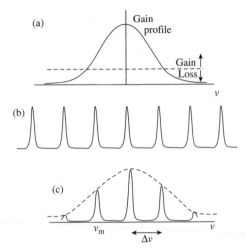

Figure 15.10 Gain profile and resonant frequencies in a cavity laser: (a) gain profile of the laser transition; (b) allowed resonances of the Fabry–Pérot cavity; (c) oscillating laser frequencies

direction. A transverse mode is an electric and magnetic field configuration at some position in the laser cavity which, on propagating one round trip in the cavity, returns to that position with the same pattern; some of these field patterns are shown in Figure 15.11.

The laser output is at one or more frequencies from this set of modes. When only one longitudinal and transverse mode is selected, in a *single mode laser* (Chapter 16), the bandwidth of the laser light is almost unbelievably small. For comparison, light from a single line of a low-pressure gas discharge lamp has a spectral width of about 1000 MHz. Non-pulsed laser light in contrast typically has a bandwidth of less than 1 MHz and may, with careful design, have a bandwidth of less than 10 Hz.

As can be seen in Figure 15.11, the transverse modes can have polar (or circular)[5] symmetry or Cartesian (rectangular) symmetry; these are known respectively as Laguerre–Gaussian modes and

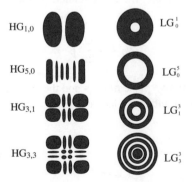

Figure 15.11 Distribution of irradiance for various transverse modes: Hermite–Gaussian (HG), where the double subscript refers to the number of nodes in the x and y directions, and the corresponding Laguerre–Gaussian (LG), where the superscript and subscript refer to cycles of azimuthal phase and the number of radial nodes respectively.

---

[5]In three dimensions, the LG modes are actually helical and carry angular momentum.

Hermite–Gaussian modes. Although most lasers are constructed with circular symmetry, the modes with Cartesian symmetry are most common; this arises when some element in the laser cavity imposes a preferred direction on the transverse electric and magnetic field vectors. The lowest order transverse electromagnetic mode (HG_{00} or LG_0^0) is labelled TEM_{00}. This is the fundamental mode with the largest scale pattern across the laser beam. The zero subscripts indicate that there are no nodes in the x and y directions, transverse to the direction of the laser beam.

The cavity mirrors are, of course, required to reflect at the laser wavelength in order to make the cavity resonant. Typically one mirror has a reflectivity as close to 100% as possible and one is arranged to have a carefully selected transmission, chosen to produce the optimum laser output power; this necessarily means that the transmission must be less than the overall laser gain. For efficient operation the deviations of the mirrors from their ideal shapes are required to be within a small fraction of the laser wavelength (usually $\sim \lambda/20$).

15.8   Beam Irradiance and Divergence

The beam of light leaving the laser is coherent in relation to both its narrow spectral linewidth and its spatial coherence over its emitted wavefront. As it leaves the laser, the beam will spread into a narrow angle by diffraction, the width depending on the field distribution across the beam. This can be viewed as the beam's cross-section acting as its own diffraction aperture. The simplest mode (the TEM_{00} mode), which has the narrowest beam, has a Gaussian radial dependence of irradiance $I(r, z)$ with peak irradiance along the axis, so that at radial distance r from the axis

$$I(r, z) = I_0 \exp\left(\frac{-2r^2}{w^2(z)}\right). \tag{15.36}$$

The radial width parameter w is referred to as a *spot size* and varies with distance along the axis. (For $r = w$ the amplitude is $1/e$ of the amplitude on-axis, but for convenience this is often referred to as the edge of the beam.) The spot size is smallest within the laser cavity, where there is a *beam waist*. Here the width w_0 (Figure 15.12) is related to the length L of the resonator and the wavelength λ as

$$w_0 = \left(\frac{\lambda L}{2\pi}\right)^{1/2}. \tag{15.37}$$

This applies for both the cavity with two plane mirrors and the symmetric confocal cavity. As we discuss below, the cavity mirrors must be significantly larger than this spot size to avoid diffraction loss.

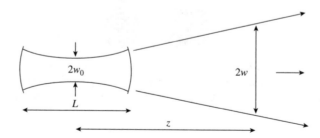

Figure 15.12   The beamwidth w_0 at the waist and at a distance z from the waist

The laser beam spreads by diffraction (Figure 15.12) both inside and outside the resonator. Analysis of the Gaussian beam solutions of the paraxial wave equation leads to the width $w(z)$ of the beam at distance z from the beam waist:

$$w(z) = w_0 \left[1 + \left(\frac{\lambda z}{\pi w_0^2} \right)^2 \right]^{1/2} \tag{15.38}$$

which approximates to

$$w(z) \simeq \frac{\lambda z}{\pi w_0} \quad \text{for} \quad z \gg \frac{\pi w_0^2}{\lambda}. \tag{15.39}$$

Note that the larger the beam waist, the smaller the angle of spread of the beam. For the TEM$_{00}$ mode, which has a Gaussian spatial profile, the half angle θ of the divergence cone for the propagating beam is

$$\theta = \frac{\lambda}{\pi w_0}. \tag{15.40}$$

As expected from Fraunhofer diffraction, the angular width is of order λ/w_0. For example, an He–Ne laser with $\lambda = 632.8\,\text{nm}$ operating with a symmetric confocal resonator of length $L = 30\,\text{cm}$ has

$$\text{minimum spot radius } w_0 = \left(\frac{\lambda L}{2\pi} \right)^{1/2} = 0.17\,\text{mm} \tag{15.41}$$

$$\text{divergence angle } \theta \simeq \frac{\lambda}{\pi w_0} \simeq 1.2\,\text{mrad} = 0.066°. \tag{15.42}$$

Note that the beamwidth w_0 is determined by the length and not the width of the laser. There is, however, a need for the resonator mirrors to be sufficiently wide, so that the beam is not lost by diffraction at each reflection. For example, consider the diffraction broadening of a beam that arrives at mirror M_2 after it reflects off M_1. Assuming initially that the beam fills mirror M_1, the diffraction half angle at mirror M_1 is $\sim \lambda/d_1$ where d_1 is the diameter of the mirror M_1 and also of the beam at M_1. If d_2 is the diameter of M_2, low loss requires $d_2 \geq d_1 + 2L\lambda/d_1$, or approximately

$$\frac{d_1 d_2}{\lambda L} > 1. \tag{15.43}$$

This is known as the Fresnel condition. For a symmetrical arrangement where $d_1 = d_2 = d$ the condition is $d^2/\lambda L > 1$; the quantity $d^2/\lambda L$ is known as the *Fresnel number* of the optical arrangement. (Note the close relationship to the Rayleigh distance (Section 10.4), which defines the boundary between Fraunhofer and Fresnel diffraction.)

The beam remains almost parallel for some distance from the laser. In equation (15.38) the width is almost constant for distances $z \leq \frac{1}{2}z_0$, where $z_0 = \pi w_0^2/\lambda$ defines the Rayleigh range, i.e. the distance over which a laser beam is effectively collimated. For example, a red-light beam from a laser with 1 mm aperture remains parallel for about 5 m. A longer but wider parallel beam can be achieved by using a beam expander, which is a telescope system used in reverse (Figure 16.3). This effectively gives a larger coherent wavefront than the laser aperture alone. A survey theodolite with a 25 mm aperture would have a parallel beam over a distance of 3 km. Over longer distances the beam expander achieves a smaller angular spread than the laser alone. Given an optical system accurate to a fraction of a wavelength, and in the absence of atmospheric turbulence, a very narrow beam can be generated. A telescope with 1 m diameter aperture can transmit a laser beam with a divergence less than 1/2 arcsecond; this would illuminate a spot only 1 km across on the Moon.

15.9   Examples of Important Laser Systems

15.9.1   *Gas Lasers*

Gas lasers may be divided into several types, depending on the active amplifying species in the gas and the excitation mechanism. The wide range of gas lasers is summarized in Table 15.1. The wavelengths of gas lasers cover a very broad range from the vacuum UV to the far IR, in continuous wave and pulsed operation, and with some lasers operating up to high powers. A mixture of gases is often used in gas lasers to enable excitation by energy transfer between the components or to enhance their operation. There are many different pumping mechanisms, including continuous, pulsed or radio frequency electrical discharges, optical pumping, chemical reactions and intense excitation in plasmas. The laser emission may be from electronic transitions in neutral atoms (e.g. the He–Ne laser) or ionized atoms (e.g. Ar^+ or Kr^+), electronic transitions in molecules (e.g. F_2 or N_2), electronic

Table 15.1   Examples of gas lasers

Laser type	Wavelength (nm)	Typical power or pulse energy	Pulsed or CW
Neutral atom			
He–Ne	632.8	1–50 mW	CW
Cu	511, 578	20 mW	Pulsed
Ion			
Ar^+	488, 515	2–20 W	CW
Kr^+	647	1 W	CW
He–Cd	441.6, 325.0	50–200 mW	CW
Molecular			
CO_2	10.6 μm	10^2–10^4 W	CW, pulsed
N_2	337.1	10 mJ	Pulsed
F_2	157	10 mJ	Pulsed
HCN	336.8 μm	1 mW	CW
CH_3F	496 μm	1 mW	CW
Excimer			
ArF	193	mJ, kHz	Pulsed
KrF	248	mJ, kHz	Pulsed
XeCl	308	mJ, kHz	Pulsed
XeF	351, 353	mJ, kHz	Pulsed
Chemical			
HF	2.6–3.3 μm	CW to kW	CW, pulsed
I	1.3 μm	CW to kW Pulsed mJ to J	CW, pulsed
Plasma			
Se^{24+}, Ar^{8+}, etc.	3.5–47	nJ to mJ, ns	Pulsed

transitions in transient excited dimer molecules (termed excimers, e.g. KrF or ArF), and vibrational or rotational transitions in molecules (e.g. CO_2, CH_3F).

Generally gas lasers are excited by an electrical discharge in which excitation of the gas atoms, ions or molecules is by collision with energetic electrons. Optical excitation of a gas is usually inappropriate since the absorption lines of gases are very narrow (in contrast to solids).

The He–Ne laser described briefly in Section 15.2 was the first gas laser to be operated (in 1960), and was the first continuously operating laser. It is still one of the most common lasers, operating on the 632.8 nm wavelength, and is used in many applications requiring a relatively low-power, visible, continuous and stable beam.

The CO_2 gas laser provides large power outputs at the infrared wavelength of 10.6 μm. The laser action involves four vibrational energy levels, as in the scheme of Figure 15.13. The broad highest level is closely equal to an excited level in nitrogen, which is an essential added gas component. The upper level of the CO_2 molecule is populated from this state by collisions with nitrogen molecules. The excitation of the nitrogen molecules is by electron collisions, and the electrons are produced in an electric or radio frequency discharge within the laser tube. The gas also contains helium, which assists the depletion of the lower levels by collisional de-excitation and stabilizes the plasma temperature. Large continuous power outputs, up to some tens of kilowatts, are obtainable; pulsed operation can give pulse energies of joules in microsecond pulses. As the gas densities are usually comparatively low, high-powered CO_2 lasers must be relatively large to contain a sufficient number of molecules.

Regarding the rare optical pumping of gas lasers, two exceptions of interest are the atomic iodine photodissociation laser and the neutral atomic mercury laser. The iodine laser is pumped by an intense flashlamp, whose light dissociates a molecule such as CF_3I to produce iodine atoms in the first electronic excited state, and stimulated emission is on the magnetic dipole transition $^2P_{1/2}$–$^2P_{3/2}$ at 1.3 μm. The iodine 1.3 μm laser may also be pumped by a chemical reaction in which excited molecular oxygen, formed in a reaction between hydrogen peroxide and chlorine, transfers energy to atomic iodine. The mercury laser operates continuously on the strong Hg 546.1 nm transition pumped by a powerful mercury lamp.

Gain at X-ray wavelengths over 3 to 47 nm has been demonstrated from highly ionized atoms. These pulsed lasers operate in a dense plasma pumped by nanosecond laser pulses or electrical discharges. Nanosecond X-ray pulses of up to 1 mJ energy (equivalent to megawatt powers) have been produced.

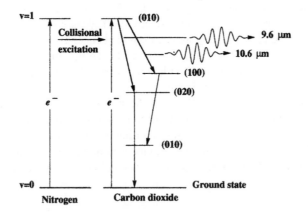

Figure 15.13 Vibrational energy levels in the CO_2 laser

Table 15.2 Examples of solid state crystal, glass and fibre lasers

Laser	Wavelength (nm)	Operation
Crystal host		
Ruby: Cr^{3+}:Al_2O_3	694.3	Pulsed
Garnet: Nd:YAG	1064	CW or pulsed
Vanadate: Nd:YVO_4	1064	CW
Titanium sapphire: Ti:Al_2O_3	670–1070	CW or pulsed
Glass		
Silicate, Nd	1064	Pulsed
Phosphate, Nd	1054	Pulsed
Fibre host		
Er-silica	1500–1600	CW
Er-fluoride	2700	CW
Yb-silica	970–1040	CW
Tm-silica	1700–2015	CW or pulsed

15.9.2 Solid State Lasers

A solid state laser, such as the ruby laser, may be in the simple form of a transparent rod with mirrors formed directly on the ends. The gain medium contains a paramagnetic ion in a host crystalline solid or glass. The active ion may be substituted into the crystal lattice or may be doped as an impurity into the glass host. There are many combinations of dopant ion and host materials which provide a wide range of laser wavelengths. The doped solids exhibit broad absorption bands which make them amenable to optical excitation from continuous or pulsed lamps or from semiconductor diode lasers. A listing of some of the more common doped crystal solid state lasers[6] is given in Table 15.2, and also includes some doped glasses where the active laser medium is a bulk glass or the central core of an optical fibre.

The dopant ion should fit readily into the crystal host by matching the size and valency of the element that it is replacing. The optical quality of the doped medium needs to be high so that there is low loss for the amplifying beam. Refractive index variations, scattering centres and absorption can contribute to loss processes. Suitable host media are garnets (complex oxides), sapphire (Al_2O_3), aluminates and fluorides. The glass hosts are easily fabricated, in large sizes and with high optical quality. The energy levels are broader in glasses than in crystals, making them more suitable for pumping by flashlamps. The lower thermal conductivity of glasses compared with crystals renders them susceptible to thermal distortion and induced birefringence. The increased linewidth leads to a reduced stimulated emission cross-section (Section 15.4) such that the pumping threshold is higher. Although pulsed and CW operation are used with various crystal hosts, pulsed operation is necessary for a glass host, especially at high power levels.

The paramagnetic dopant ions are usually from the transition metals and lanthanide rare earths. The Nd:YAG laser operating at 1064 nm is one of the most used solid state lasers; here neodymium ions Nd^{3+} provide the laser action, and yttrium aluminium garnet (YAG) is the usual crystal host. The crystal has a relatively high thermal conductivity which enables it to distribute heat efficiently following optical pumping. The laser can operate either pulsed or continuously.

---

[6]The common name for the crystal is given, with the dopant (e.g. Cr^{3+}) and host medium (e.g. Al_2O_3).

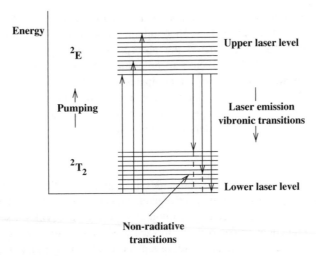

Figure 15.14   Simplified energy level diagram of titanium-doped sapphire. Absorption and laser emission bands and non-radiative transitions are shown

Semiconductor lasers, which are dealt with in Chapter 17, are derived from light-emitting diodes. They are distinct from the solid state lasers such as the ruby laser in their pumping and photon generating processes, deriving their energy from the electrical excitation of electrons within the semiconductor and emitting radiation with photon energy approximately equal to the bandgap energy.

The titanium–sapphire laser (Ti^{3+} ions doped into sapphire, Al_2O_3) has assumed much importance as it is tunable over a wide band of 670 to 1070 nm and produces CW powers up to 50 W depending on the pump power; it can also be mode locked (described in Chapter 16) to produce femtosecond pulses. The titanium–sapphire crystal has a broad optical absorption band between 400 and 600 nm and is optically excited, usually by another laser such as the argon ion laser or the frequency-doubled Nd:YAG laser. The broad emission band of width $\Delta\lambda \sim 400$ nm has a peak wavelength near 800 nm. The lower ($^2T_2$) and upper laser (^2E) levels, shown in Figure 15.14, are composed of overlapping vibrational–rotational (termed vibronic) levels. The simple energy level structure, in which there are no states with energy levels above the upper laser level, avoids excited state absorption from the upper laser level, which in some solid state lasers reduces the efficiency and tuning range. Non-radiative relaxation in the upper and lower laser levels acts to maintain population inversion.

Optical fibres in which the central core is doped with a rare earth ion, such as Er^{3+} or Yb^{3+}, can act as an efficient laser. The pump light may be fed in either from one end or from the side and is then trapped in the fibre together with the stimulated wave, thereby ensuring strong coupling between the pump and laser beams. An example of a fibre laser is shown in Plate 6.[*] The operating wavelength of the erbium-doped silica glass fibre at 1.54 μm has a value which qualifies it for use as the erbium-doped fibre amplifier (EDFA) in optical communications. The Yb-doped silica-fibre laser operating at 1.05 μm has high efficiency, and in a double-clad configuration produces output powers up to 1 kW. The double-clad fibre structure has a second concentric cladding with a diameter of typically 400 μm into which pump power can be efficiently coupled and that power is then transferred into the narrow fibre core as the light travels down the fibre.

---

[*]Plate 6 is located in the colour plate section, after page 246.

Efficient and compact solid state lasers can be made using pumping by high-power semiconductor diode lasers (described in Chapter 17). Semiconductor diode lasers with high powers have been developed with wavelengths which match the absorption wavelengths of doped solids. As an example, the 1.064 μm Nd:YAG laser is able to be pumped by the 808 nm GaAs diode laser, with substantially reduced heating of the crystal compared with broadband pumping by flashlamps, thereby conferring on it improved laser efficiency and greater optical beam quality of the laser output. The Nd:YVO$_4$ vanadate crystal also pumped by the 808 nm laser diode has a greater gain coefficient than the Nd:YAG crystal and is more tolerant of cavity losses.

15.9.3 Liquid Lasers

The major liquid lasers employ organic molecules in solution as their amplifying medium. The characteristic absorption and emission spectra of organic molecules derive from their molecular structure of a backbone of carbon atoms with conjugated double bonds; this provides a set of p-state electrons (π electrons) with wavefunctions spread over the molecule. The electronic energy states of the molecule are determined by the π electrons and have a set of singlet (total spin zero) and triplet (total spin unity) states. Each electronic state has associated vibrational and rotational modes which form a continuous energy band. These molecules have broad absorption and fluorescence bands. Fluorescence transitions between singlet levels are allowed dipole transitions, so that the excited singlet states have nanosecond lifetimes, and emit often with high efficiency. Triplet–singlet transitions are not allowed as dipole transitions, so that the lifetimes are greater than microseconds in the lowest triplet state. Excitation to the long-lived triplet level may therefore lead to loss in the laser due to absorption to higher triplet levels.

The broad absorption band can be pumped by flashlamps or by another laser such as the argon or krypton ion, frequency-doubled Nd:YAG, excimer or copper vapour lasers. The emission band is also broad so that tunable radiation can be achieved over a bandwidth of about 30 nm from a single molecule, and over the range of 320 to 1500 nm from a set of molecules. The large fluorescent bandwidth enables mode-locking techniques (described in Chapter 16) to be used to generate ultrafast pulses with durations down to a few femtoseconds.

Problem 15.1

(i) For a continuous wave laser, and ignoring photon losses by absorption and scattering, calculate the rate at which photons are being produced by stimulated emission in (a) a 1 watt laser at wavelength 600 nm and (b) a 1 milliwatt maser at a frequency of 3000 MHz.

(ii) Assuming a pulse length of 100 ns, calculate the total energy available and estimate the peak power in a single pulse from (a) a $\lambda = 694$ nm solid state laser in the form of a rod 10 mm in diameter and 0.1 m long containing 3×10^{19} active ions per cm^3 and (b) a CO$_2$ gas laser of wavelength 10.6 μm, 30 mm in diameter and 2 m long containing gas with 6×10^{18} molecules per cm^3.

(iii) Calculate the longitudinal mode separation in the cavity of a 633 nm He–Ne laser with mirrors separated by 0.3 m. How many of these modes could oscillate if the width of the gain curve is 2×10^9 Hz?

Problem 15.2
Show for a blackbody that the energy density u per unit frequency interval is related to the radiance (brightness) R as

$$u = \frac{4\pi R}{c}.$$

Problem 15.3
For a system in thermal equilibrium calculate the temperature at which the rates of spontaneous and stimulated emission are equal for a wavelength of $10 \, \mu m$.

Problem 15.4 A proof that photons are bosons
In thermal physics, three kinds of particles are considered: bosonic, fermionic and classical (the high-temperature, or low-density, limit shared by the other two kinds). Particles with integer spin, such as photons $(s = 1)$, act as bosons and can have any number of particles per quantum state or mode. Particles with half-odd-integer spin, such as electrons, protons or neutrons $(s = \frac{1}{2})$, behave as fermions, with only 0 or 1 in each state.

In equilibrium at temperature T, photons must occupy states with a mean density per mode $= [\exp(h\nu/kT) + \Delta]^{-1}$, where the constant Δ depends on which kind of particle they are: $\Delta = +1$ (fermions), 0 (classical), -1 (bosons). In all three cases, the density of modes per unit volume per unit frequency (including two polarizations) is $D(\nu) = 8\pi\nu^2/c^3$.

(a) Following Einstein's approach, derive relations for the A and B coefficients including the constant Δ. Then show that:

(b) photons cannot be classical, since there would be no stimulated emission;

(c) photons cannot be fermions, since Einstein's model of radiative transitions would fail.

Problem 15.5
Compare the Doppler-broadened linewidth of the He–Ne laser with that of the argon ion laser given the following data:

	He–Ne	Ar^+
Atomic mass	20 (Ne)	40
Wavelength (nm)	633	488
Gas temperature (K)	400	5000

Problem 15.6
In the He–Ne laser operating at 633 nm the Einstein A coefficient of the upper laser state is $3 \times 10^6 \, s^{-1}$. The upper state has a degeneracy of 3 and a population of $10^{16} \, m^{-3}$ and the lower state a degeneracy of 5 and a population of $10^{15} \, m^{-3}$. The mirror reflectivities are 1.0 and 0.95, the losses are 3% per round trip, and the gas temperature is 400 K (as in Problem 15.4). The laser transition has a Doppler inhomogeneously broadened spectral line profile. Calculate the minimum length required for the gain medium to achieve laser operation.

Problem 15.7
An He–Ne laser operating at 633 nm in the TEM_{00} mode has an output power of 1 mW and a minimum spot radius of 0.3 mm. Find:

(a) The beam divergence angle.

(b) The laser radiance or brightness $R = P(A\Omega)^{-1}$, where P is the power, A the spot area, and Ω the solid angle subtended $\simeq \pi\theta^2$ for divergence angle $\theta \ll 1$.

(c) The temperature of a blackbody with the same brightness.

Problem 15.8
Calculate (a) the threshold gain coefficient and (b) the population inversion $n_2 - (g_2/g_1)n_1$ for a ruby laser operating at 694.3 nm. The spontaneous lifetime of the upper laser level is 3 ms, the linewidth of the transition is 150 GHz and the ruby crystal refractive index is 1.78. The laser transition is homogeneously broadened, and the degeneracies of the upper and lower laser levels are $g_2 = 2$ and $g_1 = 2$. The laser cavity has reflectivities 1.0 and 0.96; the length is 5 cm and other losses are negligible.

Problem 15.9

Calculate the fraction of the beam power and the average photon flux within the beam waist for a 1 watt argon ion laser operating at 515 nm, and with a cavity length of 1.5 m.

Problem 15.10

Determine the number of longitudinal modes in an argon ion laser of length 80 cm if the laser wavelength is 515 nm. The laser transition is Doppler broadened and the gas temperature is 5000 K (see Section 12.2 and Appendix 4 for the linewidth). In this case the loss coefficient is one-third the peak gain value.

Calculate the maximum length of the laser cavity for only one longitudinal mode to oscillate.

Problem 15.11

Explain why the frequency spacing of modes for a laser in the form of a ring is twice that for a standing wave cavity of the same length.

Problem 15.12

A carbon dioxide laser operating at the 10.6 μm transition has a gas pressure of 1 atmosphere and gas temperature of 400 K.

By estimating the contribution to the linewidth from Doppler and pressure broadening determine if the transition is broadened by homogeneous or inhomogeneous effects (see Appendix 4 for the linewidth equations).

16 Laser Light

How far that little candle throws his beams!/ So shines a good deed in a naughty world.

William Shakespeare, *Merchant of Venice*.

Stimulated emission, which is at the heart of laser action, produces a multiplicity of photons, identical in frequency, phase and direction. This *coherence* in laser light contrasts sharply with the chaotic nature of light from spontaneous emission, and gives laser light its extraordinary properties of narrow spectral linewidth (i.e. temporal coherence) and directionality (spatial coherence). Some of these properties are familiar; a laser beam can be pencil-sharp over a large distance, and the *speckles* in a spot of laser light on a surface distinguish it at a glance from incoherent light. Coherence allows laser light to be focussed to a spot only a few wavelengths across, with an intensity that is useful in heating and cutting many different materials. The highest power flux is obtained in lasers operating with short pulses, and techniques exist for producing pulses only a single cycle or a few femtoseconds (10^{-15}s) long. High intensity also means high electric fields; dielectrics may behave non-linearly at such high fields, producing effects such as the generation of shorter wavelength laser light at a harmonic of the laser frequency.

In this chapter we examine the temporal and spatial coherence properties of laser light, including its directionality and radiance (or brightness), and the theoretical and attainable limits on linewidth, focussing and pulse width. We also consider the effects of the extremely high electric fields in short laser pulses, including the non-linear behaviour of dielectrics and the generation of harmonics.

16.1 Laser Linewidth

In a laser cavity the frequency of the oscillation is determined by a resonance in the cavity rather than by a natural resonance in an atom or ion. The process of stimulated emission usually leads to laser light with a considerably narrower linewidth than that of the spontaneously emitted radiation, and the laser beam consequently has a very pure colour. The beam will, however, contain a small proportion of spontaneously emitted photons, which add to the beam with random phase. It is this addition of incoherent photons that ultimately limits the coherence of the laser light. We start with a simple

Optics and Photonics: An Introduction, Second Edition   F. Graham Smith, Terry A. King and Dan Wilkins
© 2007 John Wiley & Sons, Ltd

situation when only the ratio between stimulated and spontaneous emission limits the coherence, which occurs when there is a large population inversion. The theoretically attainable limit of bandwidth then depends only on the laser power and the width of the cavity resonance.

Consider a laser oscillating in a single mode, populated by $\bar{n}$ photons. Nearly all these photons have been produced by stimulated emission, and are completely coherent. There is also a population of photons generated by spontaneous emission from the excited atoms in the laser, and a small number of these must be in the same mode as the coherent population. The population of these incoherent photons relative to the coherent photons limits the coherence of the laser beam. We therefore need the relative rate of stimulated and spontaneous emission in a single mode. This ratio is simply the mean population $\bar{n}$ in the mode;[1] recalling that the A_{21} and the B_{21} coefficient represent incoherent and coherent photons respectively, it follows that there is on average only one incoherent photon and $\bar{n} - 1$ coherent photons (or $\bar{n}$ to a very good approximation). The single incoherent photon contributes on average $1/\bar{n}$ of the power, and thus $1/\sqrt{\bar{n}}$ of the electric field. The photons in the population last for a certain time in the cavity before they are either emitted from the cavity or absorbed; this time, τ_{cav}, is the *decay time* for the cavity. The phase relation between the coherent and incoherent components changes randomly on this time scale.

The effect of the vector addition of this incoherent field component is shown in Figure 16.1, which shows two successive phasors of the incoherent photons added to the coherent component. The tip of the resultant phasor follows a random walk as these changes accumulate. A single step of this random walk can modify the phase of the main field by the ratio of amplitudes $\Delta\phi \simeq \bar{n}^{-1/2}$ radians. The phase change builds up randomly; when it reaches approximately 1 radian the original phase is effectively lost. This occurs after $\bar{n}$ changes, i.e. after a time $\tau_{\mathrm{cav}}\bar{n}$. This *phase diffusion time* determines the frequency width $\Delta\nu_{\mathrm{L}}$, from the bandwidth theorem (Section 4.12):

$$\Delta\nu_{\mathrm{L}} \simeq \frac{1}{2\pi\bar{n}\tau_{\mathrm{cav}}}. \tag{16.3}$$

The number of photons in the beam at any time is related to the power P and the cavity decay time by

$$\bar{n}h\nu = P\tau_{\mathrm{cav}} \tag{16.4}$$

---

[1] The ratio of the transition rates is, from Section 15.3,

$$\frac{B_{21}u(T, \nu)}{A_{21}} = \frac{u(T, \nu)c^3}{8\pi h\nu^3} \tag{16.1}$$

where $u(T, \nu)$ is the energy density and A_{21}, B_{21} are the Einstein coefficients. The classical density of modes $\rho(\nu)$ in a blackbody cavity is

$$\rho(\nu) = \frac{8\pi\nu^2}{c^3} \tag{16.2}$$

(see for example F. Mandl, *Statistical Physics*, 2nd edn, John Wiley & Sons, 1988). The required ratio is

$$\frac{u(T, \nu)/h\nu}{\rho(\nu)},$$

which is $\bar{n}$, the number of photons per unit volume and unit frequency interval divided by the number of modes per unit volume and unit frequency; the ratio of rates is therefore the average number of photons per mode.

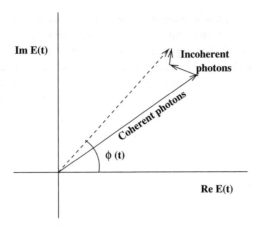

Figure 16.1.   Phasor diagram showing the small addition of an incoherent spontaneously emitted photon to coherent laser light. Further additions occur at random phase at intervals of τ_{cav}, the decay time of the laser cavity. The size of the incoherent components is greatly exaggerated in this diagram

giving the laser frequency width as

$$\Delta v_L = \frac{hv}{2\pi\tau_{cav}^2 P}. \tag{16.5}$$

The decay time of the cavity resonance τ_{cav} is related to the width of the resonance Δv_{cav} by

$$\tau_{cav} = \frac{1}{2\pi\Delta v_{cav}}. \tag{16.6}$$

The laser linewidth may therefore be related to the width of the cavity resonance by

$$\boxed{\Delta v_L = \frac{2\pi hv}{P}(\Delta v_{cav})^2.} \tag{16.7}$$

Note that this theoretical minimum linewidth decreases as the power increases.[2]

The theoretical limit is hard to attain in practice in gas lasers. A typical calculation from equation (16.7), shows that an ordinary He–Ne laser should give a linewidth of order 10^{-2} Hz, while in practice a width of a few kilohertz is commonly observed. The difference is due to thermal and mechanical instabilities which cause random changes in the cavity length.

Theoretical linewidths are much larger in semiconductor lasers, because the small cavity length (typically $300\,\mu$m) leads to a very short decay time (5 ps) and a large cavity resonance linewidth (typically 10^{10} Hz). The linewidth Δv_L of a semiconductor laser is usually around 10^6–10^7 Hz.

---

[2]A more rigorous calculation takes into account the numbers n_2, n_1 of species in the upper and lower laser levels and the population inversion $\Delta n = n_2 - (g_2/g_1)n_1$ where g_2, g_1 are the degeneracies of the levels. Then

$$\Delta v_L = \frac{2\pi hv(\Delta v_{cav})^2}{P}\frac{n_2}{\Delta n}.$$

The further progress of the phasor diagram of Figure 16.1 is shown in Figure 16.2, where the essential difference between ordinary incoherent light and laser light is shown in the probability distribution of amplitude and phase. A plane quasi-monochromatic electromagnetic mode may be described by the electric field $\tilde{E}(t) = E_0 \exp(-i\omega_0 t) a(t) \exp[i\phi(t)]$. The electric field consists of a carrier wave at frequency ω_0 with random amplitude and phase modulation, represented by $a(t)$ and $\phi(t)$. In ordinary light the arrival of photons follows Gaussian statistics; the probability distribution

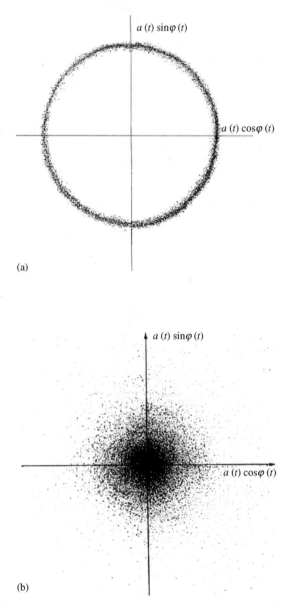

Figure 16.2 Probability distributions of amplitude and phase of the electric field for (a) laser light (b) ordinary light. The magnitude of the probability is proportional to the density of shading. (R. Loudon, *The quantum theory of light, 2nd ed.*, Oxford University Press, 1983.)

expressed as a function of amplitude and phase has a peak at zero, and all phases are equally probable. For laser light the distribution of amplitude is Poissonian about a mean; all phases are again equally likely, but on a long time scale during which the phasor slowly wanders round the circle.

For chaotic light the probability distribution of amplitude is Gaussian; it has the highest probability at the origin and all phases are equally probable. For this reason chaotic light is referred to as Gaussian light. Note that the *frequency* distribution (i.e. the spectrum) of chaotic light can have a Lorentzian, Gaussian or Voigt profile (see Chapter 18) and is not to be confused with the Gaussian amplitude distribution described here. A distinction between chaotic light and laser light can be observed in a photon-counting experiment in which the rate of arrival of photons at a detector is measured. In chaotic light the arrival of photons shows photon correlation (or bunching); this is discussed also in Section 13.8. The probability $p(n)$ of detecting n photons in a certain time interval, for a chaotic (thermal) source with mean number $\bar{n}$, is a Bose–Einstein distribution:

$$p(n) = \frac{1}{1+\bar{n}} \left(\frac{\bar{n}}{1+\bar{n}} \right)^n. \tag{16.8}$$

For coherent light from a single mode laser, the photon arrival times are statistically independent and follow a Poisson distribution:

$$p(n) = \frac{\bar{n}^n}{n!} \exp(-\bar{n}). \tag{16.9}$$

In this book the semi-classical theory is adopted in which the photon energy and atomic energy levels are quantized, while the electromagnetic field is classical, and non-quantum-mechanical. It is appropriate here to draw attention to the extension from this approach in which the electromagnetic field is treated quantum-mechanically. In that case, a single mode laser operating well above threshold denotes a coherent state corresponding to a classical stable electromagnetic wave. Briefly in the quantized field description the energy of a mode is quantized. This implies that the lowest energy of a radiation mode, corresponding by definition to the vacuum state, is a zero-point energy $W_0 = \frac{1}{2}\hbar\omega$. There is also an associated fluctuation in the background field, which contributes noise in a measurement. The phases of the zero-point fluctuations can be influenced to reduce the noise, giving *squeezed light*. A property of squeezed light is that it has a variance in photon number which is reduced (squeezed) below that for a coherent state, and is termed sub-Poissonian or non-classical. Several methods have been devised to generate squeezed light using laser sources and parametric down conversion. The reduced noise in squeezed light leads to its potential applications in digital optical communications, interferometry and precision measurements.

16.2 Spatial Coherence

The term *spatial coherence* usually refers to coherence transverse to the laser beam, and is also termed transverse or lateral coherence. Spatial coherence is described in Section 13.1. When operating on a single transverse mode laser radiation has a high degree of spatial coherence, while a laser operating on more than one transverse mode has reduced spatial coherence. For a laser the transverse coherence length L_t is dependent on whether the laser is operating in a single transverse mode or in multiple transverse modes. The beam divergence angle θ_t is related to L_t as

$$\theta_t \sim \lambda/L_t \tag{16.10}$$

The high degree of collimation of lasers results from the high effective values of L_t that are able to be established.

The spatial coherence of the beam can be measured using Young's double slit interferometer described in Chapter 8.

The quantitative measure of spatial coherence is described by the coherence function $\gamma^{(1)}(r_1t_1, r_2t_2)$ (equation (13.18)) which includes the spatial and temporal coherence dependence of a light beam at space and time points r_1t_1 and r_2t_2. Measurement at two points r_1 and r_2 across a beam at the same time $(t_1 = t_2 = t)$ yields the first-order degree of spatial coherence

$$\gamma^{(1)}(r_1, r_2) = \frac{\langle E(r_1, t)E^*(r_2, t)\rangle}{[\langle |E(r_1t)|^2\rangle \langle |E(r_2t)|^2\rangle]^{1/2}}. \tag{16.11}$$

This quantity has values $0 \leq \gamma^{(1)} \leq 1$. A beam with $\gamma^{(1)}(r_1, r_2) = 1$ has perfect spatial coherence.

The irradiance of the TEM_{00} transverse laser mode has a radial dependence which is a Gaussian function (equation (15.36)). A beam with minimum spot diameter $2w_0$ has spatial coherence across that dimension, the divergence angle is $\theta_t \simeq \lambda/\pi w_0$, and the transverse coherence length is $L_t \sim \pi w_0$. There is therefore a direct connection between the spatial coherence, the transverse coherence length and the divergence of the beam. The beam divergence can be reduced by expanding the beam with a telescope as shown in Figure (16.3).

A particular feature of laser light is that it can have a very high degree of spatial coherence. When combined with the high irradiance obtained from lasers this confers special properties on laser light. The spatial coherence of laser radiation is dependent on the transverse mode structure of the laser beam. A laser which is operating in the fundamental TEM_{00} mode has complete spatial coherence. When the laser operates on more than one transverse mode the spatial coherence is reduced because of the loss of coherence between the modes. A laser operating on multiple longitudinal (frequency) modes can still have high spatial coherence. A single transverse mode in a laser can be selected by reducing the transverse cross-section of the beam, e.g. by placing an aperture in the laser cavity. The diameter of the aperture is selected to achieve a single transverse mode but has the consequence of reducing the output power of the laser. The spatial coherence determines the divergence of the laser beam, as described in Section 15.8. It also determines in part the size of the focused beam, and the formation of the speckle pattern observed with laser beams.

The Gaussian dependence of laser irradiance, $I(r, z) = I_0 \exp(-2r^2/w^2(z))$ (equation (15.36)), results from the spherical resonator used with the laser. It is possible to generate laser beams which have other transverse irradiance distributions. One example of these is for the electric field to be described by a Bessel function. For a monochromatic wave propagating in the z direction

$$E(r, z, t) = KJ_0(k_r r) \exp[-i(\omega t - k_z z)] \tag{16.12}$$

with J_0 the zeroth-order Bessel function, k_r and k_z the radial and longitudinal components of the wave vector **k**, and K being a constant. A remarkable property of a wave described by the ideal zeroth-order Bessel function J_0 is that it is planar, and the irradiance $(I \propto EE^*)$ is independent of the propagation distance z, i.e. the irradiance is the same for all positions along z. This means that the beam does not spread out, i.e. it is non-diffracting, so that its size and irradiance remain constant. Several methods have been demonstrated to produce beams which are close to a J_0 Bessel beam. In practice, the perfect plane wave Bessel beam cannot be produced since it is required to have infinite radial dimension. However, close approximations to Bessel beams can be created which have an equivalent Rayleigh range much greater than that of a Gaussian beam.

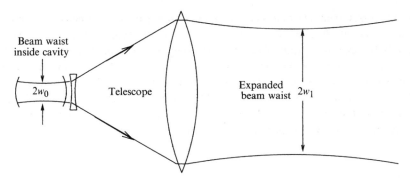

Figure 16.3   A beam-expanding telescope, used to increase the diameter of the coherent beam and reduce its angular divergence. The beam waist is increased by the ratio of focal lengths of the two telescope lenses. In this diagram the beam is outlined at the $1/e^2$ irradiance level; the curvatures of the emergent beam are greatly exaggerated

16.2.1   Laser Speckle

An extended spot of laser light on a rough surface may easily be recognized by its grainy appearance, with a pattern of individual light and dark spots (speckles) which changes as the eye moves. This arises from random variations in phase of the reflected light at the rough surface: light is scattered from each point of the surface with a phase depending on the height. In any direction in front of the surface, light from the component sources combines coherently, but the combined amplitude depends on the addition of their random phases.

A bright speckle observed by eye occurs in a direction in which the component sources are adding more or less in phase; this can be anywhere within the direction of the spot. As in any diffraction problem, we may think of the light leaving the surface as an angular spectrum of plane waves; each point on the retina responds to a small range of these, covering an angle which is the angular resolution of the eye. In the direction of a bright spot this small range of component waves happens to add in phase. Outside this range, and at a different point on the retina, the response is the sum of an unrelated set of waves, whose phases are very unlikely to add in the same way. The bright speckles therefore have an angular width which is the angular resolution of the eye; the width does not depend on the scale of the surface roughness, provided that it is sufficiently rough to introduce phase changes of at least one wavelength, and that the lateral scale of the roughness is also at least one wavelength.

If the eye is at distance D from the surface, and the full diameter d of the pupil is illuminated, the scale size of the speckles as seen on the surface is approximately

$$S_E \simeq D\frac{\lambda}{d} \tag{16.13}$$

where λ/d is the angular resolution. This is easily tested by moving to different distances D and by squinting to reduce d. The pattern changes with small movements of the eye, giving a shimmering effect. Remarkably the pattern does not disappear when the eye is defocussed, as when one's spectacles are removed. The pattern also changes when the scattering surface moves or changes: if instead of a surface, the scatterer is a liquid suspension of particles, e.g. milk micelles or chalk dust in water, the speckle pattern becomes dynamic, giving an easily observed demonstration of Brownian motion.

When speckle is observed by eye it is due to an interference pattern formed on the retina of the eye. There is, however, an interference pattern in the whole space in front of the scattering surface, as may be found simply by holding a piece of paper or exposing a photographic plate at a fixed position. A point on the plate is now receiving contributions from the whole of the illuminated surface and the luminance depends on the superposition of these contributions. Their relative phases change significantly at an adjacent point on the plate when the path difference between the two edges of the illuminated spot changes by λ. If the spot diameter is s, this requires an angular movement of λ/s; at a plate distance D this gives a speckle scale S_P on the plate, where

$$S_P \simeq D\frac{\lambda}{s}. \tag{16.14}$$

Note the similarity of these two equations, and that the scale is proportional to D for both, as may easily be tested experimentally. Perhaps the most unexpected feature of speckle is that the scale is independent of the scale of the roughness of the surface.

16.3   Temporal Coherence and Coherence Length

The temporal coherence of the laser output is directly related to the spectral bandwidth. A spread of frequencies in a laser output having a bandwidth $\Delta\nu_L$ leads to a changing phase relation between the components in the spread, changing randomly the amplitude and phase of their sum (see Chapter 13). The temporal coherence is characterized by the *coherence time* τ_c, during which the frequency components maintain a fixed phase relation. Assuming a Gaussian line profile, τ_c is

$$\tau_c = \frac{1}{\Delta\nu_L}. \tag{16.15}$$

The *coherence length* l_c is the distance $c\Delta\tau$ travelled in the coherence time, so that

$$\boxed{l_c = c\tau_c = c/\Delta\nu_L = \lambda^2/\Delta\lambda_L.} \tag{16.16}$$

Very large coherence lengths are often encountered with lasers; even for a comparatively large linewidth of 1 MHz the coherence length is 300 m. Interferometric measurement of such narrow linewidths requires interferometers with correspondingly long optical paths.

The line emission from a low-pressure discharge lamp has a comparatively small coherence length; for example, the 546.1 nm mercury emission line would have a linewidth of about 2.5×10^{-2} nm and a coherence length of about 1 cm. In contrast the stable single mode He–Ne laser with a bandwidth of 1 kHz gives a coherence length of 300 km. Even the normal He–Ne laser, which usually operates on several modes simultaneously and consequently has a larger bandwidth, gives a coherence length of about 50 cm.

The quantities τ_c and l_c can be measured using a Michelson interferometer or a Mach–Zehnder interferometer as described in Chapters 8 and 12. The open-path Michelson interferometer is suitable for laser outputs with linewidths above 1 GHz, since these correspond to path length differences less than 30 cm. A laser source with a linewidth of 1 MHz observed with a Michelson interferometer would require a path difference between the two interferometer arms of about 300 m. Such long path differences can be accommodated using a long optical fibre in one of the arms (Section 8.6).

Lasers can be used for measuring distances of several kilometres or more in terms of the wavelength of the laser light; it must be emphasized, however, that the wavelength is determined by the resonant cavity and is not a fundamental physical parameter, so that only comparative measurements can be made. A change in distance of a fraction of a wavelength can be detected; for example, the gravitational wave detector (Chapter 9) is an interferometer with a path length of several kilometres, designed to detect a periodic fractional change as small as 1 part in 10^{20} (as may easily be seen, this represents a very small fraction of a wavelength).

16.4   Laser Pulse Duration

Commonly encountered lasers such as the He–Ne laser and semiconductor lasers produce a continuous beam of light, although for communications purposes a semiconductor diode laser may be switched electrically at high rates. Other lasers operate predominantly or only as pulsed sources, usually because there is not an effective pumping mechanism to sustain CW operation. Where the laser is pumped by a pulsed source, e.g. a flashlamp, the gain is driven above the threshold value, and a pulse of laser radiation is emitted. The pulsed laser emission often shows wide fluctuations in irradiance due to the dynamics of the excitation mechanism, gain and laser output; this is referred to as laser spiking or relaxation oscillations.

Many lasers are designed to produce individual very short pulses at extremely high intensity. Phenomenally high irradiances can be achieved in these pulsed lasers; the mean power may, however, be kept to a manageable level by using a low pulse repetition rate. We describe here two techniques for producing these ultrashort pulses.

16.4.1   Q-switching

Intense short-duration pulses, in the nanosecond range, may be obtained by the technique of *Q-switching*. In normal laser operation the resonant cavity has low losses: it is a resonator with high quality factor Q. Laser action can be inhibited by reducing the Q of the cavity, e.g. by rotating one of the mirrors out of alignment. If the excitation (pumping) of the laser is maintained during the time the cavity is in a low-Q state, the population inversion can build up to a very high value, as shown in Figure 16.4. Suddenly restoring the original high Q by realigning the mirrors then gives an intense burst of laser light. The build-up of the pulse is very rapid, since the gain at the instant when the Q is restored is very much larger than the threshold value. Laser action then removes the excitation in a very short pulse undergoing only a few passes through the laser medium.

The duration of the Q-switched pulse is approximately the same as the cavity lifetime τ_{cav} described in Section 16.1. This depends on the length L of the resonator, the refractive index n and the irradiance reflection coefficient R of the mirrors. As the initial laser pulse builds up within the resonator, at each mirror reflection a fraction $(1 - R)$ of the energy is lost from transmission. The pulse makes $1/(1 - R)$ passes of the resonator, which occurs in the characteristic cavity lifetime τ_{cav}:

$$\tau_{\mathrm{cav}} \leq \frac{nL}{c(1 - R)}.\tag{16.17}$$

More accurately, in addition to transmission loss τ_{cav} takes into account all losses in the cavity.

Q-switching can be achieved in several ways, most simply by rotating one of the mirrors, typically at about 10 000 rpm, giving a pulse for each rotation. Alternatively the cavity mirror can be replaced by a roof-top prism or a combination of a rotating prism and a mirror, Figure 16.5; here the cavity alignment is achieved for a small angular range of the prism.

Rather than mechanical Q-switching, active Q-switches are more generally used based on electro-optic or acousto-optic modulators in the laser cavity, which act as shutters. The electro-optic modulator in the form of a solid state Pockels cell has been previously described in Section 7.11; it is inserted into the laser cavity as shown in Figure 16.5(b) together with a polarizer. Application of a voltage to the Pockels cell induces birefringence proportional to the applied voltage. The orientation of the Pockels cell is such that the induced birefringence is in the plane orthogonal to the axis of the resonator and the polarizer is set at 45° to the birefringence axis. Applying the bias voltage, a pulse transmitting the Pockels cell twice with reflection at the mirrors has its plane of polarization rotated by 90° and is switched out of the cavity by the polarizer. A switching time of a few nanoseconds can be achieved.

The acousto-optic modulator (Figure 16.5c) consists of a crystal or glass (e.g. fused silica) in which an ultrasonic wave is propagated, inducing refractive index variations at the acoustic frequency. These

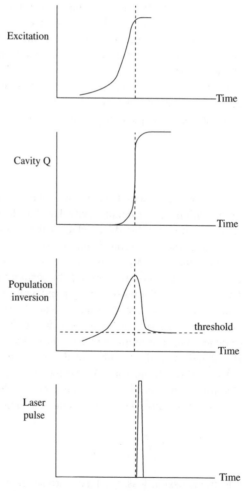

Figure 16.4 Q-switching, showing (a) the growth of excitation, (b) the step increase of Q in the laser cavity, (c) the growing population inversion and (d) the short laser pulse

periodic variations act as an optical Bragg grating with an effective period equal to the acoustic wavelength; the laser beam is diffracted out of the resonator by this phase grating. The ultrasonic waves are driven by a piezoelectric transducer attached to the crystal which is typically operated at frequencies in the range of 100 MHz to 1 GHz.

A passive modulator *Q*-switch is also often used in which a cell is inserted into the laser cavity containing a medium, e.g. a dye solution or a solid state absorber, which absorbs at the laser wavelength. The medium is selected to provide an absorption which can be saturated at relatively low irradiance, and hence become transparent. At saturation, the medium is bleached and becomes transparent at the laser wavelength. The switch is then effectively open and the laser intensity may rapidly build up to a high value.

As an example, the solid state Nd:YAG laser can be flashlamp pumped in normal operation to produce a pulse of about 1 ms duration, typically with a peak power of order 1–10 kW. When *Q*-switched the pulse duration can be reduced to 10–100 ns, with a peak power of several megawatts.

16.4.2   Mode Locking

As we have seen in Chapter 4, a short pulse must contain components over a wide bandwidth. A laser oscillation has an inherently narrow bandwidth, but it may be able to oscillate in several modes with frequencies spaced within the bandwidth of the resonator and the linewidth of the lasing medium. If these modes can be excited simultaneously, with a suitable relation between their phases, the effect of

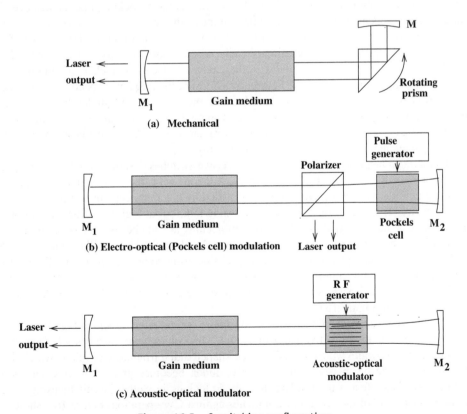

Figure 16.5   *Q*-switching configurations

a broad bandwidth is obtained in a regular train of pulses which may individually be shorter than one picosecond (10^{-12} s).

Consider a cavity resonator with two mirrors a distance L apart. The separation of the N modes in angular frequency is $\Delta\omega = \pi c/L$. If they are all excited simultaneously so that the different modes maintain the same relative phase, i.e. are *mode locked*, and with equal amplitude, their complex electric fields add as

$$\tilde{E}(t) = E_0 \sum_{n=1}^{N} \exp(i\omega_n t) \tag{16.18}$$

where $\omega_n = \omega + n\Delta\omega$. The sum of this series is already familiar in the context of diffraction gratings:

$$\tilde{E}(t) = E_0 \exp(i\omega t) \exp[i(N+1)\pi ct/2L] \frac{\sin(N\pi ct/2L)}{\sin(\pi ct/2L)}. \tag{16.19}$$

The output irradiance is (apart from a constant factor)

$$I(t) = E_0^2 \frac{\sin^2(N\pi ct/2L)}{\sin^2(\pi ct/2L)}. \tag{16.20}$$

Such coherent oscillation in the set of N modes, known as mode locking, may be achieved by the rapid and repetitive opening of an electro-optic shutter in the laser cavity. The output is a train of pulses uniformly spaced in time at the period $t = 2L/c$, which is the round-trip transit time of a pulse in the laser cavity, Figure 16.6. The duration of each pulse is approximately $(\Delta v)^{-1}$, where Δv is the bandwidth of the set of longitudinal modes in the cavity. Under mode-locked conditions the time-dependent amplitude of the output is the Fourier transform of the frequency spectrum. For a Gaussian frequency spectrum with linewidth Δv, i.e. a set of modes with amplitudes having a Gaussian distribution, the mode-locked pulses have a Gaussian time profile with width (FWHM) $\Delta\tau_p = 2\ln 2/(\pi\Delta v) = 0.441/\Delta v$. The maximum irradiance is $N^2 E_0^2$; note that without coherence between the modes the maximum would have been only NE_0^2.

Mode locking can be produced by the active or passive switches described for Q-switching. Active mode locking may be achieved by amplitude modulation using the electro-optic Pockels cell or acousto-optic modulator. The passive switch is provided by placing a cell containing a *saturable dye* in the laser cavity. The dye absorbs over a wide bandwidth, but at high intensities the absorption is reduced because a large proportion of the dye molecules are in the excited state. If the laser is oscillating with several modes covering the wide range Δv, the pulse shape will contain structure with width $(\Delta v)^{-1}$, within a complex shape whose length is determined by a single mode (Figure 16.6(a)). The highest peak will be amplified more than the lower peaks at each pass through the cell (Figure 16.6(b)); repeated passages through the dye cell eventually amplify this peak into a single sharp pulse as seen in Figure 16.6(c).

Mode locking may also be achieved using a non-linear optical effect in which high laser irradiance produces an increase in the refractive index of a solid. This consequently induces self-focusing of the beam since, for a beamwidth which has a Gaussian-shaped radial irradiance profile in which the beam is more intense at the beam centre, the refractive index becomes greater on-axis, and acts as a converging lens. This effect is used in Kerr lens mode locking in which the self-focusing selects the pulsed mode-locked set of modes and discriminates against CW operation. Figure 16.7 shows a laser gain medium, particularly titanium–sapphire $Ti:Al_2O_3$, in a cavity containing an aperture whose

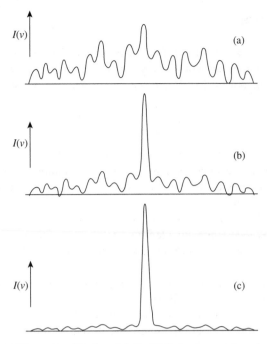

Figure 16.6 Mode locking with a saturable dye. (a) Oscillation at several modes simultaneously produces a complex pulse. (b) The highest peak is amplified selectively. (c) After many passes through the dye cell the peak becomes a single narrow pulse

function is to create high loss for CW operation. However, a laser pulse will suffer self-focusing in the laser medium and will have higher transmission through the aperture.

The pulse of a mode-locked laser can be compressed in time by factors of 20 or more to generate pulses as short as 1fs. This is achieved by a technique which induces a linear change in frequency along the pulse (known as a frequency *chirp*), followed by propagation through an optical system with dispersion in group velocity. The propagation delay for the back of the pulse is thus made less than for the front, and the pulse is correspondingly compressed. This may be achieved using a combination of two diffraction gratings. In a different spectral regime, this technique is also used to make short radio pulses for high-resolution radar systems.

Ultrashort laser pulses find many applications. Very fast processes in atoms, molecules and materials can be excited and probed. The short pulses, suitably amplified, are used as the pump source for X-ray lasers and to study high-temperature and high-density plasmas.

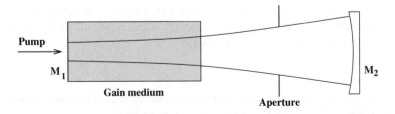

Figure 16.7 Kerr lens mode locking

16.5   Laser Radiance

The radiance[3] even of low-power lasers is often many orders of magnitude greater than the radiance of incoherent sources of light, because of the very high directionality of the laser beam. We recall that radiance R is defined as the power flow P per unit area A and per unit solid angle:[4]

$$R = \frac{P}{A\Delta\Omega} \text{ W m}^{-2} \text{ sr}^{-1}. \tag{16.21}$$

We recall also that no optical system can increase the radiance of a light source (provided that object and image are in media with the same refractive index); for example, by focussing with a lens it is possible to create an image with smaller area than the source but with light flowing over a correspondingly larger solid angle.

As an example of a bright non-laser source, the radiance of the Sun is about $5.0 \times 10^6 \text{ W m}^{-2} \text{ sr}^{-1}$; this cannot be increased by focussing with a lens or a mirror. In contrast even an ordinary low-power, e.g. 1 mW, He–Ne laser operating at 632.8 nm has a radiance $R \sim 10^9 \text{ W m}^{-2} \text{ sr}^{-1}$ which is brighter than a hundred Suns.

An ultrashort pulse laser, such as a mode-locked 1.06 µm Nd:YAG laser producing 1 mJ pulses with a pulse duration of 50 ps, has a power of 20 MW during the pulse; this is equivalent to a radiance $R = 2 \times 10^{19} \text{ W m}^{-2} \text{ sr}^{-1}$. High-power pulsed lasers followed by a train of amplifiers can achieve a radiance approaching $10^{22} \text{ W m}^{-2} \text{ sr}^{-1}$; furthermore the coherent wavefront from a laser can be focussed into a very small area. Focussed mode-locked pulses can attain extremely high power densities (of order TW cm^{-2}) in the focal region. These have widespread application in the processing of materials.

16.6   Focusing Laser Light

A laser beam may be focussed to very small focal spot, not much more than a wavelength across, giving extremely high power densities. Since diffraction from a circular aperture with diameter D uniformly illuminated by a plane wave gives a beam with angular radius

$$\theta = \frac{1.22\lambda}{D}, \tag{16.22}$$

the spot produced by a lens with focal length f has a diameter

$$\text{Focused spot diameter} = \frac{2.44\lambda}{D}f = 2.44\lambda F \tag{16.23}$$

where F is the focal ratio f/D. Even allowing for some lens aberration, spots of a few wavelengths in diameter can easily be achieved, giving power densities high enough for cutting and welding metals. For example, CO_2 lasers operating at 10.6 µm wavelength with a power of 500 W can be focussed to a spot 50 µm across, giving a power density of 250 kW mm^{-2}.

---

[3]The *radiance* of a source is frequently termed brightness in earlier and some current literature; we adopt the term radiance to conform with international convention (see Appendix 1).

[4]Note that *spectral* radiance (spectral brightness) also includes 'per unit bandwidth'.

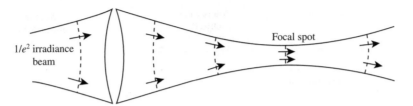

Figure 16.8   Focussing a laser beam by a converging lens

The action of a lens with a short focal length f is illustrated in Figure 16.8. Here the diameter d of the lens has been chosen to match the width of the wavefront of a beam at distance z from the waist, where the beam has expanded according to equation (15.38). Then

$$d = 2w_1 = 2\frac{\lambda z}{\pi w_o} \tag{16.24}$$

where $2w_o$ is the beamwidth at the waist. The wavefront emerging from the lens converges to form a focal spot with width $2w_f$, limited by diffraction of the wavefront to

$$w_f = \frac{2f\lambda}{\pi d} = \frac{2}{\pi}\lambda F \tag{16.25}$$

where F is the F-number of the lens. Provided that the lens diameter matches the width of the laser beam, the spot size is limited only by the F-number and the wavelength of the light. A practical low value is $F = 1$, giving a smallest spot size approximately equal to the wavelength of the laser light.

A 1 mW He–Ne laser focussed by a lens with $F = 1$ has a focal radius of $r_f = (2/\pi)(6.3 \times 10^{-7}) = 4 \times 10^{-7}$m. The power per unit area at the focus is 2×10^9 W m^{-2}.

16.7   Photon Momentum: Optical Tweezers and Trapping

The precision of focussing and the spectral purity of laser light have led to two remarkable applications of photon momentum, which we now describe.

16.7.1   Optical Tweezers

Optical tweezers use focussed laser light to manipulate microscopic objects and even individual atoms by trapping them in a focal spot. The mechanism is illustrated for a small transparent dielectric sphere in Figure 16.9. In Figure 16.9(a) a ray is refracted through the sphere, and the angular deviation of the ray transfers momentum to the sphere in the opposite direction. Figure 16.9(b) shows rays converging on a focal spot above the centre of the sphere, with the corresponding reaction forces combining to give a net upwards force, towards the focal point. Similar diagrams can be drawn for a sphere below or to one side of the focal spot; in each case the net force is towards the focal spot, which forms a trap for the dielectric sphere.

The force on a microscopically small dielectric sphere may be measured in nanonewtons (nN). Such a sphere may be attached to a biological molecule, such as DNA or a molecular motor, allowing measurements to be made of their strength and elasticity.

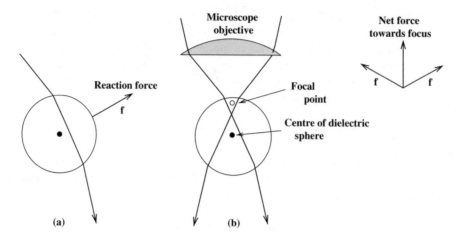

Figure 16.9 Forces on a dielectric sphere at the focus of laser light. (a) A ray is refracted and deviated, transferring momentum to the sphere in the opposite direction. (b) The reactions from rays converging from opposite sides combine to force the sphere towards the focal point. The converging rays are from a laser, focussed by a microscope objective lens

16.7.2   Laser Cooling

A dilute atomic gas in a vacuum chamber may be cooled by a laser beam which acts selectively on atoms with large thermal velocity, slowing them and thus cooling the gas. The selection is achieved by tuning the laser to a frequency immediately above a resonant frequency of the atom. The effective cross-section of the atom is maximum at resonance, but the Doppler effect of a thermal velocity towards the laser source shifts the resonance into coincidence with the laser frequency. Radiation pressure therefore slows the atoms moving towards the laser source. A laser illuminating the gas from the opposite direction acts similarly on atoms moving away from the first laser beam; two further pairs of laser beams on the orthogonal axes deal similarly with the other components of motion.

The interaction is best considered in terms of the transfer of momentum from photons by absorption in the atoms. Taking a sodium atom as an example, the r.m.s. thermal velocity at $300\,\mathrm{K}$ is about $570\,\mathrm{ms}^{-1}$. At a sodium D-line (wavelength $589\,\mathrm{nm}$), the laser must be tuned to a shorter wavelength, calculated from the Doppler shift for an atom travelling towards the laser, which is nearly $10^9\,\mathrm{Hz}$. A single collision with a photon transfers momentum $p = h/\lambda$, reducing the speed of the atom by about $0.03\,\mathrm{ms}^{-1}$. The 20000 collisions required to bring the velocity to zero occur typically within milliseconds.

Laser cooling was first achieved in 1985 by S. Chu, who reduced the temperature of a cloud of sodium atoms to below 1 millikelvin.

16.8   Non-linear Optics

Before the discovery of the laser the propagation of a light wave travelling in a medium could be described by a linear dependence of the polarization on the electric field of the light wave, $\mathbf{P} = \epsilon_0 \chi \mathbf{E}$. With laser beams the light irradiance can readily be large enough that the polarization response of the medium is non-linear on its dependence on the electric field. This has opened up the dramatic new subject of non-linear optics.

From Section 5.3 the r.m.s. electric field E in any electromagnetic radiation field is related to the irradiance I by $I = n\bar{E}^2/377\,\mathrm{W\,m}^{-2}$ (in a non-magnetic medium with refractive index n). The peak field E_{max} in a dielectric with refractive index n is therefore

$$E_{\mathrm{max}} = 27.4\left(\frac{I}{n}\right)^{1/2}\,\mathrm{V\,m}^{-1},\qquad(16.26)$$

where I is measured in $\mathrm{W\,m}^{-2}$. A peak field reaching $10^{12}\,\mathrm{Vm}^{-1}$ is attainable in an ultrashort pulse from a high-powered laser. This is greater than the typical internal field strength of a dielectric, or the field binding the electron to a proton in the hydrogen atom.[5] A laser pulse can therefore completely disrupt a dielectric medium. Expensive optical components have been destroyed in a few picoseconds in this way!

At lower fields, in the range 10^7 to $10^9\,\mathrm{V\,m}^{-1}$, the dielectric may respond non-linearly to the field and generate harmonics. We have previously treated the polarization of a dielectric as proportional to the electric field; we must now include further terms and write

$$\mathbf{P} = \epsilon_0(\chi\mathbf{E} + \chi^{(2)}\mathbf{E}^2 + \chi^{(3)}\mathbf{E}^3 + \ldots)\qquad(16.28)$$

where χ is the normal linear susceptibility of the dielectric, and $\chi^{(2)}, \chi^{(3)}$, etc., are second, third and higher order terms; $\mathbf{P}$ and $\mathbf{E}$ represent (signed) components along any given direction. The origin of the non-linear response is from the non-linear movement of the outer, more loosely bound electrons in the medium. In the Lorentz model for the interaction of electromagnetic radiation with a dielectric, described in Chapter 19, electrons are harmonically bound to an ionic core. In the linear model the outer electrons respond to the electric field of a light wave experiencing a force $F = -m\omega_0^2 x$, where m and x are the mass and displacement of the electron. The classical model is modified under strong electric fields with the addition of an anharmonic force proportional to x^2, leading to a non-linear equation of motion for the electron of the form

$$\ddot{x} + \omega_0^2 x + \alpha x^2 = \frac{e}{m}E\cos\omega t\qquad(16.29)$$

where damping has been omitted.

A light wave with a field $E = E_0\cos\omega t$ induces a polarization

$$\begin{aligned}P &= \epsilon_0(\chi E_0\cos\omega t + \chi^{(2)}E_0^2\cos^2\omega t + \chi^{(3)}E_0^3\cos^3\omega t + \ldots)\\ &= \epsilon_0[\chi E_0\cos\omega t + \tfrac{1}{2}\chi^{(2)}E_0^2(1 + \cos 2\omega t)\\ &\quad + \tfrac{1}{4}\chi^{(3)}E_0^3(3\cos\omega t + \cos 3\omega t) + \ldots].\end{aligned}\qquad(16.30)$$

The polarization P is therefore oscillating at harmonics $2\omega, 3\omega$, etc., and radiating waves at these higher frequencies. Frequency doubling, i.e. the generation of the second harmonic, is commonly

---

[5]The field of a point electric charge e at distance $r_0 = 0.1\,\mathrm{nm}$ is

$$E = \frac{e}{4\pi\epsilon_0 r_0^2} \simeq 10^{11}\,\mathrm{V\,m}^{-1}.\qquad(16.27)$$

achieved in non-isotropic materials; harmonics above second order may also be produced at higher field strengths in isotropic materials. The term $\frac{1}{2}\chi^{(2)}E_0^2$ is time independent and describes the creation of a constant field across the medium. This effect is known as *optical rectification*.

There is a distinction between the second- and third-order processes. For materials that are isotropic or centrosymmetric, $\chi^{(2)} = 0$ and no second-order processes occur. A medium has a centre of symmetry if an electron at position **r** relative to that point experiences the same field when at position $-\mathbf{r}$. If we imagine reversing the sign of E, the sign of the total polarization must also reverse. However, since $P^{(2)} \propto \chi^{(2)}E^2$ this can only occur if $\chi^{(2)} = 0$. Hence $P^{(2)}$ only occurs in materials without a centre of symmetry, i.e. non-centrosymmetric. Certain crystals are non-centrosymmetric, while gases and liquids are centrosymmetric. Third-order processes occur for both centrosymmetric and non-centrosymmetric materials. Typical values for the non-linear susceptibilities are $\chi^{(2)} \sim 2 \times 10^{-11} \text{m V}^{-1}$ and $\chi^{(3)} \sim 4 \times 10^{-23} \text{ m}^2 \text{ V}^{-2}$. In general for anisotropic materials, $\boldsymbol{P}$ and $\boldsymbol{E}$ are not in the same direction. The non-linear polarizability $\chi^{(2)}$ depends on the polarization of the electric field, the orientation of the optic axis of the crystal and the direction of propagation. This requires $\chi^{(2)}$ to be a tensor, such that the second-order non-linear polarization is

$$P_i^{(2)} = \epsilon_0 \sum_{ijk} \chi_{ijk}^{(2)} E_j E_k. \qquad (16.31)$$

Here i, j, k represent the coordinate directions x, y, z. (Equation (16.31) includes isotropic materials as a special case.)

An interesting aspect of these processes is their interpretation in terms of photons. Two identical photons arriving nearly simultaneously at a molecule in a crystal lattice can emerge from the encounter as a single photon with twice the energy: this is frequency doubling. The probability of such close encounters depends on the flux of photons, since two must be found close to the same molecule for the interaction to occur; this is equivalent to the power-law dependence on the field strength in equation (16.30).

Frequency doubling is important as a way of producing coherent light at new or shorter wavelengths; a laser beam at frequency v_1 traversing a medium for which $\chi^{(2)} \neq 0$ can be converted into a beam at frequency $v_2 = 2v_1$. A practical problem is that the original laser light and its second harmonic must travel along the ray path through the dielectric with the same velocity; if they are different, the second harmonic light generated from different parts of the path will not add correctly in phase. Most dielectrics are sufficiently dispersive for this to be a serious limitation on the thickness of a harmonic generator. In some birefringent materials it can be arranged that the fundamental and second harmonic waves are polarized as ordinary and extraordinary waves (see Chapter 7), and a propagation direction can be chosen in which the two refractive indices are equal. A commonly used material for this purpose is potassium dihydrogen phosphate, known as KDP; the efficiency of frequency doubling can exceed 50% with this material.

The non-linear crystal may be placed outside the laser resonator, or inside where the fundamental irradiance is greater; the latter generally leads to higher efficiency. A common application is the frequency doubling of the pulsed or CW Nd:YAG laser at 1.064 μm to its second harmonic at 532 nm. Coherent radiation in the UV down to ~ 200 nm can be obtained by second harmonic generation in $\beta\text{-BaB}_2\text{O}_4$ which transmits in the UV.

The irradiance of the second harmonic at frequency 2ω grows as the fundamental wave at frequency ω propagates in the crystal (Figure 16.10). For the propagation direction z

$$\frac{dI^{(2)}}{dz} \propto P^{(2)}(z). \qquad (16.32)$$

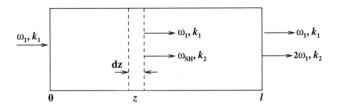

Figure 16.10   Second harmonic generation in a crystal

The induced second harmonic dipole moment per unit volume $P^{(2)}$ is proportional to E^2, with angular frequency $2\omega_1$ and wave vector $2k_1$. However, the second harmonic propagates with wavevector k_2. Because of the dispersion in the refractive index of the crystal, $k_2 \neq 2k_1$. For a crystal of length l (Figure 16.10) the second harmonic irradiance produced from the element dz at position z is

$$dI^{(2)}(l) \propto dI^{(2)}(z) \exp[ik_2(l - z)]dz$$
$$\propto \exp[i(2k_1 - k_2)z] \exp[i(k_2 l - 2\omega_1 t)]dz. \qquad (16.33)$$

If we assume that the conversion of the fundamental into the second harmonic is small, so that the incident fundamental irradiance is undepleted, equation (16.33) can be integrated to give

$$I^{(2)}(l) \propto \frac{\sin[(2\pi/\lambda)(n_2 - n_1)l]}{(2\pi/\lambda)(n_2 - n_1)}. \qquad (16.34)$$

The second harmonic irradiance is a maximum when $l = \lambda/4(n_2 - n_1)$. The length over which conversion of fundamental to second harmonic occurs is the coherence length for second harmonic generation. This length can be greatly extended by ensuring that $n_2 = n_1$, i.e. the refractive index at the second harmonic is equal to the refractive index of the fundamental. This is the *phase match* condition. It may be achieved by using the birefringence of an anisotropic crystal.

Figure 16.11 shows the angular dependence of the refractive indices for ordinary and extraordinary waves in a negative uniaxial birefringent crystal. If the fundamental wave at wavelength λ_1 is incident as an ordinary ray, there is coincidence with the refractive index of the second harmonic generated at $\lambda_2 = \lambda_1/2$ for a certain angle θ_m if the second harmonic is propagating as an extraordinary ray. Under these conditions the two waves are phase matched.

An alternative method for efficient second harmonic generation is to create a material in which the orientation of a ferroelectric domain is alternated after each coherence length. Then successive elements add to the irradiance of the second harmonic and quasi-phase matching is achieved. This structure may be realized by periodic application of an electric field to the crystal (called periodic poling) such as $LiNbO_3$ or $KTiOPO_4$, in a manner similar to microelectronics fabrication.

Two laser beams with different frequencies ω_1, ω_2 propagating in a non-linear dielectric may induce polarization oscillating at the difference and sum frequencies $\omega_1 - \omega_2$, $\omega_1 + \omega_2$. This is known as *optical mixing*. Again these processes are valuable in generating new coherent wavelengths. The efficiency of the process depends on matching refractive indices.

Which optical mixing process is dominant is determined by the phase-matching condition. In difference frequency mixing in which $\omega_1 \rightarrow [(\omega_1 - \omega_2), \omega_2]$ the frequencies ω_2 and $(\omega_1 - \omega_2)$

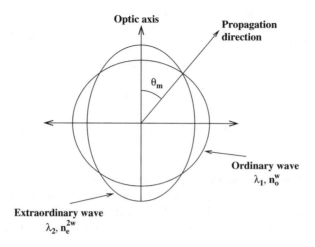

Figure 16.11   Phase matching with a negative uniaxial crystal ($n_e < n_o$)

are generated. By placing the crystal in a resonator which selectively resonates ω_2, the wave at this frequency can be amplified. This is the basis of the optical parametric oscillator. The intense beam at ω_1 is designated as the pump, the amplified wave at ω_2 is the signal and the difference frequency is termed the *idler*. Importantly the parametric oscillator is a tunable source in which the signal frequency ω_2 can be varied by rotating the crystal or changing its temperature.

In frequency doubling the two conditions apply: on photon energy $\omega_{SH} = 2\omega$ and on photon wave number $k_{SH} = 2k_\omega$. These equations are clearly consistent with the conservation of energy $\hbar\omega_{SH} = 2\hbar\omega$, and the conservation of momentum $\hbar k_{SH} = 2\hbar k_\omega$.

The third-order non-linear susceptibility $\chi^{(3)}$ provides the interaction for the generation of the third harmonic of the fundamental beam. It also enables the non-linear process of optical phase conjugation via a four-wave mixing interaction. In this process a wave $E_1 = E_0 \exp[i(\omega t - kz)]$ incident on a phase conjugate cell can be converted into a reflected counter-propagating wave $E_r = aE_0^* \exp[i(\omega t + kz)]$ where E_0^* is the conjugate of E_0, so that E_r is exactly the phase conjugate of the incident wave, with a change in amplitude through the reflectivity coefficient a. The reflected wave retraces the path of the incident wave and its spatial phase distribution replicates the phase distribution of the incident wave. As an example of its usefulness, consider a plane wave which traverses a medium in which phase distortion occurs, e.g. from aberrations in an optical system or thermal aberrations in a laser amplifier, as illustrated in Figure 16.12(a). On reflection from a phase conjugate mirror, the reflected wave retraces its path such that the original phase distortion is removed.

Phase conjugation acts as a real-time adaptive optical system, able to compensate for beam propagation in a turbulent or distorting medium. A phase conjugate mirror can be formed by four-wave mixing, in which a signal of amplitude E_3 interacts with two counter-propagating waves E_1 and E_2 in a third-order non-linear medium illustrated in Figure 16.12(b). The induced non-linear polarization is proportional to E_3^*. The induced electric field is then proportional to the complex conjugate of the input electric field. A practical phase conjugate medium is a gas cell containing carbon disulphide CS_2 for which $\epsilon_0\chi^{(3)} \sim 4 \times 10^{-32}$ SI units (CmV^{-3}).

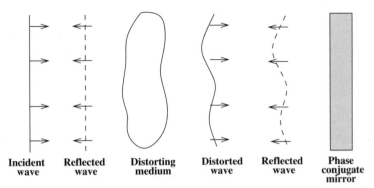

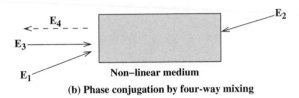

(a) Phase conjugate mirror acting to correct wavefront distortion

(b) Phase conjugation by four-way mixing

Figure 16.12 Phase conjugation. (a) A phase conjugation mirror acting to correct wavefront distortion. (b) Phase conjugation by four-wave mixing

---

Problem 16.1

A 1 watt laser beam is focussed onto a spot 10 μm in diameter. Calculate the irradiance (see Appendix 1) and the mean electric field in the spot (see Section 5.5). What is the maximum temperature attainable in the spot?

Problem 16.2

A solid ruby rod laser 0.2 m long with refractive index 1.76 and coated end faces to form the resonator produces mode-locked pulses. What is the time interval between the pulses?

Problem 16.3

A collimated He–Ne laser beam, wavelength 632 nm, is required for surveying over a distance of 10 km. The beam will be expanded optically: what waist diameter will be needed?

Problem 16.4

Summarize the properties of a Gaussian light beam. Explain why a Gaussian light beam remains Gaussian after passing through a lens.

Problem 16.5

From equation (16.7) calculate the minimum linewidth Δv obtainable from a 1 mW He–Ne laser ($\lambda = 633$ nm) if the cavity decay time is 10^{-7}s. Why is this theoretical limit never attained? If the laser length is 1 m, what change in length would give a frequency shift Δv equal to the linewidth? If the coefficient of thermal expansion of the cavity is 10^{-6} K^{-1}, what temperature change would change the length by this amount?

Problem 16.6

Compare the coherence lengths of the following sources: (a) a heated filament lamp with a white light output over the wavelength range 400 to 700 nm; (b) a stabilized CW Nd:YAG laser operating on a single mode with a linewidth of 20 kHz; (c) an He–Ne laser with a resonator length of 30 cm oscillating in three longitudinal modes.

Problem 16.7

A 3 mW helium–neon laser ($\lambda = 633$ mm) has an emission linewidth $\Delta v = 8$kHz.

(a) If the beam diameter is 0.34 mm, find, and compare with that of the laser, the power emitted by an equal area of the following:

  (i) The Sun over all frequencies. Assume it radiates like a blackbody at temperature $T = 5800$ K.

  (ii) The Sun over a frequency range equal to Δv of the laser, and centred on the same frequency. (Hint: The spectral irradiance $I(x)$ of a blackbody (power per unit area per unit frequency), where $x = hv/kT$, is given in Section 5.7.)

(b) Comment on the preceding.

Problem 16.8

The radiation pressure P_{rad} of blackbody radiation is related to its energy density u by $P_{rad} = u/3$.

(a) At what temperature T will the radiation pressure be 10^{-2} bar? (1 bar $= 10^5$ Nm$^{-2} \approx 1$ atm).

(b) The pressure supporting stars is the sum of gas pressure and radiation pressure. Models of the Sun's interior predict that at the centre of the Sun, the temperature is $T = 1.55 \times 10^7$K, and the total pressure is $P_{tot} = 3.4 \times 10^{11}$ bar. Assuming that the interior acts like a blackbody cavity, find out the relative importance of the radiation in the pressure balance at the Sun's centre.

Problem 16.9

A laser of power Π is used to focus a spot of diameter 2 μm on a totally reflective mirror surface. Find the value of Π such that the spot exerts a pressure of 10^{-2} bar.

Problem 16.10

An argon ion laser has a resonator length of 100 cm and a Doppler broadened linewidth $\Delta v_D = 3.5$ GHz. In this laser the magnitude of the loss coefficient is half that of the peak value of the small-signal gain coefficient. The refractive index of the laser medium can be assumed to be unity. Determine (a) the frequency spacing of the longitudinal resonator modes, (b) the number of longitudinal modes that the laser can sustain.

Problem 16.11

An He–Ne laser operating at 633 nm generates a Gaussian beam with a minimum spot diameter $2\omega_0 = 0.2$ mm. Determine (a) the angular divergence of the beam, (b) its depth of focus, (c) the radius of curvature of the wavefront with distance z along the propagation distance for $z = 0$ and $z = z_0$, where z_0 is the Rayleigh range, (d) the diameter of the laser beam after travelling across the city of London, assuming a distance of 25 km.

Problem 16.12

The beam from an Nd:YAG laser ($\lambda = 1.06$ μm) has an initial diameter of 5 mm and is required to be focused to a diameter of 0.5 mm. Calculate the focal length of the lens required. With the assumption that the focal region can vary by up to 10%, what is the depth of focus?

Problem 16.13

Consider the conversion of a fundamental wave to its second harmonic when propagating over a length L in a non-linear crystal. If k_1 and k_2 are the wave vectors for the fundamental and second harmonic waves show that the irradiance I of the second harmonic is

$$I \propto \left[\frac{\sin(k_1 - k_2/2)L}{(k_1 - k_2/2)L}\right]^2. \tag{16.35}$$

Estimate the propagation distance in a KDP crystal, under conditions without phase matching, for the highest conversion from a fundamental wave of 800 nm to its second harmonic. For KDP the refractive index at 800 nm is 1.5019 and at 400 nm is 1.4802.

Explain how phase matching can increase the conversion efficiency.

Problem 16.14

Two lasers of high irradiance have wavelengths of 0.5μm and 0.75μm. What non-linear optical processes could be used to generate light at (a) 0.3μm and (b) 1.5μm? How can one of the processes be made to dominate the other?

Problem 16.15

The momentum transfer exploited by optical tweezers can be illustrated by considering the interaction between a light beam and a lens.

(a) A uniform monochromatic light beam of power P falling normally on the vertex of a thin lens of focal length f is brought to a focus on-axis. The incident light consists of N photons per unit volume, each carrying momentum p and energy pc parallel to the optic axis ($+ z$ axis). If the beam's radius is r, show that the power can be written as $P = Npc^2\pi r^2$.

(b) By considering the deflection of the photons by the lens, find the average change of photon momentum along the axis, and show it is $-(p/4)(r/f)^2$. Deduce that the total refractive force on the photons is $F_{z(\text{phot})}^{\text{refr}} = -(r/f)^2 (P/4c)$.

(c) In addition to the refractive force, the photons also experience a scattering force when reflected at the glass–air interfaces. From equation (5.36), the reflectance is $R = [(n_2 - n_1)/(n_2 + n_1)]^2$ in going from medium 1 to 2, or from 2 to 1. For a typical glass–air interface with $n_2 = 1.5, n_1 = 1, R$ is only 4% and our neglect of this in part (b) was justified. Ignoring multiple reflections, the two surfaces of the lens give $R \approx 2[(n-1)/(n+1)]^2$, where $n = n_2/n_1$. Write an expression for the scattering force on the light.

(d) Deduce an expression for the total reaction force on the lens, including both types of force.

(e) Evaluate the total force on the lens for $P = 6$ mW, $r = 0.5$ mm, $f = 8$ cm, $n = 1.5$.

17 Semiconductors and Semiconductor Lasers

How bright these glorious spirits shine!

Isaac Watts, 1674–1748.

Semiconductors play a vital role in optics both as sources and as detectors of light. The light-emitting diode (LED) and laser diode are widely used, as are the various forms of photodiode detector.

The semiconductor laser in its many forms is the most numerous of all lasers. It has widespread application, e.g. in optical fibre communication systems, barcode scanners, laser printers and the compact disc player. Semiconductor lasers, like the gas and solid state lasers considered in Chapter 15, depend on stimulated emission to produce coherent light in a resonator. They have, however, different pumping and photon generating processes, which we now consider.

Semiconductor lasers have many valuable properties. They have high efficiencies (defined as laser power output/electrical power input) of typically 30 to 50%, which is higher than most other lasers. They are very small, typically with dimensions of less than 1 millimetre, and require only modest power supplies, operating typically at a few volts and currents of 10 mA to a few amps. Semiconductor lasers use direct electrical pumping; modulation at frequencies typically up to 20 GHz makes them very suitable for optical communications. The wide range of semiconductor lasers at many wavelengths and power levels and their particular radiation characteristics lead to many other applications, such as in spectroscopy, sensing and optical data storage.

In this chapter we review the basic physics and radiative mechanisms of semiconductors and LEDs. We describe the structures and operation of practical forms of semiconductor lasers, including heterostructures and quantum well diodes. We briefly review the radiation characteristics of semiconductor lasers, which are distinct from those of other lasers.

17.1 Semiconductors

An isolated atom, of atomic number Z, consists of a positively charged nucleus with charge $+Ze$, surrounded by Z electrons of charge $-e$. The electron energies are quantized into discrete levels, and

Optics and Photonics: An Introduction, Second Edition   F. Graham Smith, Terry A. King and Dan Wilkins
© 2007 John Wiley & Sons, Ltd

without thermal excitation the electrons occupy the Z lowest energy states of the atom. When a large number N of such atoms are brought together to form a solid, the interactions between the atoms spread the allowable energy levels into bands, each containing $2N$ energy states (the factor 2 results from the two-fold degeneracy of the atomic levels due to electron spin). Without thermal excitation, electron energies fill the lowest possible bands. For electrons to provide conductivity by moving through the material, their energies must be in higher levels. Thermal excitation can raise electrons into higher energy states; however, if Z is even, as in silicon, the topmost level of the allowed band of states is full at low temperatures, and the electron can only be excited into an empty energy state by surmounting a gap between the full and empty bands. The last fully occupied band is the *valence band* and the first empty band is the *conduction band*. The bands are separated by the *bandgap E_g*; for silicon the bandgap is 1.1 eV.

Silicon is called a semiconductor because of the relative ease of exciting electrons from the valence band to the conduction band. For a much larger bandgap, the solid would be an insulator. If the bandgap does not exist (i.e. the lowest energy of the upper band is less than the highest energy of the lower band), or if Z is odd giving a half-filled upper band, then the solid is a conductor, i.e. a metal. The schematic energy bands and their occupancy are shown in Figure 17.1 for a metal, a semiconductor and an insulator.

At absolute zero temperature there are no electrons in the conduction band of a pure semiconductor and the material is a perfect insulator. Thermal excitation raises the energies of a small number of electrons into the conduction band. This leaves a corresponding number of unoccupied energy states in the valence band; both the free electrons and the vacancies are important in the behaviour of a semiconductor.

When the energy of an electron takes it into the conduction band, the unoccupied state, or *hole*, in the valence band allows some movement among the remaining electrons. This movement of the valence electrons is best understood by regarding the hole as a positively charged particle which has its own mobility and mass and which can contribute to the conductivity of the semiconductor. If an external electric field is applied, the electron and the hole move in opposite directions, the electron moving faster than the hole. The thermally excited electrons in the conduction band and the holes in the valence band are *carriers* and provide conduction in the semiconductor. Electrons may also be excited into the conduction band by absorbing the energy of a photon; this is the basis of a *photoconductor*, in which photons with sufficient energy to excite electrons directly from the valence band into the conduction band are detected by an increase in conductivity. For metals the conduction band is part filled with electrons which, with application of a potential, are able to move to provide

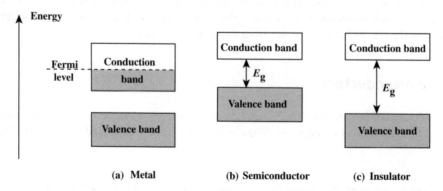

Figure 17.1   Electron energy levels in (a) a metal, (b) an intrinsic semiconductor, (c) an insulator

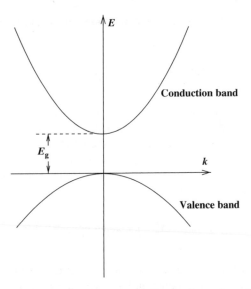

Figure 17.2 Parabolic electron energy–wave vector diagram for a direct bandgap semiconductor

conduction. In the insulator the valence band is filled with electrons and the conduction band is empty, such that conduction is not possible. The bandgap for the semiconductor is less than for the insulator so that electrons are more readily able to be excited from the valence band to the conduction band.

The energy E of an electron excited into the conduction band is measured from the bottom of this band. Electrons of energy E move as waves with wave vector magnitude k. The allowed energies of the electrons are related to the wave vector as $E = \frac{1}{2}m^*v^2 = \hbar^2k^2/2m^*$, where m^* is the effective mass of the electron.[1] E and k for an electron in the conduction band are therefore related as shown in the upper curve of Figure 17.2. The energy of an electron in the valence band is measured downwards from the top level of the valence band, giving the lower curve.

The concentration of electrons n_e in the conduction band and the concentration of holes in the valence band are determined by the density of available states as a function of energy, $\rho(E)$, and the probability $f(E)$ in the conduction band and $[1 - f(E)]$ in the valence band of the states being occupied. The number of electrons and holes per unit volume, n_e and n_h, within the energy range dE is

$$n_e dE = f(E)\rho(E)dE$$
$$n_h dE = [1 - f(E)]\rho(E)dE. \tag{17.1}$$

The density of states as a function of wave vector $\rho(k)$ is $\rho(k)dk = k^2dk/\pi^2$. Substituting for k, the density of states as a function of energy is

$$\rho(E)dE = \frac{1}{2\pi^2}\left(\frac{2m_e}{\hbar^2}\right)^{3/2}E^{1/2}dE. \tag{17.2}$$

[1] The effective mass differs from the free electron mass because of interactions between the electron wave and the crystal lattice.

The probability at temperature T of an electron being found in an energy state E follows Fermi–Dirac statistics and is

$$f(E) = \frac{1}{1 + \exp\left[(E - E_F)/kT\right]}. \tag{17.3}$$

In equation (17.3) E_F, the Fermi energy, is that energy value for which the probability of the state being occupied is $\frac{1}{2}$. At temperature $T = 0$ all the energy states below E_F are completely filled and above E_F they are completely empty. In equilibrium the electrons and holes have a common Fermi energy. To obtain the total number of electrons per unit volume in the conduction band we integrate over the range of energies. With E_c being the lowest energy in the conduction band, the concentration of electrons in the conduction band is then

$$n_e = 2\left(\frac{2\pi m_e kT}{h^2}\right)^{3/2} \exp\left[(E_F - E_c)/kT\right]. \tag{17.4}$$

From equation (17.4) we see that the concentration of electrons in the conduction band markedly increases as E_F moves closer to the conduction band. (A change in E_F by 0.5 eV corresponds to a change of about 5×10^8 in the concentration of electrons in the conduction band.) The concentration of holes in the valence band can be calculated in a similar way.

Equation (17.4) is invalid if the concentration of electrons in the conduction band is low and we regard the electrons in the conduction band as forming a gas of classical particles obeying Boltzmann statistics. The quantity before the Boltzmann term, $2(2\pi m_e kT/h^2)^{3/2}$, is the effective concentration of levels in the conduction band.

There is a discrete set of allowed wave vectors for an electron in the crystal lattice. Figure 17.3(a) shows an electron transition from an allowable state in the valence band, leaving a hole, into the conduction band. Transitions without a change in wave number, as shown in Figure 17.3(a), occur in *direct gap* semiconductors, e.g. GaAs. Figure 17.3(b) shows a transition in an *indirect gap* semiconductor, in which the energy minimum of the conduction band is not at the same value of k as that of the valence band. For the indirect bandgap case, e.g. in Si or Ge, the transition of the electron from the conduction band to the valence band must involve a change in wave vector, which contravenes the selection rule for an allowed electron transition. The transition can occur only if there

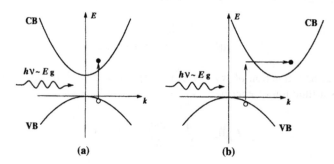

Figure 17.3 Electron excitation from the valence to the conduction band in (a) a direct gap and (b) an indirect gap semiconductor. Occupied and vacant allowable states are shown as filled and empty dots on the parabolic E/k curve

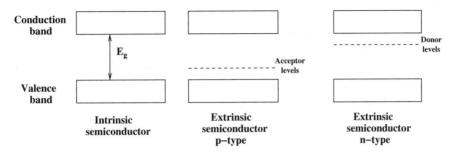

Figure 17.4 Electron energy levels in a doped (extrinsic) semiconductor

is also an interaction with a quantized unit of mechanical lattice oscillation, or *phonon*,[2] within the crystal, in order to conserve momentum. For this reason only the direct bandgap materials are efficient light emitters.

The energy band structure of the semiconductor may be modified by introducing impurity atoms into the crystal lattice; this is known as *doping*. At small concentrations the impurity may replace atoms without changing the crystal structure, and increase the conductivity either by releasing extra electrons or by creating holes.

The crystal lattice may be able to accept a replacement of 1% or more of its atoms, but doping usually extends only to the order of 1 in 10^3; a lightly doped semiconductor may have only one impurity atom in 10^6 of the host atoms. Silicon atoms are in group IV of the Periodic Table and so have four outer electrons. A donor impurity from group V with five outer electrons has one more electron than required for covalent bonding with neighbouring silicon atoms. This additional electron is much more easily lost to the conduction band, with an excitation energy of only 0.1 eV. Such an impurity is a *donor*. On the other hand an *acceptor* impurity, such as boron (group III), has three outer electrons and so contributes a hole to the valence band by allowing an electron from the valence band to be localized at the boron atom. A silicon semiconductor with a group V donor is known as *n-type*, and with a group III donor as *p-type*.

The new dopant-induced impurity energy levels are full, and are situated within the bandgap. The donor energy levels of n-type are close to the conduction band, and the p-type acceptor levels are close to the valence band, as shown in Figure 17.4. Increasing the electron concentration in the conduction band moves the Fermi level close to the conduction band as given by equation (17.4). Semiconductors with added donor or acceptor dopants are known as *extrinsic*, in contrast to those which contain no dopants, which are known as *intrinsic*.

17.2 Semiconductor Diodes

A semiconductor diode is a junction between the two types of doped semiconductor, n-type in which the dopant produces extra electrons, and p-type in which there are extra holes. When the p–n junction is formed, electrons and holes diffuse across the junction forming a contact region which is depleted

---

[2]Phonons are discussed in detail in J. R. Hook and H. E. Hall, 2nd edn, *Solid State Physics*, John Wiley & Sons, 1991. Their important property at issue here is that they have a larger wave vector for a small energy compared with photons; hence a transition for a phonon in Figure 17.3 is almost a horizontal line.

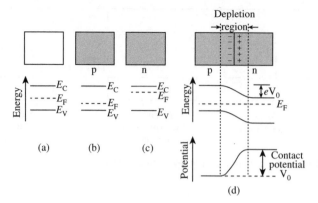

Figure 17.5 The development of a contact potential across a p–n junction, showing the Fermi energy levels (a) in the pure material; (b) in p-type (with added acceptor impurities); (c) in n-type (with added donor impurities); (d) when the junction is made

of charge carriers, known as the *depletion layer*. The depletion layer has a high resistance and a large contact potential; an electric field develops across it due to the dipole layer of positively charged donors in the n-region and negatively charged acceptors in the p-region, preventing further diffusion. In photodiode detectors (Chapter 20) photons are absorbed and generate electrons and holes within the depletion layer of the junction; the intrinsic electric field between the n- and p-type regions then transports these free charge carriers to give a current in an external circuit.

The development of the contact potential across the junction may be understood in terms of the Fermi energy levels[3] in the two components of the junction. In the intrinsic semiconductor, with no doping, the Fermi level is midway between the valence and conduction band; thermal excitation is sufficient for the energies of a small number of electrons to reach the conduction band. The impurity bands of the doped material extend the valence band upwards for the p-type and downwards for the n-type, and the Fermi levels are displaced upwards and downwards as shown in the diagrams of Figure 17.5. The valence and conduction bands of the pure semiconductor are shown in Figure 17.5(a), with the Fermi energy level between. The effect of doping is seen at (b), for the p-type, and at (c), for the n-type; the Fermi levels are lowered and raised as shown. When the two types of semiconductor are in contact, electrons and holes can flow across the junction until the Fermi levels are equalized. The energy levels adjust to give the same Fermi level, and the contact potential V_0 is developed.

This potential difference develops in the depletion layer at the junction. The n-type becomes positively charged, and the p-type negatively charged. An externally applied potential making the p-type more positive is a *forward bias* (Figure 17.6), which increases the flow of electrons from the n-region and holes from the p-region; the diode then has a low resistance. With *reverse bias* there is only a small reverse current, and the resistance is high. With biassing the conduction band electrons and valence band holes have different Fermi levels. Figure 17.7 shows the voltage–current characteristic of a typical semiconductor diode.

The exponential form of the diode characteristic at low voltages follows from the probability that a charge carrier can surmount the potential barrier at the junction; this is proportional to

---

[3]In a metal at zero temperature the Fermi level is the energy of the highest occupied state, as shown in Figure 17.1. As temperature increases, electrons move from below the Fermi level to higher states, providing electrical conduction.

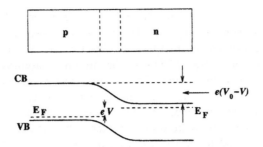

Figure 17.6   A diode junction with forward bias V

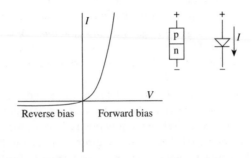

Figure 17.7   The voltage–current characteristic of a typical semiconductor diode. The diode symbol and the current flow are shown in the inset

$\exp[-e(V_0 - V)/kT]$. The diode characteristic relating current I to applied voltage V takes the form

$$I = I_0[\exp(eV/kT) - 1]. \tag{17.5}$$

17.3   LEDs and Semiconductor Lasers

The simplest light-emitting semiconductor diode is a p–n diode in a material such as gallium arsenide (GaAs), illustrated in Figure 17.8. The active region of the diode is at the junction between layers of p- and n-doped GaAs. The n-doped side of the junction contains mobile electrons in the conduction

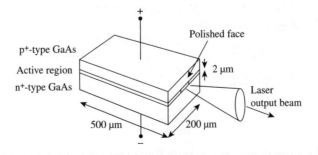

Figure 17.8   Schematic illustration of a semiconductor homojunction diode laser

band, and the p-doped side contains mobile holes in the valence band. Typical dimensions are submillimetre, as shown in the diagram. The active laser material is grown on a substrate selected so that the lattice spacings of the two materials are closely matched. In a laser diode the parallel end faces of the crystal are cleaved and polished to form a laser resonant cavity. The surfaces are often not given reflective coatings, since the reflection coefficients can be large enough due to the high refractive index (for GaAs, $n = 3.6$, giving a reflectivity, calculated from the Fresnel equation (5.30), of 32%). A simplified diagram of the energy band structure is shown in Figure 17.9. The donor and acceptor concentrations are sufficiently large that the Fermi level is in the conduction band for the n-type material and in the valence band for the p-type material. The electrons in the conduction band and the holes in the valence band act as degenerate gases, and equation (17.4) no longer strictly applies.

When a current passes in the forward direction, electrons from the n-doped side of the junction are injected at high density into the p-region of the junction, and holes from the p-region into the n-region. The electrons and holes recombine to emit photons; this mechanism of *radiative recombination* is the basis of the LED and the semiconductor laser. The electron–hole recombination time ($\sim 10^{-9}$s) is equivalent to the radiative lifetime of an atom or molecule in a gas or an ion in a doped crystal laser material.

For the semiconductor diodes used in LEDs and lasers the p- and n-regions are heavily doped ($\sim 0.1\%$) to give a large population inversion; the n-type material is more heavily doped than the p-type material and may be denoted by n^+. When the junction is formed, the movement of electrons and holes causes the n-region to be depleted of majority electron carriers and the p-region to be depleted of majority hole carriers. The contact potential V_0 creates a barrier to further electrons moving from the n- to p-region or to further holes moving from the p- to n-region.

When the junction is forward biassed (by giving the p-region a positive potential V with respect to the n-region), carriers are injected, the band energies are modified and the junction potential is reduced to $(V_0 - V)$. Filled electron states in the conduction band have energies above those of hole (empty electron) states in the valence band as shown in Figure 17.9. In a heavily doped p–n junction the concentration of electrons in the bottom of the conduction band can be much greater than in the top of the valence band. This is equivalent to a *population inversion* and stimulated recombination radiation can occur; this is the basis of gain and laser action in the diode laser.

Current flows in the p–n junction by injection of minority carriers – electrons into the p-region and holes into the n-region. To maintain electrical neutrality in the n- and p-regions the concentrations of mobile electrons in the n-region and holes in the p-region rise to balance the injected excess carriers.

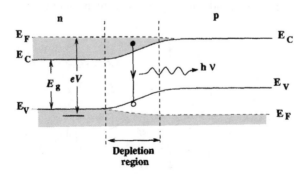

Figure 17.9 Radiative recombination in a strongly forward-biassed p–n junction. An electron undergoes a transition from the conduction band to the valence band, providing a photon

The injected excess carriers are removed by recombination of electrons and holes; an electron with energy in the conduction band falls into an empty electron state of lower energy in the valence band. Either this produces an emitted photon or the energy is lost non-radiatively.

The concentration of electrons in states at the bottom of the conduction band can be much greater than the concentration of electrons in states at the top of the valence band. With n_e electrons per unit volume in the conduction band and n_v electrons per unit volume in the valence band, then $n_e > n_v$. The gain of the laser from equation (15.25) is

$$\gamma(v) = \frac{\lambda^2 A_{21}}{8\pi n^2} g(v) \left(n_2 - \frac{g_2}{g_1} n_1 \right). \tag{17.6}$$

For the semiconductor laser $n_2 \equiv n_e$ and $n_1 \equiv n_v$, A_{21} is the electron–hole radiative recombination rate, $g(v)$ is the lineshape function and n is the refractive index of the medium. The normal situation is that there is a high density of holes in the valence band such that $n_v \approx 0$. Then the gain coefficient is

$$\gamma(v) = \frac{\lambda^2 A_{21}}{8\pi n^2} g(v) n_e. \tag{17.7}$$

The radiative recombination transition is homogeneously broadened with a Lorentzian lineshape and linewidth Δv. At line centre v_0 and assuming $g(v_0) = 2/\pi\Delta v$ the gain coefficient is

$$\gamma(v_0) = \frac{\lambda^2 A_{21} n_e}{4\pi^2 n^2 \Delta v}. \tag{17.8}$$

From equation (15.32) the threshold gain for a gain length L and k losses per unit length, excluding reflector losses, is

$$\gamma_{\text{thr}} = k + \frac{1}{2L} \ln\left(\frac{1}{R_1 R_2} \right) \tag{17.9}$$

where R_1 and R_2 are the reflectivities of the laser cavity mirrors. The inversion required to reach laser threshold is when $\gamma(v_0) = \gamma_{\text{thr}}$ such that

$$(n_e)_{\text{thr}} = \frac{4\pi^2 n^2 \Delta v}{\lambda^2 A_{21}} \left[k + \frac{1}{2L} \ln\left(\frac{1}{R_1 R_2} \right) \right]. \tag{17.10}$$

The loss coefficient k in the laser diode is mainly from scattering. Forward biasing of the diode produces a threshold injection current which enables the population $(n_e)_{\text{thr}}$ to be established and the gain then depends on the current flowing in the diode. In equilibrium the rate of injection of carriers must equal the rate R_e at which they are lost by radiative recombination and non-radiative processes. In the main form of diode laser, the double heterojunction described later, almost all the injected carriers recombine in the junction region. For injection current I and gain region with depth d, width w and length l, giving an active volume dwl, the rate of loss of carriers $= R_e n_e = I/e(dwl)$. The threshold current I_{thr} is then

$$I_{\text{thr}} = e(dwl) \left(\frac{R_e}{A_{21}} \right) \frac{4\pi^2 n^2 \Delta v}{\lambda^2} \left[k + \frac{1}{2L} \ln\left(\frac{1}{R_1 R_2} \right) \right]. \tag{17.11}$$

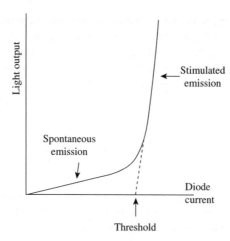

Figure 17.10   Light output from a semiconductor laser diode with variation of injection current

It is useful to write the threshold current in terms of the current density, defined as the flow of electrons per unit area per unit time, $J_{thr} = I_{thr}/wl$. The threshold current density is

$$J_{thr} = \frac{4\pi^2 n^2 ed\Delta v}{\lambda^2} \left(\frac{R_e}{A_{21}} \right) \left[k + \frac{1}{2L} \ln \left(\frac{1}{R_1 R_2} \right) \right]. \tag{17.12}$$

The active region shown in Figure 17.8 is the region in which laser gain is possible and its height d is determined by the diffusion of the charge carriers. The stimulating laser radiation inside the laser cavity occupies a mode volume whose height D may be greater than d. Where the radiation mode depth D is greater than the gain depth d the threshold current increases in the ratio D/d. The ratio $A_{21}/R_e = $ (radiative recombination rate)/(total rate of recombination) is the internal quantum efficiency of the laser. The gain coefficient of semiconductor lasers is typically $10\,000\,\mathrm{m}^{-1}$ and losses are typically $1000\,\mathrm{m}^{-1}$. The recombination radiation is homogeneously broadened with a linewidth of about 20 nm and about 5 nm for the quantum well laser described later.

The light output power from a semiconductor laser as the diode current increases is shown in Figure 17.10. Up to a threshold current light is emitted by spontaneous emission, and the diode acts as an LED. Above the threshold current laser action starts, and stimulated emission begins to dominate spontaneous emission. Beyond the threshold the efficiency of conversion of electrical energy into light increases rapidly. A fully efficient laser would produce one photon for each injected electron. The rate of carrier injection above threshold is $(I - I_{thr})/e$. The formal definition of efficiency η is as follows: for a laser with drive current I and a threshold current I_{thr}, the output power of the laser at wavelength λ is

$$P = \eta \frac{hc}{e\lambda} (I - I_{thr}), \tag{17.13}$$

where $\eta < 1$ and accounts for the fraction of injected carriers that combine radiatively and generate laser photons.

The beam from a semiconductor laser such as that in Figure 17.8 is emitted in the plane of the junction; here the path through the lasing material is greatest. The laser cavity is formed by polishing the ends of the diode. The beam is usually elliptical in cross-section, since the emitting area is rectangular. As expected from diffraction theory, small dimensions produce large divergence angles.

The emission photon energy of the semiconductor laser is close to the bandgap energy E_g. In GaAs $E_g = 1.43\,\text{eV}$ and the wavelength of the GaAs diode laser is about $870\,\text{nm}$; the typical spontaneous linewidth is $20\,\text{nm}$, due to the energy distribution of electrons and holes in the conduction and valence bands respectively.

As an example, for the homojunction GaAs laser diode, $\lambda = 870\,\text{nm}, \Delta v \approx 10^{13}\,\text{Hz}, n = 3.6$. It is found that the laser diode is highly efficient so that we may assume that $A_{21}/R_e \approx 1$. Typical diode dimensions are $l = 0.5\,\text{mm}, d = 2\,\mu\text{m}, w = 0.2\,\text{mm}$ and $k = 1\,\text{mm}^{-1}$. Then the theoretical value is for $J_{\text{thr}} \approx 500\,\text{A cm}^{-2}$ and the threshold diode current is $I_{\text{thr}} = 0.5\,\text{A}$. A more rigorous calculation would take into account the band structure more precisely. Practical values of the threshold current density for the homojunction laser are about $10^5\,\text{A cm}^{-2}$. The higher practical current density arises from the relatively large thickness of the active region and the spread of the beam into the p- and n-regions where there is absorption. This value may be reduced by operating the laser at the low temperature of liquid nitrogen to reduce the population in higher levels. The homojunction laser normally has to be operated in pulsed mode to minimize temperature rise from the high current density and resistive heating in continuous operation. The junction region has relatively high resistance since the charge carriers are neutralized compared with the neighbouring p- and n-regions. The active volume can be reduced by reduction in the depth or width of the active region. The addition of Al to the active layer forms GaAlAs which is also a direct bandgap material, and increases the bandgap. GaAlAs lasers are able to generate wavelengths from 750 to $850\,\text{nm}$; the most usual wavelength of $780\,\text{nm}$ is used in the CD player, laser printers and other common applications.

The efficiency of light emission from the LED depends on the relative rates of radiative recombination to the non-radiative mechanism of conversion to phonons. The presence of total internal reflection in the device also affects the emission by reflecting back some of the emitted light. The critical angle for total internal reflection from a medium of refractive index n_2 to a medium of refractive index n_1 is $\theta_c = \sin^{-1}(n_2/n_1)$. Light emitted from the active layer at an angle greater than θ_c will be reflected back. For the GaAs LED, for which $n_2 = 3.6, \theta_c = 16°$. The fraction of light able to escape into air is $[1 - (1 - n_1^2/n_2^2)^{1/2}]$. Then for GaAs a fraction 0.05 can escape. This fraction is increased by placing a hemispherical dome of high refractive index on the LED to reduce the total internal reflection.

17.3.1 Heterojunction Lasers

More efficient diode lasers employ layers of different materials at the junction: they are known as *heterojunction* and have largely replaced the *homojunction* lasers we have described so far. The heterojunction is formed between two different semiconductors with different bandgap energies; typical materials are GaAs and AlGaAs. The double heterojunction is the most common form in which the active layer of one semiconductor is sandwiched between two cladding layers of another semiconductor.

One form of double heterojunction is shown in Figure 17.11; here the active region is a thin layer of GaAs which is sandwiched by p- and n-regions of AlGaAs. In the heterostructure the crystal lattice periodicities should be closely matched to avoid interface dislocations; this is achieved in GaAs/AlGaAs where the lattice periods are respectively $564\,\text{pm}$ and $566\,\text{pm}$.

The heterojunction has three significant advantages over the homojunction. First, the cladding layer, e.g. AlGaAs, has a larger bandgap than GaAs, so that it traps the charge carriers in the central region where recombination is more probable; this is known as carrier confinement. Second, an active

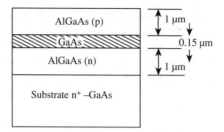

Figure 17.11 Diagram of a double heterostructure laser in which the active GaAs layer (shown hatched) is sandwiched between p- and n-regions of AlGaAs

region of higher refractive index can be formed which acts as a light guide, concentrating the light and increasing the efficiency of stimulated emission; this is termed optical confinement. Third, the laser emission is only weakly absorbed in the adjacent regions so that the losses are minimized. These three effects result in a much reduced threshold current density ($\sim 10^3 \mathrm{A\,cm^{-2}}$) compared with the homojunction, allowing continuous wave operation at room temperature. The threshold current may be further reduced by confining the current in the active region to a narrow stripe along the length of the diode.

An important extension of the double heterostructure is made by reducing the thickness of the central active region so that it becomes comparable with the electron or hole de Broglie wavelength given by $\lambda = h/p$, where p is the electron or hole momentum. Since in the double heterostructure electrons and holes are confined to the central region, where the bandgap E_g is smaller than that for the cladding region, the electrons and holes are confined within a potential well. The energy levels for electrons and holes in the potential well are quantized to values dependent on the well dimensions. These structures are referred to as *quantum wells*. The quantum well lasers based on these types of structures have increased gain and reduced current threshold and have become the predominant structure for semiconductor lasers. Further confinement of the charge carriers has been enabled by the use of a stripe geometry in which the injection current is confined to a narrow width. In this way the current flows through a smaller area and a certain current density can be achieved for a lower total current. The stripe laser diode is able to operate with threshold currents down to $50\,\mathrm{mA}$.

A semiconductor quantum well laser operating on a fundamentally different principle is the *quantum cascade laser*. It works on electron transitions between discrete levels in the conduction band which arise from the quantization of electron motion perpendicular to the plane of the active layer. This mechanism involves only the electrons in the conduction band and is to be contrasted with the normal semiconductor laser based on electron–hole recombination. The quantum cascade lasers produce radiation over a range of long infrared wavelengths.

17.4   Semiconductor Laser Cavities

The fabrication of diode lasers involves the deposition of multiple layers of single crystals with lattice matching and precise thicknesses. The substrate is also selected to match closely the lattice spacing of the deposited layer. In the semiconductor laser the optical resonator is often made by directly using the cleaved ends of the crystal as mirrors. A more efficient alternative is to use reflection from a periodic variation of refractive index within a layer coated on the ends of the active region, forming a multi-layer mirror known as a *Bragg reflector*. Reflection at this periodic structure is wavelength

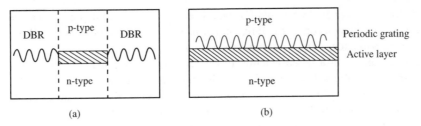

Figure 17.12   Distributed Bragg reflectors in resonant laser cavities: (a) as mirrors at either end of a cavity; (b) distributed throughout the length of the cavity, in the DFB laser

sensitive; components reflected at each step are in phase if the periodic spacing Λ satisfies the Bragg condition $2n_{\mathrm{eff}}\Lambda = m\lambda$, where n_{eff} is the effective refractive index and m is an integer. A suitable choice of Λ selects a single mode of oscillation for the laser. An example is shown in Figure 17.12(a), where a distributed Bragg reflector (DBR) is positioned at each end of the active region. A similar effect is obtained if the periodic variation in refractive index extends throughout the active gain region, when the resonant modes of the cavity are restricted to wavelengths satisfying the Bragg condition; this is called a *distributed feedback* (DFB) laser. In the DBR and DFB devices the periodic variation in refractive index is achieved by periodic variation of the thickness of one of the cladding layers to change the effective refractive index. The selection of wavelength which is available in the DFB lasers is important in the use of multiple wavelengths in optical communications.

Semiconductor lasers which generate laser light parallel to the surface of the diode junction interface, as shown in Figure 17.8, are called *edge emitting lasers*. An important alternative type is the *vertical cavity surface emitting laser* (VSCEL), shown in Figure 17.13, in which the laser beam is emitted parallel to the junction surface. In this structure, since the length of the active region is very small in the output direction, high reflection coefficients are required to form the laser cavity. This is achieved by cladding with Bragg reflectors on both sides. VCSELs operate with a high efficiency and a low current threshold. They can be packed into two-dimensional arrays in which the individual lasers can be individually addressed. The broad area facet of the diode laser shown in Figure 17.13 provides multiple transverse mode output. Single transverse mode can be achieved by reducing the width of the active region to $\leq 5\mu$m using a narrow-stripe electrical contact, and also by reduction in the threshold current, and is able to give output powers up to about 100 mW. Greater output power up to 5 W is obtained using a linear diode array of stripes on a single substrate, with the individual stripes sufficiently close such that they are phase locked and emit coherently. Several diode arrays can be

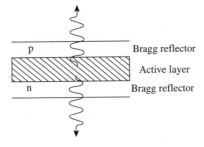

Figure 17.13   Diagram of a vertical cavity surface emitting laser (VCSEL)

combined to form a linear diode bar, and for the highest power several diode bars can be arranged to form a stack in a two-dimensional structure. Output powers from diode bars up to 20 W are produced and output powers in excess of 1 kW from diode stacks.

The beam emitted from a heterojunction diode laser at the output facet (the near field) has an elliptical shape with typical dimensions in the directions perpendicular and parallel to the junction plane of 1 and 5 μm. In propagation the beam size expands in the perpendicular (fast-axis) and parallel (slow-axis) directions by diffraction. Assuming a Gaussian beam profile, the divergence half angle $\theta = 2\lambda/\pi d$, the beam divergence perpendicular to the junction is typically $\theta \approx 20°$ and is larger than the beam parallel to the junction where $\theta \approx 5°$.

17.5   Wavelengths and Tuning of Semiconductor Lasers

The direct gap semiconductor GaAs is typical of the large group known as III–V lasers based on a combination of elements from the third group of the Periodic Table (e.g. Al, Ga and In) and from the fifth group (N, P, As and Sb). The ternary alloys AlGaAs and InGaAs and quaternary alloys such as InGaAsP also fall in this group. These produce emission wavelengths over the range 610 to 1600 nm, covering the range of optical communications as well as for reading and writing CDs, and DVDs, metrology and laser pointers. Table 17.1 lists some common semiconductor lasers.

As we remarked in Chapter 15, laser action becomes more difficult to induce at shorter wavelengths: covering the whole visible spectrum requires special attention to lasers producing blue light. Semiconductor lasers based on nitride compounds, e.g. InGaN, provide continuous

Table 17.1   Emission wavelengths of various semiconductor lasers

Wavelength region	Active semiconductor material	Wavelength range
Ultraviolet/blue	Group III nitride: GaN	370–490 nm
	GaInN	380–490
	ZnSe	460–490
	ZnCdS	300–500
	ZnCdSe	500–700
	Frequency-doubled AlGaAs	460–480
Visible	Heterojunction, quantum well or VCSEL structures:	
	CdSeS	500–700 nm
	GaAsP	600–900
	AlGaInP	620–580
	AlGaAs	700–920
	InGaAsP	700–900
Near-infrared	GaAs	0.8–0.9 μm
	GaAsSb	1.0–1.7
	InGaAs	1.0–3.2
	InGaAsP	1.2–1.6
	InAsSb	3.0–5.5
Mid-infrared	Lead salt (PbS, PbTe, PbSe, PbSnTe)	3.0–30 μm
	HgCdTe	3.2–15
	Cascade lasers	3.0–24, 65–87

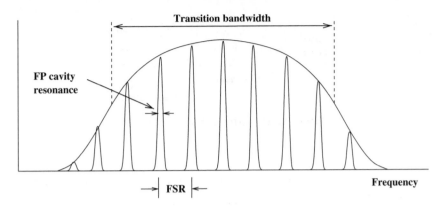

Figure 17.14 The tuning range of a semiconductor laser. The free spectral range limits the range achievable by tuning the cavity resonances

emission in blue and ultraviolet light, at 380–450 nm, with particular application in optical data storage, spectroscopy and biophotonics. Lasers based on the combination of elements from the second group (Cd, Zn) with those from the sixth group (S, Se), called II–VI lasers, give wavelengths in the blue–green region due to the larger bandgap compared with the III–V compounds. Wavelengths in the mid–infrared (4 to 30 μm) are produced by the Pb salts of IV–VI compounds (S, Se and Te)–however, these sources must be operated at low temperatures, and it may be preferable to use the quantum cascade laser described above.

The wavelength of oscillation of a semiconductor laser must lie within the comparatively broad band determined by the electronic band structure, within which there can be a number of cavity resonances (Figure 17.14).

The bandwidth of the recombination radiation spontaneous emission is determined by the distribution of density of states for the conduction and valence bands and is influenced by temperature. The typical width of spontaneous recombination radiation from a heterojunction laser at room temperature is 20 nm and is reduced to about 5 nm in a quantum well laser. The linewidth of the laser emission is much smaller due to the narrowing from the laser gain and the laser cavity. The resonances are often well separated, since there is only a small spacing between the polished faces of the crystal forming the Fabry-Pérot resonator. The laser can operate with many longitudinal modes covering the spectral width of the spontaneous emission, but since the transition is homogeneously broadened a single mode near line centre will dominate at low power. The separation between resonances is the *free spectral range*, which we have already encountered in the performance of spectrometers (Chapter 12). It is possible to tune the cavity resonance by mechanical pressure to change its length, by heating to change the refractive index ($d\lambda/dT \sim 0.2$ nm K^{-1}), or by varying the junction current; the available range of laser action of a single mode is, however, limited to the free spectral range, since the laser oscillation will otherwise jump to an adjacent mode.

The free spectral range depends on the dispersion due to the refractive index of the semiconductor crystal as well as its length. For the Nth longitudinal mode in a resonator with length L and refractive index n

$$N = \frac{2nL}{\lambda}.$$ (17.14)

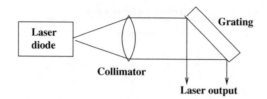

Figure 17.15   External cavity diode laser

The spacing $\delta\lambda$ between resonant modes, where $\delta N = -1$, is found by differentiation:

$$\delta N = -1 = \frac{2L}{\lambda}\frac{dn}{d\lambda}\delta\lambda - \frac{2nL}{\lambda^2}\delta\lambda \tag{17.15}$$

$$\delta\lambda = \frac{\lambda^2}{2nL}\left(1 - \frac{\lambda}{n}\frac{dn}{d\lambda}\right)^{-1} \tag{17.16}$$

or in terms of frequency the tunable bandwidth is

$$\delta v = \frac{c}{2nL}\left(1 + \frac{v}{n}\frac{dn}{dv}\right)^{-1}. \tag{17.17}$$

For example, an infrared diode at wavelength $1\,\mu m$ ($v = 30000\,\text{GHz}$) may have a resonator with $L = 0.5\,\text{mm}$, $n = 2.5$ and dispersion $(v/n)dn/dv = 1.5$. The free spectral range is then 24 GHz. Note the comparatively large contribution of the dispersion in the refractive index.

Selection of the precise operating wavelength of the diode laser can be made by several techniques: by the use of a secondary coupled cavity, by DFB or DBR, or by injecting light from another laser, termed injection locking. Tuning of the wavelength of the diode laser across its gain profile can be made by mounting the diode laser in an external cavity, Figure 17.15. The cavity contains a grating in a Littrow mounting and diffracts about 0.15 of the incident power back in first order into the laser diode. The wavelength-selective feedback modifies the gain of the laser diode, and laser linewidths down to 10 kHz can be generated.

At low powers the diode laser typically oscillates in many longitudinal modes. With increase in injection current the laser changes into one or a few longitudinal modes. The theoretical spectral linewidth of a single longitudinal mode of a diode laser may be calculated using equation (16.7), $\Delta v_L = 2\pi h v(\Delta v_{\text{cav}})^2/P$. A GaAs laser operating at 870 nm at a power of 5 mW may use a typical Fabry-Pérot resonator of length 0.5 mm, refractive index 3.6 and operate with end reflectors with reflectivities of 0.3. From equations (16.6) and (16.17), the cavity linewidth is $\Delta v_{\text{cav}} = c(1-R)/2\pi nL$, from which $\Delta v_{\text{cav}} \approx 2 \times 10^{10}\,\text{Hz}$. The theoretical minimum linewidth is then $\Delta v_L \simeq 10^5\,\text{Hz}$. Practical operating linewidths are much larger, typically $\geq 10\,\text{MHz}$.

17.6 Modulation

Communications via optical fibres may achieve large bandwidths, usually using pulse sequences formed either by switching on and off a laser source or by using an external modulator. A directly modulated semiconductor laser requires a switched current source, which for low-power lasers can easily be provided for a bit rate up to $10^9 s^{-1}$ ($1G\,bit\,s^{-1}$), and local communication networks often use this simple system. The main disadvantage is that the laser frequency changes during the short pulse, so losing the advantage of narrow bandwidths in avoiding the dispersion problems discussed in Chapter 6.

The bit rate in pulse modulation is limited by the delay in the laser output as the current is switched on and while the population inversion is built up. There is also an initial short-duration oscillatory instability or relaxation oscillation, similar to those that appear in other pulsed lasers such as the flashlamp pumped solid state laser, due in this case to a resonance between the carriers and photons in the cavity. Pulse durations of a few nanoseconds can be produced, with rise and fall times less than 1 ns.

External modulation of a free-running laser is used for long-distance communications, routinely using a bandwidth of $10\,GHz$. The modulation may be either by a semiconductor in which the transmission of light may be switched on and off by applying a voltage, or by an interferometer similar to the Mach–Zehnder (Section 9.3). The semiconductor device is an *electro-absorption modulator*, or EAM. Light is guided through an absorber region located between p-doped and n-doped semiconductor regions, in which the population of absorbing electrons is switched by an electric field. An EAM is often incorporated in the same structure as the laser diode itself.

The Mach–Zehnder modulator is used for the higher power lasers which are necessary for long-distance communications. The laser light is split between two paths, one or both of which contain a dielectric material, such as lithium niobate, whose refractive index may be changed by applying an electric field (the Pockels effect, see Section 7.11). The phase difference between the light emerging from the two paths depends on the applied field, and their sum may be switched between constructive and destructive interference. The bit rate may be $10\,G\,bit\,s^{-1}$ or higher.

Lasers and LEDs may also be modulated in analogue mode, as a continuous oscillator with a modulated injection current. In lasers the maximum attainable modulation bandwidth is limited by the carrier lifetime, determined by the spontaneous and stimulated emission processes, and the cavity (photon) lifetime, determined by the cavity losses. The very short cavity lengths of semiconductor lasers lead to low values of the cavity and photon lifetimes. Depending on the type of semiconductor laser, the carrier lifetime is typically $10\,ps$ and the photon lifetime about $2\,ps$, giving bandwidths greater than 1 GHz. An increased bandwidth is obtained for shorter cavity lifetimes and at higher laser powers.

The carrier lifetime in a LED is considerably longer than it is in a laser, where it is shortened by stimulated emission recombination. The available bandwidth is correspondingly smaller; a double heterostructure LED has a typical response of about $100\,MHz$.

17.7   Organic Semiconductor LEDs and Lasers

Organic materials are usually encountered as electrical insulators, and not as sources of light. Fireflies show us that there are other possibilities: they use the chemical excitation of organic molecules to produce light in a process with very high overall efficiency. *Electroluminescence*, which is described in Section 18.11, is the emission of light following electrical excitation, is achieved in certain solid organic materials by applying a high electric field to a thin layer of organic polymer. Electrons and holes are injected as in semiconductor diodes; recombination then leads to photon emission with photon energy determined by the energy of a bandgap. This LED source may also act as a laser, with laser action achieved by using reflecting electrodes which form a resonant cavity.

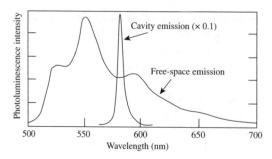

Figure 17.16 Photoluminescence spectra of a thin film of PPV (a) free-space emission (b) within a resonant cavity, showing laser action. (Friend et al, Nature 397, 121, 1999).

Electroluminescent organic materials offer considerable practical advantages in the production of matrix displays, which may be used for example in computer displays, instrument panels and motorway indicators. A matrix display might require many millions of individually controlled light-emitting elements, often using three colours. The development of such complex devices is a major task; we are concerned here only with the basic physical processes of a single element of such an array.

An organic polymer which is sufficiently conducting for electrical excitation may still require a field of order $10^7 \, \text{Vm}^{-1}$ for efficient operation. Thin films only $100 \, \mu\text{m}$ thick are therefore used, requiring only 2–3 volts between electrodes, one of which is a thin transparent conductor such as an alloy of magnesium and silver. The conducting polymer is usually a so-called π-conjugated polymer, in which there is a long chain of alternate double and single carbon–carbon bonds. The conductivity depends on electrons which are 'delocalized' along this backbone. Electrons emitted by one electrode, and holes emitted by the other, recombine forming an excited molecule known as an 'exciton': if this does not decay and lose its energy by interaction with adjacent molecules it will emit a photon at a wavelength determined by its molecular structure.[4] Efficient electroluminescence was demonstrated in 1990 by Richard Friend in Cambridge University, using the π-conjugated polymer poly(p-phenylene vinylene), known as PPV. Based on this material, a luminous efficiency of over 20 lumens per watt was achieved a few years later.

Figure 17.16 shows the luminescence spectrum of PPV when optically excited. Optical excitation is achieved without metal electrodes, so that the natural wide bandwidth of luminescence is observed. If the thin film is contained within a resonant cavity, which may be formed by electrodes or by tuned dielectric mirrors, laser action may produce a sharp spectral line as shown.

Problem 17.1

(a) Why is the spectral width of a semiconductor laser much less than for an LED, and much greater than for a typical gas laser?

(b) Why is laser output not observed on all the allowed longitudinal frequencies within the spontaneous spectral profile?

(c) What are the main differences between the output characteristics of an LED and a semiconductor laser? How do these affect the coupling of light into an optical fibre?

---

[4]The resonance of conjugated polymers at optical wavelengths is significant in their occurrence as opsins in vision and as chlorophyll in photosynthesis–see Chapter 21.

(d) How may a semiconductor laser be configured to produce (i) the fundamental single transverse mode, or (ii) the single frequency longitudinal mode?

Problem 17.2

At 860 nm GaAs has a refractive index of about $n_s = 3.6$. (a) Determine the critical angle for GaAs. (b) Assuming that the emission from an uncoated planar and thin surface of a sample of GaAs is into air with refractive index n_a, show that the fraction of light emitted is approximately $n_a^2/2n_s^2$. Estimate this fraction for GaAs. Discuss how this emitted fraction might be increased.

Problem 17.3

An injection laser has a loss coefficient per unit length k and a region of gain of length L. The two crystal faces have reflectances R_1 and R_2. Show that the threshold gain per unit length γ_{th} is $k + 1/2L \ln(1/R_1 R_2)$.

Problem 17.4

The threshold gain γ_{th} of an injection laser is related to the threshold current density J_{th} as $\gamma_{th} = \beta J_{th}$, where β is the gain factor. Consider an injection laser with a gain factor $\beta = 2.1 \times 10^{-2} A^{-1} cm^3$, a loss coefficient $10 \, cm^{-1}$ and an optical cavity of width $100 \, \mu m$ and length $250 \, \mu m$. The semiconductor medium has a refractive index of 3.6 and the cleaved end faces of the cavity are uncoated. Assuming that the injection current is confined to the optical cavity, determine the threshold current of the laser. (This indicates that typically the current in an injection laser is low.)

Problem 17.5

The output power of an injection laser may be expressed as $P_0 = A(J - J_{th})\eta_i(h\nu/e)[(1/2L)\ln(1/R_1 R_2)]/[k + (1/2L)\ln(1/R_1 R_2)]$. Here J and J_{th} are the actual and threshold current densities, η_i is the internal quantum efficiency and k is the loss coefficient.

A GaAlAs semiconductor laser has an active length of $300 \, \mu m$, width $10 \, \mu m$ and depth $0.5 \, \mu m$. The reflectances of its coated emission surfaces are both 0.9. It has a loss coefficient from scattering of $1000 \, m^{-1}$ and its internal quantum efficiency is 0.7. Estimate the efficiency of the laser. (This indicates that the efficiencies of injection lasers can be high.)

Problem 17.6

The threshold current density of an injection laser is found to vary exponentially with temperature as $J = J_0 \exp(T/T_0)$, where T_0 is the temperature coefficient. A GaAs injection laser has a threshold temperature coefficient $T_0 = 160 \, K$. When operated at $20°C$ the threshold current density is $2500 \, A \, cm^{-2}$. Determine the threshold current density when operated at $80°C$. (This indicates that the semiconductor devices are strongly temperature dependent.)

Problem 17.7

Show that the minority carrier lifetime of an LED is $\tau = (ew/JB)^{1/2}$, where e is the electronic charge, w is the width of the active region, J is the injection current and B is the electron–hole recombination coefficient.

Describe the mechanism determining the modulation bandwidth of a light-emitting diode (LED).

Problem 17.8

A light-emitting diode has a Lorentzian spontaneous emission spectrum with a full width at half maximum (FWHM) of $10^{13} Hz$ and a centre wavelength of $680 \, nm$. Calculate (a) the linewidth of the emission spectrum in wavelength, (b) the coherence time, (c) the coherence length of the LED light.

Problem 17.9

The frequency response of an LED with minority carrier lifetime τ and operating with an optical power $P(f)$ at frequency f is of the form $P(f) = P(0)/(1 + 4\pi f^2 \tau^2)^{1/2}$.

A GaAs LED has an effective recombination region of width $0.2\,\mu m$ and a recombination coefficient $B = 7 \times 10^{-16}\,m^3 s^{-1}$ and is operated at a current density of $2 \times 10^7\,A\,m^{-2}$. Calculate the 3 dB modulation bandwidth of the LED. Comment on why this should be regarded as an upper limit.

Problem 17.10

What effect do the dimensions of the active optical region of a semiconductor laser, including the size of the emitting aperture, have on the radiation characteristics of the laser?

Problem 17.11

A thin disc of GaAs, refractive index 3.6, intercepts a 10 mW light beam of near monochromatic light of wavelength 630 nm for which the absorption coefficient is $5 \times 10^4 cm^{-1}$. The bandgap of GaAs is 1.42 eV. For a disc of 1 μm thickness determine: (a) the power absorbed by the disc, (b) the power deposited in the crystal, (c) the rate of photon emission from the disc, (d) power radiated from the disc. Assume that the efficiency for the generation of recombination radiation is 0.6.

Problem 17.12

Light from an LED with an active area A_s is to be coupled from a medium with refractive index n_m into an optical fibre which has a numerical aperture NA and a fibre core area A_c, with $A_s < A_c$.

Show that the coupling efficiency $\eta^c = (NA)^2/n_m^2$.

Problem 17.13

For an acousto-optical modulator show that the first-order deflection angle for light of wavelength λ is given by $\sin \theta = \lambda/2\Lambda$, where Λ is the wavelength of the sound in the modulator. What factors determine the power in the deflected beam?

A lithium niobate (LiNbO$_3$) acousto-optical modulator is to be used to impress a 50 MHz signal on the light beam from an AlGaAs laser ($\lambda = 830$ nm) operating in the fundamental TEM$_{00}$ mode. The velocity of sound in LiNbO$_3$ is $6.6\,km\,s^{-1}$. Estimate the maximum diameter of the laser beam which can be modulated. Comment on how the irradiance of the modulated beam can be optimized.

Problem 17.14

The operating wavelength of a semiconductor laser can be changed by changing the temperature of the laser. Three relevant temperature-dependent parameters which affect the wavelength are refractive index, cavity dimensions and position of the peak of the gain profile. Make an assessment of their relative magnitudes.

18 Sources of Light

Fiat lux.

Genesis 1:3 (Vulgate).

Light visible to the human eye, covering wavelengths from 380 to 780 nm, is generated by a variety of mechanisms, which we may classify broadly as classical electromagnetic, thermal and quantum. These processes extend beyond the visible spectrum, the quantum processes becoming relatively more important at higher energies. In this chapter we consider the electromagnetic radiation from an accelerated charge, and the quantized model of the atom which leads to the formation of spectral lines. We also describe some of the practical sources of light, both natural and artificial.

18.1 Classical Radiation Processes

Although the interaction of light with matter must ultimately be considered in terms of photons, there are many circumstances when a classical electromagnetic theory of radiation in terms of accelerated charges provides a simple and remarkably complete description.

In the first part of this chapter we consider examples of classical radiation processes in which an isolated electron is accelerated either in the electric field of a positive ion, as in a dilute ionized gas, or when its motion is deflected by a magnetic field, as in *cyclotron* and *synchrotron* radiation. Both processes are enhanced in an interesting way when the electron has a very high energy. Classical theory can also be applied to the scattering of light by individual free electrons (*Thomson scattering*) and by individual atoms and molecules (*Rayleigh scattering*); the quantum approach becomes essential when there is an exchange of energy between an atom and a photon. The various classical and quantum scattering processes are described in Chapter 19.

18.1.1 Radiation from an Accelerated Charge

We first discuss briefly the basic equation for the radiation from a single accelerated charge. In classical electrodynamic theory the electric field generated by a single charge has three components. The first is the electrostatic field, whose strength varies as the inverse square of the distance from the

Optics and Photonics: An Introduction, Second Edition   F. Graham Smith, Terry A. King and Dan Wilkins
© 2007 John Wiley & Sons, Ltd

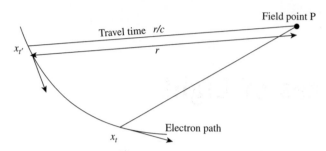

Figure 18.1   Radiation from an accelerated electron. The radiated electric field at the field point P at a time t depends on the acceleration of the electron at time $t' = t - r/c$

charge. A uniformly moving electron generates a second field component which is proportional to its velocity; this field also falls away as the square of the distance from the charge. The third component of the field is only present when the charge is accelerated. This is the radiation field, which falls away only as the first power of the distance.

The strength of the radiation field at a point some way from the accelerating charge depends on the component of the acceleration perpendicular to the line of sight. Since radiation reaches the field point after a finite travel time, the acceleration must be measured at the time the radiation leaves the charge rather than when it arrives (Figure 18.1). The component of the field perpendicular to the line of sight is given by

$$\mathbf{E}_\perp = -\frac{q}{4\pi\epsilon_0 c^2 r} \times \mathbf{a}_\perp \tag{18.1}$$

where by $\mathbf{a}_\perp$ we mean:

1. The component of acceleration projected onto a plane perpendicular to the line of sight from the field point to the charge.

2. The acceleration is measured at time r/c earlier, where r is the distance from the charge and c is the phase velocity of the radiated wave. The radiated electric field is therefore always transverse and perpendicular to the line of sight. The magnetic field B is also transverse; it is perpendicular to the electric field E.

The total power radiated instantaneously by a non-relativistic particle with charge q and acceleration a is given by the *Larmor* formula

$$P = q^2 a^2 / 6\pi\epsilon_0 c^3. \tag{18.2}$$

18.1.2   The Hertzian Dipole

A dipole is a pair of equal and opposite charges separated by some distance. The product of charge and distance is the *dipole moment*. A dipole whose moment oscillates sinusoidally, and whose dimensions are small compared with the corresponding radiated wavelength, is known as a Hertzian dipole. It radiates in the same way as a single oscillating charge, since the two opposite charges oscillate with opposite accelerations and their radiated fields must therefore add.

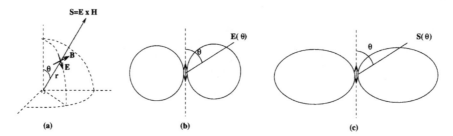

Figure 18.2 Radiated field from a Hertzian dipole. (a) The field is aligned along the surface of a sphere, with strength proportional to $\sin\theta/r$. (b) The variation of field strength E with θ (at constant distance from the dipole) can be represented as a 'polar diagram'. (c) The power polar diagram, showing the variation of Poynting flux S with angle θ

We therefore consider only one charge q, moving linearly according to the complexified displacement $\tilde{z}$

$$\tilde{z} = z_0 \exp(i\omega t). \tag{18.3}$$

The component of acceleration as seen from the field point P at a distance r in free space, and in a direction θ from the axis of the dipole (Figure 18.2), is then easily found from equation (18.1), giving a longitudinal component of the field

$$E(\theta) = -\frac{\omega^2 q z_0}{4\pi\epsilon_0 c^2 r}\sin\theta \exp[i\omega(t - r/c)]. \tag{18.4}$$

The variation with θ is known as the *radiation pattern* or *polar diagram*. This is shown in Figure 18.2 (b). The power polar diagram shown in Figure 18.2 (c) is the Poynting flux as a function of θ; note that the radiation pattern of a radio antenna is often plotted as a variation of power (E^2) rather than field (E). The term r/c in the oscillatory function is already familiar in a propagating wave as the phase term $\phi = \omega r/c$.

The total time-averaged power $\bar{P}$ radiated by the dipole is found by integrating over a sphere the time-averaged power flux $\epsilon_0 c E(\theta)^2$, giving

$$\bar{P} = \frac{1}{12\pi\epsilon_0 c^3}\omega^4 q^2 z_0^2. \tag{18.5}$$

The radiated power $\bar{P}$ depends on the square of the dipole moment (qz_0), and on the fourth power of the frequency.

18.2 Free–Free Radiation

Light may be emitted by the electrons of an ionized gas either in discrete spectral lines or as a continuum caused by their acceleration in the Coulomb field of ionized nuclei. The radiation from a single electron in an ionized gas, accelerated as it encounters a positive charge, is known as *free–free radiation*, to distinguish it from processes involving quantized energy levels. In the encounter the

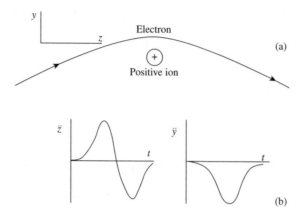

Figure 18.3 Free–free, or bremsstrahlung, radiation from an electron accelerated in its encounter with a charged nucleus. Two components of acceleration are shown: in the direction of travel z the electron speeds up and slows down, while there is a single impulse normal to z

electron loses energy; the radiation is therefore also known as *bremsstrahlung* (German for 'braking radiation'). (The acceleration of the nucleus need not be considered as it is smaller than that of the electron by the ratio of their masses.)

The acceleration of the electron is in the form of an impulse, shown in Figure 18.3; it therefore has a broad spectrum, in contrast to the narrow-bandwidth radiation from an oscillating dipole. The closest encounters give the narrowest pulses, containing the highest frequencies: a full calculation of the overall spectrum requires an integration over a range of impact parameters and electron energies. An upper limit to the emitted frequency is set by the corresponding photon energy: this cannot be greater than the original electron energy. This limit is encountered in an X-ray discharge tube, where electrons are accelerated by a high voltage V. The minimum wavelength of the X-rays emitted when the electrons reach the target anode is

$$\lambda_{\min} = \frac{hc}{eV},\qquad(18.6)$$

and there is a continuum of radiation emitted at lower frequencies. The emissivity of the gas is proportional to $n_e n_i Z^2$, where n_e and n_i are the electron and ion densities and Ze is the nuclear charge.

Free–free radiation is encountered over a wide range of the electromagnetic spectrum, including radio emission from interstellar clouds and the continuum emission from X-ray tubes; it is also important in high-energy laboratory physics, since it is responsible for the energy loss of high-energy electrons in solid matter.

18.3   Cyclotron and Synchrotron Radiation

An electron moving at right angles to a steady and uniform magnetic field follows a circular path with a centripetal acceleration, and consequently emits electromagnetic radiation. When the electron speed is $v \ll c$ this is *cyclotron radiation*; when the electron is moving at relativistic speed v near c it is *synchrotron radiation*.

Consider first an electron moving with $v \ll c$ perpendicular to a magnetic field B. It follows an orbit satisfying

$$evB = mv^2/R \tag{18.7}$$

where R is the radius of the orbit. The angular frequency of the electron in the orbit is $\omega_{cyc} = eB/m$, known as the *cyclotron frequency*. The electron acceleration is

$$a = v\omega_{cyc} = veB/m. \tag{18.8}$$

From the Larmor formula, equation (18.2), the total power radiated by one electron is

$$P = q^2 a^2/6\pi\epsilon_0 c^3 = e^2\omega_{cyc}^2 v^2/6\pi\epsilon_0 c^3. \tag{18.9}$$

The radiation is at the cyclotron frequency, $\omega_{rad} = \omega_{cyc}$.

In a synchrotron light source high-energy electrons are injected into a storage ring which confines the relativistic electrons in a circular orbit by a magnetic field. The continuous inward acceleration of the electrons produces synchrotron radiation, first observed in particle ring accelerators used in high energy physics.

We consider high-energy electrons travelling at relativistic speed v near c and kept in a circular orbit by moving perpendicular to a magnetic field. Special relativity theory shows that the only change from equations (18.8) and (18.9) required near the light speed is to replace m by γm_0, where m_0 is the rest mass and $\gamma \equiv 1/\sqrt{1 - v^2/c^2}$. We can also set $\gamma = E/m_0 c^2$ where E is the total relativistic energy (kinetic energy plus $m_0 c^2$). The electron angular frequency is

$$\omega_{sync} = c/R = eB/\gamma m_0 = \omega_{cyc}/\gamma. \tag{18.10}$$

Observed perpendicular to the orbit the acceleration is towards the centre of the circle, and a circularly polarized wave is emitted along the magnetic field. A much more intense radiation is emitted in directions close to the plane of the orbit; this must be calculated at distance r using the retarded time $(t - r/c)$, which has a large effect when the electron velocity is close to c. The effect is to compress the time during which the electron moves towards the observer and stretch the time during which it is receding (the relativistic Doppler effect; see Section 4.7). Since the electrons are moving at speeds near c the radiation appears like a shock wave in the direction of the path of the electrons. Each time the electron travels towards the observer a brief pulse of radiation is observed. The observed field therefore consists of a regular series of pulses, separated by the orbital period of γ/ω_{cyc}.

In its own frame of reference the accelerating electron emits a characteristic electric dipole radiation pattern. As v approaches c, when viewed in the laboratory frame, the pattern becomes strongly peaked in the forward direction, as shown in Figure 18.4(a). The light is emitted in the forward direction, tangential to the electron's orbit, and is linearly polarized with E in the plane of the orbit.

The angular width of the radiation pattern is $\theta \simeq 1/\gamma = m_0 c^2/E$. For $E = 1\text{GeV}$ and $m_0 c^2 \simeq 0.5\text{MeV}$ the angular width is $\theta \simeq 5 \times 10^{-4}$ radians, indicating that the radiation is highly collimated in the forward direction.

The observed pulse duration in the electron frame is $\tau = R\theta/v \simeq R\theta/c$, and in the laboratory frame is $\tau = (R\theta/\gamma^2 c)$. Substituting for R and θ, the pulse duration is

$$\tau = m_0/\gamma^2 eB = (m_0/eB)(m_0 c^2/E)^2. \tag{18.11}$$

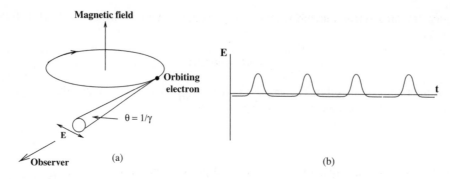

Figure 18.4 Synchrotron radiation. (a) An electron gyrating in a magnetic field is viewed from a point in the plane of its orbit. The radiated electric field (b) has a sharp maximum each time the electron travels towards the observer.

The waveform of the radiation can now be Fourier analysed. It contains harmonics of the frequency $v_{sync} = \omega_{sync}/2\pi$, usually closely spaced so that the spectrum is effectively a continuum up to a frequency of order $1/\tau \sim \gamma^3 v_{sync}$.

The relativistic analysis shows that for circular motion at arbitrary velocity, the power radiated is still given by the Larmor formula, equation (18.2), but with an extra factor γ^4 on the right. Setting $v = c$ in the centripetal acceleration, $a = c^2/R$, we obtain the total power radiated by the synchrotron source as

$$P = \gamma^4 (e^2 c / 6\pi\epsilon_0 R^2). \tag{18.12}$$

Synchrotron sources provide intense broadband radiation which is well collimated. For many purposes a specific wavelength band can be selected; storage rings typically provide radiation with bandwidth $\Delta\lambda/\lambda = 0.1\%$, with spectral radiance $\sim 10^{22}$ photons $s^{-1}mm^{-2}mrad^{-2}$. A smaller orbit and a high electron energy promote higher power. The storage ring has a sequence of bending magnets connected by linear sections which contain radio frequency cavities to replenish the energy radiated by the synchrotron emission. To compensate for defocusing of the electron beam due to Coulomb repulsion, quadrupole focusing magnets are used. The electrons orbit in bunches of a few hundred particles; the bunches are a few picoseconds long, and are separated by a few nanoseconds. Storage rings for X-ray production have a bending radius of 5 to 100 m, a magnetic field of 1 T and electron energies of 0.5 to 8 GeV, giving wavelengths of 0.1 to 1 nm. Synchrotron machines with energies up to 10 GeV provide powerful sources of X-rays for diffraction analysis of many types of organic and inorganic materials.

There are many sources of synchrotron radiation observed in radio astronomy, involving very high electron energies and very low magnetic fields. Synchrotron radiation from some astronomical sources produces visible light and even X-rays: for example, the light from the Crab nebula is generated by electrons with cosmic ray energies, probably up to 10^{12}eV, moving in a magnetic field of the order of 10^{-6}T. Interstellar gas contains a magnetic field of only 10^{-10}T, but this is still sufficient to accelerate cosmic ray electrons and produce measurable radio waves. By contrast, a terrestrial electron synchrotron might have a field of 1 T, and visible light would be emitted by electrons with energies of 10^9eV.

18.4 Free Electron Lasers

A free electron laser (FEL) converts some of the kinetic energy of a relativistic electron beam into coherent electromagnetic radiation, providing a tunable, coherent, high-power source at wavelengths from millimetres through infrared to visible, and potentially to ultraviolet and X-rays. This is achieved by passing the electron beam through an alternating transverse magnetic field produced by an array of magnets, known as an *undulator* or *wiggler* (Figure 18.5). The magnetic field reversals are typically spaced around 1 centimetre apart, and may extend for about 1 metre. The resulting alternating transverse motion of the electrons produces an electromagnetic wave, which travels with the electron beam. The wave reacts back on the electrons, and induces a coherent motion, which gives rise to coherent radiation similar to laser radiation. The emission wavelength depends on the energy of the electron beam and the undulator periodicity.

For small deviations of the electron beam the lateral oscillation produces radiation at a wavelength $\lambda \simeq \lambda_0/(2\gamma^2)$, where λ_0 is the periodicity of the undulator and $\gamma m_0 c^2$ is the relativistic energy of the electron. The factor γ^{-2} derives from two Doppler shifts: one as the electron beam encounters the undulator and the second as the oscillating beam is observed in the laboratory frame. If the beam deviations are large, the bending also produces synchrotron radiation, which may contain high harmonics of the fundamental oscillation.

The description of the FEL as a laser is somewhat misleading, as it does not involve stimulated emission between quantized levels as in the conventional laser. However, the radiation does feed energy back into the electron stream, leading to enhanced radiation from the coherent motion of many electrons. The electrons form bunches within the undulator, which is due to a velocity modulation as they enter. The effect may be enhanced by forming a resonant cavity round the undulator, consisting of mirrors at the ends (shown as M_1, M_2 in Figure 18.5). The efficiency of conversion of beam kinetic energy to radiation is typically around 1% without a cavity, rising to over 10% with a cavity.

Radiation from the relativistic electron stream is confined to an angle $\sim \gamma^{-1}$ in the direction of the beam, and it is coherent over a large wavefront, as in a conventional laser. The spectral purity of the line radiation, $\nu/\delta\nu$, is approximately equal to the number of magnetic poles in the undulator.

An FEL can generate radiation at millimetre wavelengths from an electron beam with energy of a few megaelectronvolts. A storage ring, generating electrons with energy 0.1 to 10 GeV, may be used for radiation at wavelengths in the micrometre to nanometre range. The emission has a time dependence which follows that of the electron beam, which can be continuous or in pulses with

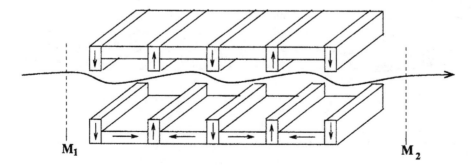

Figure 18.5 A free electron laser, showing the undulator and the lateral oscillation it produces in the electron beam. The optical resonator cavity is formed by the mirrors M_1, M_2 and the magnetic field direction is shown by the arrows.

durations down to femtoseconds. The FEL has the potential to produce high average power up to kilowatts, and high peak power. These characteristics make the FEL suitable for a wide range of applications involving physical, chemical and biological materials.

18.5   Cerenkov Radiation

A charge moving with high relativistic velocity in a transparent dielectric may radiate without any applied acceleration, if its velocity is higher than the local phase velocity of light; this is *Cerenkov radiation*, which is important, for example, when a cosmic ray particle enters the Earth's atmosphere.

It is easy for a boat to travel faster than the waves it generates, and it is commonplace for an aircraft to travel faster than the speed of sound. In both cases a sharp wavefront travels out in a V-shape, with half angle whose sine is the ratio between the group velocity of the waves and the velocity of the vehicle. By analogy, it might be expected that electromagnetic waves can be generated by a charge moving faster than the wave velocity c/n. Light emitted from an electron moving with a relativistic velocity in a medium with $n > 1$ was first observed by P. Cerenkov, and is named after him.

We note first that the electron itself has no acceleration normal to its velocity which can account for the radiation. It is in fact possible to regard the radiation as generated by the electron, by expressing the field in terms of the electron charge, its velocity and its acceleration, at a suitable retarded time $(t - nr/c)$, but it is simpler to regard the electron as exciting the dielectric medium as it progresses, and then to consider the dielectric as the source of the radiation (Figure 18.6). The dielectric becomes polarized, the electron attracting positive charge and repelling negative, so that as the electron passes each part of the dielectric acts like a small dipole giving only one impulsive oscillation. The radiation from all impulses along the track adds in phase where the Huygens' wavelets coincide. For an electron moving below the critical speed, Cerenkov radiation is understandably absent because successive wavelets (each of negligible amplitude) do not overlap and reinforce one another.

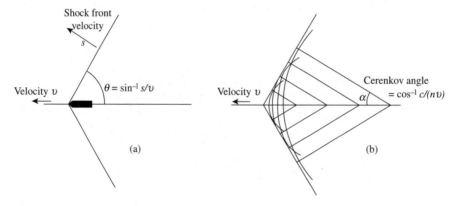

Figure 18.6   Comparison of shock wave in air with Cerenkov radiation from an electron. (a) Shock wave from a bullet, forming a cone with semi-angle θ. (b) Cerenkov wave from an electron in a medium with refractive index n. Huygens' wavelets form the wavefront

What is the angle α between the Cerenkov ray and the electron track? Consider a single wavelength component of the impulsive radiation from a point in the medium. The track is now a one-dimensional array of radiators, phased progressively as

$$\phi(x) = \frac{2\pi}{\lambda} \frac{nvx}{c} \tag{18.13}$$

where v is the velocity of the electron. By analogy with the diffraction grating (Chapter 11) we know that this radiates a cone at angle $\alpha = \cos^{-1}(c/nv)$, provided that n/c is the *phase* velocity of the wave. There is a small paradox to resolve here, since the Huygens' wavelets of Figure 18.6 obviously travel at the *group* velocity, leaving us uncertain about the true value of α. The solution is simply that a true impulse only continues without change of form if the medium is non-dispersive. The correct value for α when the medium is dispersive is obtained by using the group velocity in the range of wavelengths which is being observed instead of the phase velocity c/n.

The high-energy particles of a cosmic ray shower generate in the Earth's atmosphere a flash of Cerenkov light, which shines within a few degrees of the direction of the shower axis. Light generated by high-energy particles as they pass through liquids or gases can be used as a sensitive detector for cosmic ray showers, or for particles from accelerators. When the Cerenkov angle α can be measured it gives a direct and simple measurement of the velocity of high-energy particles.

18.6 The Formation of Spectral Lines

Spectral lines[1] are formed in transitions between quantized energy states of atoms and molecules. The quantized states in atoms are primarily *electronic*; the transitions are between discrete energy states of an electron orbiting in the electric field of the nucleus. In molecules there are also *vibrational* states; for simple diatomic molecules these are oscillations in atomic separation. Molecules also have *rotational* states: these and the vibrational states become numerous and complex in multi-atomic molecules. Electronic transitions involve typical energy steps of order 1 eV, giving a spectral line in the visible domain; vibrational transitions typically have lower energies, of order 0.1 eV, giving lines in the infrared, and rotational transitions typically involve energies of 0.01 eV or less, corresponding to the far-infrared and radio regimes. Figure 18.7 illustrates these regimes for a diatomic molecule, giving a rough guide to the spectral regions in which they mostly apply.

The theory of atomic and molecular spectra is complex, and we consider here only the simplest atomic spectra, starting with the spectrum of the hydrogen atom.

18.6.1 The Bohr Model

Although the quantum nature of energy levels in all atoms demands a full quantum-mechanical analysis, which is described later, a useful description due to Bohr[2] uses the classical concept of an

---

[1] The term *spectral line* derives from the use of slit spectrographs, in which features appear as bright or dark lines (emission or absorption) crossing a continuous spectrum.

[2] Niels Bohr, 1885–1962, creator of the Institute of Theoretical Physics in Copenhagen, and exponent of complementarity between classical and quantum theories. His model was published in 1913, following a visit to Rutherford in Manchester in 1912.

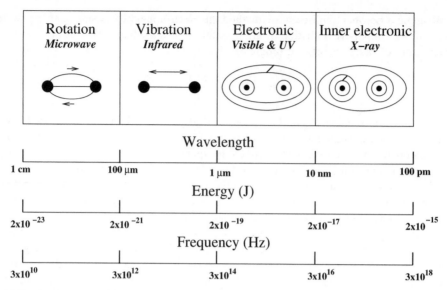

Figure 18.7 Quantum transitions in a diatomic molecule. The lowest photon energies are in transitions between rotational states and between vibrational states; the highest are due to electronic transitions in electron orbits, either in outer or inner orbits

electron orbiting a nucleus, with simple rules which confine the orbits to defined quantized energy states. The following rules were introduced by Bohr:

- Electrons can only occupy orbits in which their orbital angular momentum is an integral multiple of the quantity $h/2\pi$.

- The transition of an electron between orbits with energy separation ΔE is due to either the emission or the absorption of a photon with energy $h\nu = \Delta E$.

In Bohr's model of the hydrogen atom an electron with mass m and charge $-e$ is in orbit at radius r round the nucleus (charge $+e$) with angular velocity[3] ω. Then the quantization of angular momentum states that

$$m\omega r^2 = \frac{nh}{2\pi} \quad (n = 1, 2, 3, \ldots) \tag{18.14}$$

where n is known as the *principal quantum number*. The balance between electrostatic force and the centrifugal force for the orbiting electron is

$$\frac{e^2}{4\pi\epsilon_0 r^2} = m\omega^2 r. \tag{18.15}$$

---

[3]This is not the angular velocity of an electromagnetic wave: contrary to classical theory, the orbiting electron does not generate radiation.

Eliminating ω between these equations gives the radii of the quantized orbits

$$r = a_H n^2 \tag{18.16}$$

where

$$a_H = \frac{\epsilon_0 h^2}{\pi m e^2}. \tag{18.17}$$

The radius a_H of the smallest orbit is 5.3×10^{-2}nm.
The total energy (E) of the electron (the sum of the potential and kinetic energies) is

$$E = -\frac{e^2}{8\pi\epsilon_0 r} \tag{18.18}$$

or

$$E = -\frac{Rhc}{n^2} \tag{18.19}$$

where the constant $Rhc = me^4/8\epsilon_0^2 h^2$ is known as the Rydberg constant. This is the binding energy of the electron in the lowest energy state (the *ground state*), which is 13.6 eV.

The photon energy $h\nu$ for a transition between orbits n_1 and n_2 is

$$h\nu = Rhc\left(\frac{1}{n_1^2} - \frac{1}{n_2^2}\right). \tag{18.20}$$

There is a series of transitions for each value of n_1, which occur in different parts of the spectrum, named as follows:

- $n_1 = 1$ $n_2 = 2, 3, 4, \ldots$ Lyman series (ultraviolet)
- $n_1 = 2$ $n_2 = 3, 4, 5, \ldots$ Balmer series (visible and ultraviolet)
- $n_1 = 3$ $n_2 = 4, 5, 6, \ldots$ Paschen series (infrared)
- $n_1 = 4$ $n_2 = 5, 6, 7, \ldots$ Brackett series (infrared).

The energy levels for the hydrogen atom are shown in Figure 18.8, with these transitions. Figure 18.9 shows the Balmer series.

The spectra of the alkali metals lithium, sodium, etc., follow a similar pattern, with series of lines labelled successively *sharp, principal, diffuse* and *fundamental*.

18.6.2   Nuclear Mass

The simple Bohr model can be extended to provide some insight into the influence of nuclear mass on energy levels in hydrogen and other atoms. Equation (18.15) assumes an infinite nuclear mass; to take account of the revolution of both electron and nucleus, with mass M, about a common centre of mass, the electron mass m should be replaced in all formulae of the preceding subsection by a *reduced mass* μ given by

$$\mu = \frac{mM}{m + M}. \tag{18.21}$$

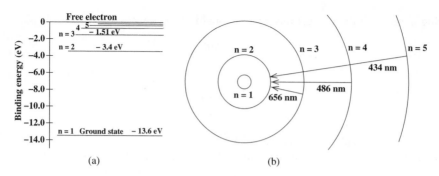

(a)   (b)

Figure 18.8   Energy levels and transitions in the hydrogen atom, showing the Lyman, Balmer and Paschen series. (a) The quantized energy levels have energies $E_n = -13.6n^{-2}$eV. (b) The orbital radii in the Bohr model: $r = na_o^2$, with $a_o = 0.0529$ nm. The wavelengths for the Balmer series are shown.

This fractional change is different for deuterium, which has twice the nuclear mass. The difference, known as an *isotope shift*, can be seen as a doubling of lines in the spectrum of a gaseous mixture of hydrogen and deuterium, which allows a precise analysis of the proportions in the mixture. (In deuterium the fractional shift in photon energies or wavelengths between ordinary hydrogen and deuterium is 1/3700, which requires more resolving power than a simple prism spectrograph.)

18.6.3   Quantum Mechanics

The *wave mechanics* theory of atomic structure which was introduced by Schrödinger, and *matrix mechanics* introduced by Heisenberg, both in the 1920s, are now subsumed into quantum mechanics. We outline Schrödinger's wave theory to demonstrate its relationship to the Bohr particle theory; the duality between the two approaches has a close similarity to the duality between photons and electromagnetic waves.

In quantum-mechanical theory the distribution of a single electron within an atomic system is described in terms of a *wavefunction* Ψ, whose distribution in space is determined by a potential field $V(x, y, z)$; in the simple hydrogen atom this is the potential energy field

$$V(r) = -\frac{e^2}{4\pi\epsilon_0 r}.\tag{18.22}$$

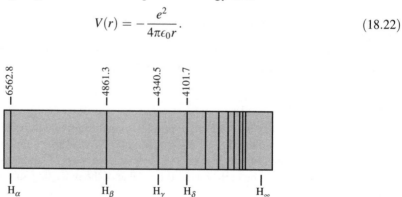

Figure 18.9   The Balmer series of spectral lines in the emission spectrum of atomic hydrogen (wavelength in Å) (G. Herzberg, *Atomic Spectra and Atomic Structure*, Dove, Publications, 1944)

The magnitude of $|\Psi|^2$ at any point is the probability density of finding an electron at that point. The general solution of the Schrödinger wave equation represents a state of uncertain energy, where measurements of the energy will reveal only a statistical distribution. But there are special solutions called *stationary states*, or *energy eigenstates*, each of which possesses a unique, well-defined energy E. For a single particle (e.g. an electron in a hydrogen atom), the wavefunction reduces to a product $\Psi(\mathbf{r}, t) = \psi(\mathbf{r}) \exp(-2\pi i E t / h)$. It turns out that the Bohr orbits correspond, roughly, to these energy eigenstates; transitions between these states occur in the emission or absorption of a photon. The evaluation of the spatial part of the wavefunction $\psi(\mathbf{r})$, and hence the probability of finding an electron with energy E at any distance from the nucleus in the hydrogen atom, is found by evaluating the time-independent Schrödinger equation

$$\nabla^2 \psi + \frac{8\pi^2 m}{h^2} (E - V)\psi = 0. \tag{18.23}$$

This famous equation is closely related to the equations for electromagnetic radiation confined to a waveguide or resonator, and for any field $V(x, y, z)$ a series of resonant modes appear in its solution. For the potential field of the hydrogen nucleus (equation (18.22)) these correspond to Bohr's quantized orbits. With increasing difficulty but remarkable success the Schrödinger equation can be applied to more complex atoms, providing the required energy levels and also the probabilities of transitions between them.

18.6.4 Angular Momentum and Electron Spin

When observed with sufficiently high resolving power, the hydrogen spectral lines are seen to consist of several closely spaced components; for example, the hydrogen line at 656 nm wavelength is split into two components separated by about 0.016 nm. This is the *fine structure*, which was discovered in many spectra by Michelson; it demands the use of a high-resolution spectrometer (Chapter 12). The explanation was given by Pauli, who famously proposed that an electron has an intrinsic angular momentum in addition to its orbital angular momentum. The combination of this spin with the orbital angular momentum obeys quantum rules, which become complex in multi-electron atoms.

In terms of the Bohr model, the energy split due to electron spin may be thought of as an interaction between the magnetic moment of the electron spin and the magnetic field at the electron generated by orbital motion of the proton relative to it.[4] A further manifestation of electron spin is a split of a quantized level into two states in which the spins of the electron and the nucleus are aligned or opposed; this gives rise to *hyperfine structure* which is observable in narrow optical spectral lines. In hydrogen this hyperfine splitting of levels is also observable as a direct transition between the split levels; this low-energy transition has a very low probability, but it is observed by radio astronomers as a radio spectral line at 21 cm wavelength (1420 MHz) in low-density clouds of interstellar hydrogen.

18.7 Light from the Sun and Stars

To a first approximation, the visible radiation from stars has a blackbody spectrum at the temperature of the surface. This temperature ranges from over 40 000 K for the youngest and most

---

[4]P.A. Tipler, *Modern Physics*, Worth 1969, sections 7–5.

massive stars, designated type O, to below 2000 K for the coolest M type[5] Blackbody curves for temperatures of 45000 K, 5800 K and 4350 K, shown in Figure 18.10, correspond respectively to white, blue and red stars of types O, G and K. These spectra fit the observed stellar spectra well at long wavelengths, but at wavelengths shorter than 100 nm there is a deficit due to absorption at the wavelength of molecular resonances. The atmospheric absorption in the ultraviolet is due to ozone; outside the atmosphere the absorption is mainly due to the hydrogen Lyman series of spectral lines.

The spectrum of light from the Sun, which is type G, is shown in Figure 18.11 together with a blackbody curve for a temperature of 5900 K, chosen to give a reasonable fit to the actual spectrum. The departure from pure blackbody radiation is mainly due to the chromosphere, a gaseous layer whose thickness is only about 1% of the solar radius. In the lower part of this layer the temperature is lower than at the surface (the *photosphere*), and the ionized atoms absorb the continuum radiation, giving rise to the Fraunhofer lines in the solar spectrum. The absorption in the chromosphere differs between the centre of the solar disc and its edge, so that the best fit blackbody curve varies across the disc. Blackbody temperatures of around 5800 K to 5900 K are often quoted for the centre of the solar disc.

The chromosphere, which is normally observed through its absorption of photospheric light, can be observed as an emitter immediately outside the edge of the solar disc when the bright photosphere is obscured in a solar eclipse. In this brief moment, the line of light at the edge of the disc may be analysed in a spectrograph, and the spectral lines of several elements then appear in emission.

18.8   Thermal Sources

The commonest example of a thermal source is the incandescent tungsten filament lamp. The spectrum corresponds nearly to a blackbody at the emitter temperature of about 2800 K, and the efficiency of converting electrical heating power into visible light is about 9%, corresponding photometrically (see Appendix 2) to 15 lumens per watt. To prevent oxidation and burn-out, the filament is enclosed in a glass bulb filled with an inert gas. To reduce the input power required to maintain the filament at a given temperature the inside may be coated to transmit visible light and reflect infrared light.

The *tungsten–halogen* lamp has a tungsten filament with the bulb filled with iodine or bromine halogen vapour. Tungsten (chemical symbol W) evaporated from the heated filament reacts with the halogen to form WI_6 or WBr_6[6] which dissociates in collision with the filament, depositing tungsten, so prolonging the life of the filament. Expressed in photometric units as the luminous output per watt of electrical power consumed, the tungsten–halogen lamps can produce up to 40 lumens per watt.

---

[5]The sequence of stellar types is O,B,A,F,G,K,M. This was originally an alphabetical sequence based on the line structure of spectra, but is now ordered in a temperature sequence.

[6]These compounds are examples of *halides* in which a halogen (such as iodine or bromine) combines with another element or group in the Periodic Table.

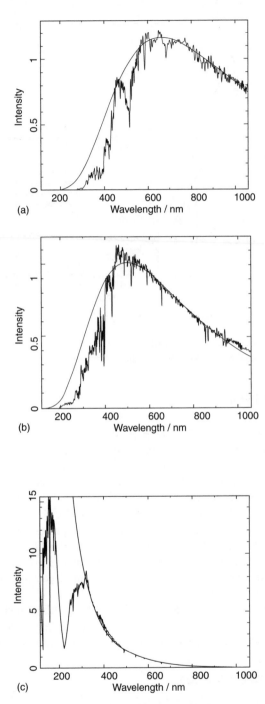

Figure 18.10   Blackbody spectra fitted to the spectra of stars with three different temperatures (spectral types (a) O, (b) G and (c) K). The short-wavelength regions of these spectra show the effects of absorption, mainly by hydrogen, in the stellar atmosphere and interstellar space. (*Courtesy T. O'Brien; spectra* from *A.J. Pickles, Publication of the Astronomical Society of the Pacific,* **110**, 1998, 863)

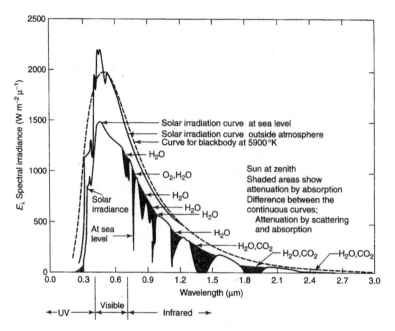

Figure 18.11   Solar spectral irradiance at the Earth's surface. (From M. Chaline et. al. 1983), *Manual of Remote Sensing*, 2nd edn. page 165. Ed. R.N. Colwell, Am. Soc. of Photogrammetry.

18.9   Fluorescent Lights

Fluorescent lights are discharge lamps in which the direct emission is absorbed in a *phosphor* and converted into new wavelengths. An example is the mercury discharge lamp which produces ultraviolet light with a strong line at 253.7 nm. By coating the inner wall of the discharge tube with a phosphor visible light is produced. Typical phosphors are calcium phosphate, zinc silicate or calcium tungstate. Discharge lamps may be excited by a radio frequency discharge using external electrodes. Arc lamps are short-length discharge lamps and operated at high current; typical gases are xenon, mercury or sodium. The high-pressure mercury lamp operates at pressures up to 10 bar and emits high-irradiance visible light. By the addition of various halides to the mercury vapour the colour distribution can be altered to produce increased white light content. A lamp based on emission from a dimer species of sulphur can closely replicate the solar spectrum. Metal–halide lamps give a light output of up to 100 lumens per watt.

Mercury discharge lamps may be designed to emit mainly the ultraviolet 253.7 nm line; these find widespread use in lithography and as germicidal lamps for killing bacteria. Line emission from discharge lamps provides standard wavelengths for instrumental calibration.

18.10   Luminescence Sources

Luminescence is the emission of light from materials that have been excited in some way other than by heating. (It is distinguished from *incandescence* where heating generates thermal radiation.) The excitation may be by light (*photoluminescence*), electric field (*electroluminescence*), electron beam (*cathodoluminescence*), chemical reaction (*chemiluminescence*), sound waves (*sonoluminescence*) and in biological matter (*bioluminescence*).

In photoluminescence light is absorbed and is followed by re-emission at a wavelength which may be equal to the exciting wavelength, or at a longer wavelength, when it is known as *Stokes radiation*. There is also a low probability of the emitted photon gaining energy by absorbing the energy of a molecular vibration or a phonon, so that the emission is at a shorter *anti-Stokes* wavelength.

In cathodoluminescence a cathode ray tube produces a visual image on a phosphor coated on a faceplate by excitation with an electron beam. Colour images can be generated based on $Eu:Y_2O_2S$ (producing red light), Ag:ZnS–CdS (green) and Ag:ZnS (blue) phosphors. Chemiluminescence may be produced in certain exothermic chemical reactions, when molecules in excited states are produced. An efficient chemiluminescence reaction is based on peroxy-oxalate, producing sufficient light to be used in light sticks. Chemical reactions in biological materials produce bioluminescence. Biological organisms such as bacteria, insects or fungi are able to act as light emitters. The transmission of high-intensity ultrasound in liquids containing dissolved gases produces sonoluminescence. In the presence of ultrasound minute inhomogeneities in the liquid grow into microscopic bubbles; these undergo rapid oscillation, expand and, finally, collapse. In this process a large amount of energy can be taken up by the bubbles, which is then released as light during the bubble collapse.

The absorption and emission properties of photoluminescent materials depend on the electronic structure of the material. The probability of emission is quantified by the ratio η of the number n_e of photons emitted to the number absorbed n_a. It is termed the *quantum efficiency*, and it is always less than 1.

The excitation energy may be dissipated by non-radiative processes such as lattice phonon vibrations in a solid, reducing the value of η. Photoluminescent materials are based on doped inorganics, organic molecules or semiconductors. Inorganic crystals and glasses doped with transition metals or rare earth ions are termed *phosphors*. Typical energy levels in an inorganic phosphor are shown in Figure 18.12.

X-ray imaging in medicine detects X-rays by using a phosphor screen from which ultraviolet or visible light is emitted and recorded on a photographic film. These phosphors are based on rare earth dopants such as indium, europium or terbium.

Luminescence from organic molecules has wide-ranging applications which include biological probes, medical diagnostics and environmental analysis; it is also used in a type of laser known as

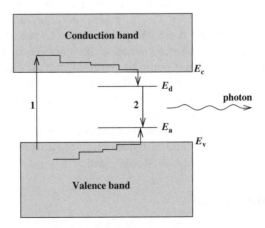

Figure 18.12 Energy levels and light emission process from a phosphor. E_v and E_c are the limits of the valence and conduction bands. Acceptor and donor dopant levels are E_a and E_d. Excitation of the phosphor (1) can lead to light emission (2)

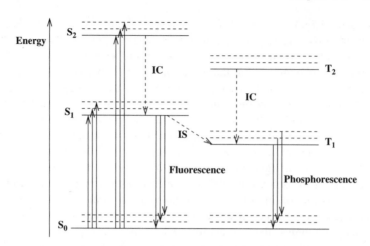

Figure 18.13   Energy levels of an organic molecule with singlet (S_0, S_1, S_2) and triplet (T_1, T_2) electronic states and vibrational levels, internal conversion (IC), intersystem crossing (IS), fluorescence and phosphorescence

the *dye laser.* Organic molecules made up of alternating single (C–C) and double (C=C) carbon bonds have a set of outer-lying electrons in molecular orbitals with angular momentum $l = 1$, designated π electrons. These electrons determine the absorption and emission properties of the molecules. The electrons fill the molecular orbitals in pairs, following the Pauli exclusion principle in which the state function is antisymmetric with respect to exchange of particles, and two electrons in a given orbital must have opposing spins. With paired spins the ground state is a singlet state with no net spin, the total spin S being zero. Excited electronic states may be singlet or triplet ($S = 1$). Electronic absorption of energy promotes one electron to a higher singlet or triplet state; singlet–singlet transitions are allowed while singlet–triplet transitions are forbidden.

For each electronic energy level there are a set of vibrational levels; the population in these levels follows the Boltzmann distribution. As shown in Figure 18.13, energy may redistribute within the molecule by vibrational relaxation (characteristic time $\sim 10^{-12}$s), internal conversion (by higher singlet to S_1 and higher triplet to T_1 transitions) and intersystem crossing (singlet–triplet transition). The emission may be by fluorescence on the $S_1 - S_0$ transition (decay time $\sim 10^{-8}$s) or by phosphorescence on the partially forbidden $T_1 - S_0$ transition (decay time $\sim 10^{-3}$s). (*Fluorescence* ceases rapidly as the excitation ceases, while *phosphorescence* can persist for a considerable time.)

Fluorescence can be excited by ionizing radiation in inorganic and organic materials; this mechanism is termed *scintillation*. (This is not to be confused with the effects, such as the twinkling of starlight, observed in the passage of light through a distorting medium, and also called scintillation.) Detectors to measure fluorescence scintillation quantitatively are called scintillation counters. In these detectors a scintillator medium is used to absorb the energy of the ionizing particle, such as X-rays or gamma rays, and the fluorescent light emission is detected by a photomultiplier, photographic film or a CCD. Suitable scintillation materials include inorganic crystals such as sodium or caesium iodide, organic crystals or liquids, doped plastics or a gas.

18.11   Electroluminescence

Inorganic phosphors may be excited by an electric current by placing the phosphor layer between electrodes. The emitting materials are doped with wide-bandgap semiconductors from groups II–VI,

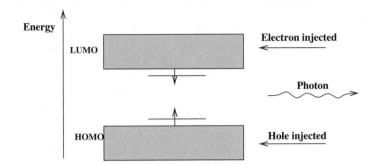

Figure 18.14 Molecular energy band structure which shows the highest occupied molecular orbital (HOMO), the lowest unoccupied molecular orbital (LUMO) and dopant levels. An electron–hole singlet spin configuration is illustrated with radiative emission

IIa–VIb and IIb–VIb, or with organic polymers. When placed in an electric field electron carriers are able to excite the activators. The ability to excite electroluminescence by an electric current is a major convenience and opens up applications in emergency lighting and backlighting for LCDs. Fabrication in thin films enables miniaturization and integration in flat-panel displays.

Electroluminescent sources based on doped ZnS, SrS or CaS powders with transparent dielectric binder and sandwiched between two electrodes provide emission wavelengths in the blue, green and red regions suitable for visible displays. The LED is an electroluminescence source using single crystal semiconductors, organic molecules or polymers. The injection of minority carriers into a forward-biassed p–n junction is followed by recombination of electron and hole carriers and light emission, as described in Chapter 17. LEDs are able to produce wavelengths across the visible region and with relatively high efficiency. LEDs based on GaP for green emission and the four-element semiconductor AlInGaP emitting in the red are able to produce up to 100 lumens per watt. Violet (370–420 nm), blue (460–490 nm) and green (500–520 nm) emitting LEDs using the nitride GaInN are able to emit $30\,\mathrm{lm\,W^{-1}}$. LEDs are suitable sources for illumination, instrumentation indicators, fibre optic communications, large-area and full-colour displays, road signs, traffic signals and vehicle lighting. LEDs can be more than 10 times as efficient as conventional light bulbs;[7] the replacement of lamps in traffic lights with LEDs gives a substantial saving in power consumption.

Solid state organic molecules and organic polymers have energy level characteristics similar to semiconductors. The energy differences between the molecular S_0 and S_1 levels (Figure 18.13) in the solid state form molecular orbital states, termed LUMO (Lowest Unoccupied Molecular Orbital) and HOMO (Highest Occupied Molecular Orbital) similar to the semiconductor bandgap (Figure 18.14). They are able to be doped to provide electron or hole semiconductivity. In organic layers the electrons and holes are mobile and form an excitation, termed an exciton, when they collide. Organic electroluminescent light sources are based on small molecules (OLEDs) or polymers (PLEDs). The organic LED is made up of hole transport and electron transport layers sandwiched between electrodes which inject carriers. They offer convenient fabrication together with high efficiency and fast response.

---

[7]Radiant efficiency is the ratio of the power emitted in the visible range, to the total input power.

Problem 18.1

(a) Find the linewidth in wavelength and frequency terms for a cooled neon (atomic mass number $M = 20$) discharge at 300 K. (b) Find the half-width of the H_α line ($\lambda = 656$ nm) emitted by atomic hydrogen in an ionized interstellar cloud at a temperature of 10^4 K assuming that the width is entirely due to the thermal Doppler effect.

Problem 18.2

The tungsten filament of a 100 W light bulb operates at a temperature of 2800 K. Assume that it radiates as a blackbody, and ignore that fraction of the power lost in the form of heat energy. (a) Determine the number of photons emitted per second. Note that for a blackbody at temperature T, the mean photon energy is $2.70kT$. (b) Use $E = mc^2$ to determine the total mass emitted by the bulb during its life of 750 h. Compare this with the mass of an *E. coli* bacterium, 6×10^{-13} kg.

Problem 18.3

A 5 mW helium–neon laser radiates $\lambda = 633$ nm with a linewidth $\Delta v = 10^4$ Hz. The beam diameter is 0.4 mm. (a) How many photons will the laser emit per second? (b) What temperature would be required for a blackbody to emit as many photons over the same frequency range and from the same area?

Problem 18.4

A synchrotron has 20 equally spaced bunches of electrons of 3 GeV energy held in circular orbits by a 2 T magnetic field. The total circulating current is 50 mA. Determine the radius and period of the orbits. Calculate the total power radiated and the characteristic energy of the emitted photons. (The total power radiated by an accelerated non-relativistic electron is given by equation (18.2).)

Problem 18.5

The Bohr model can be applied to all atoms which, like ordinary hydrogen, contain a single electron, such as deuterium, singly ionized helium and doubly ionized lithium. It predicts that the wavelength emitted or absorbed in a transition between levels n_1 and n_2 is given by $1/\lambda = R(1/n_1^2 - 1/n_2^2)Z^2$, where Z is the atomic number. For an infinitely massive nucleus, the Rydberg constant is

$$R_\infty = m_e e^4 / 8\epsilon_0^2 h^3 c = 1.097\,373\,157\ldots \times 10^{-2}\,\text{nm}^{-1}. \tag{18.24}$$

For nuclei of finite mass, R is found by replacing the electron rest mass with the reduced mass μ.

(a) For any given transition, find the fractional wavelength $\Delta\lambda/\lambda$ difference between ^{3}He and ^{4}He.

(b) What transition in ^{4}He yields a wavelength closest to that of the H_α line of the Balmer series ($n_1 = 2, n_2 = 3, \lambda = 656.3$ nm), and what is the fractional wavelength difference between them?

Problem 18.6

Bohr's correspondence principle claims that when applied to larger and larger microsystems, quantum-mechanical predictions must tend towards those of classical, macroscopic physics. This suggests, in particular, that for large n values, the electron orbits in Bohr's model of the hydrogen atom should behave increasingly as in the classical theory. Classical electrodynamics predicts that a charge orbiting in a circle with angular frequency ω will emit radiation of frequency $f = \omega/2\pi$. Consider the frequency v of the photon emitted in Bohr's model for the quantum jump $n + 1 \to n$. Show that in the limit $n = \infty, v = f$. (You can assume a hydrogen atom with infinitely massive nucleus for simplicity.)

19 Interaction of Light with Matter

Why make so much of fragmentary blue/ In here or there a bird, or butterfly/ Or flower, or weaving-stone, or open eye,/ When heaven presents in sheets the solid hue?

Robert Frost.

Although the interaction of light with matter must ultimately be considered in terms of photons and quantum mechanics, there are many circumstances when a classical electromagnetic theory provides a practically complete description. The most interesting phenomena, however, involve resonances within an atom or molecule; only quantum mechanics can account for the existence of these resonances, but given their existence it turns out to be convenient and illuminating to consider scattering, absorption and reradiation of light by individual charges and atoms through a classical electromagnetic approach. As in radiation theory (Chapter 18), however, there comes a point when interactions with matter are necessarily considered as photon processes; these interactions, such as Compton and Raman scattering, are considered in the later sections of this chapter. Some simple scattering processes, such as Rayleigh and Compton scattering, are concerned with isolated particles behaving independently of one another. The radiation scattered from many particles then adds with random phase, so that it is the power and not the amplitude which is proportional to the number of radiating or scattering events. In dense materials it is no longer appropriate to consider individual particles, and we consider instead the *polarization* of a dielectric and its effect on an electromagnetic wave. A concise guide to the various scattering processes is given in a table at the end of this chapter.

19.1 The Classical Resonator

In a neutral gas it is often appropriate to treat the neutral atoms or molecules as classical harmonic oscillators, as in Section 18.1. We derive the response of such oscillators to the electric field of the electromagnetic wave, and calculate the reradiated, or *scattered*, radiation from an individual oscillator. The scattered radiation is at the same frequency, but in directions away from the direction of the incident wave. Following the classical analysis, the atom or molecule is represented as a simple

Optics and Photonics: An Introduction, Second Edition   F. Graham Smith, Terry A. King and Dan Wilkins
© 2007 John Wiley & Sons, Ltd

harmonic oscillator in which a mass m with charge q responds to an applied field $E_0 \exp(i\omega t)$ with a complexified displacement x given by

$$m\frac{\mathrm{d}^2 x}{\mathrm{d}t^2} + \xi\frac{\mathrm{d}x}{\mathrm{d}t} + \omega_0^2 x = qE_0 \exp(i\omega t) \tag{19.1}$$

where ξ is the *damping constant* and ω_0 is the undamped resonant frequency. (The inverse of ξ is the *lifetime* $\tau = 1/\xi$.) The response[1] is an oscillation at angular frequency ω with displacement $x(t) = x_0 \exp(i\omega t)$:

$$\begin{aligned} x(t) &= \frac{qE_0}{m}\frac{\exp(i\omega t)}{(\omega_0^2 - \omega^2) + i\omega\xi} \\ &= \frac{qE_0}{m}\frac{\exp(\omega t + \phi)}{[(\omega_0^2 - \omega^2)^2 + (\xi\omega)^2]^{1/2}} \end{aligned} \tag{19.2}$$

where E_0 is the incident field. The phase difference ϕ is of no consequence, as the scattered fields from many atoms add in random phase. Figure 19.1 shows the amplitude of response for a wide range of driving frequency ω and for three values of the damping constant ξ. Note particularly:

- The full width at half power (FWHP) of the resonance is approximately ξ (for small ξ).

- At low frequencies (i.e. low photon energies) the terms in ω^4 and $(\xi\omega)^2$ can be neglected in comparison with ω_0^4. The result is that the amplitude of oscillation is independent of the wave frequency.

- For large damping the maximum amplitude is at a lower frequency than ω_0.

The radiated electric field from the electron is proportional to its acceleration, found by twice differentiating equation (19.2) with respect to time.

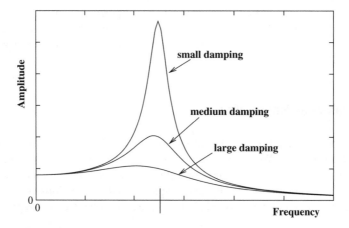

Figure 19.1   The amplitude of response of a simple harmonic oscillator for various degrees of damping

---

[1]If the damping constant is small enough ($\xi \le \omega_0$), there is also a transient oscillation with frequency $\omega' = (\omega_0^2 - \xi^2/4)^{1/2}$, which dies away exponentially according to the factor $\exp(-\xi t/2)$.

19.2 Rayleigh Scattering

At low frequencies compared with the resonant frequency, the acceleration is simply proportional to ω^2, and the reradiated power is therefore proportional to ω^4, or to λ^{-4}. This is the *Rayleigh law of scattering*. Note that there is no change of frequency in Rayleigh scattering. Its most familiar application is to the scattering of visible sunlight by air molecules in the Earth's upper atmosphere, where blue light is scattered more than the longer wavelength red light.[2] Light received directly from the Sun is unpolarized, but scattered sunlight is plane polarized. This may readily be demonstrated by using polaroid sunglasses; the plane of polarization is at right angles to the direction of the Sun, since the induced vibration of the air molecules is transverse to the illuminating rays (Figure 19.2). Rayleigh scattering only applies when the scattered radiation from individual particles adds incoherently, which is not the case for the lower atmosphere; at sea level the molecules are spaced by only about 3 nm, and the atmosphere behaves as a dense medium.

The Rayleigh law can also apply to the scattering of light in transparent solids or liquids, where the scattering elements are irregularities in density or structure rather than individual atoms. An important example is in the transmission of light along glass fibres, as used in long-distance communications (Chapter 6). Here the lowest achievable attenuation is mainly limited by Rayleigh scattering by small, widely spaced irregularities in the glass. The Rayleigh law dictates the use of the longest infrared wavelengths for which light sources and detectors are available, but avoiding the hydroxyl ion resonance at 1.39 μm; in practice the lowest attenuation is obtained at about 1.55 μm.

Rayleigh scattering applies to small particles, with diameter D less than about 0.1 λ. Scattering from larger particles, with diameters of around one wavelength and above, is more complex since there can be interference between waves separately scattered from different parts of the particle.

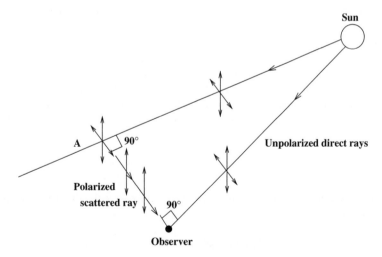

Figure 19.2 Polarization of scattered light. Light reaching the observer directly from the Sun is unpolarized. But light scattered off air molecules at A and observed from any direction perpendicular to the incident sunlight is seen to be polarized. (Notice that the Sun is so far distant, compared with the size of the Earth or Sun, that the rays from the Sun are nearly parallel, though perspective makes it appear that they converge)

---

[2]See E. J. McCartney, *Optics of the Atmosphere*, John Wiley & Sons, 1976.

There will then be a non-isotropic scattering pattern, which depends on shape as well as material and size. Scattering by spherical conducting particles of any size was analysed by G. Mie and is known as *Mie scattering*; the theory is applicable also to non-conducting particles.

19.3 Polarization and Refractive Index in Dielectrics

When the scattering particles are close together, as in a solid, they can no longer be considered as individuals. Instead we consider the effect of a continuous dielectric on an electromagnetic wave in terms of the *electric polarization* induced by the electric field of the wave. The polarization represents a charge separation within the material, forming a dipole moment which oscillates at the wave frequency: this reradiates a secondary wave, which combines with the original wave. The combination may have a different phase from that of the original wave: this means that the phase velocity of the resultant wave has been determined by the polarization properties of the dielectric; that is, by its dielectric constant.

The velocity of electromagnetic waves is given generally by $c/(\epsilon_r\mu_r)^{1/2}$, where c is the velocity in free space (Chapter 5). For a pure dielectric, whose magnetic susceptibility is unity, the refractive index therefore equals $\epsilon_r^{1/2}$, the square root of the dielectric constant. The relation between ϵ_r and the polarization P is given by simple electrostatic theory as

$$D = \epsilon_r\epsilon_0 E, \qquad \text{or} \qquad P = \epsilon_0(\epsilon_r - 1)E = \epsilon_0\chi_e E, \tag{19.3}$$

where χ_e is the electric susceptibility. The polarization P is defined as the electric dipole moment per unit volume. If there are N atoms or molecules per unit volume, each polarized by the field E to form a dipole with moment $p = \alpha E$, where α is their individual *polarizability*, then

$$\epsilon_r = 1 + \frac{\alpha N}{\epsilon_0} = 1 + \frac{P}{\epsilon_0 E}. \tag{19.4}$$

We must allow for the possibility that the atomic polarization is not in phase with the field E, which may happen when the atom behaves like a resonant atomic oscillator. The refractive index $\epsilon_r^{1/2}$ is therefore complex, having real and imaginary parts. If we denote it by $n - i\kappa$, the effect is that the wave becomes

$$E = E_0 \exp\{i[\omega t - (n - i\kappa)kx]\} \tag{19.5}$$

in which k is the wave number in free space. The imaginary part κ then represents absorption, since equation (19.5) becomes

$$E = E_0 \exp(-\kappa kx) \exp[i(\omega t - nkx)]. \tag{19.6}$$

κ is called the *extinction coefficient*. The wave proceeds at phase velocity c/n, while the amplitude decays exponentially with distance.

Equation (19.4) becomes

$$(n - i\kappa)^2 = n^2 - \kappa^2 - i\,2n\kappa = 1 + \frac{\alpha N}{\epsilon_0} \tag{19.7}$$

and if κ is small enough, as in a transparent gas or any dielectric where the attenuation is small,

$$n^2 \simeq 1 + \frac{\alpha N}{\epsilon_0}. \tag{19.8}$$

A further approximation may be made if n is close to unity:

$$n \simeq 1 + \frac{\alpha N}{2\epsilon_0}. \tag{19.9}$$

We now have the required relationships between the refractive index of a medium and the response of its charged particles to the oscillatory electric field. The response, represented by the polarizability α, will be found for three particularly interesting cases in the following sections.

19.4   Free Electrons

A particularly simple situation arises when the dielectric polarization is entirely due to electrons which are not bound to nuclei. This is the case for radio wave propagation through an ionized gas such as the terrestrial ionosphere, and also for X-rays in metals, where the electrons are only lightly bound compared with the photon energy at these short wavelengths. Neglecting the effect of collisions between electrons and ions in the ionosphere, and the effects of lattice irregularities in metals, the polarizability α for a single electron is found from the simple equation of motion

$$\frac{\mathrm{d}^2 x}{\mathrm{d}t^2} = \frac{eE}{m} = \frac{eE_0}{m} \exp(i\omega t). \tag{19.10}$$

The electrons all follow a (complexified) oscillatory displacement $x = x_0 \exp(i\omega t)$, and the polarizability α is the dipole moment per unit field:

$$\alpha = \frac{ex}{E} = -\frac{e^2}{m\omega^2}. \tag{19.11}$$

Substituting in equation (19.8) we find

$$n^2 = 1 - \frac{Ne^2}{\epsilon_0 m\omega^2}. \tag{19.12}$$

This gives the refractive index for any dielectric where the polarization is due to free electrons. It may be written as

$$n^2 = 1 - \frac{v_p^2}{v^2}, \qquad v_p = \frac{1}{2\pi} \left(\frac{Ne^2}{\epsilon_0 m}\right)^{1/2} \tag{19.13}$$

where v_p, known in gases as the *plasma frequency*, is characteristic of the medium. For the terrestrial ionosphere v_p is at radio frequencies, reaching 10 MHz where the electron density is greatest (i.e. in the F-region), while for metals it lies in the ultraviolet. For frequencies above v_p the refractive index is real and less than unity, and the phase velocity is therefore greater than the free space velocity c. It is easy to show that if equation (19.13) holds, then the product of the group velocity and phase velocity is c^2.

19.5   Faraday Rotation in a Plasma

In the Faraday effect (Section 7.11) the plane of polarization of an electromagnetic wave propagating parallel to a magnetic field B in a material medium will rotate, due to the different phase velocities of the two hands of circular polarization. The angle of rotation ψ is given by $\psi = BVl$, where V is the Verdet constant and l is the path length. In this section we add a magnetic field to the analysis of the previous section, and find the Verdet constant for an ionized gas with a magnetic field.

Consider a plane wave incident parallel to a magnetic field $\mathbf{B}\hat{z}$, in a plasma of electrons with charge $q = -e$ and mass m. For non-relativistic particle speeds, only the electric component of the light wave is significant, and we can write the Lorentz force on an electron moving with velocity $\mathbf{v}$ as $m\mathbf{a} = q\mathbf{E} + q\mathbf{v} \times \mathbf{B}$, or in component form

$$m\frac{\mathrm{d}^2 x}{\mathrm{d}t^2} = qE_x + q\frac{\mathrm{d}y}{\mathrm{d}t}B, \quad m\frac{\mathrm{d}^2 y}{\mathrm{d}t^2} = qE_y - q\frac{\mathrm{d}x}{\mathrm{d}t}B, \quad m\frac{\mathrm{d}^2 z}{\mathrm{d}t^2} = 0. \tag{19.14}$$

Defining a complex position $\tilde{r} = x + iy$ and complex electric field $\tilde{E} = E_x + iE_y$ we obtain from the x and y equations

$$\frac{\mathrm{d}^2 \tilde{r}}{\mathrm{d}t^2} + \frac{iqB}{m}\frac{\mathrm{d}\tilde{r}}{\mathrm{d}t} - \frac{q\tilde{E}}{m} = 0. \tag{19.15}$$

A circularly polarized monochromatic wave in the direction of the z axis is represented by $\tilde{E} = E_0 \exp[\pm i(\omega t - kz)]$; the upper sign refers to left-hand polarization (LHP) and the lower to RHP. The motion of the electron is represented by $\tilde{r} = r_0 \exp[\pm i(\omega t - kz)]$.

Following the analysis of Section 19.4 above, the amplitude of motion becomes

$$r_0 = \frac{-qE_0}{m(\omega^2 \pm \omega qB/m)}. \tag{19.16}$$

For an electron, $q = -e$, this leads to a displacement of the electron $\mathbf{r} = (e\mathbf{E}/m)(\omega^2 \mp \omega\omega_{\mathrm{cyc}})^{-1}$, where $\omega_{\mathrm{cyc}} = eB/m = 2\pi\nu_{\mathrm{L}}$ is the cyclotron angular frequency in the plasma (see Section 18.3). The polarization $\mathbf{P}$ of the plasma is $\mathbf{P} = -Ne\mathbf{r}$, where N is the number density of electrons in the plasma. As before, we find $n^2 = \epsilon_{\mathrm{r}} = 1 + P/\epsilon_0 E$, giving the two-valued refractive index $n^2 = 1 - (Ne^2/m\epsilon_0)(\omega^2 \mp \omega\omega_{\mathrm{cyc}})^{-1}$. Notice that this generalizes the plasma equation (19.12) simply by relacing ω^2 by $(\omega^2 \mp \omega\omega_{\mathrm{cyc}})$. It is normally the case that $\omega_{\mathrm{cyc}} \ll \omega$ so we can set the average of n^2 to $n_0^2 = 1 - (Ne^2)/(m\epsilon_0\omega^2) \approx 1$ for a thin plasma. Notice also that $n_{\mathrm{L}} < n_{\mathrm{R}}$, so that the phase velocity of the LHP wave is greater than that of the RHP wave.

A plane polarized wave may be regarded as the sum of two circularly polarized waves of opposite hands; the plane of polarization is the direction in which the oppositely rotating electric vectors intermittently coincide. After travelling a distance l in the plasma, a phase difference $\Delta\phi = kl\Delta n$ develops between the two polarizations, and the vector field of the LHP wave has rotated by $\Delta\phi$ ahead of the RHP wave. Their vector superposition has then rotated $\frac{1}{2}\Delta\phi$, i.e. half this amount, so that the plane of polarization has rotated by $\psi = \frac{1}{2}kl\Delta n$. Since for a thin plasma $\Delta n = \frac{1}{2}\Delta(n)^2$, we find the rotation formula

$$\psi = \frac{e^3 BN\lambda^2 l}{8\pi^2\epsilon_0 m^2 c^3} = (2.63 \times 10^{-13}\,\mathrm{rad})(B/T)N\lambda^2 l, \tag{19.17}$$

where $T = 1\,\mathrm{T}$. (Note that $1\,\mathrm{T} = 1\,\mathrm{Wb\,m^{-2}} = 1\,\mathrm{kg\,s^{-2}\,A^{-1}} = 1\,\mathrm{N\,m^{-1}\,A^{-1}}$.)

Faraday rotation is particularly significant in radio wave propagation through the terrestrial ionosphere and interstellar space (see Problem 19.5).

19.6   Resonant Atoms in Gases

The third case for which we can easily evaluate the polarizability, and hence obtain the refractive index, concerns neutral atoms in gases. In a low-pressure gas the atoms may be assumed to act independently. Electrons are bound within the atoms at a series of energy levels; the atoms emit and absorb light with photon energies equal to energy differences between the energy levels. The classical analogy is a simpler picture, in which the atoms are considered as oscillators which resonate at the appropriate spectral line frequencies. These resonators respond to the electric field of a wave with a forced oscillation, whose amplitude depends on the resonant angular frequency ω_0, and the sharpness of resonance, characterized by a width (FWHM) $\Delta\omega_{1/2} = \xi_0$. The approach of equations (19.2), (19.10), and (19.11) then leads (for a low-density gas) to a polarizability

$$\alpha = \frac{e^2}{m} \frac{1}{(\omega_0^2 - \omega^2) + i\omega\Delta\omega_{1/2}}. \tag{19.18}$$

Some awkward algebra is needed to relate α through equation (19.18) to the refractive index $n - i\kappa$, but if κ and $n - 1$ are both small, as is appropriate for a low-pressure gas, and using equation (19.5) a good approximation may be found:

$$n = 1 + \frac{Ne^2}{2\epsilon_0 m} \frac{\omega_0^2 - \omega^2}{(\omega_0^2 - \omega^2)^2 + (\omega\Delta\omega_{1/2})^2}. \tag{19.19}$$

$$\kappa = \frac{Ne^2}{2\epsilon_0 m} \frac{\omega\Delta\omega_{1/2}}{(\omega_0^2 - \omega^2)^2 + (\omega\Delta\omega_{1/2})^2}. \tag{19.20}$$

The form of the variation of n and κ with the wave frequency is shown in Figure 19.3. Well away from resonance, the refractive index n increases steadily with frequency, and the extinction is small.

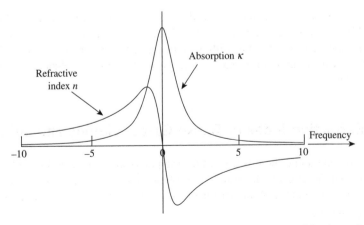

Figure 19.3   The variation of absorption and refractive index n (plotted as $n - 1$) in the region of a resonance. The width of the resonance is determined by the lifetime τ

Normal dispersion is said to apply for any frequency where $dn/d\omega > 0$. As the wave frequency approaches the atomic resonance, and the amplitude of the atomic oscillators increases, the absorption increases and the dispersion $dn/d\omega$ in the refractive index changes sign. This *anomalous dispersion*, defined by $dn/d\omega < 0$, is found at every resonant frequency of the atoms in the gas. Anomalous dispersion is most marked in gases at low pressure, since it is only when the atoms behave independently that a sharp resonance is found. When resonators of the same frequency are coupled together the combined resonance curve is broadened; in quantum terms this means that the lifetime of an excited state is reduced, while in terms of refractive index there are less marked effects of anomalous dispersion.

In real gases the atoms will have many resonances, each of which has an associated anomalous dispersion. As compared with monatomic gases, molecular gases also have rotational and vibrational resonances, occurring generally at lower frequencies. A complete curve showing refractive index and absorption for a gas therefore shows a series of the dispersion curves with the form of Figure 19.3. Away from resonances, and where the absorption is negligible, the refractive index increases with wave frequency; this is seen in glass, where the refractive index is higher at the violet than at the red end of the spectrum. When the wave frequency is very much higher than the resonant frequencies, the refractive index becomes less than unity; this is the free electron case discussed in the previous section.

A formula by W. Sellmeier provides a useful representation of refractive index as a function of wavelength:

$$n^2(\lambda) = A + \sum_j \frac{B_j \lambda^2}{\lambda^2 - \lambda_j^2}. \tag{19.21}$$

Here each λ_j represents a resonance, at frequency ν_j, with $\lambda_j \nu_j = c$. (Obviously the Sellmeier formula is inapplicable close to a resonance.)

The wave propagates with phase velocity c/n, and with an amplitude decreasing exponentially with distance. This attenuation is expressed by an *absorption coefficient* μ, such that the fraction of the intensity lost in a thin slab with thickness dx is μdx. An expansion of equation (19.6) shows that

$$\mu = \frac{2\omega\kappa}{c}. \tag{19.22}$$

The absorption may be regarded as the sum of the absorptions by all the individual atoms. Let each atom have an effective cross-sectional area σ for absorption or scattering. When a beam of area A travels a small distance Δx in a medium with N atoms per unit volume, it will pass $NA\Delta x$ atoms and suffer a fractional loss of irradiance $\mu\Delta x = -\Delta I/I = (NA\Delta x)\sigma/A$. Hence the *cross-section for absorption* $\sigma = \mu/N$.

19.7   The Refractive Index of Dense Gases, Liquids and Solids

In gases at high pressure and in liquids and solids, the polarization induced in neighbouring atoms or molecules means that the effective local polarizing field, say E_{loc}, on each atom or molecule differs from the applied external field E. A model for the local field taking into account the polarization of the neighbouring atoms gives

$$E_{loc} = E + P/3\epsilon_0. \tag{19.23}$$

For a dense medium $P = N\alpha(E + P/3\epsilon_0)$, and rearranging

$$P = N\alpha E/(1 - N\alpha/3\epsilon_0). \tag{19.24}$$

With the definitions $P = \epsilon_0\chi_e E = \epsilon_0(\epsilon_r - 1)E$, we obtain

$$N\alpha/3\epsilon_0 = (\epsilon_r - 1)/(\epsilon_r + 2). \tag{19.25}$$

This is the *Clausius–Mossotti* equation relating the relative permittivity to the polarizability of the molecules. We may use the relation between the relative permittivity and the real part of the refractive index, i.e. $\epsilon_r = n^2$, to obtain the related *Lorentz–Lorenz* equation

$$(n^2 - 1)/(n^2 + 2) = N\alpha/3\epsilon_0. \tag{19.26}$$

For cases in which the refractive index is near unity this equation reduces to equation (19.9). Equation (19.26) is found to be valid over a wide range of values of n and N.

19.8   Anisotropic Refraction

In most substances the refractive index is independent of the direction of propagation and independent of the polarization of the light. From the foregoing discussion this is obviously due to isotropy in structure of the substance: the polarizability is independent of the direction of the electric field. For some materials, however, this may not be so. In a crystal the unit cell may be anisotropic, so that the polarizability depends on the direction of the electric field in relation to the directions of the crystal axes. This leads to the birefringence already discussed in Chapter 7. Again, a liquid may be composed of molecules with a permanent dipole moment, so that they may be aligned by an external steady electric field. In this way a dielectric liquid may be made birefringent, and used to modify the polarization of light passing through it. Birefringence induced by an electric field in a normally isotropic liquid is known as the *Kerr effect* (Section 7.11): a similar effect induced by a magnetic field is known as the *Cotton–Mouton effect*. Another related phenomenon in which the existing birefringence of a crystal is modified by an electric field is known as the *Pockels effect*.

These effects relate to the refractive indices of plane polarized components of a light ray. The two components will also show anomalous dispersion for frequencies close to atomic resonances. Figure 19.4 shows a further effect in which the resonant frequency of an atomic oscillator is changed

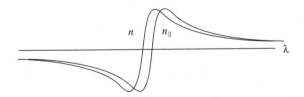

Figure 19.4 Anisotropic dispersion. Strong magnetic or electric fields can induce birefringence in some substances by changing the resonant frequency of atomic oscillators. The oscillators then become anisotropic, with resonances in the dispersion curve appearing at frequencies which differ for two planes of polarization, related to the orientation of the applied field. The dielectric then shows magnetic or electric double refraction at frequencies close to the resonance

by a strong electric or magnetic field. The resonance then appears in the dispersion curve at different frequencies for the two planes of polarization, depending on the plane of polarization in relation to the direction of the applied external field. The result is a large birefringence in the region of the resonance. This is the *Voigt effect*.

Optical activity, discussed in Chapter 7, is due to a difference in refractive index between the two hands of circular polarization.

19.9   Brillouin Scattering

In condensed matter (liquid or solid), light waves may be scattered by non-uniformities in dielectric constant created by acoustic waves. This is *Brillouin scattering*, which is often used in measurements of the elastic properties of materials; here phonons (which are quantized acoustic waves and intrinsic to the material) cause the light scattering. Figure 19.5 shows scattering by a plane acoustic wave, wavelength λ_s and travelling at velocity v_s. The wave causes variations in density, and hence of refractive index, in planes separated by λ_s. The reflected electromagnetic waves of wavelength λ_1 are only in phase and observable when the glancing angle of incidence θ for mth-order scattering satisfies the Bragg law (derived for X-rays in Chapter 11):

$$2\lambda_s \sin \theta = m\lambda_1. \tag{19.27}$$

The reflecting planes are moving at velocity v_s, causing a Doppler shift, which in Figure 19.5 changes the frequency from v_1 to a lower frequency v_2 given by

$$\frac{v_1 - v_2}{v_1} = 2\frac{v_s \sin \theta}{c}. \tag{19.28}$$

The shift in frequency is small, and does not appreciably affect the Bragg reflection condition. It is, however, sufficiently large to allow the sound velocity to be measured. Combining equations (19.27) and (19.28),

$$v_1 - v_2 = v_1 \frac{2v_s}{c} \frac{m\lambda_1}{2\lambda_s} = mv_s, \tag{19.29}$$

where we have used $v_s\lambda_s = v_s$ and $v_1\lambda_1 = c$, showing that the frequency shift can only be at integer multiples of the acoustic wave frequency. The sign of the frequency shift depends on the direction of

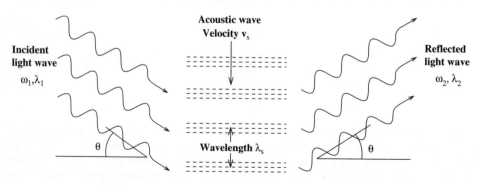

Figure 19.5   Brillouin scattering by an acoustic wave

the acoustic wave. Brillouin shifts have typical values in the range of 0.1 to 100 GHz, and may be observed with a laser probe source and a Fabry-Perot interferometer.

19.10 Raman Scattering

The scattering of light by atoms and molecules can be described by the Rayleigh law only for light with low photon energy. When the photon energy is comparable with or greater than the resonance energies in the scatterers, there may be a quantum interchange of energy, so that a photon emerges from a collision with a different energy.

The response of the damped driven classical oscillator, as in equation (19.2), is maximum at resonance, when $\omega = \omega_0$ and the photon energy is close to a resonant energy. The scattering which follows from this response corresponds to the re-emission of photons with unchanged energy. This is not always the case: the atom or molecule may absorb a photon and subsequently decay to a different excited state by emitting a photon of different energy; this inelastic scattering is known as *Raman scattering*.

Two types of Raman scattering may be distinguished. Normally the emergent photon has a lower energy than the incident photon: this is the usual inelastic scattering. If, in contrast, the scatterer is already in an excited state before the interaction, a photon may be emitted with higher energy than the incident photon: this is referred to as superelastic scattering. Analysis of both types must provide both for conservation of energy and for conservation of momentum, as for Compton scattering of free electrons.

In the quantum theory of Raman scattering from a molecule an incident photon interacts with quantized vibrational or rotational energy levels of the molecule. The photon raises the molecule to an excited electronic state, which is not an eigenstate of the system but may be an intermediate or virtual excited state. This state rapidly decays producing a scattered photon and the molecule in a different vibrational level. Where the final vibrational level is of higher energy the scattered photon has lost energy and has a longer wavelength than the incident photon; this leads to the Stokes Raman spectrum. Alternatively the photon may gain energy by interacting with a molecule in a higher vibrational level than the ground level, and will then have a shorter wavelength. This is the anti-Stokes Raman spectrum; it has lower amplitude than the Stokes spectrum, since it is partly determined by the population of molecules in the higher vibrational level given by the Boltzmann factor $\exp(-\Delta E/kT)$, where ΔE is the energy of the higher vibrational level. When the incident photon energy is close to an energy eigenstate of the molecule, i.e. an absorptive transition, this corresponds to a resonance, and in this *resonance Raman scattering* the strength of the scattering is greatly enhanced.

Insight into the Raman effect can be gained by a classical (non-quantum-mechanical) theory. A molecule in an electric field is distorted from the zero-field case with the electrons and nuclei moving in opposite directions; this creates an induced electric dipole moment or polarization in the molecule. For a molecule with polarizability α in an applied electric field E, the induced electric dipole moment is

$$p = \alpha E. \tag{19.30}$$

The electric field from a light wave oscillating at frequency ω, $E = E_0 \sin(\omega t)$, induces an oscillatory dipole moment

$$p = \alpha E_0 \sin(\omega t). \tag{19.31}$$

The normal modes of vibration of the molecule modulate the polarizability, such that for a molecular vibration at frequency ω_v the polarizability α becomes

$$\alpha = \alpha_0 + \beta \sin(\omega_v t). \tag{19.32}$$

The polarizability at equilibrium is α_0 and β is a constant. Then

$$p = [\alpha_0 + \beta \sin(\omega_v t)]E_0 \sin(\omega t). \tag{19.33}$$

Expanding the sine product we obtain

$$p = \alpha_0 E_0 \sin(\omega t) + (1/2) \beta E_0 [\cos(\omega - \omega_v)t - \cos(\omega + \omega_v)t]. \tag{19.34}$$

The oscillating dipole has components at frequency ω leading to Rayleigh scattering, at $(\omega - \omega_v)$ giving the Stokes Raman spectrum, and at $(\omega + \omega_v)$ giving the anti-Stokes spectrum.

   This classical analysis indicates a selection rule for Raman scattering: for the molecular vibration (or rotation) to be Raman active it must cause a change in the molecular polarizability.

   The Raman effect gives only a weak scattered irradiance, but a laser is an ideal light source, providing monochromatic and focusable light with high irradiance. In the Raman scattering experimental arrangement the scattered light is dispersed in frequency by a monochromator, which is most usually two monochromators in series, to discriminate against strong Rayleigh scattering at the incident frequency and stray light. Suitable high-sensitivity detection is provided by the photomultiplier or CCD detectors.

19.11   Thomson and Compton Scattering by Electrons

In Section 19.4 we considered the collective effect of free electrons on the propagation of an electromagnetic wave, by analysing the polarization of an electron gas. If the electrons instead act independently, as they will in a dilute gas where their spacing is large compared with the wavelength, they scatter electromagnetic radiation in a simple classical process, analogous to Rayleigh scattering by neutral atoms. This is known as *Thomson scattering*, for which the individual electrons have the *Thomson cross-section*

$$\sigma_T = \frac{8\pi r_0^2}{3} = 6.652 \times 10^{-29} \, \text{m}^2 \tag{19.35}$$

where r_0 is the classical electron radius.[3] In Thomson scattering the radiation is not absorbed, but reappears as radiation travelling in a different direction. There is no change of frequency; in quantum terms the photons collide elastically with the electrons. This applies when the photon energy $h\nu$ is much less than the rest mass energy of the electron $m_0 c^2$.

   Thomson scattering is a valuable technique in the characterization of plasmas produced in the laboratory. A probe beam of irradiance I_0 is attenuated in passage through a plasma according to $I = I_0 \exp(-N\sigma_T z)$, where N is the electron density and z is the distance travelled in the plasma. A

---

[3]That is, $r_0 = e^2/(4\pi\epsilon_0 m_e c^2) = 2.818 \times 10^{-15}$ m is the approximate radius of the electron viewed as a charged sphere, in order for its mass to derive entirely from its Coulomb self-energy.

splendid natural display of the scattering of light by free electrons is in the solar corona, where sunlight is scattered by electrons in the outer atmosphere of the Sun, and observable during a solar eclipse.

For higher photon energies the scattering can only be regarded as a quantum process in which a collision between an electron and a photon involves an exchange of both energy and momentum. This is the realm of *Compton scattering* (see Section 5.6), in which the photon emerging after the collision has a lower energy, i.e. a longer wavelength, the amount of the change depending on the geometry of the collision. The analysis which depends on the conservation both of energy and of momentum gives the Compton scattering formula for the increase in wavelength $\Delta\lambda$ as a function of θ, the angle through which the photon is deviated:

$$\Delta\lambda = \frac{h}{m_e c}(1 - \cos\theta). \tag{19.36}$$

Compton scattering is observed in X-rays passing through a solid or a gas. The essential interaction is between a high-energy photon and an individual electron, whether or not that electron is bound to an atomic nucleus.

19.12 A Summary of Scattering Processes

Scattering type	Initial system	Final system	Description
Compton	Photon + charge q	Photon + charge q	Photon, esp. X-ray or gamma ray, acts like particle in colliding with free charge $hv \geq m_e c^2$, $\Delta\lambda > 0$
Thomson	EM wave + charge q	EM wave + charge q	Dipole radiation from oscillating free charge, especially an electron $hv \ll m_e c^2$, $\Delta\lambda = 0$ Low-frequency limit of Compton
Rayleigh	EM wave + atom, molecule, particle or density fluctuation	EM wave + atom, molecule, particle	Dipole radiation from oscillating, bound electrons $D/\lambda \leq 0.1$, $\Delta\lambda = 0$
Mie	EM wave + spherical particle	EM wave + spherical particle	D/λ arbitrary. Reduces to Rayleigh for $D/\lambda \leq 0.1$. Forward scattering accentuated for $D/\lambda \geq 0.1$
Raman	Photon + molecule or material	Photon + molecule or material	Increase or decrease in energy of photon due to change in vibrational or rotational energy of molecule $\Delta\lambda > 0$ "Stokes" $\Delta\lambda < 0$ "anti-Stokes"
Brillouin	EM wave + phonon	EM wave + phonon	Acoustic waves in solid or liquid $\Delta\lambda > 0$ or $\Delta\lambda < 0$

Problem 19.1
Consider the complex refractive index of a gas for the case where the atomic oscillators are lightly damped: $\xi = \Delta\omega_{1/2} \ll \omega_0$. Equations (19.15)–(19.16) can be written conveniently in the form

$$n(\omega) = 1 - A(\omega^2 - \omega_0^2)/f(\omega), \quad \kappa(\omega) = A\omega\xi/f(\omega), \tag{19.37}$$

where we define $f(\omega) = (\omega^2 - \omega_0^2)^2 + (\omega\xi)^2$, $A = Ne^2/2\epsilon_0 m$. (Note: the gas density is assumed to be low enough that $A/\omega_0\xi \ll 1$.)

(a) Show that to first order in the small quantity ξ/ω_0, the frequencies where $n(\omega)$ has its local minimum and maximum are $\omega_\pm^2 = \omega_0^2 \pm \omega_0\xi$. Find the extremal values $n(\omega_\pm)$.

(b) Working to second order in ξ/ω_0, show that $\kappa(\omega)$ has its maximum at the frequency $\omega_\kappa = (\omega_0^2 - \xi^2/4)^{1/2} \approx \omega_0$. Find the peak value $\kappa(\omega_\kappa)$.

(c) Show that the extrema of $n(\omega)$ satisfy $\kappa(\omega_\pm) = \kappa(\omega_\kappa)/2$. What does this give for the width (FWHM) of the "extinction index" $\kappa(\omega)$?

Problem 19.2

X-rays of wavelength 0.712×10^{10} m undergo Compton scattering in carbon. Calculate the wavelength of the radiation scattered at $\pi/2$ if the scattering particle is (a) a free electron, (b) the whole carbon atom.

Problem 19.3

An isotopically pure sample of hydrogen chloride gas, when illuminated by light at 488 nm from an argon ion laser, shows a Raman-scattered line at 565 nm. Calculate the vibrational frequency of the molecules.

Problem 19.4

Explain why (a) all polished bulk metals have strong reflectance in the visible region and appear shiny, (b) gold has a reddish appearance, (c) white light appears blue–green after passing through a thin film of gold.

Problem 19.5

Find the Faraday rotation for a linearly polarized radio wave, wavelength 1 m, travelling for 500 light-years through the ionized gas of interstellar space in which the electron density is 10^4 m^{-3} and the magnetic field component along the line of sight is 10^{-10} T.

20 The Detection of Light

Get Thee glass eyes;/ And, like a scurvy politician, seem to see the things thou dost not.

William Shakespeare, *King Lear.*

The range of photon energies, from the negligibly small quanta of long radio waves to the overwhelmingly large quanta of cosmic gamma rays, is reflected in the wide variety of methods of detecting radiation. At the extremes, there is no obvious connection between the measurement of the oscillating electric field of a radio wave and the measurement of the momentum interchange in a collision between a gamma ray photon and a material particle. All methods lead to a measurement of the flow of energy in radiation, and when we consider the ultimate sensitivity of any measurement we specify it in terms of the smallest quantity of radiant energy that can be detected. At short wavelengths this smallest quantity is the energy in a single photon; at long wavelengths the sensitivity is limited by thermal effects in the detector and individual photons cannot be detected.

Light occupies a middle position in the electromagnetic spectrum, where most detectors are photonic. In this chapter we concentrate on photonic detectors and include only a brief section on thermal detectors. The photonic detector may involve photoemission, in which an electron is excited and detected outside the photoemitting surface, or where a photon excites an electron from the valence band to the conduction band of a semiconductor, so creating an electron–hole pair. Mobile electrons and holes within a semiconductor may be detected by applying an external voltage, as in photoconduction, or as a current in the internal electric field of a semiconductor photodiode.

20.1 Photoemissive Detectors

In all *photoemissive* detectors an incident photon frees an electron from a solid by ionization. If the electron has sufficient energy it may escape from a photoemissive surface; this is the *photoelectric effect*. The photoemitting material may be a metal or a semiconductor. The minimum photon energy for photoemission depends on the material; it is called the *work function W*. If the photon has sufficient energy the electron may leave the surface with kinetic energy E given by

$$E = h\nu - W. \tag{20.1}$$

Optics and Photonics: An Introduction, Second Edition   F. Graham Smith, Terry A. King and Dan Wilkins
© 2007 John Wiley & Sons, Ltd

The minimum frequency v for photoemission corresponds to a maximum photon wavelength λ_0 given by

$$\lambda_0 = \frac{hc}{W} = \frac{1.24 \times 10^3}{W} \, \text{nm,} \tag{20.2}$$

where W is in electron–volts.

Each photoemissive material has its own work function; for example, for gold $W = 4.5\,\text{eV}$, so that only photons of light with a wavelength shorter than 275 nm can release electrons from it. W is smaller for the alkali metals; for caesium $W = 2.1\,\text{eV}$, giving a limiting wavelength of 590 nm, while for a widely used alloy of alkali metals, Na_2KCsSb (the S20 cathode), the wavelength limit is 850 nm.

Even if an electron is given enough energy to overcome the work function, it may be trapped in the bulk of the photoemissive material. The proportion of emitted electrons to incident photons is the *quantum efficiency η*. Typically only about 10% of the photons manage to eject an electron from the surface, corresponding to $\eta = 0.1$.

Figure 20.1(a) shows the wavelength dependence of quantum efficiency for a typical S20 photocathode, with $\eta = 0.2$. Figure 20.1(b) shows the *responsivity R*, which takes into account the wavelength dependence of the photon energy; it is defined as the ratio of the photoelectric current A (amps) to the input signal power I (watts). Semiconductors such as gallium arsenide (GaAs) or caesium oxide (Cs_2O) are widely used as photoemitters; they have a lower surface reflectance than metals, and they can provide quantum efficiencies of up to 30%.

In a simple photoelectric detector (Figure 20.2) the electrons from the emitting surface (the *photocathode*) in a vacuum tube may be collected by an electrode (the *anode*); the current is then proportional to the rate of incidence of photons. At low light levels it is possible to detect the arrival of individual photons by amplification in a *photomultiplier* tube (Figure 20.3). Here the electrons emitted from the photocathode are accelerated to a second metal surface, or *dynode*, which emits several electrons each time a primary electron strikes it. An accelerating potential of typically 1 to

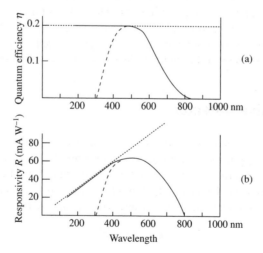

Figure 20.1   (a) The quantum efficiency η and (b) the responsivity R of a typical photoemitter, showing the fall in performance at lower quantum energies. The dotted line shows the performance expected with a quantum efficiency $\eta \approx 0.2$, independent of wavelength. The broken line indicates the effect of absorption at higher quantum energies

Anode | | Photocathode

Figure 20.2 Photoelectric tube. An electron emitted by the photocathode is collected by the anode

2 kV is required between the anode and the photocathode. A series of these dynode secondary emitting stages can be used, usually from 6 to 14, eventually multiplying the charge by a factor of 10^5 to 10^8. The photocathode may be held at negative potential so that the signal is extracted from the anode at ground potential, but for lowest noise in photon-counting mode the photocathode may be operated at ground potential. Photomultipliers may be used either to determine an average light irradiance (intensity) by recording the direct current, or to detect pulses of light, including low light levels corresponding to individually detected photons. Photomultipliers are available which detect in the short ultraviolet near 100 nm, and others through the visible range to the near infrared at 1500 nm using a GaAs photocathode.

The usefulness of all photoelectric detectors at low light levels depends both on the proportion of photons which produce a detectable output, and on any random output, or *noise*, generated inside the detector. (We consider in Section 20.5 below how to quantify the overall performance of a detector at

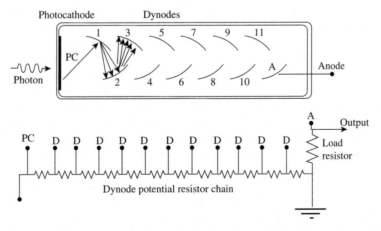

Figure 20.3 Photomultiplier tube. An electron emitted by the photocathode PC is accelerated to D_1, the first of a series of dynodes. At each dynode each electron stimulates the emission of several secondary electrons, and a large pulse of current is collected by the anode A. The potentials of the dynodes are set by the resistance chain, and the output voltage pulse is developed across the load resistor

low light levels.) In the photomultiplier most of the noise is generated by the random emission of electrons from the first photoemissive surface due to thermal excitation; there is, however, an added noise from the dynodes, mainly from the first. The background current from thermal emission is known as the *dark current*. The added noise from the photocathode depends on the temperature and the work function. The noise current may be reduced by cooling the photocathode, typically to $-20°C$; a semiconductor junction working by the Peltier effect is used for this purpose. At room temperature the thermal energy of electrons is of order $1/40$ eV, so that for a metal such as gold with a high work function this random background is negligible. Photomultipliers for high photon energies may therefore be made to be practically noise free.

A single primary photon falling on the photocathode produces an output pulse of appreciable length; this may be important since it limits the photon-counting rate which can be achieved without overlapping pulses. The output pulse length from a photomultiplier is determined mainly by the spread in travel time of electrons between the dynodes, which typically restricts the time resolution to about 10 ns. A time resolution of less than 1 ns is achieved in photomultipliers which are specially designed for a high counting rate.

Photomultipliers are widely used for their high sensitivity and time resolution, which they achieve in operation at room temperature.

20.2   Semiconductor Detectors

As we saw in Chapter 17, electrons in a semiconductor can become mobile if given sufficient energy. In a photonic detector, the energy to free the electron is derived from an incident photon. Given sufficient energy, the electron freed by a photon may escape the solid entirely, in the process of photoemission. Given a lower energy (but greater than the energy gap), the electron may be transferred from the valence band to the conduction band, creating an electron–hole pair. Semiconductor detectors may detect individual photons, or a number of mobile electrons within the crystal lattice may be created and sensed as an increase in the conductivity.

We recall from Chapter 17 that the electron energies within the crystal lattice of a pure semiconductor are almost all constrained to lie within the valence band, where they occupy almost all available energy levels. Above this band of energies is the bandgap, and above this gap again is the conduction band. The excitation of an electron into the conduction band leaves a hole in the valence band. If an external electric field is applied, the electron and the hole move in opposite directions, the electron moving faster than the hole. This is the action of a *photoconductor*, in which the rate of arrival of photons with sufficient energy is measured by an increase in conductivity.

The *responsivity* of a photoconductive detector is the output current divided by the input photon power. Figure 20.4 shows how the quantum efficiency η and the responsivity R vary with wavelength, showing the departure from the behaviour of an ideal detector. Ideally η is unity and R rises linearly up to the cut-off wavelength. In practice the photon absorption falls off at shorter wavelengths, while the cut-off is broadened by thermal excitation.

Most photoconductors use *extrinsic* semiconductor material, in which electrons are excited from an impurity energy level into the conduction band. In *intrinsic* materials the excitation is from the valence band; the energy gap is greater, so that intrinsic materials are appropriate for higher photon energies. Figure 20.5(a) shows the spectral response of silicon, an intrinsic semiconductor. The response of cadmium sulphide (CdS) is close to that of the human eye, so that it is particularly suitable for use in exposure meters for cameras, while that of the alloy HgCdTe can be adjusted to give a peak between 1 and 30 µm depending on the ratio of HgTe to CdTe in the alloy. Figure 20.5(b)

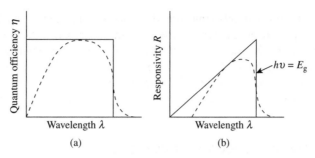

Figure 20.4   Ideal (solid line) and practical (broken line) photoconductor detector responses: (a) quantum efficiency η; (b) responsivity R

shows the response, extending well into the infrared spectrum, of three doped germanium extrinsic semiconductors. All such semiconductors with small bandgaps must be cooled to avoid excessive thermal excitation.

The current in the detector circuit of a photoconductor may be much larger than that caused by a single electron–hole pair for each absorbed photon, due to the process of *photoconductive gain*. When a detected photon creates an electron–hole pair, the electron moves under the action of the applied field to the positive terminal, and the hole to the negative terminal. The electron moves faster than the hole, since the electron mobility is greater than the hole mobility (by a factor of 200 in GaAs, for example). The electron reaches the positive terminal and is absorbed before the hole reaches the negative terminal. In order to keep charge neutrality this triggers the release of another electron from the external circuit. This electron then begins to transit the detector. This process can continue until the hole disappears, either by recombination with an electron, or on reaching the negative terminal

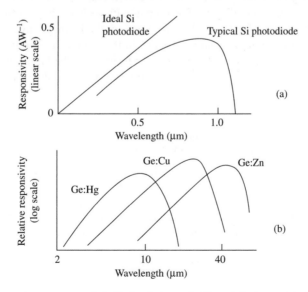

Figure 20.5   The spectral response of typical photoconductive detectors: (a) intrinsic, (b) extrinsic semiconductors

where it recombines. The effective recombination time may then be the time it takes the hole to reach the negative terminal. Thus many electron transits can be made before recombination takes place. Then, photoconductive gain $G = $ (mean time before loss of hole by recombination)/(time for electron to drift across detector), which is the ratio of the transit times for the slow and fast carriers. Photoconductive gain does not, of course, affect the quantum efficiency, which is the proportion of photons which are absorbed and produce mobile electron–hole pairs.

20.3 Semiconductor Junction Photodiodes

In the depletion layer of a junction photodiode (Section 17.2) an electric field is already present within the semiconductor. Absorption of a photon in the depletion layer generates an electron–hole pair which is separated by the internal electric field, creating a change in current and potential across the photodiode.

Figure 20.6 shows the current–voltage relation for a p–n photodiode, with and without incident light. The diode is usually operated with a reverse bias voltage; the current due to the generation of electrons and holes in the depletion region is shown as I_P in Figure 20.6. The reverse bias increases the field in the depletion region, which has two advantages: it increases the width of the depletion region, which improves photon detection, and it shortens the transit time for electrons and holes. The output is in the form of a voltage developed across a load resistance R_L, following the load line shown in Figure 20.7(a).

A photodiode can also be used in the *photoconductive* mode as a source of current, as in the short-circuit mode of Figure 20.7(b). Operation at open circuit, as in Figure 20.7(c), produces the voltage V_P; this is known as the *photovoltaic mode*. A *solar cell* is required to deliver power into a load resistor R_L, as in Figure 20.7(d). The highest power is produced for the maximum product IV. A solar cell can convert about 15% of incident radiation into electrical power.

The efficiency of a photodiode may be improved by including a layer of undoped, or *intrinsic*, material between the n- and p-type layers. Such a p–i–n diode has a larger effective depletion layer, giving it a higher quantum efficiency. A silicon p–i–n diode may have a quantum efficiency approaching unity at its maximum spectral response in the region of 0.8 μm. The p–i–n diode also has a rapid response time (some tens of picoseconds) because of the larger depletion region, which has a reduced capacitance. The response of silicon diodes, however, does not extend beyond about

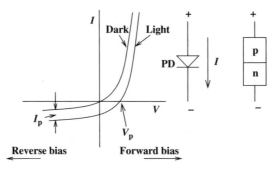

Figure 20.6 Current–voltage relation for a p–n junction showing the change with incident light. The photodiode current at reversed bias is I_P, and the open circuit voltage is V_P

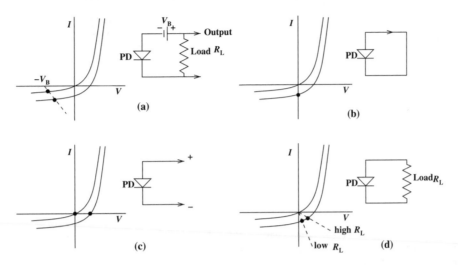

Figure 20.7 A p–n junction diode used (a) with reverse bias and load resistance R_L, (b) short circuit, (c) open circuit, (d) with a low or high load resistance R_L, as in a solar cell

1.3 µm. Diodes with a response in the important wavelength range 1.3–1.6 µm, in which fibre optic communication systems operate (Chapter 6), are often *heterojunctions*, using different materials on either side of the junction. An example is the InGaAsP/InP diode, which has a quantum efficiency approaching 0.75 over this infrared communication band.

The electric field across the depletion layer may be increased by operating a junction diode with a large reverse bias, typically ∼300 V. The electrons and holes created by photons may then acquire sufficient energy to ionize more atoms, generating more electron–hole pairs which are in turn accelerated and may create further ionization. A diode designed to operate in the voltage region immediately below this 'avalanche' level is known as an *avalanche photodiode*, or APD. The current amplification in an APD may be greater than 100; the response time is not much affected by the avalanche process and may be less than 1 ns. APDs are widely used in optical communications.

The p–n boundary is usually between layers of the two materials in a thin slab or wafer; for example, a thin layer of n-type dopant may be deposited on a substrate of p-type material, and diffused into it in sufficient quantity to form the junction. A thin metal film, forming a transparent electrode, is then deposited on the surface.

Metal–semiconductor photodiodes, also known as *Schottky diodes*, are formed by depositing a thin transparent metallic film (usually gold) on a doped semiconductor, usually n-type. The structure and the energy bands are shown in Figure 20.8. When a metal with work function W is in contact with an n-type semiconductor, equilibrium is attained by charge transfer occurring until the Fermi levels attain the same value. The electrostatic potential of the semiconductor is raised in relation to the metal and a depletion layer forms in the semiconductor near the junction. Photons with energy $E > E_g$ go through the metal film and are absorbed in the depletion layer of the semiconductor, creating electron–hole pairs. Schottky diodes have some advantages over p–n diodes: the depletion layer is close to the surface, which minimizes the loss of photons before they reach the active region of the diode, and they may also have a shorter response time, of order 10^{-11} s, giving bandwidths in the region of 100 GHz. The metal film can be sufficiently thin to transmit blue and near-ultraviolet light, giving the diode short-wavelength sensitivity.

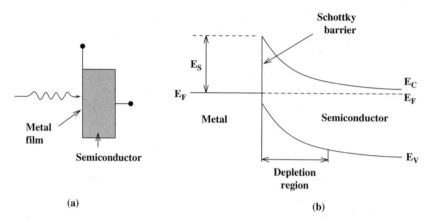

Figure 20.8 (a) Structure and (b) energy band diagram for a metal–semiconductor photodiode, in which a metal is deposited on an n-type material. The energy barrier E_S, known as the Schottky barrier, determines the lowest photon energy at which the photodiode will operate

20.4 Imaging Detectors

Photoelectric detectors have been described so far as single event detectors. There are a number of ways in which a two-dimensional array of detectors may be assembled so that an image may be recorded, as in an electronic digital camera. An array of photodiodes on a silicon chip can be used as a very efficient detector of an image, with sufficient pixels (picture elements) for use in a camera. It would be quite impractical to make a separate wired connection to each diode, so the outputs are scanned sequentially after exposure to light. An arrangement which is widely used is the *charge-coupled device* (CCD).

Three principles are involved in the CCD: photodetection, charge storage and charge transfer. The photodetectors may be photodiodes or a photo-metal–oxide–semiconductor (MOS) capacitor structure. In Figure 20.9(a) an MOS photodetector is shown as an electrode insulated from a semiconducting silicon substrate by a thin film of metal oxide. The substrate is usually p-type, and the electrodes are biassed positively. Under each electrode the positive carriers (holes) are repelled, leaving a depletion layer at a positive potential. Light transmitted to the metal electrode and oxide layer generates an electron–hole pair in the p-type silicon, and the pair separates under the electrostatic field. The electrons move towards the metal electrode but are unable to penetrate the oxide layer and are trapped in the depletion layer, which acts as a potential well. The amount of charge accumulated is proportional to the integrated light flux. Photoelectrons accumulating in this potential well are stored with very little loss for periods up to some hours, allowing long integration times for faint-light photography, as for example in astronomy.

To provide readout the accumulated charge is required to be transported to an output by the technique of charge transfer. An example of the charge transfer process in a line of MOS photodetectors is shown in Figure 20.9(b) and (c). In this example the charge is accumulated under every third electrode, where a deeper potential well is created by applying a larger voltage. Charge can be transferred along a row of electrodes by the successive transfer of this bias voltage, as shown in Figure 20.9(b). The action is often likened to a 'bucket brigade', in which buckets of water are passed from hand to hand to fill a tank. The last electrode on the line is the input to a transistor

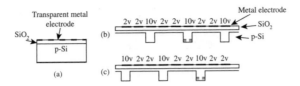

Figure 20.9   The charge transfer system in a CCD, showing the deeper potential wells under the electrodes with higher voltages. One of the wells in (a) has accumulated electrons, which are shifted to the adjacent store in (b). A three-phase cycle of voltages moves the charges progressively along the line, as shown in (c)

amplifier. The outputs of the individual rows can similarly be read sequentially. Efficient illumination of a CCD array requires the surface electrodes to be transparent: they are often made of a metallic compound such as PtSi, forming a Schottky barrier diode. Alternatively the diode array can be illuminated from the rear, when the silicon chip must be thinned to allow the photodiode action to occur close to the potential wells.

Cooled CCD cameras are efficient in infrared light, with low noise, high sensitivity, low dark current and wide dynamic range. CCDs are extensively used in digital cameras for both visible and infrared light; they provide an electronically stored image which can be read into a digital store and subsequently processed in a computer. They are very valuable in astronomy, where they provide images with several million pixels, operating with a high quantum efficiency (detective quantum efficiency DQE reaching over 50%; see the definition of DQE in the next section) and with a linear response. Schottky barrier diodes can be used to extend the wavelength coverage to about 6 μm. Diodes operating at these long wavelengths must be cooled to reduce thermal emission.

20.5   Noise in Photodetectors

Ideally the output of a photodetector is simply proportional to the input optical power. In practice, there may be an output current with no optical input: this is known as a *dark current*. Both the desired output and the dark current have a random component, known as *noise*, which limits the accuracy of measurements at low light levels. We briefly describe the various sources of noise in photodetectors, and define the parameters which specify their overall performance when a signal is to be detected in the presence of noise.

The first source of noise is inherent in the photon nature of light itself. Each detected photon gives a pulse of output current, and the fluctuations in the photon stream become electrical noise, known as *shot noise*. If a photon stream is random, the number of photons m arriving at the detector in a given time interval varies about the mean rate n according to Poisson statistics, with a probability $P(m)$ given by

$$P(m) = \frac{n^m exp(-n)}{m!}.$$ (20.3)

For Poisson statistics the standard deviation, r.m.s. noise N, is the square root of the mean, n, and since the mean n is also a measure of the signal strength S the *signal-to-noise ratio* is

$$S/N = n^{1/2}.$$ (20.4)

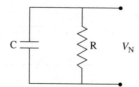

Figure 20.10   Noise voltage V_N in a simple CR circuit

The same argument applies to the photoelectron flux from a detector with quantum efficiency η, when the randomness in the photoelectron output is the statistical fluctuation of a photon stream with mean rate ηn, giving a signal-to-noise ratio

$$S/N = \eta^{1/2} n^{1/2}. \tag{20.5}$$

The degradation of signal-to-noise ratio in a detector in comparison to that of the input photon stream is specified as the *detective quantum efficiency* DQE, defined as

$$\mathrm{DQE} = \frac{(S/N)^2_{\mathrm{out}}}{(S/N)^2_{\mathrm{in}}}. \tag{20.6}$$

If the degradation is entirely due to the quantum efficiency, then $\mathrm{DQE} = \eta$.

In a communication system with a bandwidth B the noise relates to the number of detected photons arriving in a time interval $1/2B$. If the photon flux (number of photons per unit time) is Φ then

$$S/N = \left(\frac{\eta \Phi}{2B}\right)^{1/2}. \tag{20.7}$$

We have noted that detectors sensitive to low-energy photons may have to be cooled to minimize random thermal excitation of electrons. This is an example of *thermal noise*, also called *Johnson noise*, which arises from the random motion of charge carriers in resistive materials. This may be evaluated from the simple circuit in Figure 20.10.

From the classical statistical thermodynamics of a system in equilibrium,[1] an average random thermal energy $kT/2$ is associated with each degree of freedom.[2] The energy stored in a capacitor is $CV^2/2$, so that the noise voltage V_N in a single degree of freedom is given by

$$\frac{1}{2}C\langle V_N^2 \rangle = \frac{1}{2}kT. \tag{20.8}$$

The corresponding kinetic component is the Johnson noise current I_J given by

$$\langle I_J^2 \rangle = \langle V_N^2 \rangle R^{-2} = \frac{kT}{CR^2}. \tag{20.9}$$

---

[1] See for example F. Mandl, *Statistical Physics,* 2nd edn, 1988.

[2] It may be necessary at very low temperatures and very high signal frequencies to include the quantum factor $h\nu/[\exp(h\nu/kT) - 1]$.

The number of degrees of freedom is the bandwidth $\Delta f = (4CR)^{-1}$ of the circuit,[3] giving the Johnson noise current

$$\langle I_{\mathrm{J}}^2 \rangle = \frac{4kT\Delta f}{R}. \tag{20.10}$$

Thermal noise may arise in a photodetector either within the detector itself or in the electronic circuit which detects the output. Its importance depends on the photon flux: if the photon flux is large, the photon noise dominates, while at low photon flux the combined detector and circuit noise dominates.

   In practice, for any photodetector at low light levels, we want to know the smallest signal that can be distinguished from noise. This has led to the concept of the *noise equivalent power (NEP)*, which can be defined as the radiant power (in a sinusoidally modulated signal) that produces a signal-to-noise ratio of unity at the output of a given optical detector; it is quoted for a given modulation frequency and noise bandwidth (typically 1 Hz).

   The inverse of NEP is termed the *detectivity*[4] D, which is a measure of the ability to distinguish a light signal from the detector noise.

20.6   Image Intensifiers

An image on an extended photoemissive surface may be detected by focussing the emitted electrons onto a phosphor screen, as in the *image intensifier* (Figure 20.11). Here the electrons emitted from the photocathode surface in a vacuum tube are accelerated towards a phosphor screen which is at high potential. The focussing is achieved by an electromagnet, as shown in the figure, or by electrostatic field lenses. The brighter image obtained in this way is useful for night vision, especially if the photocathode has a spectral response extending into the infrared.

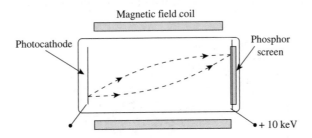

Figure 20.11   An image intensifier using magnetic focussing

---

[3]See for example Grant and Phillips, *Electromagnetism*, 2nd edn, 1990.

[4]The performance of several types of detectors varies approximately as the square root of their area A, leading to the definition of *specific detectivity*

$$D^* = \frac{A^{1/2}}{\mathrm{NEP}}. \tag{20.11}$$

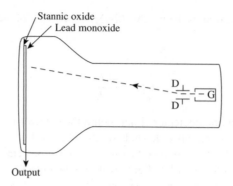

Figure 20.12   The vidicon. The conductivity of the lead oxide layer is proportional to the number of photons that fall on it. An electron beam from the gun G, deflected by a voltage applied across the plates D, scans over the layer, and a current is collected by the transparent conducting layer of stannic oxide

In television cameras an image stored on a photosensitive screen is scanned by an electron beam, providing a sequential electronic signal. In the *vidicon* camera (Figure 20.12) the material of the screen is photoconductive, such as lead oxide (PbO). The current from the scanning electron beam is collected by a transparent conductor, such as stannic oxide, on the front of the screen; the current then depends on the conductivity of the PbO at the point scanned by the electron beam.

The *image orthicon*, shown diagrammatically in Figure 20.13, depends on photoemission. Electrons emitted from the photocathode are accelerated to a thin non-conducting target plate P, which may consist of glass or magnesium oxide. The potential of P is about 300 volts above that of the photocathode, so that secondary electrons are emitted on impact; these are collected by a fine wire mesh at a potential slightly above P. The plate P therefore accumulates a positive charge density whose distribution represents the image on the photocathode. A beam of electrons now scans P, with the electron energies so arranged that the beam is scattered back more or less according to the potential of the part of P close to the beam. At the same time the beam neutralizes the charge on P.

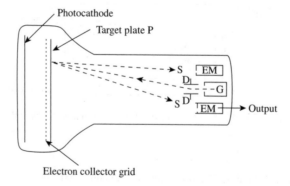

Figure 20.13   The image orthicon. Photoelectrons from the photocathode are accelerated to the target plate, where they release a large number of electrons. These are collected by a fine wire grid, leaving an image on the target plate in the form of a stored positive charge. An electron beam from a gun G is scanned by the deflector plates D. The number of scattered electrons S depends on the charge on P. The electrons are detected in an electron multiplier EM surrounding the electron gun

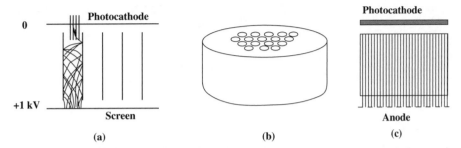

Figure 20.14 Microchannel plate. (a) An electron emitted from the photocathode enters a channel, strikes the wall and stimulates a shower of secondary electrons. Part of the shower development is shown. A large bunch of electrons reaches the fluorescent screen. (b) The channels, typically 16 μm across, are mounted in a plate several centimetres across, between the photocathode and anode (c)

The scattered electrons are detected in the secondary electron multiplier surrounding the electron gun.

Another approach to an imaging detector is the *microchannel plate*, or channel plate multiplier, shown in Figure 20.14. This is essentially a close-packed array of photomultiplier tubes, each of which is a thin glass tube coated inside with photoemissive material. The array of tubes, each about 16 μm in diameter, forms an insulating plate about 1 mm thick. As in the electronic image intensifier this microchannel plate is placed between photoemitting and phosphor plates.

The photomultiplier action is illustrated in the figure. A potential difference of around 1 kV between the surfaces of the plate provides a gradient of potential along each tube, and several stages of multiplication are effectively achieved. The phosphor plate may of course be replaced by a CCD or other array detector. The microchannel plate can then be used as a photon-counting imaging detector.

20.7 Photography

The photographic process has two important advantages over photoelectronic devices. First, it can store permanently an enormous amount of information. The linear resolution of a photographic plate is about 1 μm, while plates over 10 cm across are available for use in wide-field cameras, such as the Schmidt telescope (Chapter 3). Second, it can be used to build up an image during very long exposure times; this is a process of *integration.*

A photographic emulsion on a glass plate or embedded in a plastic film contains individual grains of silver halide crystals, each about 0.5 to 1 micron in diameter. Each crystal can be developed by a reducing agent to produce a grain of silver. The reducing agent is not, however, powerful enough to develop grains unless they contain an imperfection in the form of a single silver ion which has already been converted into a silver atom. This conversion can occur when a photon is absorbed by the crystal. The photon acts by exciting an electron into the conduction energy band, where the electron is free to move until it is trapped by a silver atom in the crystal lattice; thereby producing a silver atom. Groups of silver atoms within the crystal grain form a *latent image.* The latent image is converted into a visible image by the process of development, in which a chemical reducing agent is used to reduce the silver bromide to silver. Grains containing groups of silver atoms are more quickly reduced and form the visible image.

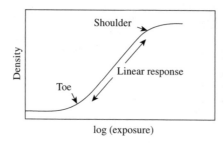

Figure 20.15   The characteristic curve for the response of a photographic film

There is a natural relaxation process in the crystal, and a developable grain is only produced if more than one photon is absorbed within some minutes of time. Further, red and green light have insufficient quantum energy for direct action on the crystal, and absorption must be arranged to occur in organic dyes which coat the grains. Photography is not therefore a process with a high DQE. The spatial resolution of the emulsion is limited by the size of the grains. More sensitive films have larger grains, and lower spatial resolution. The resolution is specified in terms of the closeness of a resolvable pattern of parallel lines: photographic films have a spatial resolution from 200 to 2000 lines per mm.

Unlike most electronic detectors the response of the photographic emulsion can be non-linear at both low and high light levels. Figure 20.15 shows a typical relation between density on the developed plate and the logarithm of the light exposure.[5] The lower non-linear portion, referred to as the *toe*, is due to the requirement for a minimum rate of arrival of photons in an individual grain. The *shoulder* is the effect of saturation, where all grains have become developable. Between these is the straight line region in which the response is more or less linear.

20.8   Thermal Detectors

For the low photon energies of long infrared wavelengths (100 μm to 1 mm), and occasionally also at shorter wavelengths, we may require a measurement of a steady flow of energy through its heating effect on an absorbing surface. The resultant temperature rise may be detected by a change of resistance, in a *bolometer*, or by a change in thermoelectric potential at a junction of two dissimilar metals, in a *thermocouple*.

In both types of detector it is essential to use a sensitive element with a small thermal capacity, well insulated from its surroundings, so that its temperature can respond rapidly to the incident energy. The response time τ is the ratio of thermal capacity C to the rate of heat loss Q:

$$\tau = \frac{C}{Q}. \tag{20.12}$$

---

[5]This *characteristic curve* is known as the Hurter and Driffield, or HD, curve, after its originators.

A response time of a few milliseconds can be achieved. A rapid response requires a small heat capacity C; however, for a very small detector thermal fluctuations may be important. The sensitivity limit may then be set by spontaneous temperature fluctuations with mean value ΔT given by

$$\Delta T = T\sqrt{\frac{k}{C}} \tag{20.13}$$

where k is Boltzmann's constant.

The response of a bolometer element depends on the temperature coefficient α of resistance in the detector material, defined as

$$\alpha = \frac{1}{\rho}\frac{d\rho}{dT} \tag{20.14}$$

where ρ is the resistivity of the material. Most metals at room temperature have α values of around $+0.005\,\mathrm{K}^{-1}$; larger values are available in so-called *thermistor* semiconductors, usually oxides of manganese, nickel or cobalt. Here the temperature coefficient depends on the bandgap in the semiconductor; values of $-0.06\,\mathrm{K}^{-1}$ are obtained at room temperature. Carbon provides a useful resistance element at low temperature; infrared astronomy from satellites uses such detectors cooled by liquid helium.

Thermocouples are generally less sensitive than bolometers, but they are more rugged and generally more convenient to use. They operate at room temperature, and their wide spectral response is useful for wideband spectroscopy. A larger and more sensitive detector can be made by connecting an array of thermocouples in series; this arrangement is known as a *thermopile*.

Pyroelectric detectors provide a wide and flat spectral sensitivity over the range 1 to $100\,\mu\mathrm{m}$ with a detectivity comparable with that of a thermopile but with a rapid (subnanosecond) response time. They are used for infrared sensing in fire detection and security alarm systems. The pyroelectric detector is made from a ferroelectric crystal which has a permanent electric dipole moment. A change in the temperature of the crystal alters the overall dipole moment, inducing a charge on electrodes on the surfaces of the crystal.

Problem 20.1
Someone who is outdoors at midday, but shielded from direct sunlight, may typically receive from the diffuse light of the sky an irradiance of about $80\,\mathrm{W\,m}^{-2}$. If the diameter of the pupil is 3 mm, and a photon of average energy has a wavelength of 500 nm, find the rate at which photons are entering each eye.

Problem 20.2
A photomultiplier detecting a weak source has a gain of 10^7 and a quantum efficiency $q = 0.1$. What photon rate will produce a current of $10^{-9}\,\mathrm{A}$? How long would it be necessary to integrate to detect a 1% change in the source?

Problem 20.3
A photomultiplier has a photocathode efficiency of 0.1, a series of 14 dynodes, an electron multiplication gain at each dynode of 4 and an anode load of resistance of 50 ohms. For a continuous photon arrival rate of 10^6 photons per second, estimate the voltage generated at the anode.

Problem 20.4

Estimate the longest wavelength which an intrinsic photodiode based on GaAs can detect. The bandgap energy of GaAs at 300 K is 1.43 eV.

Problem 20.5

An intrinsic semiconductor used as a photoemitter has an upper wavelength limit of 200 nm; when used as a photoconductor the upper limit is 1.8 μm. Find the width E_g of the bandgap and the width E_c of the conduction band.

Problem 20.6

The forward current in a p–n junction diode with applied bias V and reverse leakage current I_0 is $I = I_0[\exp(eV/kT) - 1]$ (see equation (17.5)). In the photovoltaic mode the diode is used in open circuit and there is no net current; the voltage V_{pv} generated across the diode is measured. For a photon flux Φ, show that $V_{pv} = (kT/e) \ln (\eta\Phi\lambda e/I_0 hc)$.

A photodiode is used in the photovoltaic mode to detect a 1 mW light beam at 800 nm and at ambient temperature. The quantum efficiency of the diode at 800 nm is 0.5 and the reverse bias leakage current is 50 μA. Calculate the photovoltaic output voltage.

Problem 20.7

A p–n photodiode used at a wavelength of 800 nm gives a photocurrent of 1 μA for an incident optical power of 2.8 μW. Determine the efficiency and responsivity of the photodiode.

Problem 20.8

The quantum detection mechanism in a photodiode results in shot noise in which there are statistical fluctuations in the photocurrent about its mean value. For a mean current I and photodiode bandwidth B, the r.m.s. current variation is $(i_s^2)^{1/2} = (2eBI)^{1/2}$.

A p–i–n photodiode working at 1.3 μm has a quantum efficiency of 0.5 and a dark current of 10 nA. Under conditions in which the dark current is significantly greater than the photocurrent, determine the noise equivalent power. Given that the active area of the diode is 100 μm × 150 μm, what is the specific detectivity of the diode?

Problem 20.9

Explain the following:

(a) Photomultipliers with extended infrared sensitivity have poor noise characteristics compared with photomultipliers designed for blue or UV response.

(b) A photodiode used in photovoltaic mode has a non-linear irradiance response and a slow response time.

(c) Thermal detectors which have a higher sensitivity have an extended response time.

(d) For a photographic film, why is the detective quantum efficiency in practice lower than the quantum efficiency calculated taking into account only photon absorption and reflection losses?

(e) Highly sensitive photographic films have reduced spatial resolution compared with less sensitive films.

(f) Qualitatively justify the statement: "CCD camera detection arrays have high sensitivity, low noise, limited dynamic range and limited frame rate performance".

21 Optics and Photonics in Nature

In Nature's infinite book of secrecy/ A little I can read.

William Shakespeare, *Anthony and Cleopatra.*

A lover of Nature responds to her phenomena as naturally as he breathes and lives.

Marcel Minnaert, *The Nature of Light and Colour in the Open Air*, Dover, 1954.

Nor ever yet / The melting rainbow's vernal-tinted hues/ To me have shown so pleasing, as when first/ The hand of science pointed out the path/ In which the sunbeams gleaming from the west/ Fall on the watery cloud.

Mark Akendale (eighteenth century).

In common with most academic physics, this textbook has presented an analysis of optics and photonics as a product of the human intellect. There are, however, many examples of applied optics in nature, which would be regarded as major technological achievements if they were the product of human design, but which are solely the products of evolution and natural selection. We have already referred to the outstanding example of the human eye, and to the use by insects of polarized light in navigation.

In this chapter we look in more detail at techniques of vision which occur throughout the animal kingdom, and at the photonic detectors used by both animals and plants. We also describe examples of fibre optics and polarimeters, and examples of interference phenomena involved in the coloration of iridescent insects and fish.

21.1 Light and Colour in the Open Air

We start by following Minnaert and Akendale (see the epigraph above) in celebrating two of the many non-biological phenomena that occur in the natural world, the rainbow and the aurora. These glorious spectacles are examples of dispersive refraction and of spectral line emission, respectively.

The simple geometry of the rainbow (as set out in Problem 1.3) gives a single arc at an angular distance of 42° from the anti-solar point. Plate 7[*] shows two bows: the primary at 42° and a secondary

---

[*]Plate 7 is located in the colour plate section, after page 246.

Optics and Photonics: An Introduction, Second Edition   F. Graham Smith, Terry A. King and Dan Wilkins
© 2007 John Wiley & Sons, Ltd

at 52°. This secondary bow, which, although fainter, is often seen, is accounted for by two reflections inside the waterdrop instead of one.[1]

The colours of the bow are due to dispersion: the refractive index for red light is less than for violet, so that the primary bow is red on the outside, while in the secondary bow (in which the rays cross over) the sequence is reversed. The colours are not a pure spectrum; each angle has a cusp-like maximum at one wavelength with additional light from longer wavelengths. Note that the space between the bows is dark: this shows that each bow is tracing a limiting value of angular deviation, allowing light inside the primary and outside the secondary but not vice versa.

Our second example of atmospheric physics is the aurora (Plate 8).[*] This is an airglow in which molecules are excited by energetic particles, mostly electrons, streaming from the Sun and channelled by the Earth's magnetic field into the two auroral zones round the north and south magnetic poles. Collisions with the atmospheric molecules lead to their excitation and dissociation and light is emitted in a series of spectral lines due to de-excitation and recombination of several different molecular species. The colour of the aurora depends on the balance between these different lines. The spectacular shapes follow the paths of the solar particles, typically forming rays or sheets at heights of a few hundred kilometres.

The reader is recommended to pursue the study of these meteorological phenomena through the reading list at the end of this book.

21.2   The Development of Eyes

It has long been recognized that more than one route of evolution has led to the development of eyes. All vertebrate animals have similar eyes, which must share a common ancestry. Octopus eyes, however, despite their external appearance, cannot have followed the same route of evolution, since the retina has its network of nerve connections behind the surface of photodetectors, while in the vertebrate eye they are in front. Evolution could not involve such a topological inversion, and a separate path of development must have been followed. There is also a fundamental difference in eyes that work in air and under water: in the human eye, for example, most of the converging power is at the air/cornea interface, with the lens acting as a focussing device with only half the power of the cornea, while in fish the water interface is relatively weak and focussing is achieved in a spherical lens inside the eye. Insect eyes are more obviously of different types; many are multiple eyes, with several different types of focussing systems. It appears that many different and independent evolutionary routes have been followed to produce this diversity.

---

[1]The angular deviation D for k reflections is $D = 2(i - r) + k(\pi - 2r)$. The bow occurs when dD/di is zero, giving $dD/dr = k + 1$. Using Snell's law we can then find for the minimum deviation

$$\cos i = \left(\frac{n^2 - 1}{k^2 + 2k} \right)^{1/2}. \tag{21.1}$$

The bows with $k > 2$ would occur towards the Sun rather than away from it, and cannot be seen against the light scattered and reflected directly from the raindrops.

The theory of the rainbow was first given by Descartes in 1637. An interesting account is to be found in the classic *Physical Optics* by R.W. Wood, 3rd edn, Macmillan, 1934, including a discussion of the so-called *supernumerary* bows seen immediately inside the primary in Plate 7.

[*]Plate 8 is located in the colour plate section, after page 246.

The basic element, from which all separate developments have evolved, is the photoreceptor. Many plants as well as animals have internal clocks which mark out a diurnal rhythm; this is kept in phase with day and night by simple photodetectors. Even the human eye has a special set of photodetectors dedicated to this task. The succession from a simple light detector to the wonderful optical systems now to be found in nature follows a simple and logical route. The first step is the eye-cup, which is commonly found in lower phyla such as the flatworms. If the receptors are recessed in a cup or pit, some crude directivity is achieved; the animal can then react by moving towards or away from the light. If the pit is filled with transparent tissue, the directivity is improved, again with obvious advantage. If the surface of the transparent tissue is curved, the directivity is again improved, and a simple eye has been evolved. Beyond this there is an amazing diversity of optical systems, some of which we now describe.

21.3   Corneal and Lens Focusing

Eyes which depend on the focussing power of the front surface of a cornea are universal among the land vertebrates, and are also found in spiders. The sharp focussing which is so important in human eyes requires an optical system which is relatively free from chromatic and spherical aberrations. There is no known system in nature for correcting chromatic aberration; the only relief is the restricted wavelength range of the photoreceptors, and especially those that are involved with high-acuity vision. Spherical aberration, however, is reduced in two ways, by shaping the surface of the cornea and by grading the refractive index inside the lens. As we have seen in Chapter 2, axial and peripheral rays refracted at a spherical lens surface focus at different distances. Camera lenses correct for this by using non-spherical surfaces, which are more difficult to make, but biological processes lead naturally to such a solution. In many types of eye, including human eyes, the front surface of the cornea is non-spherical, with flattened outer edges; the radius of curvature of the outer parts is typically twice that of the centre.

A second type of correction for spherical aberration is a graded refractive index in the lens itself. This is the main source of correction in fish eyes, but it also plays a smaller part in the eyes of terrestrial animals. Together these corrections provide a precision of focussing which, on-axis, matches the diffraction limit set by the diameter of the eye pupil. In daylight the diameter of the pupil of the human eye is about 2 mm, and the image of a point source in the centre of the field of view is about 1 minute of arc across.

Fish and other aquatic animals cannot follow the same focussing system as eyes operating in air, since there can be only a small step in refractive index at the front surface of the eye. A simple lens, either at the surface or internally, would necessarily have too long a focal length, since the maximum refractive index in transparent biological materials is around 1.56. Focussing is instead achieved in a spherical lens with graded refractive index, the *fish-eye lens* first described by Maxwell as an example of a perfectly stigmatic system (Chapter 2).

A spherical lens with a single ungraded refractive index (Figure (21.1(a)) has serious spherical aberration. The fish-eye lens (Figure 21.1(b)), with the continuous gradation of refractive index which bends rays throughout the lens, gives a short focal length as well as a perfect focus. The spherical geometry also gives a wide field of view over which sharp images are obtained. The ratio of focal length to radius is around 2.5, giving a compact and very efficient eye.

It is not surprising, in view of the excellent performance of the graded-index lens, that it is found in many unrelated marine animals. It is very surprising, however, to find a marine animal which uses instead a totally different lens system, with multiple elements resembling those of the cameras

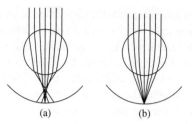

Figure 21.1   Ray tracing through a spherical lens: (a) with homogeneous refractive index; (b) with graded refractive index

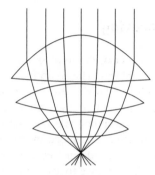

Figure 21.2   The multi-element eye of *Pontella* (after land, 1984)

described in Chapter 3. The animal is the copepod crustacean *Pontella*.[2] The lens, which is sketched in Figure 21.2, has three elements with refractive index 1.52. How such a complex system evolved, and with what advantage, is a mystery.

Many animals have multiple eyes, which may operate independently in different directions and with different resolving powers. An extreme form of multiple eye is found in the brittlestar *Ophiocoma wendtii*,[3] which has thousands of simple light detectors within its skeletal structure and distributed over its whole body. It has no obvious eyes or brain, and yet uses a sensitivity to light so as to avoid predators. The sensitive elements are calcite crystals 40 to 50 μm in diameter forming lenses which correct for spherical aberration, each focussing on a nerve bundle. Moreover, these microlenses are so aligned as to avoid any doubling up of images from the normal birefringence of calcite.

21.4 Compound Eyes

The fly's eye (Figure 21.3) is a well-known example of the compound eyes which are found in more than half of the species of the animal kingdom. Instead of the single camera system, there are in the various types of compound eye many optical units packed together over a spherical surface. The advantage is the wide field of view, which may extend almost to 180°. In the simplest types the units act independently, each with its own lens and photoreceptor, and each covering a limited field of view. The whole eye covers a large field of view, with an angular resolution depending on the spacing of the separate units. These are known as *apposition* eyes (Figure 21.4(a)).

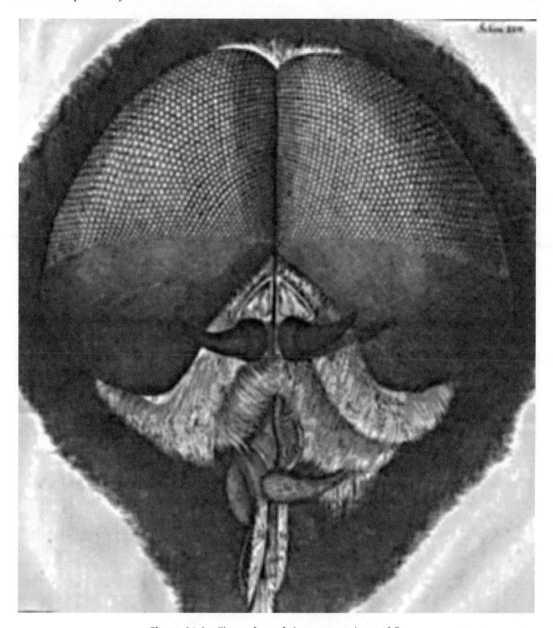

Figure 21.3 The surface of the compound eye of fly

The resolution of this simple compound eye is typically about 1 degree, which is the field of view of each unit; this contrasts with the high resolution, typically 1 arcminute, provided by camera eyes such as the human eye. Light from the whole of this field is concentrated into a single detector unit, a

[2]M.F. Land, *Photoreception and vision in invertebrates*, Plenum Press, 1984, pp. 401–38.

[3]J. Aizenberg, A. Tkachenko, S. Wiener, L. Addadi and G. Hendler, *Nature*, **412**, 819, 2001.

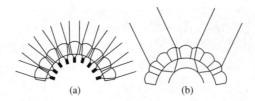

(a)   (b)

Figure 21.4 Compound eyes: (a) the apposition eye, in which each lens acts independently; (b) the superposition eye, in which many facets contribute to each image point

rhabdom, in which the light is transmitted through a light guide to receptor cells. A rhabdom may taper down to a cylinder only a few wavelengths across, so that propagation is determined by wave optics rather than simple geometry. Here nature has anticipated the development of fibre optics! Propagation down an asymmetrical rhabdom can be sensitive to polarization, providing insects such as bees with the polarimeter which they use for navigation by detecting the polarization angle of scattered sunlight.

An entirely different and more complex use of a multiple lens system is shown in Figure 21.4(b). Here the separate facets do not each have their own individual receptors, but in the *superposition* eye they combine to produce a single image on a retina, as does the lens in the human camera eye. Unlike the normal camera eye, however, the multiple units combine to give an erect image, and the separate facets are redirecting light so that an emergent beam is travelling at an opposite angle to the axis of the facet, as shown in the figure. This is achieved in two distinct ways, by refraction and by reflection. A separated pair of lenses would behave in this way, as in a telescope (Figure 3.16), but the units of the insect superposition eye each have instead a single nearly cylindrical elongated lens with a radially graded refractive index, which has the same effect. Each detector on the retina receives light from many facets of the multiple eye, giving a large increase in sensitivity; eyes of this superposition type are common in moths and other nocturnal insects, while the apposition eyes are found in diurnal insects such as flies and bees. The retina is placed at the curved focal surface, well separated from the multiple lens.

The second type of superposition eye, encountered in the multiple eyes of shrimps and lobsters, uses reflection optics to reverse the angle of travel of the incident rays (Figure 21.5(a)). The multiple lens surface is replaced by an array of radially placed mirrors, forming a box-like structure. Within each square box a ray can be twice reflected, focussing light onto the curved retina. (The figure shows reflection in one plane only; reflection also occurs on the sides of the box for rays out of the plane of the diagram.) This amazing application of reflection optics in the 'lobster eye' has provided the inspiration for a design of an X-ray telescope system,[4] which uses an array of long mirror boxes in which the X-rays are reflected at grazing incidence.

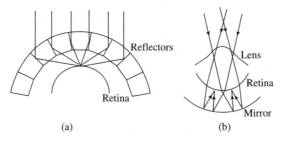

(a)   (b)

Figure 21.5 Reflection optics in eyes: (a) the radial mirror system of the lobster eye; (b) the spherical mirror system of the scallop eye

21.5 Reflection Optics

Although there is no biological process which can make a metallic mirror, there are many examples in nature of reflecting surfaces made from steps in refractive index. The reflectivity is often enhanced by thin coatings, as described later, and is sufficient for reflection optics to be used in the reflecting superposition eye described above. Another form of eye using reflection optics, shown in Figure 21.5(b), is a camera eye, similar to the Schmidt telescope system (Figure 2.25). It uses a concave mirror to form an image on a retina facing away from the light source. This reflector eye is found in scallops and some other crustacea. An obvious disadvantage is that light must pass through the retina to reach the reflector, with a consequent loss of sensitivity. As with the Schmidt telescope, the advantage of the spherical optics is the large available field of view; correction for spherical aberration is achieved, as in the Schmidt, at the centre of curvature of the mirror where the weak corneal lens is shaped as seen in the figure.

Reflectors in biological organisms such as the shiny surfaces of fish scales, and the glint of reflected light in cats' eyes, are familiar examples of remarkably good specular but non-metallic reflectors. Their high reflectivity derives from multiple layers of materials with alternating high and low refractive indices. The layers are thin, and their reflectivity depends on the thin-film interference phenomena described in Chapter 8. The reflectivity is therefore wavelength dependent, and the reflected light is coloured, often showing a marked variation in colour and intensity with angle.

The multiple layers of a reflecting surface take many different forms. The shiny surface of fish skin is due to stacks of plate-like cells in the scales; the same effect is used in the scallop's eye (Figure 21.6), in which the plates are of guanine (refractive index 1.83) separated by body fluid (refractive index 1.33). The layers of guanine repeat at intervals of a quarter wavelength at the centre of the visible spectrum, requiring a spacing of order 0.1 µm. Where there are several layers, the reflectivity can exceed 90% over a restricted wavelength range. A smaller step in refractive index, or a difference in thickness of the two types of layer, decreases the wavelength range of high reflectivity, resulting in colours which may vary with the angle of reflection.

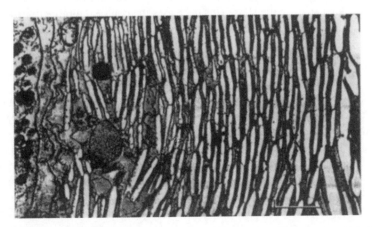

Figure 21.6 A biological reflector; the multi-layer reflector in the scallop's eye, in which layers of guanine (n = 1.83) alternate with body fluid (n = 1.33). The scale bar is 1 µm. (From M. land, *progress in Biophysics & Molecular Biology*, **24**, 75-106, 1972)

[4]R. Angel, *Astrophysical Journal*, **233**, 364, 1979.

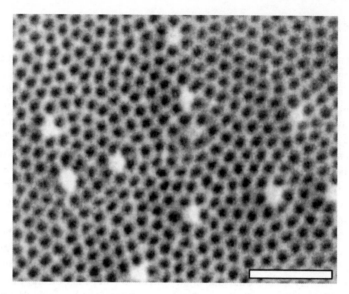

Figure 21.7   The photonic crystal slab (PCS) in a wing scale of the *P. nireus* butterfly. This is a scanning electron microscope image of the underside of the PCS. The length of the scale line is $\sim 1.4\,\mu$m. (Vukusic P and Cooper I., (2005), *Science*, **310**, 1151).

The iridescent colours in butterfly wings, and in particular the bright blue colours of many tropical butterflies (Plate 9(a) and (b)),[*] are due to light reflected from scales on the wing surfaces. Here the thin-film effect is achieved by the structure shown in Plate 9(c)–(e).[*] The cells on the surface form an array of deep slots, which contain serrations at quarter-wave intervals. The projections form a multi-layer of chitin (the main structural material of insects) and air in which the refractive index alternates between 1.54 and 1.0. The regular pattern of slots also acts as a diffraction grating. The maximum reflection is in blue light, where natural pigments are rarely encountered. The interference nature of this structural colour in the butterfly wing can be demonstrated by placing a drop of liquid such as acetone ($n = 1.36$) on the wing, displacing the air in the slots. This decreases the contrast in refractive index, and changes the colour of the wing from blue to green. The blue colour returns when the acetone evaporates.

21.6   Fluorescence and Photonics in a Butterfly

Swallowtail butterflies in the *Princeps nireus* group add two further optical technologies, fluorescence and photonic arrays, to the diffraction gratings described above. Fluorescence, in which short-wavelength (ultraviolet) light stimulates emission at visible wavelengths, is also known in some shrimps and birds. In butterfly wings, as the insect flies it provides a flashing signal at ~ 505 nm, the peak sensitive wavelength of the butterfly eye.[5] The efficiency of the fluorescent emission from butterfly wings is enhanced by two photonic devices which are remarkably similar to those developed recently for this purpose in LEDs: these are the distributed Bragg reflector (DBR) and the photonic crystal slab (PCS). The fluorescence occurs within the PCS, which has a cellular

---

[*]Plate 9 is located in the colour plate section, after page 246.

[5]P. Vukusic and I. Hooper, *Science*, **310**, 1151, 2005.

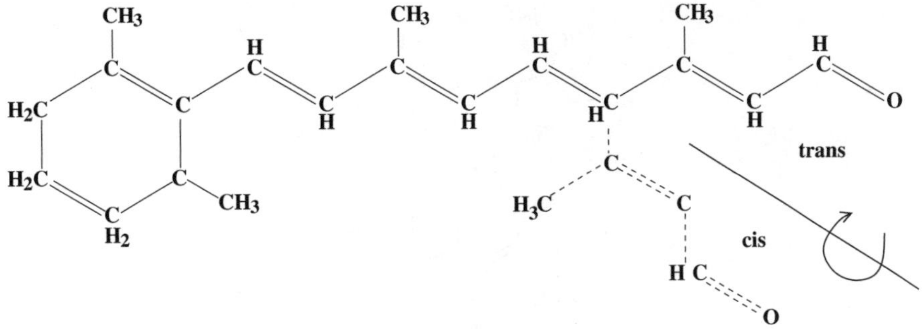

Figure 21.8 The chromophore 11-*cis*-retinal, and the *trans* configuration to which it is converted when it absorbs a photon of light with a wavelength near 500 nm

structure forming a photonic array. This is arranged to inhibit unwanted propagation in the plane of the slab while allowing propagation perpendicular to the slab. The DBR is below the slab, reflecting light upwards. Figure 21.7 shows the quasi-periodic structure of the PCS slab; it is about 2 μm thick, and comprises an array of hollow cylinders, diameter ~240 nm, spacing ~340 nm. The three-layer reflector DBR is immediately below this slab.

21.7 Biological Light Detectors

The process of detection of light is very similar over the whole of the animal world. Although the evolution of the geometric optics of eyes has evidently followed several different routes, all have followed the same basic system, involving a protein molecule with a photonic absorption in the visual range and a signalling system to convey a nerve impulse to the brain. The photoreceptor molecule correspondingly has two parts, a *chromophore* and an *opsin*. In contrast to most proteins, which are colourless with no specific absorption of light in the visual range, the chromophore is a pigment with a resonant response to light. The basic chemical structure which achieves this both in eyes and in plants is a carbon chain with alternating single and double bonds, the *conjugated* chain. The universal animal chromophore is *retinal*, shown in Figure 21.8; it has a similar structure to vitamin A. The combination with opsin forms the detector molecule *rhodopsin*.

Electrons in the conjugated carbon chain of retinal are held only loosely, and the absorption of a photon can temporarily disrupt a double bond allowing it to rotate before reforming in an *isomeric* form. Figure 21.8 shows the two spatial forms, 11-*cis*-retinal and its *trans* configuration to which it is converted when it absorbs a photon of light with a wavelength near 500 nm. The elongated rhodopsin molecule is dipolar, and consequently the photon absorption is sensitive to the polarization of light. Parts of many insect eyes contain aligned arrays of rhodopsin, allowing them to detect the angle of polarization of the sky, particularly in ultraviolet light.

The opsin signals a nerve impulse to the brain; the retinal then breaks from the opsin, recovers to the *cis* isomer and attaches again to form rhodopsin. The process takes a few milliseconds, after which the receptor is ready to receive the next photon. Prolonged exposure to bright light results in failure to recover, which is experienced in 'after images' of complementary colours. There are also random nerve impulses from the eye, the 'noise' of the detection process, and there is a very complex recognition process involved in distinguishing real signals from noise. Recognition of a light stimulus requires the

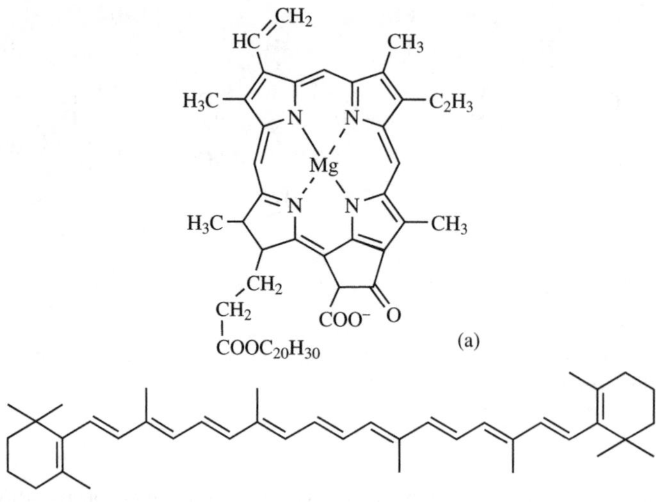

Figure 21.9   The structures of (a) chlorophyll *a* and (b) *β* carotene, the two principal agents of photosynthesis

simultaneous firing of several cells; in fact the most sensitive condition in the eye occurs when light falls on a considerable area of the retina, corresponding to at least one square degree in the field of view. The most sensitive detection corresponds to about 50 photons falling on this area of retina, which corresponds to about 100 photons on the cornea, since half of the light is lost on its way to the retina.

In Chapter 20, we defined the quantum efficiency η as the ratio of detections to the number of incident photons. In the human eye η is only about 10^{-2} at best; however, the astonishing adaptability of the eye is such that η remains above 10^{-3} over a range of $10^{7}:1$ in light intensity. The adaptation involves several processes, including a change of up to a factor of 10 in the area of the pupil; the most important is a change between two types of detecting cells, the rods and the cones. The cones are concerned with higher intensities, and in the human eye (as in other primates) they contain three forms of rhodopsin, absorbing at three different wavelengths: this provides for colour vision in sufficiently bright light. The rods are the most sensitive elements, but they are not colour selective; in consequence, colours cannot be seen in faint illumination, as for example in moonlight. Many stars are coloured, but, except for the brightest, they can only be detected by rods and appear colourless.

Rhodopsin occurs with an entirely different function in the *halobacteria*, which are microbes which live in very salty water. They actually need an external concentration of about 20% sodium chloride, and the protoplasm inside the cell contains about 5%. But their life style also needs potassium chloride, which must be present inside the cell at about 30% even though the environment may contain 1000 times less. The chemical engine that performs the necessary sorting of ions is powered by sunlight, which is absorbed by patches of rhodopsin in the cell membrane. In the process of generating ATP, the universal source of biological energy, ions are exchanged between the interior and exterior, leaving sodium outside and potassium inside. In halobacteria photon energy is used solely to drive this ion pump, and not to detect light or to use the energy to synthesize complex molecules.

21.8   Photosynthesis

Resonance in a conjugated carbon chain is also at the heart of the green pigment *chlorophyll*, which is responsible for photosynthesis in plants. Chlorophyll *a* (Figure 21.9(a)), found in all eukaryotes, has a resonance at 680 nm; chlorophyll *b*, found in vascular plants, has a very similar structure with a resonance at 650 nm. Figure 21.9(b) shows the simpler molecule *β carotene*, which is associated with chlorophyll in many plants and contributes to the photosynthetic process. Carotene, which gives colour to many plants including carrots, is probably a stage in the synthesis both of chlorophyll and of rhodopsin; it is one of the group of *xanthophylls* which contribute to the colours of a wide variety of proteins in diverse substances such as egg yolks, shrimps, corn and fruit.

Photosynthesis in plants, and the process of vision, both involve the movement of an electron in response to light. In chlorophyll the process is very complex. The free electron goes to a receptor molecule which uses the energy to synthesize ATP. The positively charged chlorophyll then removes an electron from other molecules; four such actions result in the breakdown of water molecules into oxygen.

One of the most remarkable aspects of all these processes is the complexity of the molecules. Chlorophyll has the chemical formula $C_{55}H_{72}MgN_4O_5$, 11-*cis*-retinal is $C_{20}H_{28}O$, and carotene is $C_{40}H_{56}$. Despite their complexity, it seems that each of these proteins has evolved along several different evolutionary routes, which have converged on the same complex but identical molecules. Given the diversity of the natural world, it is remarkable that the solutions of the problems of detecting light have been arrived at within such a small range of molecular structures. The evidence is all around us in the universal chlorophyll green of plants, with carotene and its relatives contributing to the colours of leaves in autumn and in fruit.

Appendix 1: Answers to Selected Problems

Full solutions to all problems are available to instructors by emailing their request to the Wiley website.

1.1 $R = 0.087$ mm.

1.8 $\theta \approx 12$ arcminutes.

2.3   The ratio is $(n - 1)/n$.

2.4   The change is a factor -7.7.

2.19 $P_C = 4.65\,D$, $P_F = -2.33\,D$. The radius of the concave side of the flint lens is -267 mm.

2.21   The plate is thicker by $1.3\,\mu$m at the edge.

3.1   Diameter 1.7 mm.

3.5 $u_1 = 20$ m.

3.6   (a) $\times 8$; (b) 1.9 cm; (c) 5 mm; (d) 1.57 cm.

3.7   19 mm closer to the objective.

3.8   Additional length 5 cm.

4.5   Group velocities $v/2$, $3v/2$, $2v$, c^2/v.

4.7   1.2 Hz.

4.10   25 days.

4.11   0.006 Hz.

4.12   Aberration 21 arcseconds. Parallax 0.33 arcseconds. Parallax and aberration are in quadrature.

5.2   (i) $R = 32\%$, $T = 68\%$. (ii) The total fraction reflected is 2.6×10^{-3}, half that of the two-layer system. (iii) $B = 5 \times 10^{-8}$ T. $U = 9.96 \times 10^{-10}\,\mathrm{J\,m^{-3}}$. (v) 6×10^8 N.

5.3   (b) Case (i): $(1 - T_{12}) = 0.006\,38$, and $(1 - T_{1f2}) = 0.003\,20$. The interposed film gas cut the loss by 50%. Case (ii): $(1 - T_{12}) = 0.1696$, and $(1 - T_{1f2}) = 0.090\,67$, and the loss is cut by 46.5%.

5.4   The intensity reflected is 4%. A coating with refractive index 1.225 is needed.

5.5 10^{-7} m.

Optics and Photonics: An Introduction, Second Edition   F. Graham Smith, Terry A. King and Dan Wilkins
© 2007 John Wiley & Sons, Ltd

5.6 $7.7\,\mathrm{V\,m^{-1}}$.

5.7   Irradiance $= 5 \times 10^7\,\mathrm{W\,m^{-2}}$. Peak electric field $= 1.9 \times 10^6\,\mathrm{V\,m^{-1}}$. Peak magnetic field $= 6.3 \times 10^{-3}\,\mathrm{T}$.

5.12 $1.7 \times 10^{48}\,\mathrm{kg}$. The corresponding black hole radius is $2.5 \times 10^{21}\,\mathrm{m}$.

6.1   (i) $\mathrm{NA} = 0.3$, $\Delta = 2\%$, (ii) $1.05\,\mu\mathrm{m}$, (iii) $49.6\,\mathrm{km}$, (iv) $2\,\mathrm{Mbit\,s^{-1}}$, (v) $2\,\mathrm{ns\,km^{-1}}$ and $0.1\,\mathrm{ns\,km^{-1}}$

7.9 $0.85\,\mu\mathrm{m}$.

7.10 $(1/8)\,I_0$.

7.11 $6.3\,\mathrm{mm}$.

7.15   Major axis at $52°$; axial ratio $= \cot 28° = 1.88$.

8.1   (i) 6×10^{-5} radians, or about 0.2 arcminutes. (ii) Radii 0.54 mm and 0.94 mm. (iii) $N = 500$.

8.10 $N_1 = 33\,333$; $N_2 = 33\,332$. $\theta_1 = 4.5\,\mathrm{mrad}$; $\theta_2 = 6.8\,\mathrm{mrad}$.

8.11 $2.62\,\mathrm{cm}$.

8.12 $(1 + F) = 16$.

8.13 $h = 0.216\,\mathrm{mm}$.

9.1   (i) 540 nm, (ii) 170 metres, (iii) $5\,000\,630 \pm 1$ nm.

9.3 $\Delta N = 0.23$.

9.6   Approximately 10 cm.

9.8   2.7 atmospheres.

10.1   (i) $\theta = \lambda n / d = 7 \times 10^{-3}\,n$ radians for red light and $4.5 \times 10^{-3}\,n$ radians for blue light. (ii) 20 fringes. (iii) 20 interference fringes between zeros of the single slit pattern. (iv) 4 arcseconds.

10.9   4.5 m.

10.11 (i) 2 arcminutes, (ii) 0.5 arcminutes, (iii) 0.03 arcseconds, (iv) 10^{-3} arcseconds, (v) 4×10^{-3} arcseconds.

10.12 450 km.

10.15 (i) 15 m.

11.1   (i) (a) $16.0°$, $33.4°$, $55.6°$. (b) $d = 1.98 \times 10^{-10}$ m, $\theta = 22.9°$, $51.1°$.

       (ii) The spectrograph would distinguish a wavelength separation of 50 nm in the first order and 25 nm in the second order.

12.1   The dispersion $\delta n / \delta \lambda = 3.3 \times 10^{-4}\,\mathrm{nm^{-1}}$, giving resolving power $\lambda / \delta \lambda \approx 1.7 \times 10^4$. This easily resolves the sodium doublet but not the hydrogen doublet. The angular separation of the sodium lines, using equation (12.2), is 4×10^{-5} radians $= 0.0023°$.

12.2   15 000.

12.3   (i) 4000, (ii) 18,000, (iii) 1.5×10^6.

12.4 (i) $m = 4 \times 10^4$, (ii) 100 arcseconds, (iii) 1.6×10^6.

13.1 The transverse coherence length for the Sun is approximately $120\lambda = 70\,\mu m$; for starlight it is approximately 20 cm.

13.2 1.1 m; 6 kHz; 8×10^{-9} nm.

13.3 (a) 10^{10} Hz; (b) 8.84×10^{-3} nm; (c) 0.03 m.

13.6 (ii) 0.1 nm.

13.7 $\tau = 20\,\mu s$, $N = 100$.

14.1 1570 fringes per mm.

15.1 (i) (a) $3 \times 10^{18}\,s^{-1}$, (b) $5 \times 10^{20}\,s^{-1}$.

(ii) (a) total pulse energy 67.5 J and peak power 675 MW, (b) total pulse energy 159 J and peak power 1.59×10^9 W.

(iii) Mode separation 500 MHz; four modes lie within the gain curve.

15.3 For $\lambda = 10\,\mu m$, $T = 2078$ K.

15.5 For He–Ne laser $\Delta v_D = 1.52$ GHz. For Ar^+, $\Delta v_D = 4.9$ GHz.

15.6 Minimum length is 0.136 m.

15.7 (a) $0.67\,mrad = 0.038°$. (b) Radiance $R = 2.50 \times 10^9\,W\,m^{-2}\,sr^{-1}$. (c) $T \approx 2.7 \times 10^4$ K.

15.8 (a) Threshold gain coefficient $\gamma_{thres} = 0.41\,m^{-1}$. (b) Population inversion $4.78 \times 10^{22}\,m^{-3}$.

15.9 Ratio is $1 - \exp(-2) = 86.5\%$. Photon flux $= 5.8 \times 10^{24}\,s^{-1}\,m^{-2}$.

15.10 24 modes are present. To ensure oscillation in a single mode the cavity must be less than 3 cm long.

15.12 $\Delta v_D \sim 61$ MHz, $\Delta v_P \sim 790$ MHz. The transition is homogeneously broadened.

16.8 (a) 2.51×10^3 K. (b) $P_{rad} = 1.46 \times 10^{13}\,N\,m^{-2} = 1.46 \times 10^8\,bar = 4.3 \times 10^{-4}\,P_{tot}$.

16.9 0.47 W.

16.10 (a) 1.5 GHz.

(b) Only three modes can oscillate.

16.11 (a) 2 mrad; (b) 0.1 m; (c) 10 cm; (d) 50 m.

16.12 1.9 m, 0.085 m.

17.4 Reflectance $R = 0.32$.

Threshold gain $= 2.65 \times 10^3\,A\,cm^{-2}$. Threshold current $= 663$ mA.

17.6 Threshold current density $= 3650\,A\,cm^{-2}$.

17.8 (a) $\Delta\lambda = 15.4$ nm; (b) $\tau_c = 3.2 \times 10^{-14}$ s; (c) $l_c = c\tau_c = 9.6\,\mu m$.

17.9 182 MHz.

17.11 (a) For GaAs–air interface $R = 0.32$. Power absorbed $= 6.75\,\text{mW}$.

(b) Power deposited into crystal $= 4.05\,\text{mW}$.

(c) Rate of photon emission $= 1.28 \times 10^{16}\,\text{photons s}^{-1}$.

18.1 (i) For neon ($M = 20$) the Doppler line width is $1.66 \times 10^{-3}\,\text{nm}$. The frequency bandwidth is $1.2\,\text{GHz}$. (ii) For the H_α line the half-width is $0.017\,\text{nm}$.

18.2 (a) $9.58 \times 10^{20}\,\text{s}^{-1}$; (b) $3.0 \times 10^{-9}\,\text{kg}$.

18.4 $\gamma = 6000$. $\omega = 5.9 \times 10^7\,\text{rad s}^{-1}$. $T = 107\,\text{s}$. Radius $R = 5.12\,\text{m}$. Power $= 2.28 \times 10^{-6}\,\text{W}$ per electron. Total power $= 76.2\,\text{kW}$. Characteristic energy of emitted photons $= 1.97 \times 10^{-15}\,\text{J} = 12.3\,\text{keV}$.

18.5 (a) $\Delta\lambda/\lambda \approx 4.5 \times 10^{-5}$; (b) $\Delta\lambda/\lambda \approx 4.1 \times 10^{-4}$.

19.2 Compton wavelength for the electron $= 2.43 \times 10^{-12}\,\text{m}$.

Compton wavelength for the carbon atom $= 1.09 \times 10^{-16}\,\text{m}$. For electron, $\lambda' = 0.736 \times 10^{-10}\,\text{m}$. For carbon, $\lambda' \approx 0.712 \times 10^{-10}\,\text{m}$.

19.3 $\Delta\lambda = 565 - 488 = 77\,\text{nm}$; $\Delta\bar{\nu} = 3233\,\text{cm}^{-1}$.

19.5 Rotation angle is 1.24 radians.

20.1 The average photon carries an energy $3.97 \times 10^{-19}\,\text{J}$. No. photons $\text{s}^{-1} = 1.4 \times 10^{15}$.

20.3 $G = (4)^{14} = 2.67 \times 10^8$. Electron arrival rate at the anode $= 2.67 \times 10^{13}$ per second. $V_a = 21\mu\text{V}$.

20.4 $\lambda_c < 8.7 \times 10^{-5} < 0.87\,\mu\text{m}$.

20.5 $E_g = 0.7\,\text{eV}$; $E_c = 5.5\,\text{eV}$.

Appendix 2: Radiometry and Photometry

The science of measurement of energy in electromagnetic waves over the entire spectrum is *radiometry*, and its application to the human visual response is *photometry*. The system of units in radiometry is based on SI units; for example, energy density is measured in units of joules per cubic metre. In photometry, however, there has been a plethora of units, arising from the need to define the illumination or visibility of a surface in terms which depend on the spectral characteristics of the human eye. We start with the basic concepts of radiometry.

The *radiant flux* Φ of a source is the rate of emission of energy, measured in watts. The radiant flux per unit area is termed either the *radiative exitance M*, when the radiation is leaving the surface, or *irradiance E*,[1] when radiation is incident on it. The *radiant intensity I* is the radiant flux per unit solid angle

$$E \text{ (or } M) = \frac{\mathrm{d}\Phi}{\mathrm{d}A}, \qquad I = \frac{\mathrm{d}\Phi}{\mathrm{d}\Omega}. \tag{A2.1}$$

For an isotropic radiator, e.g. a spherical star, $\Phi = 4\pi I$.

Suppose we observe radiation (reflected or emitted) from a small planar surface of area $\mathrm{d}A$ at angle θ off its normal direction through a small aperture subtending at the source a solid angle $\mathrm{d}\Omega$. We can define the *radiance L* (along some direction of interest) as the radiant flux per unit solid angle (steradian) per unit projected area of the source ($\mathrm{d}A \cos \theta$):

$$L = \frac{1}{\cos \theta} \frac{\mathrm{d}^2\Phi}{\mathrm{d}\Omega \mathrm{d}A} = \frac{1}{\cos \theta} \frac{\mathrm{d}I}{\mathrm{d}A} = \frac{1}{\cos \theta} \frac{\mathrm{d}M}{\mathrm{d}\Omega}, \tag{A2.2}$$

which is measured in $\mathrm{W\,sr^{-1}\,m^{-2}}$.

By definition, a perfectly diffusive surface has a radiant intensity proportional to the projected area, or $\mathrm{d}I(\theta) = \mathrm{d}I(0) \cos \theta$. This is *Lambert's cosine law*. Equivalently, such a surface shows constant radiance when viewed from any direction: $L(\theta) = L(0) = L$. Integrating equation (A2.2) with the help of $\mathrm{d}\Omega = 2\pi \sin \theta \mathrm{d}\theta$ yields $M = 2\pi L \int_0^{\pi/2} \cos \theta \sin \theta \mathrm{d}\theta = \pi L$. Blackbodies, for example, radiate like perfectly diffusive surfaces.

The irradiance of a flat surface under a hemispherical surface with radiance L is πL; it is useful to check that this is the same for a flat surface under an infinite parallel slab.

In radiometry we are often also concerned with the spectrum of the radiation, and each of the terms already defined may require the restriction 'per unit frequency band', or occasionally 'per unit

---

[1]This appendix features standard nomenclature, such as E_e for irradiance and I_e for radiant intensity. But in the rest of the book, for the sake of simplicity, we denote irradiance by I and intensity by N.

Optics and Photonics: An Introduction, Second Edition   F. Graham Smith, Terry A. King and Dan Wilkins
© 2007 John Wiley & Sons, Ltd

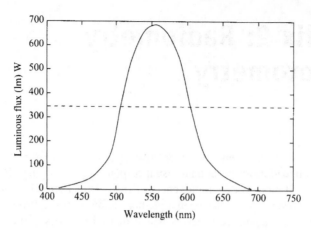

Figure A2.1 The standard relation between luminous flux and radiant flux, over the visible spectrum, as defined by the CIE. The peak sensitivity of the eye is at 555 nm, where 1 watt is equivalent to 683 lumens

wavelength range'. For example, the irradiance may be the integral of a spectrum of electromagnetic radiation with a *specific radiant intensity* $I(v)$ or $I(\lambda)$:

$$I = \int I(v)\mathrm{d}v \quad \text{or} \quad I = \int I(\lambda)\mathrm{d}\lambda. \tag{A2.3}$$

Radiance and irradiance may also be expressed as a flow of photons instead of energy. The conversion factor depends on the spectrum; at the peak sensitivity of the eye, in the yellow–green region (wavelength 555 nm) in the visible spectrum, the energy of a photon is 3.6×10^{-19} joules, so that 1 watt is equivalent to 2.8×10^{18} photons per second.

Photometric units involve the response of the human eye, usually expressed in terms of wavelength. A standard response curve has been defined by the CIE (Commission Internationale d'Éclairage); this allows the relation between luminous flux (including the response of the eye) to radiant flux to be plotted as in Figure A2.1. (The actual bright-light response of an individual eye may differ from this standardized curve; the low-light response is markedly different, with a peak at 510 nm rather than 555 nm.) The photometric equivalents of the radiometric quantities already defined include a factor $V(\lambda)$, the *luminous efficiency*; for example, the luminous flux Φ_v is related to the spectral distribution of the radiant flux Φ_e by

$$\Phi_\mathrm{v} = K \int V(\lambda)\Phi_\mathrm{e}(\lambda)\mathrm{d}\lambda. \tag{A2.4}$$

The constant K depends on the system of photometric units, of which there have been many. The most useful photometric units are shown in the table below, with their radiometric equivalents. Radiometric quantities are often written with a subscript "e" (for electromagnetic) and photometric with a "v" (for visual).

Radiometric system	Unit	Photometric system	Unit
Radiant flux Φ_e	W	Luminous flux Φ_v	lumen (lm)
Irradiance	$E_\mathrm{e}\,\mathrm{W\,m}^{-2}$	Illuminance E_v	
(or radiant exitance M_e)		(or luminous exitance M_v)	$\mathrm{lux} = 1\mathrm{mm}^{-2}$
Radiant intensity I_e	$\mathrm{W\,sr}^{-1}$	Luminous intensity I_v	candela (cd) $= \mathrm{lm\,sr}^{-1}$)
Radiance L_e	$\mathrm{W\,m}^{-2}\,\mathrm{sr}^{-1}$	Luminance L_v	$\mathrm{lm\,m}^{-2}\,\mathrm{sr}^{-1} = \mathrm{cd\,m}^{-2}$

The word candela for the SI unit of luminous intensity is a reminder that light was originally measured in terms of candlepower. Since 1979 the candela has been defined as the luminous intensity corresponding to $(1/683)$ watts per steradian being emitted in monochromatic radiation of frequency 5.40×10^{14} Hz (note that the corresponding wavelength is 555 nm, the peak of visual sensitivity, where $V(\lambda) = 1$). But since a candela is also defined as 1 lumen per steradian, this implies that the constant K in equation (A2.4) is 683 lumens per watt.

A2.1 Further Reading

M. Born and E. Wolf, *Principles of Optics*, Cambridge University Press, 1999.
W. R. McCluney and R. McCluney, *Introduction to Radiometry and Photometry*, Artech House, 1994.

A2.2 Some Examples

1. Lambert's law: the Sun and the Moon. To a first approximation, the luminance of the Sun's disc is uniform, while that of the Moon falls from centre to limb. If the surface obeyed Lambert's law the luminance would remain constant. (Halstead[2] urges a useful distinction between the terms "brightness" for the subjective visual impression, and "luminance" for its objective, measurable correlate. In this spirit, we avoided saying "the Sun's disc is uniformly bright".)

2. Solar photometry: illuminance and luminance (the photometric equivalents of irradiance and radiance). Suppose a luminous flux $\Delta\Phi$ of sunlight falls on Earth, which has a projected area ΔA_E. If the Earth and Sun are separated by a distance r, the Earth subtends at the Sun a cone of angular diameter θ_E and solid angle $\Delta\Omega_E = \Delta A_E/r^2 = \pi(r\theta_E/2)^2/r^2 = \pi\theta_E^2/4$. Likewise the Sun subtends at the Earth $\theta_S = 1\,\text{rad}/120$ and $\Omega_S = \pi\theta_S^2/4 = (\pi/4)(1/120)^2$ sr. The luminance at Earth can be written

$$L_v = \frac{\Delta\Phi}{\Delta\Omega_E \Delta A_S} = \frac{\Delta\Phi}{\Delta\Omega_E r^2 \Delta\Omega_S} = \frac{\Delta\Phi}{\Delta A_E \Delta\Omega_S}. \tag{A2.5}$$

Since the illuminance of the Earth at normal incidence is $\Delta\Phi/\Delta A_E = 10^5\,\text{lm m}^{-2}$, we find a luminance $L_v = (10^5\,\text{lm m}^{-2})(4/\pi)(120)^2\,\text{sr}^{-1} = 2 \times 10^9\,\text{cd m}^{-2}$.

3. Irradiance: moonlight versus sunlight. Assuming the Moon's surface is perfectly diffusive, we compare the radiant flux (falling on the Earth) from moonlight at full Moon with that from sunlight.

    Let S be the irradiance of full sunlight at the Moon (assumed the same as at the Earth). With radius R_M, the Moon presents a projected area of πR_M^2 to the Sun, thereby receiving a flux of $S\pi R_M^2$. The full Moon will reflect a fraction $a_M = 0.12$ of this, where a_M is the average reflectance or *albedo* of the Moon's surface. Sunlight falling on each spot of a perfectly diffusive surface is scattered isotropically into the hemisphere (2π radians) over the ground. If the earth is at distance d from the Moon, it subtends a solid angle of $\Delta\Omega_E = \pi R_E^2/d^2$ and will capture a fraction $\Delta\Omega_E/2\pi$ of the scattered sunlight. The flux of moonlight falling on the Earth is therefore

---

[2]C.P. Halstead, "Brightness, luminance, and confusion", *Information Display*, March, 1993.

$\Delta\Phi_M = a_M S\pi R_M^2 (R_E^2/2d^2)$. But the flux of sunlight falling on the Earth, $\Delta\Phi_E = S\pi R_E^2$, and with the Moon's apparent diameter as seen from the Earth of about 1/110 radian, we find

$$\Delta\Phi_M/\Delta\Phi_E = (a_M/2)(R_M/d)^2 = (0.12/2)(1/220)^2 = 1.2 \times 10^{-6}. \tag{A2.6}$$

Since the Earth presents equal projected areas to the Sun and Moon, this is also the predicted ratio of irradiances. (The observed ratio is about 2×10^{-6}. The Moon evidently back-scatters almost twice the light it would if it had a Lambertian surface.)

4. A burning glass: the illuminance of an image made with a simple converging lens. The angular width of the Sun is ($\alpha = 1/120$ radians). Suppose the lens has focal length $f = 100$ cm and diameter $D = 10$ cm. Rays traced through the vertex of the lens do not bend but form in the focal plane an image of the Sun with a diameter $f\alpha = 8$ mm. If the illuminance of the Sun at normal incidence is $S = 10^5$ lm m^{-2}, the lens will capture a flux of $S\pi D^2/4$. The illuminance of the image is therefore $\Delta\Phi/\Delta A = S(\pi D^2/4)/(\pi f^2\alpha^2/4) = S(D/f\alpha)^2 = 10^7$ lm m^{-2}, a hundred-fold increase over normal sunlight.

5. Radiance: comparison of a laser and a thermal emitter. A high-pressure mercury lamp may emit some hundreds of watts from an area of about 1 cm^2, giving a source radiance of order 2×10^6 W m^{-2} sr^{-1}. A typical 5 milliwatt He–Ne laser with a waist diameter of 0.5 mm concentrates light into a beam with angular diameter 1.6 mrad (see Section 15.8). The solid angle is 2×10^{-6} steradians and the emitting area is 2×10^{-7} m^2, giving a radiance of 1.2×10^{10} W m^{-2} sr^{-1}, i.e. 6000 times larger than the thermal lamp.

Appendix 3: Refractive Indices of Common Materials

The values in the table are for wavelength 589 nm.

Material	n
Gases	
Air (0°C, 1 atm)	1.000 29
Hydrogen (0°C, 1 atm)	1.000 13
Liquids	
Carbon dioxide	1.20
Water (0–20°C)	1.333
Ethyl alcohol	1.36
Olive oil	1.47
Sugar solution (30–80%)	1.38–1.49
Benzene	1.50
Solids	
Ice	1.31
Fused quartz	1.458
Plexiglas	1.51
Sodium chloride	1.544
Crown glasses	1.52–1.89
Flint glasses	1.58–1.89
Diamond	2.417
Iodine crystal	3.34

Optics and Photonics: An Introduction, Second Edition   F. Graham Smith, Terry A. King and Dan Wilkins
© 2007 John Wiley & Sons, Ltd

Appendix 4: Spectral Lineshapes and Linewidths

The spectral widths of emission lines from atoms or molecules observed by a spectrometer are dependent on the nature of the source, its operating conditions and also on the resolution of the measuring spectrometer. Even if the spectrometer has an infinitely fine resolution the recorded spectral width of the emission line is not extremely narrow but has a finite width. The broadening of spectral lines can be attributed to three mechanisms: natural broadening, collisional or pressure broadening and Doppler broadening.

A4.1 Natural Broadening

Excited states in atoms and molecules have a finite lifetime which, through the uncertainty principle, confers an uncertainty in the energy of the state and hence a finite frequency width to the transition. This natural broadening is determined by the radiative emission process. Natural broadening is usually not the dominant broadening mechanism but sets the minimum value to the linewidth.

We can gain an insight into natural broadening by considering the classical (i.e. non-quantum-mechanical) model of the electron in the atom. The electron may be considered as a harmonic oscillator which is damped by the energy lost by radiation. The equation of motion of the electron written in one dimension for movement along the x axis is of the form $md^2x/d^2t + m\gamma dx/dt + m\omega_o^2 x = 0$. Here γ is the damping decay rate, $\gamma = e^2\omega_0^2/6\pi\epsilon_0 mc^3$. The initial total energy of the electron is $\mathcal{E}_0 = m\omega_o^2 x_0^2/2$, where x_0 is the amplitude of the oscillation. A part of the energy is lost radiatively in each period of the oscillation. The total power P radiated is given by the Larmor formula discussed in Chapter 18; from equation (18.2), $P = e^2 a^2/6\pi\epsilon_0 c^3$, where a is the acceleration. The rate of energy loss is given approximately by $-d\mathcal{E}/dt = (e^2\omega_0^2/6\pi\epsilon_0 mc^3)\mathcal{E}$. This has the solution $\mathcal{E}(t) = \mathcal{E}_0 \exp(-\gamma t)$.

This is the spontaneous decay rate on the classical model. The lifetime of the state is $\tau = 1/\gamma = 6\pi\epsilon_0 mc^3/e^2\omega_0^2$. The time dependence on this model of the electric field radiated by the atom is $E(t) = E_0 \exp(-\gamma t/2) \cdot \exp(i\omega_0 t)$.

The frequency dependence is obtained by taking the Fourier transform. The irradiance distribution $I(v)$ is proportional to $|E(v)|^2$ giving

$$I(v) = \frac{I_0}{2\pi} \frac{\gamma/2\pi}{(v - v_0)^2 + (\gamma/4\pi)^2} \tag{A4.1}$$

$$I(v) \propto 1/[(v - v_0)^2 + (1/4\pi\tau)^2] \tag{A4.2}$$

Optics and Photonics: An Introduction, Second Edition   F. Graham Smith, Terry A. King and Dan Wilkins
© 2007 John Wiley & Sons, Ltd

where $I_0 = \int_0^\infty I(v)\mathrm{d}v$. This line profile has the Lorentzian lineshape shown in Figure 12.3. The full width at half maximum (FWHM) is $\Delta v_N = \gamma/2\pi = 1/2\pi\tau$.

This linewidth is also obtained from applying the uncertainty principle $\Delta E.\Delta t \geq \hbar$. Radiation from an atom which has a certain radiative lifetime has an uncertainty in the energy of the level $h\Delta v \geq \hbar/\Delta t$, leading to the natural linewidth.

In terms of the linewidth Δv_N, $I(v) \propto 1/[(v - v_0)^2 + (\Delta v_N/2)^2]$, and when normalized,

$$g(v)_N = \frac{2/\pi\Delta v_N}{1 + [2(v - v_0)/\Delta v_N]^2}. \tag{A4.3}$$

In general both the upper and lower levels of a transition have finite lifetimes and linewidths as shown in Figure 12.2. The uncertainty in the frequency of the emitted radiation involves the uncertainties in the energies of both levels. The frequency distribution remains a Lorentzian function modified to include the lifetimes of both levels. For the levels labelled 1 and 2 in Figure 12.2 the linewidth (FWHM) Δv_N is related to the lifetimes of the two levels, $\Delta v_N = 1/2\pi(1/\tau_2 + 1/\tau_1)$.

Natural broadening is an example of a *homogeneous broadening* mechanism, in which all the atoms are influenced in the same way. Every atom making the same transition has an identical line profile and linewidth. The natural lifetimes of atomic electric dipole transitions are $\tau \sim 10^{-8}$ s, which gives natural linewidth values $\Delta v \sim 15$ MHz. In terms of wavelength, $\Delta\lambda = \lambda^2\Delta v/c \sim 1.4\times 10^{-14}$ m $= 1.4 \times 10^{-5}$ nm. Generally the natural linewidth is much less than the linewidths due to collisional or Doppler broadening.

A4.2   Collisional or Pressure Broadening

Collisional or pressure broadening is due to the interactions between an emitting atom or molecule and its surroundings. An example is collisions between atoms in a gas at a certain pressure. In liquids and solids a similar broadening arises from interactions between the emitting species and its close neighbours and with acoustic phonon vibrations. In gases at pressures above about 0.1 bar collisions and interatomic forces become significant. Collisional broadening is the dominant mechanism in high-pressure lamps and in plasmas. The interatomic interactions include resonant and non-resonant interactions between dipoles, longer range van der Waals forces and shifts and broadening of the energy levels by Stark broadening due to the local electric fields. Since an excited atom can undergo interaction with several other perturbing atoms in a brief instant of time, the overall effect is an average over the various perturbations, each with varying collision times. For an atom emitting a steady wavetrain the collisions may be *soft*, leading to a small change of frequency or phase of the emission. Those collisions in which there is termination of the emission or in which the phase of the emission after the collision is completely out of phase with that before the collision are termed *hard* collisions. The emitted wavetrain perturbed by a collision is described by a damped harmonic oscillator in the classical electron in an atom model. In a gas the mean time between collisions t_c is inversely proportional to the gas density N and the mean relative velocity v of the perturbing atoms, $t_c = 1/N\sigma v$, where $\sigma = \pi\rho^2$ is the collision cross-section for collision radius ρ. Collisional broadening is also a homogeneous broadening mechanism and is always accompanied by natural broadening. Together the spectral distribution is Lorentzian,

$$I(v) = \frac{I_0}{\pi}(\Delta v_n + \Delta_c)\frac{1}{[(v - v_0)^2 + (\Delta v_n + \Delta v_c)^2]} \tag{A4.4}$$

where the natural broadened linewidth is $\Delta v_n = 1/2\pi(1/2\tau_1 + 1/2\tau_2)$ and the collisional broadened linewidth is $\Delta v_c = 1/\pi t_c$. The linewidth is increased over that for natural broadening, and with both natural and collisional broadening the FWHM linewidth is

$$\Delta v = (1/2\pi)[(1/\tau_1 + 1/\tau_2) + 2/t_c] = (1/2\pi)[1/\tau_1 + 1/\tau_2 + 2N\sigma v] \qquad (A4.5)$$

The collisional linewidth is proportional to the density of colliding atoms. Hence it is the dominant broadening mechanism in high-pressure and highly ionized gases.

A4.3   Doppler Broadening

In a gas the absorption or emission spectral lines of moving atoms are broadened due to the motion of the atoms; this is termed Doppler broadening. The observed frequency of an emitting atom moving with a velocity $\mathbf{v}$ relative to an observer, for $v \ll c$, is $v' = v(1 - \mathbf{v} \cdot \mathbf{r}/c)$ where $\mathbf{r}$ is a unit vector in the direction from the observer to the emitting atom. Setting a coordinate system in which r is along the x axis, the probability of an atom of mass M having a velocity between v_x and $v_x + dv_x$ at temperature T is given by the Maxwell distribution

$$P(v_x)dv_x = (M/2\pi kT)\frac{1}{2}\exp(-Mv_x^2/2kT)dv_x. \qquad (A4.6)$$

The probability of an emitted wave having a frequency between v_0 and $v_0 + dv_0$ is a Gaussian function. The normalized Doppler-broadened line profile is

$$g(v) = (c/v_0)(M/2\pi kT)^{1/2} \exp[(-M/2kT)(c/v_0)^2(v - v_0)^2]. \qquad (A4.7)$$

The Doppler linewidth (FWHM) is

$$\Delta v_D = 2v_0[2kT \ln 2/Mc^2]^{1/2}. \qquad (A4.8)$$

Doppler broadening is an example of an *inhomogeneous broadening* mechanism. In this the conditions of the emitting species are different, and the line profile is derived from many atoms with differing velocities and centre frequencies, so that the shifts in frequency differ between the atoms. When expressed in terms of the linewidth Δv_D the normalized line profile becomes

$$g(v) = (2/\Delta v_D)(\ln 2/\pi)^{1/2} \exp\{-4 \ln 2[(v - v_0)/\Delta v_D]^2\}. \qquad (A4.9)$$

The Doppler linewidth is proportional to the observing frequency v_0 and $\sqrt{T}$, and hence may be reduced by measurement at long wavelength, e.g. at microwave or radio wave frequencies, or at low temperature. The effect of Doppler broadening on a measurement may also be reduced or almost eliminated by collimating the moving source and observing at right angles.

A4.4   Observed Lineshape

The actual lineshape of a particular observation of emission from a gas will have a contribution from natural broadening and may have a contribution from collisional or Doppler broadening, and hence there will be a combination of Lorentzian and Gaussian line profiles; these functions can be combined mathematically by the convolution procedure. The combination is termed a Voigt profile. In addition the wavelength response of the measuring instrument, termed its instrumental profile, has the effect of increasing the apparent (observed) width.

Appendix 5: Further Reading

Chapter 1

Introductory, Historical and General

A.R. Hall, *All was Light - an Introduction to Newton's Opticks*, Clarendon Press, 1993.
M.I. Sobel, *Light*, University of Chicago Press, 1987. [Thoughtful tour through the many ramifications of light in physics (including its history), technology, color vision, astrophysics, etc.]
D. Park, *The Fire Within the Eye. A Historical Essay on the Nature and Meaning of Light*, Princeton University Press, 1997.

Semi-popular. Prerequisites: Basic Algebra and Trigonometry

J.H. Mauldin, *Light, Lasers and Optics*, Tab Books, 1988. [Wide-ranging survey of classical and modern optics.]
G. Waldram, *Introduction to Light*, Prentice Hall, 1983. [Well described by its subtitle, 'The physics of light, vision, and color'.]

College Level. Texts Comparable to the Present One. Prerequisites: Calculus and 1–4 Years of Physics

G.R. Fowles, *Introduction to Modern Optics*, Holt, Rhinehart and Winston, 1975 (Dover reprint, 1989). [Concise text for the student familiar with Maxwell's equations.]
O.S. Heavens and R.W. Ditchburn, *Insight into Optics*, John Wiley & Sons, 1991. [Compact text with some 470 sections squeezed into its 300+ pages.]
E. Hecht, *Optics*, 4th edn, Addison-Wesley, 2002. [Comprehensive exposition.]
J.R. Meyer-Arendt, *Introduction to Classical and Modern Optics*, 4th edn, Prentice Hall, 1995. [Emphasizes the vergence approach to geometrical optics.]
F.L. Pedrotti, L.S. Pedrotti and L.M. Pedrotti, *Introduction to Optics*, 3rd edn, Pearson Prentice Hall, 2007. [Careful, detailed treatment.]
V. Ronchi, *The Nature of Light*, Harvard University Press, 1971.
J. Simmons and M. Guttmann, *States, Waves and Photons: A Modern Introduction to Light*, Addison-Wesley, 1970.

Reference Works

M. Bass *et al.* (eds), *Handbook of Optics*, vol. I (1995), vol. II (1995), vol. III (2001), vol. IV(2001), Optical Society of America and McGraw-Hill.
C. Webb and J. Jones. (eds), *Handbook of Laser Technology and Applications*, IOP Publishing, 2004.

Optics and Photonics: An Introduction, Second Edition   F. Graham Smith, Terry A. King and Dan Wilkins
© 2007 John Wiley & Sons, Ltd

Th.G. Brown *et al.*, *The Optics Encyclopedia*, Wiley-VCH, 2003.

M. Born and E. Wolf, *Principles of Optics*, 7th edn, Cambridge University Press, 1999. [Authoritative and theoretical classic account of optics.]

E. Wolf (ed.), *Progress in Optics*, multi-volume series, Elsevier.

Chapter 2

G.A. Brooker, *Modern Classical Optics*, Oxford University Press, 2003. [Didactic approach to classical optical phenomena and applications.]

P. Mouroulis and J. Macdonald, *Geometrical Optics and Optical Design*, Oxford University Press, 1997. [Introductory treatment of geometrical optics leading to computer-aided design.]

M. Mansuripur, *Classical Optics and its Applications*, Cambridge University Press, 2002.

D. G. O'Shea, *Elements of Modern Optical Design*, John Wiley & Sons, 1985.

F. Roddier. *Adaptive Optics in Astronomy*, Cambridge University Press, 1999.

W.T. Welford, *Aberrations of Optical Systems*, Adam Hilger, 1986.

Chapter 3

G.B. Stewart, *Microscopes: Bringing the Unseen World into Focus*, Lucent Books, 1992. [Development of microscopy from ancient Greece and Rome, up to various modern devices using non-light radiations. Major microscopic discoveries in medicine and biology enliven the account.]

L. Levi, *Applied Optics*, John Wiley & Sons, 1968.

D. Malacara, *Geometrical and Instrumental Optics*, Academic Press, 1988.

W. J. Smith, *Modern Lens Design*, McGraw-Hill, 1992.

C.J.R. Sheppard, D.M. Hotton and D. Shotton, *Confocal Laser Scanning Microscopy*, Springer-Verlag, 1995.

T. Wilson, *Confocal Microscopy*, Academic Press, 1990.

Chapter 4

A.P. French, *Vibrations and Waves*, MIT Introductory Physics Series, Chapman and Hall, 1971. [Well-established undergraduate text.]

R.N. Bracewell, *The Fourier Transform and its Applications*, 2nd edn, McGraw-Hill, 2000. [Classic reference on Fourier transforms.]

R.N. Bracewell, *Fourier Analysis and Imaging*, Springer-Verlag, 2004. [Authoritative description of Fourier transforms in the formation and manipulation of images.]

D.C. Champeney, *Fourier Transforms and their Physical Applications*, Academic Press, 1973.

J.W. Goodman, *Introduction to Fourier Optics*, McGraw-Hill, 1968.

E.G. Steward, *Fourier Optics: an Introduction*, John Wiley & Sons, 1987.

Chapter 5

H.A. Haus, *Waves and Fields in Optoelectronics*, Prentice Hall, 1984.

D.H. Staelin, A.W. Worgenthaler and J.A. Kong, *Electromagnetic Waves*, Prentice Hall, 1994.

A. Liddle, *An Introduction to Modern Cosmology*, John Wiley & Sons, 1999.

Chapter 6

Basic Texts

A. Ghatak and K. Thyagarathan, *An Introduction to Fiber Optics*, Cambridge University Press, 1998. [Introductory and general description of optical fibres and applications in sensors and communications.]

J.A. Buck, *Fundamentals of Optical Fibres*, 2nd edn, Wiley Interscience, 2004. [A succinct description of the main elements of optical fibres.]

A.H. Cherin, *An Introduction to Optical Fibers*, McGraw-Hill, 1987. [Describes electromagnetic theory of optical fibres and some aspects of optical fibre technology.]

A.W. Snyder and J.D. Love, *Optical Waveguide Theory*, Chapman and Hall, 1983. [A classic text giving a comprehensive description of waves and wave optics in optical waveguides and multi-mode fibres.]

Comprehensive and More Specialized Accounts

G. Kaiser, *Optical Fiber Communications*, 3rd edn, McGraw-Hill, 2000. [Comprehensive account of optical fibre communications, including fibre theory, optical fibre measurements and non-linear effects.]

K.T.V. Grattan and B.T. Meggitt, *Optical Fiber Sensor Technology*, Chapman and Hall, 1995. [Extensive account of optical fibre sensors and applications.]

J.M. Senior, *Optical Fibre Communications: Principles and Practice*, 2nd edn, Prentice Hall, 1992. [Thorough treatment of the fundamentals of optical fibre communications.]

D. K. Mynbeer and L.L. Scheiner, *Fibre-optic Communications Technology*, Prentice Hall, 2001. [Introductory broad coverage of optical fibre communications.]

Chapter 7

J.M. Bennett, *Handbook of Optics: (a) Polarization Part 1 ch. 5, (b) Polarizers Part 2 ch. 3*, McGraw-Hill, 1995.

E. Collett, *Polarized light: Fundamentals and Applications*, Marcel Dekker, 1992.

G.R. Fowles, *Introduction to Modern Optics*, 2nd edn, Holt, Rinehart and Winston, 1975 (also Dover Publications, 1989). [Chapter 6 on optics of solids gives further analyses of anisotropic wave propagation.]

A. Gerrard and J.M. Burch, *Introduction to Matrix Methods in Optics*, John Wiley & Sons, 1975.

D.S. Kliger, J.W. Lewis and C.E. Randall, *Polarized light in Optics and Spectroscopy*, Academic Press, 1990.

W.A. Shurcliff and S.S. Ballard, *Polarized Light*, Van Nostrand, 1964.

Chapter 8

W.H. Steel, *Interferometry*, Cambridge University Press, 1983.

P. Hariharan, *Optical Interferometry*, Academic Press, 2003.

P. Hariharan, *Basics of Interferometry*, Academic Press, 1992.

H.A. Macleod, *Thin-film Optical Filters*, IOP Publishing, 2001.

G. Hernandez, *Fabry-Perot Interferometers*, Cambridge Series in Modern Optics, vol. 3, 1988.

M.J. Vaughan, *The Fabry-Perot Interferometer: History, Theory, Practice and Applications*, IOP Publishing, 1989.

Chapter 9

P. Hariharan, *Optical Interferometry*, Academic Press, 2003.

R. Hanbury Brown, *The Intensity Interferometer: its Applications to Astronomy*, Taylor & Francis, 1974.

K.J. Gasvik, *Optical Metrology*, John Wiley & Sons, 1995.

Chapter 10

M. Francon, *Optical Image Formation and Processing*, Academic Press, 1979.
S.G. Lipson, H. Lipson and D.S. Tannhauser, *Optical Physics*, 3rd edn, Cambridge University Press, 1995. [Introduction to wave aspects of optics and some modern topics.]
W.H. Steel, *Interferometry*, Cambridge University Press, 1983.
O.S. Heavens and R.W. Ditchburn, *Insight into Optics*, John Wiley & Sons, 1991.

Chapter 11

M.C. Hutley, *Diffraction Gratings*, Academic Press, 1982.
E.G. Loewen, 'Diffraction gratings: ruled and holographic', in *Applied Optics and Optical Engineering*, vol. IX, ed. R.P. Shannon and J.C. Wyant, Academic Press, 1983.
E.W. Palmer, M.C. Hutley, J.F. Verrill and B.Gale, 'Diffraction gratings', in *Reports on Progress in Optics*, vol. 38, Pt 2, p. 975, IOP Publishing, 1975.

Chapter 12

R.J. Bell, *Introductory Fourier Transform Spectroscopy*, Academic Press, 1972.
W. Demtroder, *Laser Spectroscopy*, 3rd edn, Springer-Verlag, 2002.
C.H. Banwell and E.M. McCash, *Fundamentals of Molecular Spectroscopy*, McGraw-Hill, 1994. [Elementary introduction to molecular spectroscopy with emphasis on descriptive understanding.]
P.R. Griffiths and J.A. de Haseth, *Fourier Transform Infrared Spectroscopy*, John Wiley & Sons, 1986.
S. Svanberg, *Atomic and Molecular Spectroscopy: Basic Aspects and Practical Applications*, 3rd edn, Springer-Verlag, 2001.
J. F. James and R.F. Sternberg, *The Design of Optical Spectrometers*, Chapman and Hall, 1969.
S.P. Davis, M.C. Adams and J.W.E. Brault, *Fourier Transform Spectroscopy*, Academic Press, 2001.

Chapter 13

R. Loudon, *The Quantum Theory of Light*, 3rd edn, Oxford University Press, 2000.
L. Mandel and E. Wolf, *Optical Coherence and Quantum Optics*, Cambridge University Press, 1995. [Extensive and authoritative description of coherence.]
S.G. Lipson, H. Lipson and D.S. Tannhauser, *Optical Physics*, 3rd edn, Cambridge University Press, 1995.
J. Perina, *Coherence of Light*, Reidel, 1985.
R. Hanbury Brown, *The Intensity Interferometer: its Application to Astronomy*, Taylor & Francis, 1974.
R. Hanbury Brown and R.Q. Twiss, 'Correlations between photons in two beams of light', *Nature*, **177**, 27, 1956.

Chapter 14

D. Gabor, 'A new microscopic principle', *Nature*, **161**, 777, 1948.
D. Gabor, 'Microscopy by reconstructed wavefronts', *Proceedings of the Royal Society*, **A197**, 454, 1949.
P. Hariharan, *Optical Holography: Principles, Techniques and Applications*, Cambridge University Press, 1996. [Theoretical and practical description with various applications.]
P. Hariharan, *Basics of Holography*, Cambridge University Press, 2002.

R. Jones and C. Wykes, *Holographic and Speckle Interferometry*, 2nd edn, Cambridge University Press, 1989.
H.M. Smith, *Principles of Holography*, John Wiley & Sons, 1975.
C.M. Vest, *Holographic Interferometry*, John Wiley & Sons, 1979.

Chapter 15

O. Svelto, *Principles of Lasers*, 4th edn, Plenum Press, 1998.
W.T. Silvast, *Laser Fundamentals*, 2nd edn, Cambridge University Press, 1998. [Extensive and clear description of the main lasers.]
A.E. Siegman, *Lasers*, Oxford University Press, 1986.
J.P. Harbison and R.E. Nahory, *Lasers, Harnessing the Atoms Light*, Scientific American Library, 1998.
J. Hecht, *Understanding Lasers*, H.W. Sams 1998. [Introductory, descriptive and non-mathematical description.]
C.C. Davis, *Lasers and Electro-Optics*, Cambridge University Press, 1996. [Comprehensive and advanced text with theoretical and experimental descriptions.]
A. Yariv, *Optical Electronics in Modern Communications*, 5th edn, Oxford University Press, 1997.
A. Miller and D.M. Finlayson, *Laser Sources and Applications*, IOP Publishing, 1996.
P.W. Milonni and J.H. Eberly, *Lasers*, John Wiley & Sons, 1988.

Chapter 16

O. Svelto, *Principles of Lasers*, 4th edn, Plenum Press, 1998.
P.W. Milonni and J.H. Eberly, *Lasers*, John Wiley & Sons, 1988.
R.W. Boyd, *Non-linear Optics*, Academic Press, 1992.
R. Loudon, *The Quantum Theory of Light*, Oxford University Press, 2000.
M.O. Scully and M.S. Zubairy, *Quantum Optics*, Cambridge University Press, 1997.
Y.R. Shen, *The Principles of Non-linear Optics*, John Wiley & Sons, 1984.
P.N. Butcher and D. Cotter, *The Elements of Non-linear Optics*, Cambridge University Press, 1990.
M. Francon, *Laser Speckle and Applications in Optics*, Academic Press, 1979.
A. Yariv, *Introduction to Optical Electronics*, Holt, Rinehart and Winston, 1996.

Chapter 17

B.E.A. Saleh and M. Teich, *Fundamentals of Photonics*, 2nd edn, John Wiley & Sons, 2006. [Comprehensive treatment at introductory level.]
W.T. Silfvast, *Laser Fundamentals*, 2nd edn, *Cambridge University Press*, 2004. [Extensive and clear description of the main lasers.]
M. Fox, *Optical Properties of Solids*, Oxford University Press, 2001.
C.C. Davis, *Lasers and Electro-optics*, Cambridge University Press, 1996. [Comprehensive and advanced text with theoretical and experimental descriptions.]
J.R. Hook and H.E. Hall, *Solid State Physics*, 2nd edn, John Wiley & Sons, 1991.
B. Mroziewicz, M. Bugajski and W. Nakwaski, *Physics of Semiconductor Lasers*, North-Holland, 1991.
P.W. Milonni and J.H. Eberley, *Lasers*, John Wiley & Sons, 1988.
G.H.B. Thomson, *Physics of Semiconductor Lasers*, John Wiley & Sons, 1982.
A. Yariv, *Optical Electronics in Modern Communications*, 5th edn, Oxford University Press, 1997.
D. Wood, *Optoelectronic Semiconductor Devices*, Prentice Hall, 1994.
J. Piprek, *Semiconductor Optoelectronic Devices*, Elsevier, 2003.

Chapter 18

Z.H. Malacara and M.R. Arquimedes, 'Light sources', in *Geometrical and Instrumental Optics*, ed. D. Malacara, Academic Press, 1988.

G.J. Zissis and A.J. Larocca, 'Optical radiators and sources', in *Handbook of Optics*, ed. W.G. Driscoll and W. Vaughan, McGraw-Hill, 1978.

P.J. Duke, *Synchrotron Radiation: Production and Properties*, Oxford University Press, 2000.

G. Margaritondo, *Introduction to Synchrotron Radiation*, Oxford University Press, 1988.

H. Winick and S. Doniach, *Synchrotron Radiation Research*, Plenum Press, 1980.

C.A. Brau, *Free Electron Lasers*, Academic Press, 1990.

E.L. Saldini, E.A. Schneidmiller and M.K. Yurkov, *The Physics of Free Electron Lasers*, Springer-Verlag, 2000.

B. Blasse and B.C. Grabmaier, *Luminescent Materials*, Springer-Verlag, 1994.

J.R. Coaton and A.M. Marsden, *Lamps and Lighting*, Arnold, 1997.

Chapter 19

H.C. van der Hulst, *Scattering of Light by Small Particles*, Dover, 1984.

C.N. Banwell and E.M. McCash, *Fundamentals of Molecular Spectroscopy*, 4th edn, McGraw-Hill, 1994. [Elementary introduction to molecular spectroscopy with emphasis on descriptive understanding.]

J.R. Ferraro and K. Nakamoto, *Introductory Raman Spectroscopy*, Academic Press, 1994.

A.G. Michette and C.J. Buckley, *X-ray Science and Technology*, IOP Publishing, 1993.

J. Als-Nielsen and D. McMorrow, *Elements of Modern X-ray Physics*, John Wiley & Sons, 2001.

D.T. Attwood, *Soft X-rays and Extreme Ultraviolet Radiation: Principles and Applications*, Cambridge University Press, 1999.

Y. Band, *Light and Matter*, John Wiley & Sons, 2005.

Chapter 20

G.H. Rieke, *Detection of Light from the Ultraviolet to the Submillimeter*, 2nd edn, Cambridge University Press, 2003. [A comprehensive overview of detectors.]

R.W. Boyd, *Radiometry and the Detection of Optical Radiation*, John Wiley & Sons, 1983.

G.C. Holst, *CCD Arrays, Cameras and Displays*, 2nd edn, SPIE Press, 1998.

R.J. Keyes, *Optical and Infrared Detectors*, 2nd edn, Springer-Verlag, 1980.

E.L. Dereniak and D.G. Crowe, *Optical Radiation Detectors*, John Wiley & Sons, 1984.

W.R. McCluney and R. McCluney, *Introduction to Radiometry and Photometry*, Artech House, 1994.

S.M. Sze, *Physics of Semiconductor Devices*, 2nd edn, John Wiley & Sons, 1981.

D. Wood, *Optoelectronic Semiconductor Devices*, Prentice Hall, 1994.

N. Schuster, *Sensors and Camera Systems for Scientific, Industrial and Digital Photography Applications*, *Proc. SPIE*, vol. 1669, SPIE Press, 2002.

Chapter 21

M.F. Land and D.E. Nilsson, *Animal Eyes*, Oxford University Press, 2002. [A comprehensive account of the amazing variety of eyes, and their relation to conventional optics.]

M.G.J. Minnaert, *Light and Colour in the Outdoors*, Centenary edn, Springer-Verlag, 1993. [The classic original account.]

D.K. Lynch and W. Livingston, *Color and Light in Nature*, Cambridge University Press, 1995. [With beautiful photographs of many natural optical phenomena.]

R. Greenler, *Rainbows, Halos and Glories*, Cambridge University Press, 1980. [Description of atmospheric optical effects with many colour plates.]

I. Glynn, *An Anatomy of Thought: The Origin and Machinery of the Mind*, Weidenfeld and Nicolson, 1999.

L. Stryer, 'The molecules of visual excitation', *Scientific American*, **157**, 32, 1987.

R. Wehner, 'Polarized light navigation by insects', *Scientific American*, July 1976, 105–15.

Index

Optics and Photonics: An Introduction, Second Edition   F. Graham Smith, Terry A. King and Dan Wilkins
© 2007 John Wiley & Sons, Ltd